CONSTRUCTION INSPECTION HANDBOOK

CONSTRUCTION INSPECTION HANDBOOK

Second Edition

JAMES J. O'BRIEN, P.E.

LEONARD P. SCHAEFER, P.E.
Electrical Section

RITA F. GIBSON
Editorial Assistant

VNR VAN NOSTRAND REINHOLD COMPANY
New York

Manufactured in the United States of America

Published by Van Nostrand Reinhold Company Inc.
115 Fifth Avenue
New York, New York 10003

Van Nostrand Reinhold Company Limited
Molly Millars Lane
Wokingham, Berkshire RG11 2PY, England

Van Nostrand Reinhold
480 La Trobe Street
Melbourne, Victoria 3000, Australia

Macmillan of Canada
Division of Canada Publishing Corporation
164 Commander Boulevard
Agincourt, Ontario M1S 3C7, Canada

15 14 13 12 11 10 9 8 7 6 5

Library of Congress Cataloging in Publication Data

O'Brien, James Jerome, 1929-
    Construction inspection handbook.

    Includes index.
    1. Building inspection—Handbooks, manuals, etc.
I. Schaefer, Leonard P.   II. Gibson, Rita F.
III. Title.
TH439.O27   1983   690   82-8416
ISBN 0-442-25741-4

PREFACE

This book is written for construction inspectors—a categorical description with meanings about as diverse as the number of people it applies to. Depending upon contractural arrangements, the inspector may work for the owner, a governmental building agency, the designer, or the contractor. He may or may not have a professional education, and even with an academic background in architecture or engineering, he typically will have gained most of his construction know-how on the job.

There have always been two types of inspectors: one really inspects and has an interest in the job, and the other is just on the job making the appearance of inspection. This book is written for the inspector who wants to inspect—and either wants to broaden his knowledge, or confirm that which he has learned the hard way.

Just as the name implies, the principal role of inspection is observation of quality—quality of materials, and quality of workmanship. However, the role of management in the construction field is shifting from a passive to an active one, and the successful inspector should be part of the management team. This work is aimed at enhancing the understanding of this more active role—and presenting to the inspector what is (or perhaps ought to be) expected of him by higher management.

The book is arranged in categories, so that it can be used in handbook fashion; each section stands independent of the others. The arrangement is also organized generally in the same format as building construction specifications (following the CSI format) so that easy reference can bring the information to bear on the problems at hand.

The author has had the fortunate opportunity of experiencing all project phases from planning through design and bidding into all phases of construction as a project engineer first in the petrochemical industry, then aerospace, and finally in the building construction field.

The experience presented in the handbook, however, is not limited by the author's own experiences, but represents information garnered from many sources and contributed by many acquaintances in the construction industry. Some of these sources are acknowledged formally, but many more bosses, colleagues, foremen, and craftsmen have made contributions which are well appreciated, but which must practically go unrecognized. This is perhaps just as well, since Bill the millwright, or Charlie the steamfitter tend to be characters known to everyone on a given project—but part of the passing scene.

Cherry Hill, New Jersey

JAMES J. O'BRIEN, P.E.

ACKNOWLEDGMENTS

Many sources in the construction industry have freely contributed information for this handbook. Manufacturers' information is acknowledged within the book itself, but specific recognition is given here to the singular contributions made by prior handbooks published in the public field, of which free utilization was made in the preparation of this handbook:

American Society for Testing and Materials, list of ASTM Standards used in Building Codes (1980)*

American National Standards Institute, Catalog of American National Standards (1981)*

City of New York, "Building Code of the City of New York,"

Construction Specifications Institute, Masterformat (Master List of Section Titles and Numbers) (1979).*

Corps of Engineers, Handbook for Construction Inspectors.

Environmental Protection Agency, "Construction Inspection Guide."

General Services Authority (now Department of General Services), Pennsylvania, Construction Inspector's Handbook.

Highway Department, State of California, Construction Engineer's Handbook.

New York Transit Authority, Construction Inspection Handbook.

State of New Jersey, Department of Building Construction, "Standard General Conditions."

*Used with permission.

CONTENTS

PREFACE/v

A.
THE ROLE OF THE INSPECTOR

1 Inspection/3
2 The Anatomy of a Project/13
3 Contract Documents/22
4 Codes and Standards/39
5 Job Mobilization/51
6 Surveying/59
7 Temporary Construction/75
8 Safety/99
9 Field Testing/112

B.
PROJECT CHARACTERISTICS

Divisions

1 General/131
2 Site Work/139

3 Concrete/211
4 Masonry/273
5 Metals (structural and miscellaneous)/288
6 Wood and Plastic/311
7 Thermal and Moisture Protection/343
8 Doors and Windows/344
9 Finishes/361
10 Specialties/393
11 Equipment/402
12 Furnishings/412
13 Special construction/416
14 Conveying systems/421
15 Mechanical/427
16 Electrical/465

C. PROJECT MANAGEMENT

1 Scheduling/505
2 Change Orders/526
3 Progress Payments/542
4 Documentation/558

APPENDIX A - ASTM Standards/581

APPENDIX B - ANSI Standards/633

APPENDIX C - Organizations and Associations/651

INDEX/663

CONSTRUCTION INSPECTION HANDBOOK

A.
THE
ROLE OF THE
INSPECTOR

1. INSPECTION

It is an unfortunate commentary on human nature and the psychology of the individual that we need construction inspection—but need it we do. Even more unfortunate is the apparent drive for conformity that makes companies often do things they would not do as individuals, again underscoring the need for inspection.

Without a doubt, there is a requirement for inspection, and those being inspected will usually be quick to affirm this. Our entire economy is built on the roles of buyer and seller, and we all play our part. The rule of *caveat emptor* (let the buyer beware) is not only philosophical, but legal. In construction, where the product is built in place and often takes years to develop, the buyer has an unusual opportunity to look over the shoulder of the seller. In turn, the seller puts enough in his cost (he hopes) to pay for this interference.

In the face of the real and accepted need for inspection, the inspector falls into another unfortunate general attitude—that of the adversary. With the seller (contractor) placed diametrically opposite to the buyer (owner, designer, agency) sides are clearly taken at the beginning of a construction project. The less intuitive inspector assumes that anything he can do to slow down, impede, or control the contractor is to the advantage of the owner. Nothing could be further from the truth. Without exception, jobs which go smoothly cost the owner less, and those which are dogged with problems cost everybody more. It is vital that the inspector control quality, but if he impedes progress, he inevitably costs the owner money. This is not always obvious to the owner. The experienced inspector, realizing that, after all, it is the contractor who ultimately builds the job, must tread a careful line between the owner's interest and the contractor's interest.

WHO ARE INSPECTORS?

There are biblical references to inspectors, and early monumental projects such as the pyramids had working scribes as well as overseers. In fact, the term "clerk-of-the-works" is probably a carryover from the early materials counters.

The role implied by the title "clerk-of-the-works" is as archaic as the term itself. No longer (if in fact it were ever true) is the inspector ever able to stand by and count materials coming into the job. Perhaps 100 years ago when labor was cheap and materials were scarce, there were widespread instances of shortchanging through the use of shoddy materials. In today's construction industry, it is more usual to find that good material is used, but installed improperly, with connections not made or controls not operable. This is where inspection is important— making certain that the labor and materials bought and paid for are properly applied.

In most cases, mistakes made provide benefit to no one; everyone is the loser. The craftsman reduces his productivity, the contractor is placed in the position of disputed reputation, while the owner (at some point) has to dig deeper in his pocket to make up for problems created by installation mistakes.

THE ROLE OF THE INSPECTOR

Historically, government agencies tend to view the role of inspection in the traditional and ultra-narrow sense. Nevertheless, this role is a base for the things which an inspector can and should do. The following description is from the Manual on "Construction Inspection Procedures" prepared by the General State Authority of Pennsylvania, discussing the role of the inspector:

"The primary function of the field personnel of the construction division is inspection, and the persons assigned to this task are designated as inspectors. There are three classifications of field inspectors: general, mechanical, and electrical. . . . The Inspector must be able to look upon and view critically the particular phase of the construction project to which he is assigned. This requires some degree of experience in the construction field. In addition to experience, the Inspector must also have the ability to evaluate and analyze what he is inspecting. Therefore, a most important and necessary requirement is that the Inspector be able to fully read, comprehend, and interpret the contract plans and specifications. It is also very important that the Inspector have the ability to maintain records that will fully reflect the inspections performed.

The Inspector must closely follow the progression of each stage of construction. He must be alert to existing conditions and be able to foresee future problems. When the Inspector notices through his daily inspections that certain phases of the work are not being done in accordance with the plans and specifications, or when other problems occur, he is to immediately report these errors, violations, or problems to 'management' for further action.

In effect, the Inspector is not authorized to revoke, alter, substitute, enlarge, relax, or release any requirements of any specifications, plans, drawings or any other architectural addenda. In addition, the Inspector must not approve or accept any segment of the work which is contrary to the drawings and specifications. At no time is the individual Inspector allowed to stop the construction work or interfere with the contractor's employees.

It must be recognized that the title 'Inspector' creates a barrier between him and the contractors on the job. How effective he can be in his role depends mainly on how he handles himself in this relationship. He must display knowledge, experience, integrity, ability and the use of good judgment."

The role described is a narrow one of quality control. It suggests that the Inspector cannot give up any of the prerogatives of the owner, and at the same time, cannot delay the work. This is mitigated in many cases by the assignment of resident engineers with concomitant duties to inspect. Even in these cases, where the Inspector represents management, work may not be disrupted except in accordance with the contract. In a cost-plus contract, the owner often has prerogatives to stop the work—if any such stoppage will be paid for directly by him. In fixed price contracts and negotiated .contracts, the owner's representatives must be much more careful, or they will be liable for a claim of interference.

The Corps of Engineers has the following suggestions to the Inspector on his role:

"An Inspector should at all times be thoroughly familiar with all the provisions of the contract which he is administering. This includes familiarity with the plans and specifications including all revisions, changes, and amendments. In addition, the Inspector must be thoroughly familiar with pertinent Corps of Engineers, individual District, and supervisor's administration policies."

An Inspector has different responsibilities and authorities, dependent upon the organizational set-up under which he is working, and his own capabilities. Each Inspector should know his part in the organization, and should be aware of the importance of high-quality construction. He should understand his own level of technical knowledge, and accept his responsibilities without overstepping his authority.

In order to do this, the Inspector must be aware of the extent of his authority. To that end, he always has the authority to require work to be accomplished in accordance with the contract plans and specifications. Procedures and policies on stopping work for safety violation or construction deficiencies should be reviewed with appropriate supervision before being employed.

In dealing with contractors, the Inspector should be impersonal but friendly, fair, and firm. He should be businesslike and cooperative with the contractors, and should attempt to have a clear, accurate, and appreciative understanding of the contractor's problems. Within this viewpoint, the Inspector's decisions and instructions should provide the greatest latitude possible to the contractor without prejudice and without waiving contract requirements in choice of equipment, material, or methods.

Further, not only must the Inspector make the right decisions and issue proper instructions, but he must issue them and see that they are executed promptly and at the right time. Any such decisions or instructions should be based on detailed knowledge of all the facts, good judgment, and protection of the owner's interests. Conversely, the Inspector must consider whether or not his instructions may have any adverse consequences—and if so, be certain that appropriate management be informed first. The Inspector should be cautious and avoid dictating methods of construction to the contractor, unless methods are clearly spelled out in the contract. Further, he should be careful not to discuss items directly with the contractor's craftsmen or with subcontractors unless this is done with the contractor's knowledge, and even then it is a questionable practice. Any controversies with the contractor should be settled promptly.

CONSTRUCTION MANAGEMENT

Where the owner's role in construction was once only passively to observe, many owners have now decided that the real way to achieve optimum result is to manage their own construction. To implement this viewpoint, the owner either has his own construction management staff, or retains professional assistance—or a combination of both.

The role of the Inspector in construction management is important. The construction management team does not replace inspection, but incorporates it. Further, the inspector is management's contact with the job. Through the inspection process, he develops not only knowledge of specific problems, but a general awareness of the attitudes of the contractors in the various trades. He can identify frictions, problem areas and should have a knowledge of situations which may require management attention.

The inspection team becomes the five senses of the construction manager on the job site. Further, the inspection team provides a natural training ground for new construction management staff. Recent emphasis on construction management should work in the long range to reduce the adversarial role between contractor and owner, by closing the communications gap. The application of construction management offers an opportunity for recognition of good inspection techniques— with appropriate potential rewards for the effective inspector, as well as a means of disclosing the ineffectual freeloader.

SPECIFIC DUTIES

The responsibility of the resident inspection team include the following:

1. Ensuring *compliance* by the contractor with the plans, specifications, and contractual provisions for the project.

2. Monitoring, and, where appropriate, ensuring that the *project progresses* according to schedule. Where it cannot, management must be kept aware of problems.

3. Coordination and monitoring of *reviews, approvals,* and *tests* as required by the specifications and contract.

4. *Interpreting contract drawings* and specifications, and where disputes occur, documenting these and arranging solutions.

5. Rejecting work which is not within *contractual quality* or which fails to meet contract requirements.

6. Stopping work and progress when *safety concerns* override basic contractual commitments, or when continuation will result in the inclusion of substandard work.

7. Approval, or revision so that they can be approved, contractor *estimates of progress*—which result in progress payments. This is perhaps one of the greatest areas of inspector authority, since the name of the contractor game is money, and delay in payment costs money itself. While this should not be used capriciously, it is important that the contractor recognize the willingness of the inspection team to delay payments where work is either not in place, or is not up to contract quality.

8. Approval of *shop drawings, materials, samples* and other approval items submitted by the contractor—with approval contingent either on the inspection team reports, or reports of appropriate laboratories or consultants.

9. The inspection team has the responsibility for the owner to avoid *labor situations* caused by jurisdictional problems, or contractor violations of working conditions as described within the contract. On the other hand, the inspection team cannot impose conditions which are not already included in or inherent in the contract. Nevertheless, the inspection team can utilize a far-reaching, industry-wide standard such as the Occupational Safety and Health Act implemented by the Department of Labor.

PROJECT ENVIRONMENT

The work of the inspection team is necessarily influenced by the environment within which the team must function. For instance, any of the following factors may call for different attitudes and vigilance on the part of the inspection team:

1. *Contractor Attitude,* either where a contractor has a habit of working as close to the letter of the contract as he can, or more usually where a contractor believes that he has submitted too low a bid and is constantly attempting to make up the difference through his liberal interpretations of the plans and specifications. The inspection team

must provide a careful balance—not overreacting, but not being complacent either.

2. *Contractor capabilities* cause different reactions from the inspection team. The contractor's attitude may be excellent, but he may have reached too far in bidding a project, or may have assigned project supervision that are "way over their heads." Either situation may cause problems, and the inspection team must be aware and react appropriately.

3. The general *trade atmosphere* varies from place to place, and the inspection team can be prepared for more problems in certain areas than others.

4. The *ability of the field team* itself is important to the individual Inspector, and affects his ability to perform. It also sets the scene for the way in which he will perform. If the balance of the team is skilled, highly motivated, and well organized, an inspector can emphasize his own areas of expertise. On the other hand, if the team is limited in resources, unskilled, and lacks experience, the individual Inspector may have to press himself as far as possible, operating to the maximum of his abilities, and informing management when a sufficient inspection job cannot be accomplished.

5. The practicality and completeness of the *project plans* and specifications will have an impact upon the inspection effort. In some cases, the plans and specifications will result in an excellent structure, but buildability was not one of the considerations, and many construction problems will occur. This is an unfortunate circumstance, but better than the case in which the plans and specifications include many errors and omissions, and result in many negotiations with the contractors in order to complete. Further, they leave many doorways open for extras, or arguments. Changes are inevitable in a field job, since no set of plans and specifications is perfect. Nevertheless, different design teams achieve different levels of success in the feasibility of presentation of intent. The inspection team is fortunate indeed to have a clean set of plans and specifications, describing a practical structure with the reasonable and accepted techniques of construction able to cope with the construction. (Similarly fortunate in this case is the owner who does not have to pay a lot of extras.)

6. *Field conditions* can be such, particularly when unanticipated, that they greatly influence the attitude of all parties at the job site. A designer working in an air conditioned office may recognize field problems, but may still have failed to appreciate the inhibiting attitudes resulting from extremes in temperature, either very wet or very dry working conditions, very hot or very cold working conditions, high noise levels and other environmental facts. The inspection team faces the same environment that the work force does, and, unfortunately, tends to "lose its cool" just when it needs it most.

JUDGMENT AREAS

It is not to the best interests of the owner to have the inspection team become nitpickers. When a contractor has to rework a low-value item at high cost, when that item has no significant functional or dollar impact upon the building, the Inspector has lost some of his "green stamps." Later, when an item with functional value or a high cost item requires work which might conceivably be the responsibility of either the owner or the contractor, we can assume that the owner will have to pay the bill. Unfortunately, the traditional definitions of the role of inspection appear to give the inspection team little or no latitude. There is latitude, and it can be applied positively.

The Inspector must take care to conduct himself carefully and without any potential conflict of interest. He can take no favors, gifts, nor other rewards from the contractor for doing his job in a cooperative fashion. The Inspector's reward has to be the successful progress of the job, at no real cost in quality to the owner. In doing this, the Inspector must constantly keep in mind that any delays while he reaches a fair evaluation of what is acceptable are more than delays to the contractor —they are costs, for time is money to the contractor. Accordingly, an Inspector who makes judicious and proper decisions, but does not make them in a timely fashion, can indeed be the cause of a poor attitude or resentment on the part of contractors. The Inspector who fails to inspect, or who allows the widest range of latitude to the contractor, in turn, does favor to neither the contractor nor the owner. Thus a balance must be achieved.

The inspection team should be willing to call on its own management and/or consultants when in doubt. The designer offers a pool of available information, although the inspection team should be aware of the contractual agreement between the designer and the owner. In most cases, the design team is required to at least furnish interpretation of intent, and almost inevitably will be willing to cooperate by phone call, if not site visits. In those cases in which the design team is not available, the Inspector must not be reluctant to call upon appropriate supplemental professional help such as surveyors, testing labs, manufacturer's representatives, and others who could add to his understanding of potential problems.

The construction Inspector is truly the man in the middle. Many owners, concerned that they will receive full value for their money, are encouraged by a "take charge" Inspector—one who pounds the table, shakes his finger, and threatens the contractor. Unfortunately, it is this individual who creates problems, and may cause job hang-ups which reduce the ultimate product, delay completion, or result in higher costs to the owner through extras.

The enlightened owner would recognize that his best interests are realized when the contractor makes a reasonable profit on his job, with a minimum of delays and problems. The owner that encourages overzealous inspection is usually an inexperienced one, and is depending upon the resiliency and ability of the contractor to overcome this kind of pressure. In a large job, or in a smaller job where a medium-sized contractor has reached to the limits of his ability to take the job, owner resistance can well be the straw which breaks the contractor's back. The result may be a bankrupt contractor, with the job run by the surety, and then the owner can really discover how tough a job can be.

On the other hand, the inspection team must achieve that reasonable level of quality which meets all of the literal requirements of the plans and specifications without being dogmatic or harassing.

Further, the Inspector has his own stability as a human being to consider. The over-detailed nitpicker will find it a tough row to hoe. One fussy chemical engineer who found a few too many painter's holidays returned to the construction shack one day painted with five gallons of the company color, spilled "accidentally"—but accurately—from 20 feet above. This same engineer, a slow learner, was sprayed wtih a tank lining plastic after his sharply delivered advice to the contractor's tank lining subcontractor.

Craftsmen have their own pride and dignity—and the role of Inspector is not supervisor. The careful Inspector must be able to distinguish those situations which can be left to develop without real injury to the personnel, contractor, or owner from those which require a definite work stoppage because of safety problems or because an obvious mistake is about to be made.

SUMMARY

The role of inspection requires experience, knowledge and talent. Further, it requires the development of a fine judgment in regard to the workings of a construction project. The smooth project has its own coordination, rhythm, and procedures. The uninformed Inspector is fortunate when he has a good project, a good contractor, and a good team to break in with. If he has to start on a new job alone, with tough contractors, a demanding owner, or other negative factors, his best attributes may be patience and a good attitude.

2. THE ANATOMY OF A PROJECT

While inspection occurs only during the construction phase, construction, while the most visible portion of a project, takes about one-half or in some cases only one-third of the time consumed from original concept and approval of the project through move-in and utilization. Most construction duties are confined to the construction period, but it is helpful to the Inspector to understand the roles and attitudes of the various characters within the project cast. By the time the job reaches the field, the designer, owner and various government reviewing bodies—such as building code specialists—have developed certain definite opinions. The project itself may be popular within the community or highly unpopular. This background and set of attitudes carries forward into the construction phase, and will affect the inspection role.

PROJECT STRUCTURE

There are four major categories of project progress:

1. *Pre-design activities*, with the owner having primary responsibility for progress.

2. *Design phase*, with Architect, Engineer, A-E, or in-house staff having primary responsibility for progress.

3. *Construction*, with contractor or in-house construction force responsible for progress.

4. *Furnish and move-in*, with owner or contractor having primary responsibility.

PRE-DESIGN

One of the least defined, intangible and time-consuming phases of a project is the pre-design portion. During this phase, the owner, his technical staff, or consultants should be very busy with a number of important roles. Typically, these are accomplished by omission or default, rather than carried out in a rigorously managed approach. The seeds of many project problems are planted in this field of neglect.

Most projects seem to appear from nowhere, the result is an evolutionary manner of aggregate thinking from many sources which gathers pressure, both political and personal, until the project has been articulated. Programs which evolve in this manner include schools, hospitals, public buildings, industrial plants, highways, and almost any project which can be identified. Key characteristics are power structure and consensus. Actually, the project should go through the following stages: goals; resource evaluation; decision-making; budgeting and funding.

These components make up what has been termed PPBS (Planning, Programming and Budgeting System) by the federal government. Operated in its best sense, the project is defined as a result of needs of the organization, rather than justified as a result of the personal feelings of the owner or management involved.

In order to operate PPBS, needs and goals must be preestablished, preferably as policy. Projects which can be utilized to meet the goals of the organization are reviewed against the available resources. Usually this is an intuitive relationship, making selections and establishing priorities on intuition and personal preferences, rather than through any quantitative approach.

The decision to go ahead requires identification of specific projects and the development of preliminary cost estimates. These are usually accomplished in-house on the basis of square footage costs, or they may be prepared by a consultant.

After the project has been placed into the budget, and funding is available to meet that budget, the project—still under the direction of the owner—enters the pre-design phase, but implementation has begun.

Site

When the project has been identified and approved, a site must be selected. In some cases, the site decision has been a part of the basic decision to go ahead with the project, such as a hospital addition, or a school replacement, and other finite location situations. In the majority of cases, however, a new site must or can be considered. Usually the site consideration precedes the selection of the designer, since the design must be a function of the siting. A number of nontechnical factors may funnel the choice of a site in a specific direction. However, where there are a number of possibilities, the owner or his consultant must consider a number of factors including:

Encumbrances—are there tenants who will have to be relocated?

Land costs—what are the economic factors?

Transportation—is the location adequately served?

Utilities—what is the availability, what are the potential problems?

Neighborhood—is the environment suitable for the facility?

Zoning—does the local zoning conform with the use intended?

Community—how will the community react to the new facility?

Subsurface conditions—will the project foundation require unusual support?

Topographic—is the topographical nature suitable for the project?

There are other factors, but it is clear that in choosing between more than one site, many factors must be evaluated and considered carefully. Unfortunately, many of these factors are considered from a viewpoint of hindsight rather than at the proper time in the project.

Programming

The last pre-design activity should be the development of a specific program which defines the intent of the owner in regard to the functional uses of the project. This philosophy is not only important, but is often lost during the design phase of the project. It is clearly the owner's responsibility to make his requirements known, and to interpret these in terms of cost impact *prior* to the selection of the designer. Programming is a unique talent, and usually requires a combination of the owner's knowledge, and consulting expertise.

A functional programming effort should reconfirm the budgetary estimate, and establish the material to be included in the project. The functional program incorporates policy, and should be approved by appropriate owner authority.

The architectural program is a result of the functional program, and must necessarily follow it, since it utilizes the functional program for source data. The architectural program may be incorporated in the schematic design phase by the architect, or may be furnished to the architect.

Projects typically do not have formal program documents, perhaps deliberately. The result is uncertainty at the beginning of the design phase. Since the designer does not get compensated for uncertainty, his only defense is to proceed slowly at the early portion of the project, developing program-type statements which can be affirmed or revised by the client. Unfortunately, clients typically offer a tremendous proclivity to change their minds almost constantly. From the designer's viewpoint, this is not only time-consuming, but highly expensive.

DESIGN

Design of a project is a relatively complex series of activities which become increasingly detailed as the project proceeds through the various design phases. The following is an outline of design phases.

Schematic Development

This is also called the sketch phase. Concept plans are developed by the architect, and basic engineering systems analysis is made. Design criteria are specified, and schematic drawings are prepared. A perspective is prepared. Various systems are defined in terms of performance and, where appropriate, by single line diagram or sketch.

Preliminary Design

This is also called the design development. After approval of the sketch or schematic phase, the drawings are refined to a degree sufficient to permit the development of dimensioned space layouts, heating and ventilating systems main feeders, electrical main feeders, as well as definite development of the structural framework. Requirements for utilities are also definitely developed, and specific equipment require-ments are determined. The budgetary type cost estimate is revised, and a more firm estimate made.

Working Drawings

These are also referred to as contract drawings and specifications. This phase comprises about two-thirds of the work, but fewer of the decisions. The design as defined in the prior stages is developed in complete detail, including dimensions, so that it can be specifically priced by prospective contractors. The contract documents include both drawings and specifications.

As the project proceeds, each change becomes more difficult to implement. Each revision requires many more reviews and changes of related items. The range of changes which can be accepted gracefully narrows down in the funnel effect, with the maximum changes being more acceptable early in the project, and more and more costly as the project proceeds. These factors can be quantified, but vary with each project.

The design phase often proceeds essentially unscheduled and uncoordinated. This is partially understandable, since specific interconnections between the phases, such as the structural, mechanical, and electrical, are difficult to express. Conversely, the design phase must be closely coordinated and interconnected at all stages and steps. To schedule this at the daily liaison level of detail would result in a fantastically large scheduling network. The usual compromise is the scheduling of concurrent activities with an implied understanding that continual liaison is carried out.

During the design there is a constant interplay between the designer and the owner. The owner reviews the project at the major points, and also is available on a daily basis for information. Quite often, the owner is furnishing or specifying special equipment which requires his attention. The owner and/or architect are involved in various agency reviews. The design phase offers a great potential for time gains. Many activities can actually occur concurrently, and the scheduling problem is usually one of resource allocation of the design staff, rather than a problem in sequential planning. Clear identification of the design alternates can be a benefit to the designer as well as the owner in anticipating review requirements, and in placing responsibility on the owner to expedite his own decisions. (It is typical that the owner delays decisions assuming that the designer can work around problems. Conversely, the designer utilizes patience to wait until external pressures force a decision from the owner.)

With so many people concerned with the project, all the time required, with the exception of the building time, is suspect. Referred to as red tape, inefficiency, delays, and similar derogatories, the review and planning stages have their definite roles. Very often, these are overplayed, as each individual tends to see his own activity as the most important, and is therefore willing to take more than his share of project time. Also, in the early planning stages, individuals do not look upon planning time taken as affecting the delivery date of the facility.

To understand the role of the participants in project management, it is first necessary objectively to consider the nature of a project, and the stages involved. This is surprisingly universal to most projects.

CONSTRUCTION

Construction scheduling is usually the responsibility of the contractor, and the concern of the owner. Where a single construction contract is involved, the contractor is the key to all scheduling problems and solutions. In certain cases, however, the owner is required (or chooses) to undertake contracts with several prime contractors. In this case, the owner himself may legally become the coordinating or general contractor.

Some owners have a highly developed construction management team as part of their internal structure. Typical would be major school districts, cities, industrial firms, competitive builders, states, and federal agencies. Organizations which need a highly developed project management team, but usually do not have the on-going staff include: universities, hospitals, airports, and medium to small industrial firms.

Contractors schedule in a broad range of ways and degrees. The successful ones utilize the best techniques, and manage aggressively. The worst ones don't stay in business long—but while they do, they cause problems. The majority are somewhere in between.

It is usual that the contractor does not pre-plan or schedule a project when he is bidding on it. The reason is economic. The contractor wins between 10% and 20% of the jobs he bids, and money spent on planning the jobs not acquired is totally wasted. However, pre-planning can be useful—and many owners are utilizing a *pre-construction working plan* as a convenience to the contractors. This tends to present problems for what they are, and sets an environment of thoughtfulness in regard to the construction planning and scheduling.

The owner can also utilize the pre-construction study to set a reasonable schedule—where the usual case is the establishment of a hopeful schedule.

After the contract has been awarded, the *working plan* or schedule should be developed. The use of network-based scheduling is usually recommended for any medium or major-sized project, and is now relatively well received by contractors.

This working schedule should be developed as soon after the selection of the successful contractor as possible, but there is always a certain degree of initial mobilization activity accomplished strictly on the basis of experience. This should include ordering of long lead items, if they have not already been ordered by the owner. In no case should a project be delayed while it is being scheduled. If later scheduling points out that false starts have been made, these can always be revised.

Project scheduling functions in a time dimension, and the principal benefits are best measured in terms of time. Unfortunately, both owners and contractors often fail to equate time and money. The value of a project to an owner depends on the specifics of the project, and returns something in relationship to its investment value. In most cases, earlier completion means earlier use of the facility in a meaningful way.

To the contractor, a shorter project means better returns, because overhead is reduced. The time advantage which can be realized through a well-organized project scheduling plan is necessarily in proportion to the capability of the contracting team. To a lesser degree, the impact of the intangibles such as weather, site conditions, and other exterior influences tends to obscure the results of good planning.

There is no one correct approach to the preparation of the project schedule. A conference approach, in which key persons involved form the conference, is particularly effective; however, the size of the group must be small enough to represent key areas. The plan developed by the scheduling group must be the one which the contractor expects to use. The planning network can point out the weaknesses of a poor plan just as well as it can develop the strength of a good one. An important ingredient of the successful planning effort is leadership. If the meetings are not well directed, they become disorganized and nonproductive.

The single most important result of the planning efforts is the development of a specific sequence for the project. A secondary but important area is the forecast of projected completion times. Normally, the initial projections which utilize optimum sequences result in too long a duration. Replanning is then required to bring the project back into the acceptable time zone.

The contractor has included a key staff organization as part of his planning and prebid. This group is rapidly expanded in the field through the addition of craftsmen and supervisory personnel.

The project is broken down into its normal phases which would include:

Site preparation and utilities
Foundations
Structure
Close-in
Interior HVAC and electric systems
Interior finishing work
Final finishing work

Naturally, there can be substantial overlap in these areas, and this can assist in successful scheduling and implementation of the project.

During the course of the project, it is almost inevitable that certain areas will be revised or omissions will be noted. Change orders will develop, and can have a substantial impact on the relationship between the contractor and the owner. Another important but often neglected area is the coordination of shop drawing reviews between the owner and the contractors. Delays in returning shop drawings can affect the completion time; conversely, however, contractors will often submit imperfect shop drawings as a ploy to screen problems and delays.

Contractors should have an effective project reporting system for both internal use and reports to the owner. This is often neglected, and difficult to set up post facto.

The coordination of *purchasing activities* with the contractor or with the owner in negotiated contracts can be a very difficult but important task. Again, well-organized planning and scheduling can be a substantial assist to the purchasing department in the timely order and delivery of key materials.

FURNISHING AND MOVING

Very often, the completion of the project appears to be the turnover of the facility from the contractor to the owner. This turnover usually involves a *punch list* prepared by the inspection team. The punch list may be trivial, or could go on for years. The nature of the relationship between the owner and the contractor by the conclusion of the project may directly relate to the length and difficulty of the punch list.

The owner often fails to fully appreciate his responsibilities once the contractor has phased over the building. For one thing, personnel must be provided, having been previously selected and trained, to operate the equipment in the building. The owner must staff and orient personnel far enough in advance to effect a smooth phaseover.

The *move-in* to the facility should be planned, and may be rather complex. In high-rise buildings in an area such as New York City, the move-in requires intensive planning, including studies of vertical loading of elevators, security of equipment, provision of sufficient moving personnel, truck loading dock space as required, waste disposal, and other factors.

The final decorating is often left to the owner's discretion or an interior decorator. This phase must be handled in a timely fashion. Even more important in complex installations is the ordering of equipment. In hospitals, industrial situations, and similar areas, long lead equipment is required, and must be ordered many months before the actual move-in date. Warehousing or drop shipping must be arranged and coordinated.

The actual delivery of the building is only a milestone in the project progress, and the move-in phase will have a substantial effect on the philosophical relation of the owner to his facility in the future.

SUMMARY

The construction phase is only one part of the project. While the most important in many ways, its ability to be successful is determined by the level of planning, by events beyond the control of the parties to the project, and by many other factors. While the role of the Inspector is basically quality control, the climate for inspection is influenced by many factors and considerations, in most cases not related to the inspection itself.

The Inspector has to accept the project cards as they are dealt. Nevertheless, an understanding of the attitudes of the participants, co-workers, and other organizations can help to rationalize and work with the problems generated.

The successful Inspector must be concerned with job coordination and progress, as well as quality. Cooperation and coordination are interdependent, and depend upon the personal relationships, as well as professional relationships, which the inspection team builds with the owner, designer, and contractors. An understanding of the nonconstruction aspects of the project can help the Inspector in maintaining an appropriate perspective.

3. CONTRACT DOCUMENTS

The contract documents describe the contract between the contractor and the owner, and form the basis for inspection. They generally include the contract itself, plus working drawings and specifications.

The working drawings describe in graphic form with dimensions the location, size, arrangement, layout, and spatial relationships of the work to be installed. Their organization is determined to a large degree by the organization of the disciplines within the architect-engineering organization that prepares them. Typically, the *drawing set* is divided into the following sections:

Architectural
Structural
Mechanical (Heating, Ventilating, Air Conditioning)
Mechanical, Plumbing
Electrical

In many public jobs, the local or State law requires the awarding of separate contracts, so that this breakdown is also useful in subdividing the work into the separate prime contracts.

The *architectural drawings* are generally shown in plan view at a scale of either $\frac{1}{8}''$ to 1 ft or $\frac{1}{4}''$ to 1 ft, depending on the size of the project. For a large area project, the plans may be subdivided into several sheets, using major geometric shapes as dividing points, or match lines. Typically, the column lines are marked with letter designations along one axis, and numbers along the opposite to form a location grid pattern for reference purposes. Another approach is the use of an alpha-numeric grid on the drawing margin itself.

The plans are further developed by appropriate elevations showing the outside of buildings or structures, and sections showing a cutaway view. At strategic locations, special sections may be cut to show details. These are shown at a substantially larger scale such as $\frac{1}{2}'' = 1$ ft.

In the *mechanical-electrical* sections, it is usual to use the same outline plan with the same scale as the architectural to present the various systems. The systems are shown in heavy line, superimposed upon the light-line architectural format. As in the architectural section, the mechanical-electrical sections show elevations and cross-sections. An important addition in the mechanical-electrical areas is the use of single line diagrams to indicate functional flow of fluids, gases, or current.

Notes are utilized on the drawings to explain special conditions, and owners having large building programs may reference standard details previously established within their design organization.

SPECIFICATIONS

The role of specifications is the description of the quality of the materials to be placed in the project. Where the drawings present scope in terms of quantities, dimensions, form, and building details, the specifications provide the qualities of materials for construction. Specifications, even for a small project, tend to be voluminous, and in large projects can extend to several thousand pages. Because of the legal connotations involved, the specifications tend to be loaded with all possible information.

Designers tend to evolve their own specification clauses or to use computer-generated draft specifications. In either case there is a tendency to put in all previous applicable clauses, just in case they are needed. Accordingly, something as straightforward as 3000 psi concrete may require several pages of description, including factors such as size and gradation of sand and aggregate, origin of the cement, detailed information describing the reinforcing bar, and other factors not under the direct control of the field group. Since we are all human, some omissions become apparent in the course of construction. To counter this, specifications often have substantial escape or coverall clauses to protect the designer. Further, the specifications typically include the massive amounts of boiler plate— that is, standard insert material used by the particular architectural or engineering organization, often carried forward from prior projects, and quite often including ambiguous or even incorrect material.

Format

Before 1960, many of the major organizations such as the Corps of Engineers and the Navy Bureau of Yards and Docks had their own format for specifications in an attempt to get similarity between the various specifications. These outline or guide specifications were often used by architects and engineers, even when not specifically required. The *Concrete Specifications Institute (CSI)* has done much to develop and promulgate a standardized divisional breakdown of specifications, as follows:

DIVISION	NAME
1	General requirements
2	Site work
3	Concrete
4	Masonry
5	Metals
6	Carpentry
7	Moisture protection
8	Doors, windows, and glass
9	Finishes
10	Specialties
11	Equipment
12	Furnishings
13	Special construction
14	Conveying system
15	Mechanical
16	Electrical

This breakdown lends itself more to building specifications than to heavy construction. Also, the emphasis on architectural items is obvious—with mechanical and electrical being lumped into two major headings, with the remaining 14 applying to architectural and structural.

In 1966, the Association of General Contractors developed a "Uniform System for Building Construction Specifications, Data Filing, and Cost Accounting." This format is based upon the 16 major CSI divisions and provides a breakdown under each division into sections. It is often difficult for the Inspector to dig out just what is meant by the contract specifications; the CSI format at least gives the Inspector an appropriate place to look. With the same purpose in mind, the discussion of technical specification content is organized in this book along CSI format lines. That section of the book (B. Project characteristics) includes CSI Division 2 through 16.

GENERAL CONDITIONS

Traditionally, the "General Conditions" section preceded the technical portion of the specifications. This section included definition of responsibilities of the owner, architect, contractor, and other parties, such as the construction manager. Organizations with an ongoing building program, such as the State of New Jersey or the New York State Dormitory Authority, have a standard set of General Conditions. The American Institute of Architects (AIA) publishes a series of recommended General Conditions (the A201 series) with versions tailored to federal contracts or construction management contracts. A typical set of General Conditions (paraphrased from the Standard General Conditions published by the Division of Building and Construction of the State of New Jersey) follows:

1. Definitions—Notices

A. The Contract Documents consist of the Agreement, Instructions to Bidders and General Conditions of the Contract, the Drawings and Specifications, Bulletins, and Addenda, including all modifications thereof incorporated in the documents before their execution. Whenever the word "Contract" is used herein it means all of the above documents or such part of them as are clearly indicated.

B. Whenever the word "Owner" is used herein, it means the John Doe Company.

C. Whenever the word "Director" is used herein, it means the Director of Construction of the John Doe Company, or his duly authorized representative.

D. Whenever the word "Contractor," "Prime Contractor," or "Single Contractor" is used herein, it means the individual or firm, undertaking to do all work contracted for under the Contract.

E. Whenever the word "Architect" or "Engineer" is used herein, it means the Architect or Engineer engaged by the Owner and acting as the duly authorized representative of the Director.

F. The term "Subcontractor," as employed herein, includes individual or firm having a direct contract with the Contractor, Prime Contractor, or

Single Contractor and it includes one who furnishes labor or material worked to a special design according to the drawings or specifications of this work, but does not include one who merely furnishes material not so worked.

G. When the term "acceptable" or "approved" is used herein, it means that the material or work shall be acceptable to or approved by the Director.

H. The term "Work" of the Contractor, Prime Contractor, Single Contractor, or Subcontractor as used herein, includes labor, materials, plant, and equipment required to complete the contract.

2. Architect's Status

The Architect shall interpret the drawings and specifications, and as agent for the Director shall judge the quantity, quality, fitness, and acceptability of all parts of the work. He shall certify contractor's invoices for work performed and materials delivered to the site, and shall be given access to part of the work for inspection at all times. The Architect shall not have authority to give approval nor order changes in work which alter the terms of conditions of the Contract, nor which involve additional cost. He may, however, make recommendations to the Director for such changes, whether or not costs are revised, and the Director may act, at his discretion, on the basis of the Architect's recommendations.

3. Intention

A. The drawings and specifications, with the Contract of which they form part, are intended to provide for and comprise everything necessary to the proper and complete finishing of the work in every part, notwithstanding that each and every item necessary may not be shown on the drawings or mentioned in the specifications.

B. Each Contractor shall abide by and comply with the true intent and meaning of all the drawings and the specifications taken as a whole, and shall not avail himself of any unintentional error or omission should any exist.

C. Should any error discrepancy appear or should any doubt exist or any dispute arise as to the true intent and meaning of the drawings or of the specifications, or should any portion of same be obscure or capable of more than one construction, the Contractor shall immediately apply to the Architect for the corrections or explanation thereof, and, in case of dispute, the Director's decision shall be final.

D. Determinations will be in the form of Bulletins to the Specifications which will be forwarded to all affected contractors.

E. The sequence of precedence pertaining to contract documents is as follows:

1. Agreement
2. Addenda, Bulletins
3. Specifications (including General Conditions)
4. Details
5. Figured Dimensions.
6. Scaled Dimensions
7. Drawings

F. Any provision in any of the Contract Documents which may be in conflict or inconsistent with any of the paragraphs in these General Conditions shall be void to the extent of such conflict or inconsistency unless the provision is specifically referenced as a supplement or change thereto.

G. Each and every provision of law and clause required by law to be inserted in this contract shall be deemed to be inserted herein and the contract shall be read and enforced as though it were included herein and if, through mistake or otherwise, any such provision is not inserted, or is not correctly inserted, then upon the application of either party the contract shall forthwith be physically amended to make such insertion or correction.

4. Contract Limit Lines

A. Whenever the words "contract limit lines" are used herein, they refer to the lines shown on the drawings, surrounding the contract work beyond which no construction work shall be performed unless otherwise noted on the drawings or specified. Each Contractor shall check and verify conditions outside of the contract limit lines to determine whether or not any conflict exists between elevations or other data shown on the drawings and existing elevations or other data outside of the contract limit lines.

B. Whenever the words "construction site" or "project site" are used herein, they refer to the geographical grounds of the entire Institution, college campus or other Using Agency property at which contract work is performed.

5. Responsibility for Work

A. Each contractor shall be responsible for all damages due to his operations; to all parts of the work, both temporary and permanent; and to all adjoining property.

B. Each Contractor shall protect the Joe Doe Company from all suits, actions, damages and costs of every name and description resulting from the work under this contract.

C. Each Contractor shall provide, in connection with his own work, all safeguards, rails, night lights, and other means of protection against accidents.

D. Each Contractor shall make, use and provide all proper necessary and sufficient precautions, safeguards, and protections against the occurrence of happening of any accident, injuries, damage or hurt to any person or property during the progress of the work.

E. Each Contractor shall, at his own expense, protect all finished work liable to damage, and keep the same protected until the project is completed and accepted by the Director.

6. Superintendence—Supervision—Laying Out

A. At the site of the work the Contractor shall employ a construction superintendent or foreman who shall have full authority to act for the Contractor. It is understood that such representative shall be acceptable to the Architect/Engineer and the Director and shall be one who is to be continued in that capacity for the particular job involved unless he ceases to be on the Contractor's payroll.

B. The various subcontractors shall have competent foremen in charge of their respective part of the work at all times. They are not to employ on the work any unfit person or anyone not skilled in the work assigned to him.

C. Each Contractor shall give the work his special supervision, lay out his own work, do all the necessary leveling and measuring or employ a competent licensed engineer or land surveyor satisfactory to the Director to do so.

D. If, due to trade agreement, additional standby personnel are required to supervise equipment or temporary services used by other trades, the Contractor providing such standby services shall evaluate requirements and include the cost thereof in his bid.

E. All Contractors and Subcontractors shall rely on their own measurements for the performance of their work.

7. Subcontractor—Material Approval

A. Each Prime Contractor shall submit through the Architect or Engineer to the Director for approval a list of all subcontractors, manufacturers, materials, and equipment, whether specified or not, within thirty (30) days after award of contract. No contract shall be entered into with any Subcontractor before his name has been approved in writing by the Director.

8. Subcontractors—Equipment and Materials

A. Each Contractor shall, within thirty (30) days after award of the contract, notify the Director through the Architect or Engineer in writing of the names of subcontractors proposed for the principal parts of the work and for such others as the Director may direct and shall not employ any without prior written approval of the Director or any that the Director may within a reasonable time reject.

B. Each Contractor agrees that he is as fully responsible to the John Doe Company for the acts and omissions of his subcontractors and of persons either directly or indirectly employed by them, as he is for the acts and omissions of persons directly employed by him.

9. Material—Workmanship—Labor

A. All material and work shall conform to the best practice. Only the best of the several kinds of materials shall be used, and the work carefully carried out in strict accordance with the general and detail drawings, under the supervision of the Director or his accredited representative who shall have full power at any time to reject such work or material which does not in his opinion conform to the true intent and meaning of the drawings and specifications.

B. All work, when completed in a substantial and workmanlike manner, to the satisfaction of the Director, shall be accepted by him in writing. Unless otherwise specified all materials used shall be new.

C. Each Contractor shall furnish and pay all necessary transportation, scaffolding, forms, water, labor, tools, light and power, mechanical appliances, and all other means, materials, and supplies for properly prosecuting his work under contract, unless expressly specified otherwise.

10. Defective Work and Materials

A. Any materials or work found to be defective, or not in strict conformity with the requirements of the drawings and specifications, or defaced or damaged through the negligence of any Contractor or his Subcontractors or employees, or through action of fire or the weather or any causes, shall be removed immediately and new materials or work substituted therefor without delays by the Contractor involved.

11. Inspection of Work

A. The Director or his authorized representative shall at all times have access to the work whether it is in preparation or in progress and the Contractor shall provide proper facilities for such access and for inspection.

BUSINESS CONSIDERATIONS

Business items are also included, providing the basis for payment and progress payments—perhaps one of the best means of controlling contractor cooperation. Some typical information would include:

1. Unit Schedule Breakdown

The Contractor shall file with the Director a unit schedule breakdown in sufficient detail to include the following which will be used as the basis for determining the amount of payment to be made on a periodic basis for work completed and installed in accordance with Contract Documents:

(1) Description of material or equipment and number of units involved.

(2) Lump sum price for labor and lump sum price for equipment and/or material listed.

(3) Lump sum allowances included in the specifications.

The total of items shall equal the lump sum contract price.

2. Payment

A. The basis for computing monthly progress payments shall be the Unit Schedule breakdown. The Unit Schedule breakdown shall be submitted for approval to the Director through the Architect within ten (10) days from date of written notice to proceed by the John Doe Company.

B. In making such partial payments for work, there will retained 10% of estimated amount until final acceptance and completion of all work covered by the contract: Provided, however, that after 50% of the work has been completed, if the Director determines that Contractor's performance and progress are satisfactory, the John Doe Company may make remaining partial payments in full for work subsequently completed.

3. Additions—Deductions—Deviations

The Director at his discretion, may at any time during the progress of the work authorize additions, deductions, or deviations from the work described in the specifications as herein set forth; and the contract shall not be vitiated or the surety released thereby.

A. Additions, deductions, deviations may be authorized as follows at Director's option:

(1) On the basis of unit prices specified.

(2) On a lump sum basis.

(3) On a time and material basis.

(4) Standby time or overtime.

4. Payments Withheld

A. The Director may withhold or, on account of subsequently discovered evidence, nullify the whole or a part of any certificate for payment to such extent as may be necessary to protect the owner from loss on account of:

(1) Defective work not remedied.
(2) Claims filed, or reasonable evidence indicating probable filing of claims.
(3) Failure of any Contractor to make payments promptly to subcontractors or for material or labor.
(4) A reasonable doubt that the contract can be completed for the balance then unpaid.
(5) Damage to another contractor.

B. When all the above grounds are removed, certificates will at once be issued for amounts withheld because of them.

5. Construction Progress Schedule

A. The Contractor for the General Construction Contract shall be responsible for preparing and furnishing, before the first contract requisition date, a coordinated single progress schedule which incorporates progress schedules of all Prime Contractors engaged on the project. The schedule shall be an arrow diagram network, bar chart, or other recognized graphic progress schedule in a form and in sufficient detail satisfactory to the Director.

OTHER TOPICS

Temporary Facilities are also described in the General Conditions and include items such as:

Construction sign
Temporary drives and walks
Temporary buildings and sanitary facilities
Temporary water
Temporary light and power
Temporary heat
Temporary enclosures, glass breakage and cleaning

Hoisting facilities
Fire protection

The General Conditions include a description of the manner in which the owner will utilize basic approaches to control the job, with sections as follows:

Drawing; specifications; shop drawings; as-built drawings
Samples
Testing of Equipment
Concrete and other structural testing
Photographs
Job Meetings

There are special sections on business matters such as:

Federal taxes
Guarantees
Right of the owner to terminate contract
Lands and rights of way
Disputes, claims, appeals
Withholding of payments
Payrolls and basic records
Minimum wages
Anti-kickback act
Statute prohibiting discrimination—Civil rights
Equal employment opportunity
Certification of non-segregated facilities

SPECIAL CONDITIONS

Particularly where the General Conditions are a standard package evolved over time and through experience, it may be necessary to add a section entitled "Special Conditions" or "Supplemental Requirements." This section modifies the General Conditions to reflect the specific and special circumstances of the project.

GENERAL REQUIREMENTS

The CSI Division One entitled "General Requirements" can be utilized to describe all, or part, of the General and/or Special Conditions. The format for this division follows:

DIVISION 1: GENERAL REQUIREMENTS

NUMBER   **TITLE**

01010   **SUMMARY OF WORK**
          Work Covered by Contract Documents
−11   Contracts
−12   Work by Others
−13   Future Work
−14   Work Sequence
−15   Contractor Use of Premises
−16   Occupancy
          Owner Occupancy
          Partial Occupancy
          Continued Occupancy
          Maintenance of Operation
−17   Pre-Ordered Products
−18   Owner-furnished Items

01020   **ALLOWANCES**
−21   Cash Allowances
−22   Inspection Testing Allowances
−23   Contingency Allowance

01030   **SPECIAL PROJECT PROCEDURES**
−31   Alterations Project Procedures
−32   Hospital Project Procedures
−33   Industrial Project Procedures
−34   Nuclear Project Procedures

01040   **COORDINATION**
−41   Project Coordination
−42   Mechanical and Electrical Coordination
−45   Cutting and Patching

01050   **FIELD ENGINEERING**
−51   Grades, Lines, and Levels

34

01060   **REGULATORY REQUIREMENTS**
  −61   Building Codes
  −65   Mechanical Codes, Fees, and Lateral Costs
  −66   Electrical Codes, and Fees

01070   **ABBREVIATIONS AND SYMBOLS**
  −71   Architectural Abbreviations and Symbols
  −75   Mechanical Reference Symbols
  −76   Electrical Reference Symbols

01080   **IDENTIFICATION SYSTEMS**
  −85   Control Identification Systems
  −86   Piping Identification Systems
  −87   Electrical Identification Systems

01100   **ALTERNATES/ALTERNATIVES**

01150   **MEASUREMENT AND PAYMENT**
  −51   Unit Prices
  −52   Applications for Payment
  −53   Change Order Procedures
01160−01199 (Reserved)

01200   **PROJECT MEETINGS**
  −01   Preconstruction Conferences
  −02   Progress Meetings
  −03   Job Site Administration
01210−01299 (Reserved)

01300   **SUBMITTALS**
  −10   Construction Schedules
  −11   Network Analysis Schedules
  −20   Progress Reports
  −30   Survey Data
  −40   Shop Drawings, Product Data, and Samples
  −50   Mock-ups
  −60   Layout Data
  −70   Schedule of Values
  −80   Construction Photographs
01390−01399 (Reserved)

01400   **QUALITY CONTROL**
  −10   Testing Laboratory Services
  −20   Inspection Services

01500   **CONSTRUCTION FACILITIES AND TEMPORARY CONTROLS**

  −510   **TEMPORARY UTILITIES**
  −11   Temporary Electricity

−12	Temporary Lighting
−13	Temporary Heating, Cooling, and Ventilating
−14	Temporary Telephone Service
−15	Temporary Water
−16	Temporary Sanitary Facilities
−17	Temporary First Aid Facilities
−18	Temporary Fire Protection
−19	Temporary Construction
	Interim Bridges
	Interim Overpasses
	Interim Runarounds
	Interim Decking

−520	**CONSTRUCTION AIDS (Data Filing Only)**
−21	Construction Elevators and Hoists
−22	Temporary Enclosures
−23	Swing Staging
−24	Scaffolding and Platforms

−530	**BARRIERS**
−31	Fences
−32	Tree and Plant Protection
−33	Barricades

−540	**SECURITY**
−45	Protection of Work and Property

−550	**ACCESS ROADS AND PARKING AREAS**
−51	Access Roads
−52	Parking Areas

−560	**TEMPORARY CONTROLS**
−61	Noise Control
−62	Dust Control
−63	Water Control
−64	Pest Control
−65	Rodent Control
−66	Debris Control
−67	Pollution Control
−68	Erosion Control

01570	**TRAFFIC REGULATION**
−71	Traffic Signals
−72	Flagmen
−73	Flares and Lights
−74	Construction Parking Control
−75	Haul Routes

01580   PROJECT IDENTIFICATION AND SIGNS

01590   FIELD OFFICES AND SHEDS

01600   MATERIAL AND EQUIPMENT
 −10   Transportation and Handling
 −20   Storage and Protection
 −30   Substitutions and Product Options

01650   STARTING OF SYSTEMS
 −55   Starting Piping Systems
 −56   Disinfection of Domestic Water Lines
 −57   Starting Process Systems

01660   TESTING, ADJUSTING, AND BALANCING OF SYSTEMS
 −65   Testing Mechanical Systems
 −66   Testing Piping Systems
 −67   Balancing Air Systems
 −68   Testing Electrical Systems
 −69   Testing Process Systems

01700   CONTRACT CLOSEOUT
 −10   Cleaning
 −20   Project Record Documents
 −30   Operating and Maintenance Data
 −35   Equipment Systems Demonstrations
 −36   Mechanical Systems Demonstrations
 −37   Electrical Systems Demonstrations
 −38   Process Systems Demonstrations
 −39   Conveying Systems Demonstrations
 −40   Warranties and Bonds
 −50   Spare Parts and Maintenance Materials
01800   **MAINTENANCE (Data Filing Only)**
01850–01999 (Reserved)

RECOMMENDED ORGANIZATION OF GENERAL REQUIREMENTS (from CSI MASTERFORMAT)

SUMMARY

The inspection team must become familiar with the plans and the specifications. These form the basis upon which the inspection is made. The inspector must be prepared for the fact that errors and omissions will be discovered, and must be duly reported to management. Wherever the error or omission is substantial, it is appropriate to review it with the designer first to insure that the error noted is valid.

The Inspector is on the job to be objective, and to enforce the contract as written. When the contract does not suffice to cover the work conditions, then the Inspector should request the opportunity to prepare an appropriate clarification or change order to the contract. The question of whether a change order is to result in a credit to the owner or to the contractor is a matter of interpretation of the plans and specifications. Great care should be taken to avoid casual commitment or agreement with change items. The Inspector should first research the change or requirements in the plans and specifications to determine that it is not included in the contractual scope—either partially or completely. Next, he should collaborate with his own management team, receiving their agreement before making any commitments or carrying on any conclusive discussions with the contractor. From that point on, standard change order procedures, approvals, and documentation should be followed.

4. CODES AND STANDARDS

Construction specifications often incorporate other standards which further define the quality requirements for the materials or methods to be used in the construction process. Appendix A includes a list of codes promulgated by the American Society for Testing and Materials, which lists the ASTM standards used in building construction both by alphanumeric designation of the standards and by construction category.

Appendix B is a similar list for construction-oriented standards promulgated by the American National Standards Institute. The standards promulgated by these two organizations represent the majority of those referenced in typical construction specifications. The field inspection team will need a library of appropriate standards in order to fulfill its task.

A building code may also incorporate standards. For instance, the building code for the City of New York is 472 pages long, but has as an appendix 371 pages of building code reference standards. Many of these, in turn, reference ASTM and ANSI standards.

When construction is accomplished within a jurisdiction, the building code of that jurisdiction, such as the City of New York, usually applies. There are certain exceptions. For instance, state and federal agencies are not subject to the sovereignty of a city, and therefore are not subject to its building codes. Often as a courtesy, and also a practicality, the state or federal agency chooses to abide by the building code, and requires its designers and inspectors to do so.

A major responsibility under a building code lies with the design professionals. It is their obligation to design the facility in accordance with applicable building codes. This responsibility usually culminates in the submittal of the completed design to the licensing body authorized by the building codes for review and approval. The inspection team, in turn, starts with an approved set of documents which conform to the appropriate building code. Even in the case of a design-builder, state law usually requires that the design documents be prepared by licensed architect-engineers and appropriate submittals made on behalf of the design-builder by these professionals. Accordingly, the field inspection team should be able to rely upon the approved documents as being in conformance with the appropriate building code.

There are, however, portions of the typical building code which require specific action on the part of the inspection team.

The contents of the New York City Building Code are as follows:

TITLE A—GENERAL PROVISIONS   1

TITLE B—LICENSES

1. General Provisions   7
2. Master Plumbing License   10
3. Welder License   12
4. High-Pressure Boiler Operating License   13
5. Hoisting Machine Operator License   15
6. Rigger Licenses   17
7. Sign Hanger Licenses   20
8. Oil-Burning Equipment Installer Licenses   23
9. Concrete Testing Laboratory License   25

TITLE C, PART 1—BUILDING CONSTRUCTION

1. General Provisions   26
2. Permits   27
3. Fees   28
4. Inspections   36
5. Certificates of Occupancy   37
6. Projections Beyond Street Line   38
7. Safety in Building Operations   39
8. Unsafe Buildings and Property   42
9. Miscellaneous Provisions   53

TITLE C, PART II—BUILDING CODE

1. **Administration and Enforcement**
   100.0 General Provisions   55
   101.0 Matters Covered   56
   102.0 Continuation and Change in Use   56
   103.0 Alteration of Existing Buildings   57
   104.0 Minor Alterations: Ordinary Repairs   58
   105.0 Maintenance   59
   106.0 Materials, Assemblies, Forms and Methods of Construction   59
   107.0 Service Equipment   62
   108.0 Approval of Plans   63
   109.0 Permits   65
   110.0 Applications for New Building Permits   66
   111.0 Applications for Building Alteration Permits   68

112.0 Applications for Foundation and Earthwork Permits   69
113.0 Applications for Demolition and Removal Permits   69
114.0 Applications for Plumbing Permits   70
115.0 Applications for Sign Permits   71
116.0 Applications for Equipment Work Permits   72
117.0 Applications for Equipment Use Permits   76
118.0 Issuance of Permits   77
119.0 Conditions of Permit   79
120.0 Department Inspections   79
121.0 Certificates of Occupancy   81
122.0 Posting Buildings   83
123.0 Stop Work Order   84

2. **Definitions**
    200.0 General   85
    201.0 Definitions   85
    202.0 Abbreviations   104

3. **Occupancy and Construction Classifications**
    300.0 General   106
    301.0 Occupancy Classifications   106
    302.0 Occupancy Group A-High Hazard   108
    303.0 Occupancy Group B-Storage   109
    304.0 Occupancy Group C-Mercantile   109
    305.0 Occupancy Group D-Industrial   110
    306.0 Occupancy Group E-Business   110
    307.0 Occupancy Group F-Assembly   110
    308.0 Occupancy Group G-Educational   111
    309.0 Occupancy Group H-Institutional   111
    310.0 Occupancy Group J-Residential   111
    311.0 Occupancy Group K-Miscellaneous   112
    312.0 Doubtful Occupancies   112
    313.0 Construction Classifications   112
    314.0 Construction Group I-Noncombustible   113
    315.0 Construction Group II-Combustible   113
    316.0 Mixed Construction   116

4. **Building Limitations**
    400.0 General   117
    401.0 Fire Department Access   117
    402.0 Fire Districts   118
    403.0 Limitations Inside the Fire Districts   118
    404.0 Limitations Outside the Fire Districts   119
    405.0 Area Limitations   119
    406.0 Height Limitations   124
    407.0 General Projection Limitations   124
    408.0 Permissible Projections Beyond Street Line   125

5. **Fire-Protection Construction Requirements**
    500.0 General   128
    501.0 Fire Protection Test Procedures   128
    502.0 Fire Resistance Requirements   128

503.0 Prevention of Exterior Fire Spread — 132
504.0 Prevention of Interior Fire Spread — 139

6. Means of Egress

600.0 General — 152
601.0 Determination of Exit Requirements — 152
602.0 Location of Exits — 156
603.0 Number of Exits — 156
604.0 Access Requirements and Exit Types — 157
605.0 Exit Lighting — 174
606.0 Exit Signs — 174
607.0 Exit Signs for Existing Buildings — 176

7. Special Uses and Occupancies

700.0 General — 178
701.0 High Hazard Occupancies — 178
702.0 Occupancies Involving Spray or Dip Finishing — 179
703.0 Uses and Occupancies Involving Radioactive Materials and Radiation-Producing Equipment — 180
704.0 Boiler and Furnace Rooms — 182
705.0 Dry Cleaning Establishments — 183
706.0 Heliports — 185
707.0 Automotive Service Stations — 186
708.0 Automotive Repair Shops — 186
709.0 Public Garages — 187
710.0 Open Parking Structures — 190
711.0 Private Garages — 191
712.0 Open Parking Lots — 192
713.0 Trailer Camps — 193
714.0 Swimming Pools — 194
715.0 Radio and Television Towers — 195
716.0 Outdoor Signs and Display Structures — 196
717.0 Fences — 200
718.0 Tents and Air Supported Structures — 201

8. Places of Assembly

800.0 General — 204
801.0 Basic Requirements — 204
802.0 F-1 Places of Assembly — 216
803.0 F-2 Places of Assembly — 221
804.0 F-3 and F-4 Places of Assembly — 223

9. Loads

900.0 General — 224
901.0 Dead Loads — 224
902.0 Live Loads — 225
903.0 Live Load Reduction — 230
904.0 Wind Loads — 231
905.0 Other Loads — 231
906.0 Distribution of Loads — 232

10. Structural Work

1000.0 Scope and General Requirements — 235

1001.0 Structural Design—General Requirements   239
1002.0 Adequacy of the Structural Design   240
1003.0 Masonry   243
1004.0 Concrete   243
1005.0 Steel   250
1006.0 Wood   250
1007.0 Aluminum   258
1008.0 Reinforced Gypsum Concrete   258
1009.0 Thin Shell and Folded-Plate Construction   259
1010.0 Suspended Structures   259
1011.0 Glass Panels   263

11. Foundations
1100.0 General   266
1101.0 Soil Investigations   268
1102.0 Foundation Loads   272
1103.0 Allowable Soil Bearing Pressures   273
1104.0 Soil Load Bearing Tests   282
1105.0 Footings, Foundation Piers, and Foundation Walls 285
1106.0 Pile Foundations—General Requirements   288
1107.0 Pile Foundations—Loads   294
1108.0 Pile Driving Operations   306
1109.0 Pile Types—Specific Requirements   307
1110.0 Underpinning   314
1111.0 Stability   314
1112.0 Inspection   315

12. Light, Heat, Ventilation, and Noise Control
1200.0 General   317
1201.0 Existing Buildings   317
1202.0 Standards of Natural Light   317
1203.0 Standards of Artificial Light   319
1204.0 Standards of Heating   319
1205.0 Standards of Natural Ventilation   321
1206.0 Standards of Mechanical Ventilation   322
1207.0 Ventilation of Special Spaces   324
1208.0 Noise Control in Multiple Dwellings   328

13. Mechanical Ventilation, Air Conditioning and Refrigeration Systems
1300.0 General   336
1301.0 Inspections and Tests for Equipment Use Permits   337
1302.0 Operation and Maintenance   339
1303.0 Posting   339
1304.0 Code Requirements of Other City Departments   339

14. Heating and Combustion Equipment
1400.0 General   340
1401.0 Inspections and Tests for Equipment Use Permits   340
1402.0 Licenses and Certificates   342
1403.0 Abatement of Air Contaminants   343
1404.0 Equipment Standards   343
1405.0 Equipment Classification   343

1406.0 Equipment Foundation Mountings   344
1407.0 Equipment Clearances   344
1408.0 Combustion Air   344
1409.0 Piping Containing Steam, Hot Water or Other Fluids   345
1410.0 Residence-Type Warm Air Heating Systems   345
1411.0 Unit Heaters   346
1412.0 Floor Furnaces   346
1413.0 Boilers   347
1414.0 Unfired Pressure Vessels   347
1415.0 Gas Fired Equipment   347
1416.0 Fuel Oil Equipment   347
1417.0 Refuse Disposal Systems   357
15. Chimneys and Gas Vents
1500.0 General   362
1501.0 Chimneys   362
1502.0 Chimney Connectors   372
1503.0 Incinerator Chimneys and Refuse Charging Chutes   374
1504.0 Gas Vent Systems   377
16. Plumbing and Gas Piping
1600.0 General   384
1601.0 Water Supply Systems   388
1602.0 Drainage Systems   388
1603.0 Hospital and Institutional Plumbing   389
1604.0 Swimming Pools   389
1605.0 Existing Buildings and Installations   389
1060.0 Inspection and Tests   390
17. Fire Alarm, Detection and Extinguishing Equipment
1700.0 General   393
1701.0 Existing Buildings—Fire Extinguishing Equipment   393
1702.0 Standpipe Requirements   394
1703.0 Automatic Sprinkler Requirements   406
1704.0 Interior Fire Alarm and Signal System   415
18. Elevators and Conveyors
1800.0 General   421
1801.0 Existing Installations   423
1802.0 Tests and Test Interval   424
1803.0 Equipment Permits   425
1804.0 Equipment Operation   426
19. Safety of Public and Property During Construction Operations
1900.0 General   428
1901.0 Maintenance of Site and Adjacent Areas   430
1902.0 Protection of Adjoining Property   436
1903.0 Excavation Operations   437
1904.0 Erection Operations   441
1905.0 Demolition Operations   446
1906.0 Repair and Alteration Operations   450
1907.0 Scaffolds   450
1908.0 Structural Ramps, Runways, and Platforms   461

1909.0 Material Handling and Hoisting Equipment   463
1910.0 Explosive Powered and Projectile Tools   471
Appendix—Reference Standards
  RS2 Definitions   A1
  RS3 Occupancy and Construction Classifications   A2
  RS4 Building Limitations   A21
  RS5 Fire-Protection Construction Requirements   A26
  RS7 Special Uses and Occupancies   A29
  RS8 Places of Assembly   A31
  RS9 Loads   A31
  RS10 Structural Work   A40
  RS11 Foundations   A85
  RS12 Light, Heat, Ventilation, and Noise Control   A86
  RS13 Mechanical Ventilation, Air Conditioning and Refrigeration Systems A88
  RS14 Heating and Combustion Equipment   A10
  RS15 Chimneys and Gas Vents   A119
  RS16 Plumbing and Gas Piping   A122
  RS17 Fire Alarms, Detection and Extinguishing Equipment   A205
  RS18 Elevators and Conveyors   A320
  RS19 Safety of Public and Property During Construction Operations   A332
  New York City Charter and Other Administrative Code Amendments   A361

LICENSES

Title B under Licenses describes eight different types of licenses required in the City of New York. It is the responsibility of the inspection team to assure that those requiring licenses have, in hand, a current license before they work on the project. In certain cases, the license relates to supervision, as in the case of Master Plumber, Master Rigger, Special Rigger, Sign Hanger, or Concrete Testing Laboratory. Other of the licenses relate directly to the craftsmen involved, including: Welder (structural), High Pressure Boiler Operator, Hoisting Machine Operator, and Oil Burning Equipment Installer.

As noted previously, it is the responsibility of the designer and/or owner to have plans reviewed and approved. The New York City Code states that plans will be stamped or endorsed "Approved," and then, "One set of such approved plans shall be retained in the department office of the Borough in which the building premises or equipment is located; and after the issuance of a work permit, a second set of such approved plans shall be retained at the place where the building premises or equipment is located, and it shall be open at all times to inspection by the Commissioner and his authorized representatives until final inspection of the work is completed." Accordingly, it is the responsibility of the inspection team to be sure that such a set of approved documents is available. Further, even though the plans have been approved, typically it is still the responsibility of contractors to acquire permits for either all of or stages of the work. In New York, these include:

New Building Permits
Alteration Permits
Foundation and Earthwork Permits
Demolition and Removal Permits
Plumbing Permits
Sign Permits
Equipment Work Permits
Equipment Use Permits

The typical code describes what is required for the acquisition of permits, including plans required, notification of adjoining owners, protection of adjoining property, and similar requirements. Typically, the permit is deemed to confirm compliance with the building code, and also reserves certain rights such as right of entry and inspection on the part of the Building Department.

Usually, the Building Department reserves the right to inspect during the progress of the work, and prior to issuance of any certificate of occupancy or equipment use will require a final inspection. This final inspection should, of course, be preceded by a complete checkout and punch of the building on the part of the inspection team in residence.

The occupancy-in-construction classifications are intended to first classify the use of the structure and then delineate the safety and fire requirements in terms of egress and fire protection for its various usages. This in turn provides criteria for the design, including interior fire spread rate, access requirements, and exit types. In the New York City code, this type of criteria description is covered in Articles 3 through 8 (pages 106–223).

In C26-106.3, the New York City Code defines a special level of inspection known as controlled inspection:

(a) Controlled inspection.—All such materials which are designated for "controlled inspection" under the provisions of this code shall be inspected and/or tested to verify compliance with code requirements. Unless otherwise specifically provided by code provisions, all required inspections and tests of materials designated for "controlled inspection" shall be made and witnessed by or under the direct supervision of an architect or engineer retained by or on behalf of the owner or lessee, who shall be, or shall be acceptable to, the architect or engineer who prepared or supervised the preparation of the plans; and the architect or engineer by whom, or under whose direct supervision, the required inspections and test reports, together with his signed statement that the material and its use or incorporation into the work comply with code requirements, unless the filing of such reports and statement is specifically waived by code provisions.

(b) Semicontrolled inspection.—All such materials that are not designated for controlled inspection under the provisions of this code shall be subject to semicontrolled inspection, and, as such shall be inspected and/or tested to verify compliance with code requirements by the person superintending the use of the material or its incorporation into the work, except that all required inspections and tests may, at the option of the owner or lessee, be made and witnessed by or under the direct supervision of an architect or engineer retained by or on behalf of the owner or lessee, who shall be, or shall be acceptable to, the architect or engineer who prepared or supervised the preparation of the plans. The person superintending the use of the material or its incorporation into the work, or the architect or engineer by or under whose direct supervision the required inspections and tests are made and witnessed, as the case may be, shall file with the department signed copies of all required inspection and test reports, together with his signed statement that the material and its use or incorporation into the work comply with code requirements, unless the filing of such reports and statement is specifically waived by code provisions.

(c) Off-site inspection.—In all cases where code provisions require that the inspection and/or test of materials be made off-site, or prior to actual use or incorporation into the work, the inspector shall mark or cause to be marked for identification all units (or packages of units) of the material inspected; and the reported results of such inspection shall state that the material was so marked for identification.

Table 1. Inspection of Materials and Assemblies

Materials	Elements That Shall Be Subject to Controlled Inspection[a,b,d]	Elements That Are Not Subject to Controlled Inspection[a,c,d]
Steel	None	All structural elements and connections.
Concrete	Materials for all structural elements proportioned on the basis of calculated stresses 70 per cent or greater, of basic allowable values.	(1) All materials for structural elements proportioned on the basis of calculated stresses less than 70 per cent of basic allowable values. (2) Concrete materials for: (a) Short span floor and roof construction proportioned as per section 1004.8 (Empirical formulas) (b) Walls and footings for buildings in occupancy group J-3. (Residential) (3) Metal reinforcement.
Aluminum	None	All structural elements and connections.
Wood	None	All structural elements and connections.
Reinforced gypsum concrete	None	All structural elements.
Masonry	None	All structural elements.

[a]For general provisions relating to inspection, see section C26-106.3.

[b]All structural materials and assemblies subject to controlled inspection shall be tested and/or inspected at their place of manufacture and evidence of compliance with the provisions of this article shall be provided as stipulated in sub-articles 1003.0 through 1011.0.

[c]Mill, manufacturer's, and supplier's inspection and test reports will be accepted as evidence of compliance with the provisions of this code for all structural materials and assemblies not subject to controlled inspection.

[d]Basic allowable stress values as referenced herein shall denote allowable stress value without increase for infrequent stress conditions as established in this code or in the applicable reference standard for the material or element in its proposed use.

Table 2. Inspection of Methods of Construction

Materials	Operations on Structural Elements That Shall Be Subject to Controlled Inspection[a,b,d]	Operations on Structural Elements That Are Not Subject to Controlled Inspection[a,c,d]
Steel	(1) Welding operations and the tensioning of high strength bolts in connections where calculated stresses in the welds or bolt are 50 per cent or more of basic allowable values. (2) Connection of fittings to wire cables for suspended structures, except where cables together with their attached fittings are proof-loaded to not less than 55 per cent of ultimate capacity.	(1) Welding operations and the tensioning of high strength bolts in connections where the calculated stresses in the welds or bolts are less than 50 per cent of basic allowable values. (2) All other fabrication and erection operations not designated for controlled inspection.
Concrete	Except for those operations specifically designated in this table as not subject to controlled inspection, for all concrete.	(1) All operations relating to the construction of members and assemblies (other than prestressed members) which involve the placement of a total of less than 50 cubic yards of concrete and wherein said concrete is used at levels of calculated stress 70 per cent or less of basic allowable values. (2) Placing and curing of concrete for all: (a) Short span floor and roof construction as per section 1004.8. (Empirical design) (b) Walls and footings for buildings in occupancy group J-3. (Residential) (3) Size and location of reinforcement for walls and footings for buildings in occupancy group J-3.
Aluminum	Welding operations in connections where the calculated stresses in the welds are 50 per cent or more of the basic allowable values.	(1) Welding operations in connections where the calculated stresses in the welds are less than 50 per cent of basic allowable values. (2) All other fabrication and erection operations not designated for controlled inspection.

Table 2. (*cont.*)

Materials	Operations on Structural Elements That Shall Be Subject to Controlled Inspection[a,b,d]	Operations on Structural Elements That Are Not Subject to Controlled Inspection[a,c,d]
Wood	Fabrication of glued-laminated assemblies and of plywood components.	All other operations not designated for controlled inspection.
Reinforced Gypsum Concrete	None	All operations incident to the fabrication and placement of structural elements.
Reinforced Masonry	(1) Fabrication of prefabricated units. (2) Placement and bedding of units; sizes of members, including thickness of walls and wythes; sizes of columns; the size and position of reinforcement, in place, and provisions for curing and protection against freezing for all reinforced masonry construction unless such operations are specifically not designated for controlled inspection.	(1) All operations relating to the construction of members and assemblies which involve the placement of a total of less than 50 cubic yards of masonry and wherein said masonry is used at levels of calculated stress 70 per cent or less of basic allowable values. (2) All masonry work for buildings in occupancy group J-3. (3) All mixing of mortar. (4) All other operations not designated for controlled inspection.
Unreinforced Masonry	Placement and bedding of units and sizes of members including thickness of walls and wythes; sizes of columns; and provisions for curing and protection against freezing for all masonry construction proportioned on the basis of structural analysis as described in section 4 of reference standard RS 10-1, unless such operations are specifically not designated for controlled inspection.	(1) All operations relating to the construction of members and assemblies which involve the placement of a total of less than 50 cubic yards of masonry and wherein said masonry is used at levels of calculated stress 70 per cent or less of basic allowable values. (2) All masonry work for buildings in occupancy group J-3. (3) All mixing of mortar. (4) All other operations not designated for controlled inspection.

[a]For general provisions relating to inspection see section C26-106.3.
[b]All construction operations designated for controlled inspection shall be inspected by the architect or engineer designated for controlled inspection during the performance of such operation.
[c]Certification by the fabricator or erector, as applicable, will be accepted as evidence of compliance with the provisions of this code for all construction operations not subject to controlled inspection.
[d]Basic allowable stress values as referenced herein shall denote allowable stress value without increase for infrequent stress conditions as established in this code or in the applicable reference standard for the material or element in its proposed use.

5. JOB MOBILIZATION

At the start of a project, the contractors, whether selected by negotiation or competitive bid, must mobilize their forces. This initial period of organization is also one of organization for the inspection staff. Usually, it is good practice for the resident team, including the inspectors, to conduct a job conference. This initial conference reviews the steps which the contractor will take at the beginning of the project to get organized, and to respond to contractual commitments.

These contractual requirements include, among other things, the performance bonds, liability insurance by the contractor, equal opportunity compliance documentation, building permit, and other special requirements of the project.

In addition, the contractor at this stage submits the initial material on job schedule, payment schedule, shop drawing submission plan, and materials submission schedule. In addition, special information such as the trial concrete mix should be submitted during this mobilization phase.

The special conditions of the contract may require the contractor to furnish facilities for the inspection team if the site is in a remote location. The initial meeting will review factors such as this, and situations such as temporary heat for the building and other special requirements of the job.

BONDING

It is usual practice for the owner to require the contractor to submit a bid bond with his bid. The purpose of this bond, issued by the contractor's bonding company, is to insure that if awarded the job he will perform the work. Further, once selected, it is common practice for an owner to require the contractor to post a completion bond which permits the owner to complete the work. Should the contractor default, the bonding company will arrange to complete the work, providing the funding for this contractor or another one to carry out the work to conclusion.

Just as the borrower who has sufficient collateral—and therefore less real need to borrow—is more favorably received at the bank, so is the contractor who causes the least concern in regard to his ability to complete the work. The successful (stable in both the financial sense and business reputation) contractor has less problem in receiving a bond than the less well-known, or financially unstable contractor.

The bonding company reviews the contractor's financial statements, work experience, backlog of current work, credit references, and current status with material suppliers and subcontractors before agreeing to provide either bid or completion bonds.

The fee for bonds is a function of the risk involved, and ranges from less than 1% to several percent depending upon the situation. The bonding company is interested in the cumulative amount of work in progress, and the contractor's ability to be bonded in cumulative terms is a direct reflection of his reputation, financial standing, and general well-being.

The contractor may also be required to supply a special bond to suppliers, subcontractors, and others. Generally, the more required in this area, the less stable the contractor.

In difficult relationships with contractors, it is not unusual for the owner to threaten to turn the job over to the bonding company. This is much more easily said than done—since the bonding company specializes in not taking over jobs. In any confrontation, the inspection team can anticipate that the bonding company will be very much on the side of the contractor—since it is their preference that he finish the job on his funds and his terms. While the bonding company is committed to insure that the contractor, if in default, will be replaced or facilitated in the completion of the building, it is not unusual, even in cases of bankruptcy, for a bonding company to keep the same contractor on the job. While bonds do provide some protection, they do not generally insure that a building will be completed in a timely fashion. This also becomes a negotiable area in cases of default. Further, even in cases in which the bond does agree to a hard and fast completion date, that date can be shifted or negated by changes in scope or changes in conditions beyond the control of the owner or the contractor.

INSURANCE

The General Conditions of the contract establish the level of insurance which the contractor must carry. The owner's interest is principally to avoid any liability which may pass through the underinsured contractor to the owner. Today, it is quite useful for owners and other parties to the contract to be sued, as well as a contractor, for injuries, accidents, damages, and other mishaps on a project even when they occur within the physical and contractual limits of the contractor's field of responsibility.

The inspection team or the owner's office staff should periodically check on the validity of the insurance—specifically that it has been neither cancelled nor allowed to lapse.

Generally, as a matter of state law, the contractor must agree to standard work and compensation and employer's liability policies. There are variations in different states, and the owner's reviewing personnel must be aware of the requirements of the state in which the project is being built.

Beyond the minimal/statutory Workman's Compensation and Employer's Liability, most contracts require the contractor to carry a comprehensive general liability policy. Typically, the general liability policies are broken down into five divisions as follows:

Division 1: *Operations on Premises*: for bodily injury or property damage caused by occurrences on the premises owned or leased by the insured anywhere in the country, or as limited by the policy.

Division 2: *Elevator Liability*: covers legal liability for injury or property damage by occurrence involved in the ownership, maintenance, and use of elevators which are controlled and operated by the insured, usually not including material hoists for construction operations.

Division 3: *Contractor's Protective Liability*: covering the contractor's legal liability for bodily injury or property damage arising out of operations performed for another. Policies usually include provisions to cover secondary liabilities resulting from sublet operations or subcontracts.

Division 4: *Completed Operations*: standard policy includes protection of completed operation, and to some degree repeats coverage under Division 1. Usually, it is the contractor's responsibility to protect his completed work, but often the experienced owner will require this additional protection to preclude the necessity for litigation where serious vandalism or damage arises on a project—and the contractor is unable financially to repay the difference on his own account.

Division 5: *Contractual Liability*: covers contractor's liability for bodily injury or property damage which is assumed under written contract with other parties. This would cover incidental subcontracts, or services retained by the contractor and not coming under his Division 1 coverage.

The owner's interest in specifying insurance coverage on the part of the contractor lies solely in the protection of the owner from liability or litigation. Accordingly, specification for coverage is a consideration of the owner's legal counsel, and subsequent coverage achieved by the contractor should be reviewed by this same legal counsel for the owner. The work of the inspection team should be limited, but the owner's inspection team should be aware of the coverage enforced. Further, in the event of any accidents involving damage or bodily harm, the inspection team becomes the owner's on-site representatives—and should carefully investigate situations, functioning to mitigate difficult circumstances for all parties. However, the inspection team's first responsibility is to the owner, and care must be taken not to involve the owner in problems which properly should be limited to the contractor.

EQUAL OPPORTUNITIES

Most requirements for equal opportunity employment use as a minimum base the Civil Rights Act of 1964, which provides that it is a violation of a Federal law for an employer employing more than 24 people to discriminate in hiring, business operations, or layoffs with respect to race, color, religion, sex, or national origin.

On Federal work, the contractor as part of his bid must assure that he has or will meet all equal opportunity regulations covering construction work as described by the Executive Order of the President, and enforced by the Office of Federal Contract Clients of the Department of Labor.

While it is the contractor's responsibility to meet the requirements of the Civil Rights Act, the inspection team is responsible for collection of information on the actual job work force, including compliance. The inspection team should be familiar with the current requirements of the Executive Order and local requirements of the Office of Federal Compliance, often described in terms of the locale, such as the New York Plan, the Philadelphia Plan, etc. The attachment of labels such as these is sometimes misleading, since plans and practices do shift.

In addition to Federal law, individual states often have specific civil rights laws and programs. Since these are statutory obligations, failure to properly comply with them can result in delays and even enforcement passing through the contractor to the owner.

LABOR RELATIONS

Relations with the trades are the responsibility of the contractors, but the inspection force can have an influence. Many projects involving inspection are either wholly or partially union, and a knowledge of the local working conditions and agreements is important to the inspection function. The inspection is, after all, the owner's eyes and ears, and has the responsibility to forecast problems in all areas, particularly including labor. The building trades are comprised of more than 20 crafts, which in turn are broken down into many specialties. Often, problems evolve regarding jurisdiction between trades, in some cases, triggering work delays or stoppages in which neither the contractor nor the owner are directly involved, but from which both directly suffer.

Usually, the working agreements specify that the inspection team must communicate with the trades via their supervisory structure. This is good management; informal liaison often leads to problems.

The working agreements differ even for a given craft between local union agreements and international agreements.

The inspection team must carefully practice laissez-faire in keeping hands off of the relationships between the contractor and the labor force—carefully maintaining the role of passive observers. It is not the prerogative of the inspection team to be for the contractor or for labor—but only for their client—the owner.

On-the-job disagreements which escalate may have to be settled by the National Joint Board for the settlement of jurisdictional disputes for the National Appeal Board. Prior to this, the National Labor Relations Board has jurisdiction in attempting to settle any disputes between unions with jurisdictional discord. However, by the time these rather ponderous procedures are put into action, harmful delays will have taken place. The best place to settle jurisdictional disputes is at the local level—before they become acute enough to cause walk-offs.

Implementation of the Equal Opportunities Act by the contractors may involve the inspection team in an auditing role to insure compliance.

BUILDING PERMITS

The contract often calls upon the contractor to obtain the building permit; in the case of multiple prime contracts, this is usually the responsibility of the general contractor. Naturally, where there are defects or omissions in the plans and specifications, it is the owner's responsibility to take remedial action.

Building codes may be local, city, county, or state. While there is a recommended national building code, it usually has no effect, except on Federal projects. In turn, Federal projects built on state, city, or county lands—public or private—are considered to be essentially a Federal reservation and local laws do not apply.

The regulations and subsequent review for building codes involve categories of the building such as exits, egress and access, fire protection, setbacks or building lines, building classifications, allowable stresses and loads, foundations, structure, heating, ventilating and air conditioning, lighting requirements, noise control, vents and piping, fire alarms, elevators, and any safety factors, particularly where the building involves public assembly.

Often, although a pre-building permit or preliminary review has been conducted by the design team—or a permit to allow construction has been awarded—the builder must get separate permits allowing work in certain categories such as foundations, demolition, plumbing, equipment use, and signs.

Perhaps even more important in many areas is the requirement for an occupancy certificate before the building may be utilized. Often, the owner uses this as a condition of completion and final payment. In many cases, a beneficial occupancy certificate may be issued as a matter of normal practice—and in some cities where this is the case, final building occupancy certificates are rarely issued.

The building department often handles the licensing in the locality also. Local regulation often calls for electricians and plumbers in particular to be licensed.

JOB MOBILIZATION

Whatever the importance of the contractual requirements in terms of bonding, insurance, equal opportunity, building permits, etc., it is easy to lose track of the importance of getting the job moving.

Another activity which will absorb the contractor initially will be reestablishing his pre-bid commitments from subcontractors and other contractors. Depending on the type of contract, and the contractor himself, he may attempt to improve upon the pre-bid conditions, or simply affirm them.

JOB MOBILIZATION CHECKLIST

Material submitted should include, but not be limited to:

1. Appropriate insurance certificates.
2. Performance bond.
3. Equal Opportunity documentation.
4. Building permits.
5. Special permits.
6. Progress schedule.
7. Payment schedule.
8. Shop drawing submission plan.
9. Materials submission plan.
10. Concrete mix design.
11. Subcontractor list.

JOB SCHEDULE

The job schedule must be submitted usually within 3 months in great detail. However, it is usual for most contracts to call for submission of a 90-day schedule to encourage start of work, and to view the short cycle. In certain situations, a network plan may be appropriate, while in others a bar graph will suffice for the short-range initial schedule. Items such as layout, de-watering, excavation, sheeting, shoring, and piling will be emphasized.

The key to the preparation of a meaningful schedule is the assignment of key contractor personnel—and their willingness to undertake this important initial field work.

PAYMENT SCHEDULE

The contractor usually ties his payment schedule and breakdown to the detailed schedule. The inspection team, in accordance with the contract details, should attempt to correlate the first several requisition schedules to the work to be accomplished on a short-range schedule. Actually, the completion of the detailed payment schedule, shop drawing submission plan, and material schedule will be a function of the completion of the detailed schedule. Initiation of these activities prior to preparation of the detailed schedule is premature. The detailed schedule should be the basis of each of these items.

INSPECTION TEAM MOBILIZATION

Just as the early portion of the job is difficult for the contractor, so it is for the inspection team. Just as the contractor is actually pulling together his key field staff, so is the inspection team calling upon their resources to man the job. There tends to be a mutuality of willingness to delay, combined with the optimism that much time remains in order to get the project completed.

While every effort should be made to place the long-term team on the job, it is often expeditious to place a specialist in site preparation on the initial clearing, foundations, and heavy construction phase. The same may well apply to the inspection team. Here, one of the keys is surveying, and the other is checking excavations, soil backfilling, and foundations.

6. SURVEYING

The measurement of quantities in terms of angles and lengths on the job site falls within the province of the surveyor. Surveys, to have legal meaning, must be accomplished by either professional engineers or licensed surveyors, both of which are licensed in the state in which the job is located. However, a substantial amount of the surveying accomplished on the job site is done by unlicensed surveyors, and it is quite common for the inspection team to become part of the surveying party, or conversely, to have the surveying party double as inspectors.

Surveying is an ancient profession, and is related in many ways to navigation, and the measurement of angles for artillery fire. The more involved procedures, such as determining azimuth lines, longitude, or latitude through celestial readings are not appropriate to the typical construction project. The level of work required, and the level of accuracy needed in the typical building or highway project can be accomplished with basic equipment, and on-the-job-training is usually adequate for the development of new members of the survey party.

ACCURACY

Accuracy of the survey is a relative measurement. Accuracy for the sake of accuracy alone can be an expensive commodity. In the case of tunnel work, particularly where work is proceeding from two different tunnel faces to meet at a predetermined point, a high level of accuracy is required and is done today almost exclusively with laser surveying equipment.

In the more typical situation, the boundaries of the project have been established by appropriately qualified precision survey parties, and *benchmarks* established. Within these benchmarks, the site surveying group can work and recheck distances, angles, and dimensions.

In measuring, don't pass up the obvious. In one instance, when a fractioning tower was hoisted into place ready to be put upon preset anchor bolts cast into a large concrete foundation, measurement with the naked eye was sufficient to disclose that the bolt circle diameter was substantially less than the matching holes in the base plate of the tower. A hurried review of the drawings showed that a circle of 2′-6″ was required, but a faded print had appeared to be 26″. The only survey needed to disclose this problem would have been a standard 6′ rule applied to the base of the column before it was picked up. The mistake in the foundation may or may not have been inevitable, but there was no excuse for the lift of the tower and the expense involved.

In measuring or *checking the measurements* of others, the tolerance allowed within the specifications, as well as normal construction practice, dictates the level of precision required. Where precision parts with a tolerance of plus or minus hundredths or even thousandths of an inch are involved, measurements must be commensurate. In determining the alignment of a brick wall, variances of up to ½″ might be allowed, but cumulative error would be held within that or less.

The inspector may have to preclude an overabundance of material, as in the case of a trench poured where the ground is a form, rather than a formed trench, in which the contractor offers to provide a 2′ wall where a 6″ wall is called for. In some cases, this is appropriate—while in others, review of the drawings may disclose that the oversized wall could provide an interference with other work which is to go in place adjacent to the wall. Keep in mind that the concrete man may well be a subcontractor, and interested only in his phase, so that the surveying party must maintain an overview of the project.

In urban work, particular attention must be paid to building lines, property lines, easements, and rights-of-way. Encroachment of even fractions of an inch into property of others may cause very expensive readjustments or litigation.

TRANSIT

The principal instrument for job site surveying is the transit. In fact, the transit is the basic tool for all surveying. Its principal purpose is the measurement of angles in a horizontal plane parallel to the earth's surface. In order for the angles to be meaningful, they are relative to established or identifiable points on the surface of the earth, and in the field the transit must be set up immediately over the point in question. The alignment is done with precision through the use of either a

The Wild T0 Compass Theodolite is a compact lightweight instrument, which can be used either for observing and setting-out magnetic bearings or as a normal theodolite for measuring or turning-off angles. It is particularly useful for rapid traversing, especially in areas where visibility is limited and the traverse legs are likely to be short. The T0 can also be used for photogrammetric ground control surveys, on building sites or for reconnaissance. Horizontal (i.e. Compass) circle readings are of unusually high accuracy for an instrument of this type and size, because of the coincidence setting of the diametrically-opposite parts of the circle. Horizontal and vertical circle readings are made to 1' (2ᶜ). The telescope has a magnification of 20×.

Accessories which can be attached to the T0 include a telescope level, the Wild Objective Pentaprism and the Wild-Roelofs Solar Prism.

The Wild T1A (T1AE) Double Centre Theodolite is used for low-order triangulation, underground traverses, general traversing, tacheometry, property surveys, mine surveys, building site layouts, etc. – in fact for most survey and engineering tasks where first or second order accuracies are not essential.

The simple, automatic vertical index, which replaces the index level, makes vertical circle reading an easy matter. The liquid compensator has nothing to wear out and nothing to maintain.

Each circle is read with the aid of

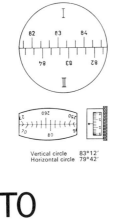

Vertical circle   83° 12'
Horizontal circle 79° 42'

TO

Fig. 1 Wild theodolites for typical construction projects. *Source: Wild Heerbrugg.*

plumbbob, or a visual plumbbob. The use of a telescopic plumbbob removes the difficulties encountered on windy days.

Angles are measured by sighting a reference point through a telescope with gunsight-like cross hairs, and then turning the telescope to a second reference point. This sighting and turning results in a measurement of the angle between the two points sighted, with the set-up point for the transit representing the apex of a triangle.

Readings are made with a micrometer-type vernier which allows the naked eye to read to within parts of a minute (within a circle of 360°, each degree has 60 minutes, and each minute 60 seconds).

The instrument portion of the transit is usually set upon a separable tripod. The tripod is established over the point, approximately in place. The plumbbob is used to exactly center the transit head over the point to be used as apex, and then clamped in place, using a built-in clamping an optical micrometer whose drum is graduated to 20″ (or 1ᶜ), with estimation made easily to 5″ (or 0.1ᶜ). Having two independent cylindrical vertical axes and two horizontal clamps (one each for the lower and upper plate and shaped differently to avoid confusion) the T1A allows the repetition method to be used for observing.

The T1A has an optical plummet and a detachable tribrach for forced-centring. A full range of accessories enlarges its usefulness.

The Wild T16 (T16E) Direct Reading Theodolite is a tacheometric theodolite suitable for all low-order triangulation, tacheometric detail and traverse surveys, mine surveys, property surveys, building site measurements, marking out, etc.

The easily read scales of the horizontal and vertical circles, with estimation to one tenth of a minute of arc (0.2ᶜ), allow work to be carried out quickly. All clamps and drive screws are placed logically so that they can be manipulated safely and comfortably. The combination of the simple circle scale reading and the operation of the instrument

itself makes the T16 a most useful instrument for use by trainees. The detachable tribrach ensures that the T16 can be used with all Wild traversing equipment and, of course, the normal accessories and attachments all provide additional uses.

The T16 ED has the extra facility of a horizontal circle with double numbering (360° circle only), allowing angles to read or set out either to the left ("North to West") or, to the right ("North to East").

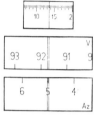

Horizontal circle 5°13′35″

T1A

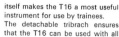
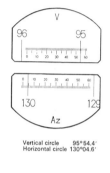

Vertical circle   95°54.4′
Horizontal circle 130°04.6′

T16

Fig. 1 (*continued*)

device in the instrument head. The transit contains leveling screws, usually either three or four, so that the transit can be established in a level or horizontal plane. The plumbbob is then rechecked to be sure that the leveling did not move the device over the point. The transit is then ready for use.

The transit can also be used to measure vertical angles, and has two matching scales for direct reading of the angle. Angles can be read between two points, or more usually between the horizontal and a point in question.

When any given angle or series of angles is read, the transit can be reset and other angles read over the same point.

Most transits utilized in construction prior to the 1950's were American-made, and utilized telescopes between 12 and 24″ in length. European *theodolites* were rarely seen on a construction site. These units

The Wild NK01 Dumpy Level is a most convenient level for the building site. The telescope rotates under a friction-braked action, thus requiring no horizontal clamp. Fine pointing is completed by means of a drive screw. Readings on the metal horizontal circle are made by eye against a fixed index mark, with estimation possible to a tenth of a degree or grade, according to the circle division. The circle itself can be turned by hand to set any desired initial bearing. Four dots around the circle indicate rapidly the right angle settings. The circular and tubular level bubbles are seen, free from parallax, in the hinged observation mirror which, when closed, protects the glass vials.

The Wild N10 (NK10) Engineers' Compact Level, small in size but big in performance, has all the qualities of a larger level and gives accurate results under all conditions. It is suitable for general levelling, route surveys, irrigation works, civil and constructional engineering. When fitted with a glass circle (NK 10) it can be used in flat terrain for tacheometric work and for measuring and setting out horizontal angles. Rotation is friction-braked and a drive screw is used for fine pointing. The line of sight is levelled with the aid of a tilting screw and the well known Wild "split-bubble" image system. The tubular level is ventilated in order

NK01   N10

Fig. 2 Precision leveling instruments. *Source: Wild Heerbrugg.*

utilize optical systems to provide the same level of magnification and accuracy without the bulk of the American equipment. Since the 1950's, these units have virtually taken over the market, and are quite often found on the job site. Figure 1 illustrates several of the Wild theodolites which could be found on a typical construction project.

The transit should be field checked when first brought to the job site or if jarred or dropped. This can be accomplished by measuring known angles and comparing the results with that measured. American units being more basic have some adjustments which can be made in the field, while the theodolites do not. Nevertheless, field adjustments of a transit are questionable, and the inspection team should be certain that instruments utilized are accurate and in good condition. A transit should never be left in the weather, and should be treated with care since it is usually worth several thousand dollars.

to avoid excessive accumulation of internal heat and it is also well protected against damage. The N10 has an extremely short minimum focussing distance of 1 m (3.3 ft.), which is most useful in restricted spaces.

It is possible to supply this instrument with an erect image telescope (models N10E and NK10E) – a feature often appreciated by those who use their level only occasionally or who find it difficult to work with an inverted image. The E models have a minimum focussing distance of 1.35 m (4.5 ft) and, of course, must be used with staves with erect numbering.

The Wild N2 (NK2) Engineers' Level is a strongly built, accurate and reliable general purpose instrument, which is easy to operate and which is the ideal level for topographic, geodetic and engineering tasks. It is unsurpassed for use on civil and constructional engineering projects, bench mark establishment, breaking down primary level networks, profiles, cross-sections, road and railway surveys and for photogrammetric height control. When fitted with a glass circle (NK2), it can be used in flat terrain

for setting out works and for measuring horizontal angles and distances. Fine pointing is made with a horizontal clamp and a drive screw and an accurate coincidence setting of the tubular level split-bubble image is made by means of a tilting screw. The telescope and tubular level can rotate simultaneously about the longitudinal axis, enabling the vertical collimation to be checked quickly and easily at one instrumental set-up and without changing the position of the target staff.

N2

Fig. 2 (continued)

LEVEL

One of the most important measurements on a construction site is that of the relative *elevation.* The difference in elevation between benchmarks or reference marks, established and accurately identified as to elevation by the precision surveying team, and other key measurement points on the job site is accomplished by relative measurement. That is, the surveyor sets up his transit and levels the telescope. He then takes a zero-degree elevation reading on a measuring rod held vertically on a reference elevation point. The distance up to or down from this benchmark is read, thus establishing the actual height of the instrument telescope center line. These readings are made using horizontal cross hairs in the transit. Once the height of instrument has been established, measurements to other points can be calculated by adding or subtracting their height relative to the height of instrument.

On a smaller job, the transit may be utilized to do both measurement of angles and leveling. Where large amounts of leveling are to be done, as in piling cut-offs, a less expensive instrument called the *level* can be utilized. The level is a precision instrument, but establishes only the horizontal plane, and need not be located over any specific point. Although more economical, the level is much more limited in its utility. A typical level is illustrated in Figure 2. In this case, also, European optics have made possible much more compact equipment.

DISTANCES

Possibly the most difficult measurement made is the measurement of distance. This is true for several reasons, but principally because it is a manual operation in most cases. In order for calculations to be accurate, distances must be measured exactly in the horizontal, and this depends upon good judgment on the part of the chainmen. Surveying *tapes* are made of invar metal which has a low coefficient of expansion and contraction. For high level work, even with invar tapes, adjustments must be made in the measured lengths to relate the distances to a set standard temperature, usually 70° F. On a hot day, the tape will expand, and the distances read will appear to be shorter than the actual comparison distance, while on a cold day, with a short tape, the apparent distances will be longer than the base.

Another inaccuracy in measurement comes in long measurements, where the *chainmen* must place pins at each 100 feet, moving the tape forward. Small errors in lining up to the pin can result in accumulative errors in the measurement. The tapes are made so that a 15-lb pull results in a standard measurement, but this will vary with the human element. The instrument man on the transit should line up the chainmen so they are exactly on line, since any variation off the line will result in diminishing the length measured. On measurements up or down a grade, the chainmen level the tape and utilize plumbbobs at the low end to move from set-up to set-up. Manual measurements are tedious and often fraught with error. For this reason, the introduction of electronic devices for the measurement of distances has been a welcome one. While not economical for a few distances, or for short distances—the devices are invaluable.

Figure 3 is the Wild Distomat which mounts on top of the standard theodolite. Its accuracy is within one centimeter irrespective of distance at a range of up to 2000 meters. Measuring time requires 15 seconds, and

Fig. 3 Distance surveying instruments. *Source: Wild Heerbrugg.*

is generated by an infrared carrier wave generated by a diode and reflected from a target at the point of measurement.

Figure 4 is the Hewlett Packard Model 3800B distance meter which measures in meters using an invisible beam of infrared light to measure distances up to 2 miles with great accuracy. The instrument is battery-powered and makes any measurement in less than 2 minutes. It weighs 17 lb and mounts on a tripod. Figure 5 is a picture of the companion model 3800A in use.

Figure 6 is a distance surveying instrument, again utilizing infrared rays, developed by the Cubic Corporation. The device measures distances from one meter to more than a mile in under 10 seconds. Called the Cubitape, it weighs less than 10 lb and operates in a temperature range from 10 to 120° F.

The popularity of the distance meter is evident from the wide range of

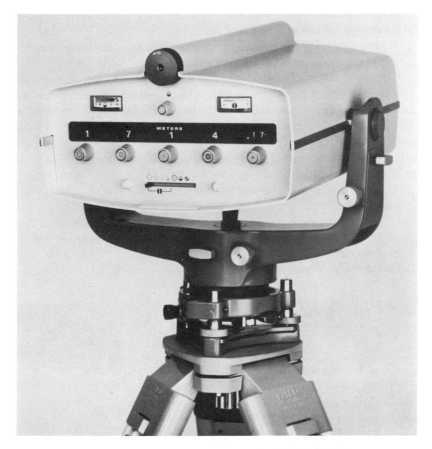

Fig. 4 Distance Meter, Hewlett-Packard Model 3800B.

units available. The AGA Corporation of Sweden distributes a unit in the same price range (Fig. 7), which measures in less than 20 seconds a range of up to two miles. In this case, the light source is a laser powered by a 12-volt battery pack.

PROCEDURES

Buildings, roads, and other major features must be correctly located within the construction limits. *Control points* must be established, usually from monuments previously established for the project, or from established USGS reference points. Control points to be used must be examined for accuracy, and should be at least the same standard of accuracy (or better) as is required for the control survey. If there are not sufficient control points available to provide a closed loop traverse, additional control points should be established.

Control surveying on complex projects is calculated using coordinate locations. Control points, alignment, and construction feature controls can be mutually coordinated prior to the time when actual construction key points must be established or checked.

Calculation of the coordinate points is done manually using logarithms, or by computer. If there is a substantial number of points to be calculated, the computer is very economical. Having sufficient control points avoids problems during construction if one or more points is lost, not an unusual situation.

Within the control points, the contractor will set stakes for rough grading and other working points. These should be checked by the inspection surveying team.

The inspection team may set its own *reference stakes* for convenience.

Fig. 5 Distance Meter 3800A. *Source: Hewlett-Packard.*

Fig. 6 Distance surveying instrument. *Source: Cubic Corp.*

These may be offset on a line parallel to the actual building line, or at right angles to the final points.

Notes should be kept on all surveying traverses and checks of key points. If any control points or reference points are lost, the situation should be recorded. Notes should always show the latest known condition of a point, and if a point has been reestablished, that fact should also be noted.

All notes should show the name of the party chief, date, personnel, and weather. Notes should be original and kept during the surveying operation.

Level notes should be kept for the establishment of key points, and traverse notes should be kept when the control loop is run periodically.

Keep in mind that monuments may be disturbed, or in some cases, may shift due to conditions such as drought or earthquake. Periodic rechecks of location and elevation of control points should be made.

SPECIAL SURVEYS

The family of *laser* measurement instruments has brought new and very practical methods of making measurements on the job site. Figure 8 illustrates the method by which the laser is attached to standard transit, transmitting a visible laser beam through the telescope to a target.

Figure 9 shows the dialgrade unit by Spectra-Physics which is used for establishing grade for pipeline installation.

Spectra-Physics also makes a rotolite which is a rotating visible laser used for measurements at various levels above a reference point. Figure 10 shows the unit in use for measuring partition wall alignment, raised floor leveling, chair rails, and ceiling grid installation. The system can also be used for establishing screeds for pouring concrete.

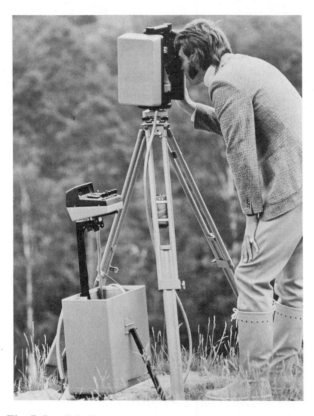

Fig. 7 Swedish distance measuring unit. *Source: AGA Corp.*

Fig. 8 Laser attachments to standard transits. *Source: Spectra-Physics.*

The unit can also be used to project a vertical line upward through a building for reference measuring.

The advantages of the laser are its relatively easy set up and its visibility. The unit is generally far superior to string lines or plumb lines which are subject to distortion by tension and/or wind.

Laser-operated equipment is subject to the cautions listed in the Occupational Safety and Health Act (OSHA):

Only qualified and trained employees may be assigned to install, adjust, or operate laser equipment, and proof of qualification of the equipment operator must be available at all times. When employees are working in areas in which potential exposure to direct or reflected laser light is greater than 5 milliwatts, they should be provided with anti-laser eye protection.

Areas in which lasers are used must be posted with standard laser warning placards, and the laser beam shall be turned off when not needed—such as at lunch time.

The laser beam may not be directed at employees. During rain, snow, dust or fog, operation of laser systems is not recommended, because of beam scatter.

Fig. 9 Laser for setting grade line. *Source: Spectra-Physics.*

SURVEYING CHECKLIST

1. Determine level of accuracy required.
2. Locate bench marks, elevation and location.
3. Check equipment:
   3.1 Transit
   3.2 Level
   3.3 Measuring tape
   3.4 Accessories
4. Set up equipment custody routine.
5. Run basic equipment tests as required.
6. Establish surveying plan.
7. Perform presurvey calculations.
8. Survey field
9. Calculate notes.

ROTOLITE IN ACTION

CEILING GRID INSTALLATION

The **ROTOLITE** is mounted on the adjustable Maxi Tripod which extends up to 14' 6". For higher ceilings, the Column Clamp is used. The plane of laser light is accessible to the installer at all times. He simply moves the grid up or down until it is in line with the red laser light, and then he ties the grid at that position. Special targets are available for improved viewing. A red laser line is also visible on the perimeter walls, which makes the installation of wall channel a great deal faster!

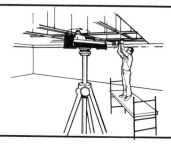

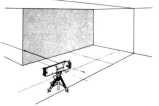

PARTITION-WALL ALIGNMENT

The **ROTOLITE** is turned on its side and set on the floor or mounted on a Mini Tripod. The plane of light is now projected vertically. Partitions and tracks are easily aligned with this light. Also, when the **ROTOLITE** is in this position, the 2 beams can be used for 90° layouts.

RAISED-ACCESS FLOOR LEVELING

Computer room floors and similar access floor systems must be installed to critical tolerances. The **ROTOLITE** is ideal for this application. The laser can rotate around a wide area at the proper elevation of the pedestal caps or a few inches higher when using the magnetic target. The same principle is used for leveling circular conveyors, floor joists, windows, tile, etc.

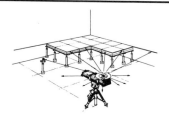

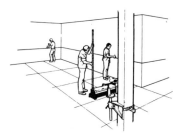

ESTABLISHING BENCHMARKS — CHECKING ELEVATIONS

Setting 4' marks is a breeze with the **ROTOLITE**. Checking the elevation of footers, subfloors, finished floors, etc., is also a natural. The rotation of the laser at high speeds gives the effect of a constant plane of light. This enables several workers to use the light simultaneously although they are working in different areas of a room.

Fig. 10 Laser used for establishing level or vertical line. *Source: Spectra-Physics.*

7. TEMPORARY CONSTRUCTION

During construction, a wide range of structures are designed and built by the contractor—often without the guidance of the structural engineer. It is in the interest of the inspection team to insure the safety of these structures. This chapter first discusses some of the general aspects of shoring, scaffolding, and temporary structures—and then reviews some basic parameters of strength of materials which the construction team should be aware of.

SCAFFOLDING

Scaffolding is used for many purposes during construction. It may be used to support working platforms, stairways, hoists, trash chutes, and more recently has been used to support shoring. The Scaffolding and Shoring Institute has defined safety rules for scaffolding. Frame shoring nomenclature is illustrated in Fig. 1.

Scaffolds are no better than the base upon which they are erected. Recommended procedure is to provide adequate sills and base plate, use adjusting screws instead of blocking to adjust to uneven grade, and plumb and level all scaffolds as the erection proceeds. Scaffolding depends very much upon the ability of tubular columns to carry load, and this ability in turn is dependent upon vertical structures. Cross bracing is important, so fasten all braces securely. Freestanding scaffold towers must be restrained from tipping by guying or other lateral restraint.

When planning scaffolds, make them to a sufficient height. Do not attempt to stack ladders or crates on top scaffolding to increase the height.

Calculate the safe load for scaffolds, preferably through the scaffolding contractor. Do not overload. (OSHA safety factor is now 4 to 1, minimum.)

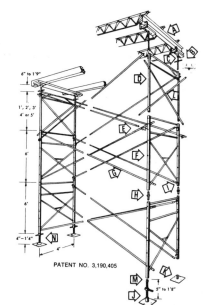

A **4″ x 8″ Steel Junior Beams.**
B **WACO Beam Clamps** (4447-00) secure the beam to the
   "J" head to eliminate the possibility of beam rotation.
C **"J" Headed Screw Jacks** (2621-00 or 2600-21+4511-14)
   give easy 15″ height adjustment.
D **Extension Frames** (2450-00—4′ x 5′4″ or 2240-00—2′ x
   4′4″) telescope into Base frames to give height adjust-
   ments of 1 ft., 2 ft., 3 ft., 4 ft.—and 5 ft. This adjustability
   eliminates the need for 2 ft., 3 ft., 4 ft., and 5 ft. frames,
   their various cross sizes, and extension sleeves.
E **Adapter Pins** (2001-00) fit into holes on legs of Base
   Frames, support the Extension Frames at the desired
   height adjustment, and provide attachment points for
   crosses. (Pat. No. 3,346,283)
F **Base Frames** (2460-00—4′ x 6′ or 2250-00—2′ x 5′) have
   "X" braced design. Holes in legs receive Adapter Pins
   at any desired level of adjustment.
G **Cross Braces.** Only one size of side cross brace is re-
   quired for both Base and Extension Frames.
H **Coupling Insert Pins** (2023-00) provide alignment of
   Base Frames and can be bolted through holes in legs of
   Base Frames to permit hoisting of assembled towers.
I **Speed-Locks** provide fast, trouble-free attachment of
   cross bracing. Pat. No. 3,174,779

PATENT NO. 3,190,405

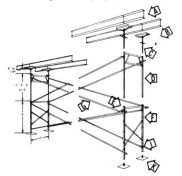

J **Fixed Screw Jacks** (2600-21+4710-00) give easy 15″
   height adjustment. (Add 2623-01 when used with 2⅜″
   frame.)
K **Fixed Base Plates** (4721-00) are plates which distribute
   loads to sills or pads.
L **End Cross Brace** (2045-00) always used at 3′-0″, 4′-0″,
   5′-0″ Extension. 4′-0″ Frame only.
M **The Cap** (2623-01) combined with the collar on the jack
   handle, provides positive alignment in the legs of the
   Base Frames.
N **Swivel-Based Screw Jacks** (2619-00), with 8 inch by 8
   inch swivel plates, compensate for uneven ground conditions, eliminating wedging, and give easy 12 inch
   height adjustment. (Patent No. 3,259,367)
   (Add 2623-01 when used with 2⅜″ frame.)

Fig. 1 Scaffolding nomenclature. *Source: Waco Scaffold and Shoring Co.*

On wall scaffolds, anchor to the structure at least every 30′ of length
and 25′ of height. If scaffolds are to be enclosed, check the load-carrying
capability due to the additional load imposed by wind and weather.

Personnel should not be permitted to climb cross-bracing, and all
working platforms should be properly provided with hand rails at a 42″
height.

The Scaffolding and Shoring Institute recommends the following
erection procedures:

1. Prior to erection, scaffolding elements should be inspected for rust,
   straightness of members, and soundness of welds. Locking devices on

frames and braces must be in good working order, and coupling pins must be well aligned with the frame and panel legs. Pivoted cross braces must have the center pivot securely in place.

2. Check the carrying capacity of the soil upon which the sill and base-plates are to be placed, and ascertain that it is not overloaded during erection, or when the design loads are applied. The ability of the scaffolding bases to carry load is roughly proportional to the size of the base.

3. The work of erection of scaffolding should be under supervision of a qualified superintendent familiar with scaffolding safety rules. Adjustment screws should be set to their approximate final adjustment before scaffolding is erected to minimize the runout required to level. After erecting the first tier of scaffolding, plumb and level all frames so that the additional tiers will also be in correct vertical alignment.

4. The maximum permissible span for $1\frac{1}{4}''$ plank is $4'$ under a medium duty loading of 50 lb/ft^2. Scaffold planks should extend over the end supports not less than $6''$ nor more than $12''$. Planking should be overlapped a minimum of $12''$ or secured from movement.

Light duty tube and coupler scaffolding has posts and bracing of nominal $2''$ OD steel tubing. Posts should be spaced no more than $6'$ apart x $10'$ along the length of the scaffold. Dissimilar metals should not be used together, to avoid galvanic action.

Medium-duty tube and coupler scaffold has posts runners and bracing of nominal $2''$ OD tubing with the posts spaced not more than $6'$ apart by $8'$ along the length with bearers of nominal $2''$ OD tubing.

Heavy-duty tube and coupler is made up of $2''$ OD steel tubing with the posts spaced not more than $6' \times 6'\text{-}6''$. Other structural metals when used must carry equivalent loads.

The specific load limits for the various categories of scaffolding are shown in Fig. 2.

Cross bracing must be installed across the width of the scaffold at least every third set of posts horizontally, and every fourth runner vertically. Longitudinal diagonal bracing on the inner and outer rows of poles is to be installed at approximately 45° angle from near the base of the first outer post upward to the extreme top of the scaffold. Where the longitudinal length permits, this bracing shall be duplicated at every fifth post.

	Light Duty	Medium Duty	Heavy Duty
Uniformly distributed load			
(not to exceed)	25 psf	50 psf	75 psf
Post spacing (longitudinal)	10'–0''	8'–0''	6'–6''
Post spacing (transverse)	6'–0''	6'–0''	6'–0''
Maximum height:			
Working level 1	125'	125'	125'
2	125'	78'–0''	—
3	91'–0''	—	—
Additional planked levels:			
Working level 1	8	6	6
2	4	0	—
3	0	—	—

Fig. 2 Scaffolding loads allowable. *Source: OSHA.*

SHORING

The recommendations of the Scaffolding and Shoring Institute apply to shoring for various purposes, principally for the support of concrete formwork. The following code by SSI applies: "Recommended Safety Requirements for Shoring Concrete Formwork."

Shoring installations must be properly designed by qualified people. The shoring layout must include details which take into account unusual conditions such as ramp, cantilevered slabs, and heavy beams, showing both plan and elevation views. The inspection team should be certain that the plan has been approved by appropriate structural engineers.

The minimum design load for any form work and shoring must be not less than 100 lb/ft² for the combined live and dead load regardless of slab thickness, and the minimum allowance for live load must be at least 20 lb/ft². When motorized carts are to be used to place the concrete, design loads must be increased 25 lb/ft². Allowable loads must be based on a safety factor, consistent with the type of shoring used.

Design stresses for form lumber shall not exceed 1,760,000 lb/in² for the modulus of elasticity, 120 lb/in² for horizontal shear, and 1450 lb/in² for extreme fiber bending. Bending and shear stresses may be increased by 25% for short-term loading.

Sound sills or bases shall be used, and the load should be applied to the center of the sill to avoid overturning of the tower.

Fig. 3 Initial Stages of Shoring Scaffolding. *Source: Waco Scaffold and Shoring Co.*

The allowable soil bearing capacity shall be studied and certified by a structural engineer.

Timber used as single post shores shall have the safety factor and allowable working load for each grade and species as recommended in "Wood Structural Design Data Book" prepared by the National Forest Products Association, Washington, D. C.

As in all vertical loads supported by frames, legs must be erected vertically when frame-type support is utilized. The maximum allowable deviation from the vertical is $\frac{1}{8}''$ per 3'.

The shoring must not be removed prematurely, and when removed, should be replaced by partial reshoring. In high-rise slabs, it is usual to replace partial shoring for seven floors below the floor in progress. Usually, slabs can be stripped on the following day, but support must be maintained for at least 3 days, and preferably 7. Some shoring is needed up to 28 days when the concrete develops its full strength. The stripping of forms is subject, also, to the live load on the new structure.

Horizontal shoring beams should be designed according to standard design procedures. Figure 3 shows the initial stages of a frame shoring. Figure 4 is frame shoring in place for a floor slab pour, while Fig. 5 shows an extra heavy duty shoring with cribbing beams and two major support beams to carry a major 600,000-lb load.

SCAFFOLDING CHECKLIST

1. Survey the conditions under which scaffolding will be erected.
2. Prepare scaffolding design under direction of professional engineer.
3. Check plank for permissible span.
4. As scaffolding goes up, check braces.
5. Free-standing towers must be restrained by guys, or equivalent.
6. Safety factor must be 4 to 1.
7. Scaffold should be anchored every 30' of length and 25' of height.
8. Working platform should be properly handrailed.
9. Check elements for soundness.
10. Check bearing capacity of soil prior to erection.
11. Check tower for plumb.

12. Cross-bracing must be installed across the width of the scaffold at least every third set of posts horizontally and every fourth runner vertically.

SHORING CHECKLIST

1. Shoring must be designed by qualified engineers with minimum load not less than 100 lb/ft^2 combined live and dead. Increase 25 lb/ft^2 for motorized buggies.

2. Design stresses to be increased 25% for short-term loads (allowable).

3. Set a soil bearing capacity before structure is erected.

4. Check for plumb—maximum ⅛ " per 3 feet.

5. Scaffolding shall not be removed prematurely.

6. Check wedges and fasteners.

MECHANICS

Since the inspection team has a varying academic background, some will have had extensive instruction in mathematics, strength of materials, physics, and similar academic subjects, while others will have a limited one. This section offers some suggestions in the practical application of mathematics in the construction field, identifying potential weak points or problem areas.

Generally speaking, the structure as designed will do the job when in place; it is not the usual responsibility of the inspection team to second-guess the structural design. However, when an apparent error is noted by the inspection team, it is quite proper to ask the design group for clarification. The result is either an explanation of what was not obvious to the inspection team, or recognition by the designer that there has been an omission or an error. The inspection team has to be willing to ask questions, and if the questions are presented in a tactful manner, there will be no comebacks from the designers.

The structural integrity of the structure is most vulnerable during *erection,* and yet this phase is often left entirely to the control of the contractors.

Complicated false work, shoring, and other temporary rigging should be approved by the design team whenever possible. Even with these approvals, failures can occur. It is up to the inspection team to have the right plans for erection sequences and temporary support, and to see that they are followed. An understanding of mechanics will help in the spotting of errors.

Fig. 4 Scaffolding in place for floor slab pour. *Source: Waco Scaffold and Shoring Co.*

Forces

Forces are usually represented for purposes of calculation as *vectors;* that is, arrows which may be drawn to scale in units of force, or may just have the forces represented. In Fig. 6, upward and downward vertical forces and the direction of horizontal forces are shown. The resolution of forces not either vertical or horizontal into their horizontal and vertical components is important for graphical representation, as well as for the understanding of forces.

The resolution works in both directions; that is, a horizontal and vertical force operating on the structure is resolved into one component which is neither completely vertical nor horizontal. It is this equivalent force or combined force which acts upon the structure, and which must be resisted in equal and opposite amounts, or the structure will be forced to move.

The scientist Sir Isaac Newton, about 300 years ago, posed his three fundamental *laws of motion.* Two of these deal with equilibrium. Newton's first law states: "A body at rest remains at rest, and a body in

Fig. 5 Heavy-duty shoring with cribbing beams. *Source: Waco Scaffold and Shoring Company.*

motion continues to move at constant speed along a straight line, unless the body is acted upon in either case by an unbalanced force." Either condition is assumed to be equilibrium, but in the case of building structures, stability and rest is the state desired. Newton's third law states: "For every action, there is an equal and opposite reaction." In other words, in a building which is still and at rest, every action, either upon the building or resulting from the weight of the building, must be resisted by an equal and opposite reaction if it is to stay in a stable or still condition. When a body is in equilibrium, its resistance to change either from rest into motion, or from motion to come to rest, is known as its inertia.

In determining the forces upon a structure at rest, or at stages of rest during erection, all forces when added up must equal zero. Thinking in terms of the components, all vertical forces must equal a net of zero, and all horizontal forces must equal a net of zero. It is therefore important to be able to convert components, and the simplest means is to visualize each angular force (i.e., neither horizontal nor vertical) as being the hypotenuse of a right triangle. The vertical and horizontal legs then become the vertical and horizontal components of that force, and when shown in relation to the structure, can replace the actual force for purposes of calculation. The method of solving or resolving forces is shown in Fig. 7.

Resolution of the forces depends upon measurement of the angles or sides, or a combination of both. For instance, in a measurement where both sides are equal, the angles must be 45° angles.

The information on angles is based upon the resolution of the force on the basis of a right triangle. Since the sum of the three angles within a triangle must equal 180°, and the right triangle identifies one of these three angles as 90°, then the sum of the other two must be 90° also. The relationship between the side of a right triangle where the longest side squared equals the sum of the squares of the opposite two sides is proven in all trigonometry books.

In order to add forces, they must be on the same axis, and signs must be given, for instance, downward forces being positive, while upward are negative, and forces to the right positive, while to the left, negative. Any sign convention will work, as long as it is applied uniformly during calculations.

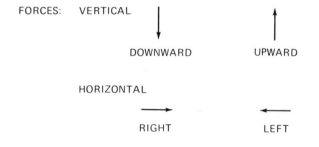

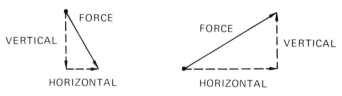

Fig. 6 Vector presentation of forces by components.

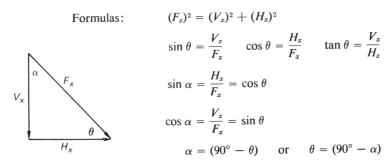

Formulas:

$$(F_x)^2 = (V_x)^2 + (H_x)^2$$

$$\sin \theta = \frac{V_x}{F_x} \qquad \cos \theta = \frac{H_x}{F_x} \qquad \tan \theta = \frac{V_x}{H_x}$$

$$\sin \alpha = \frac{H_x}{F_x} = \cos \theta$$

$$\cos \alpha = \frac{V_x}{F_x} = \sin \theta$$

$$\alpha = (90° - \theta) \quad \text{or} \quad \theta = (90° - \alpha)$$

Fig. 7 Resolution of forces.

Moments

The equilibrium of a body is not complete merely by equalization of its horizontal and vertical forces. For instance, the example in Fig. 8 shows a seesaw with no horizontal forces exerted, and a net zero upward and downward force, and yet it is obvious that the seesaw will move from its horizontal position, tilting to the left-hand side. The unbalanced force causing this motion is the moment which is equal to the arm through which it acts times the force, in this case, the arm being $2l$ times F. In Case 2 of Fig. 8, the seesaw has come to rest, and again the vertical forces must be zero, with one vertical force resisting the downward push at the left-hand side, being $F = F$. However, the uplift on the right-hand side must also be resisted, and is resisted by a vertical downward pull of the pivot. A pivot, however, can exert only a force at right angles, so that it sets up both a vertical and a horizontal component. The only place where this horizontal component can be resisted is by a pressure in the earth at the grounded end of the seesaw, which is also resisting downward press at the left.

In calculating the effect of moments, the rule for calculation states that moments about any point within the structure, or even a point taken outside of the structure must be equal and opposite. Solving for the vertical force at the center in Case 2, we see that the vertical component must be $2F$. Accordingly, since the sum of vertical forces must be zero, then the force at the left reaction in the ground must also be $2F$. To simplify calculations, the horizontal and vertical axes were shifted to be parallel, and vertical to the seesaw itself. If the force at the right-hand side is not at true right angles to the board, but actually in the vertical, and this condition is not specified, then the angle of the seesaw would have to also be considered, and the height of the pivot above the ground would have to be identified. Further, a horizontal force acting to the right at the pivot and to the left at the ground would also be incurred. These horizontal forces would be equal and opposite, but would set up a moment equal to the horizontal force times the height of the pivot from the ground, and would have to be reacted against by other moments. The other moment would be the horizontal force of F at the right end, times twice the pivot height, if force F is acting truly perpendicular to the board.

In summary, several factors are important in examining a structure:

1. All *forces* must be *equal* and *opposite,* and can best be calculated by resolving the forces into their components which can be added and subtracted arithmetically once the resolution has occurred.

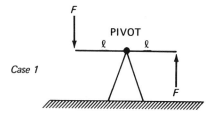

Case 1

Case 2   Axes shifted to plane of tilted board to simplify results:

Moments at left = 0

$V_c \ell = 2\ell F; V_c = 2F$

Moments at pivot = 0

$V_L \ell - F\ell - F\ell = 0; V_L = 2F$

Check $\Sigma V_{\text{forces}} = 0$

$\quad F - V_L + V_c - F = 0$

$\quad F - 2F + 2F - F = 0$

Fig. 8 Seesaw with no horizontal forces exerted.

Moment = Force × ARM

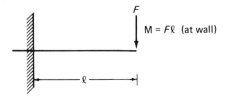

$M = F\ell$ (at wall)

Torque-twist = Force × ARM

TORQUE AT $a = F\ell$
TWIST AT $b = F\ell$

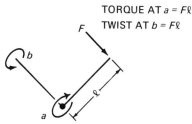

Fig. 9 Twist or torque.

2. In addition to forces being equal and opposite, *moments* about any point in relation to the body must also be *equal* and *opposite*. As demonstrated in Fig. 9, this may require the reaction to be the result of the type of structure. In this case, a cantilever wall in Case 1 is providing a reverse or reactive twist. In Case 2, a crank arm is providing a reaction torque or twist.

3. The type of *connection* is also important. In Fig. 10, various types of connections and their ability to transmit forces are illustrated.

Where two structures or members are connected only by their own relative weight, the ability to transmit force is a function of the roughness or relative smoothness of the two surfaces. The coefficient of *friction* is the measure of the relative ability of the two surfaces to develop a binding or restraining force.

STRUCTURAL FORMS

Figure 11 illustrates that not everything is what it appears to be. While the external forces must be in equilibrium in order for a structure to remain stable, similarly the structure must be used in its proper context, and must be able to have the rigidity and strength to resist the external forces applied to it. This applies to *temporary structures* as well as to the

1. Hanging Cable or Chain—Vertical downward force only, no moment

2. Taut Cable or Chain—Force in direction of pull, no moment

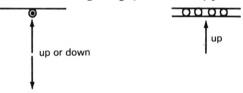

3. Pivot or Rollers: Force at right angle, no moment (up or down)

4. Pin Connection: Axial Force + or −; no moment

5. Rigid Connection

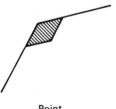

Force, any direction; moments, any direction.

6. Point

Vertical force in direction of axis; true point to surface transfer has no horizontal component, unless coefficient of friction × area is sufficient to provide equal and opposite reaction, or point "digs in" enough to permit horizontal reaction.

Fig. 10 Typical connections.

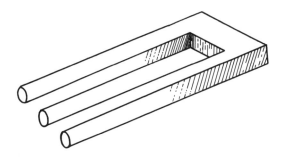

Fig. 11 Things aren't what they may appear to be, necessarily.

final structure, so that trusses or beams or derricks rigged to pick up structural pieces must in themselves have the resistive capability. This is one of the most important areas of inspection in erection safety.

For instance, a steel shape such as a pipe or H-beam can usually take much more force axially or in compression than it can in bending, if properly supported laterally. It is not sufficient for the inspection team to pull figures directly from tables in order of load-carrying capability—it is important that the conditions of those tables be understood.

Beams

A beam is a structural member used to transport or carry forces from one point to another, usually horizontally. A beam bridging between two support abutments carries weights between it, while a cantilever beam is held rigidly at its end or center, and carries forces suspended or imposed at its extremities. In carrying these forces, the beam bends or deflects in proportion to the load placed upon it, and this flecture imposes certain internal stresses upon the beam. Its ability to carry load, and its amount of deflection is a function of its material and configuration. Figure 12 shows a typical beam in what is normally called positive or downward bending. A cross section of its material is shown, indicating that the bottom fibers are being stretched or are in tension, while the upper fibers are being compressed or squeezed. the neutral axis is somewhere in the central portion of the beam, and is a balance point between the upper and lower halves of the beam, and represents a balance between their ability to resist tension or compression. Usually, the neutral axis is stable whether the beam is bent downward or upward. Using moment analysis, it is obvious that the fibers furthest from the neutral axis are those with the most ability to resist forces. Accordingly, the central portion of the beam may be made of a lighter material if the beam is a sandwich material, or can be made up of a smaller cross section, as in the case of an I beam. The ability of the beam to resist bending is dependent on the

STANDARD BEAM (IN POSITIVE BENDING)

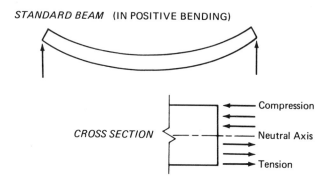

CANTILEVER BEAM (IN NEGATIVE BENDING)

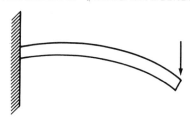

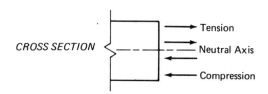

Fig. 12 Beams in bending.

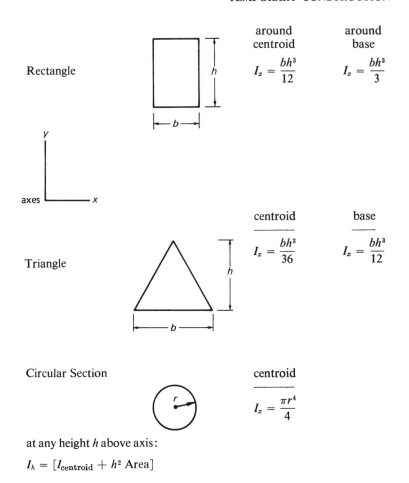

	around centroid	around base
Rectangle	$I_x = \dfrac{bh^3}{12}$	$I_x = \dfrac{bh^3}{3}$

y

axes ⎣→ x

	centroid	base
Triangle	$I_x = \dfrac{bh^3}{36}$	$I_x = \dfrac{bh^3}{12}$

	centroid
Circular Section	$I_x = \dfrac{\pi r^4}{4}$

at any height h above axis:

$I_h = [I_{\text{centroid}} + h^2 \, \text{Area}]$

Fig. 13 Typical moments of inertia.

moment of inertia of its cross section, which is symbolized as I, and equals cross section times radius of gyration, which is essentially the moment of the area above or below the neutral axis. Again, it is obvious that the greater the distance from the neutral axis, the larger the area and the greater the ability to resist. (See Fig. 13.)

Because most resistance is at the extreme section, the inspection team should refuse any temptations or requests to cut the extreme edge or *flanges* of steel beams, and similarly should allow penetrations of wood beams only in the central portion. Notching or cutting the flanges or outside edge of a timber effectively reduces its moment of inertia, and greatly weakens it. This weakening is not merely directly proportional to the depth of the cut, but is a function of the difference between the square of the distance from a neutral axis to the original outer flange, less the square of the distance to the outer edge of the cut surface.

As illustrated in Fig. 14 the resistance or load-carrying capacity of the beam is a function not only of its size, and geometry, but also of the material. The *stresses* shown are approximations of the allowable stress which may be developed in the fibers of the beam in terms of pounds per square inch. These do not represent values of the external load, or the internal stresses developed as a result of load. The factors of safety vary with the material. For instance, the usual allowable stress in steel is between 20,000 and 25,000 psi, while the steel yield value is usually on the order of 36,000 psi, and at its breaking point, the stress will be on the order of 50,000 or more psi. However, this is not really a two-to-one safety factor, but more like a 50% safety factor, since once the yield point in steel has been passed, it stretches readily, sharply reducing the area resisting load in a necking phenomenon.

Wood can only carry about 10% of the load of a steel beam. Its working stress is about $\frac{1}{3}$ of the ultimate or breaking stress, because of

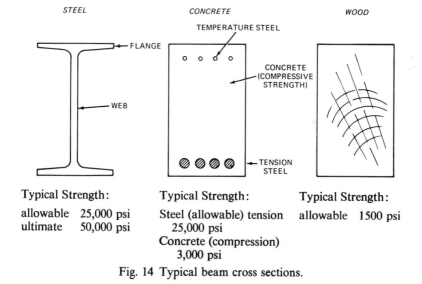

Fig. 14 Typical beam cross sections.

the uncertainties in the make-up of material. A knot hidden within a timber may cause it to break unexpectedly, thereby requiring a greater safety factor. Concrete is very strong in compression, but has virtually no tensile strength. For this reason, temperature steel is utilized to hold together the material during shrinkage or when subjected to light tensile loads. However, the reinforcing steel is placed in reinforced concrete to carry the entire tension load, while the opposite flange carries the compressive load. Obviously, the placement of the reinforcement steel is quite important. On some occasions, iron workers, not realizing the importance of the placement near the outer fibers, may pull the steel too high into the beam, seriously reducing the moment of inertia of the steel or tensile portion of the beam. In other cases, the steel is allowed to lie on the ground in the case of reinforced concrete grade beams with resulting corrosion, and ultimate loss of strength.

In beams, with the emphasis upon developing greater moments of inertia, the narrow or weak axis has limited strength if load is applied from the sideways direction, or if the beam is inadvertently carried on its side. Similarly, plates of materials or large sections, such as precast concrete wall sections, if carried flat, may crack badly, while if carried in the upright position have a high moment of inertia and can resist handling stresses. Even steel sheets or plates carried in the flat may buckle due to their own dead weight.

Handling of precast members which are to be post-stressed is a particularly critical item. The shop drawings should indicate the methods of handling, which usually involves lifting lugs placed at the ⅓ or ¼ points, so that the dead weight of the member counterbalances itself about the lifting points.

When beams are placed, the structural designer has usually arranged for crossbridging or *lateral support,* so that there is no rotation under loading of the beam from its strong toward its weak axis. If this is not done, the result is buckling or crumpling of the beam. During installation, beams may be subjected to undue lateral loading, which can cause failure.

For *lightweight beams,* bar joists may be utilized as intermediate members. The purpose of the bar joist is, again, to increase the moment of inertia of angles or similar materials by spacing them at a greater distance apart, usually on the order of 12 to 24 inches. The bar material used to space the angles has limited strength of its own.

As a rule of thumb, most steel members are about 1″ deep for each foot to be spanned, so that for a 20′ span, something on the order of a 20″ I-beam could be used. A lightweight beam might be somewhat deeper, whereas a strong heavy section might accomplish this span with 16″ of depth. For larger spans, on the order of 100′ and greater, a different form of beam is utilized—the truss. The purpose of a truss is to bridge the space between points, and this is accomplished by placing material along the flanges or outer edges through the use of spacers or chord members. Figure 15 shows some typical truss shapes. The Fink truss, for instance, might be utilized for the roof beam or span in a building with sloped roofs, while the Vierendeel is a square truss, very suitable for spanning large openings in buildings. In some cases, the dimensions of the Vierendeel truss may be made to coincide with two or three stories in height, so that lateral finished spaces might go through the truss.

Reinforced concrete beams tend to be larger in dimension when compared with steel beams of equal strength, because of the additional space required to develop the tensile reinforcing steel cross section. However, a wider beam can be utilized often because the reinforced concrete beam is incorporated into a floor structure.

In *loading* of any structure, it is important to place the load slowly and gently. It can be demonstrated that a load quickly released on to a structure has the effect of two times the dead weight of the load imposed. In order to avoid this type of dynamic loading, movements such as the final placement of a beam should be very slow and deliberate, gradually releasing the lifting power so that the dead weight of the beam does not cause dynamic loading.

A beam is better able to carry a load which is uniformly distributed across it, rather than placed at concentrated points, because the moment developed by a uniformly distributed load is less than the load imposed by concentrated weight at the center or middle portions of a beam. For this reason, during erection, it may be advantageous to use timbers on top of beams in certain temporary loading situations to spread the load uniformly. The indiscriminate use of chocks or sleepers may impose point loading when it is not intended. Similarly, the use of installed beams as tie-back points for alignment and other purposes should be carefully considered, and temporary connections should be made near the ends of the beams where they are more rigid, and not at the center or middle section. Of course, in the case of cantilever beams, any such tie-offs should be made near the rigid connection to the wall or column.

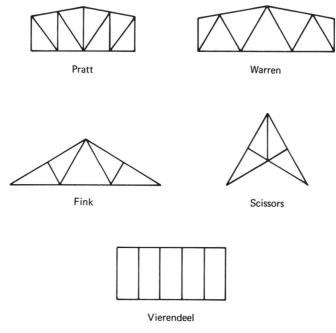

Pratt

Warren

Fink

Scissors

Vierendeel

Fig. 15 Truss types.

Columns

Columns are compression members normally installed in a vertical position. Because their primary purpose is the resistance of compressive force, they can support substantial loads with relatively small cross sections. Theoretically, a short column of substantial cross section should be able to support a load equal to the allowable stress times the cross-sectional area. However, columns fail usually not in direct compression, but by buckling. Buckling occurs because the column is relatively slender in terms of its cross section versus length, and when subjected to any small moments, stresses may be set up as a result of the moments which cause a failure. Accordingly, lateral support of columns is an important consideration. This type of support is gained, for instance, by the tie-in of beams at floor levels, and in the case of derricks or gin poles, is achieved by guy wires.

The traditional solution of column loading is the Euler equation which indicates that there is a specific axial load which will hold a column in equilibrium in a bent or curved position when the ends are pin-loaded. This relationship or allowable load is as follows:

$$P = \frac{\pi^2 EI}{l^2}$$

where

E = modulus of elasticity
I = least moment of inertia about any cross-sectional axis
l = length

This relationship indicates that the allowable load is directly proportional both to the strength as represented by the modules of elasticity, and the moment of inertia of the cross section. Since the moment of inertia is the least moment, a tube provides the maximum effective placement of material, and the larger the diameter, the more resistant the material (however, if the material is too thin, local buckling can occur).

The load is inversely proportional to the square of the length, indicating that the dimension or relationship often considered is the slenderness ratio as a controlling factor. This is the ratio between the length l to the radius of gyration of the cross section.

Column sizes are calculated not only using the traditional Euler formula, with appropriate variations for end conditions, but also by considering various size and load factors, utilizing families of curves similar to those presented in the *AISC Handbook*.

From the inspector's viewpoint, the important consideration in columns, whether permanent or temporary, is *lateral support* and a realization that a column should be a uniform member, so that its least radius of gyration or least moment of inertia does not create any particular area of weakness. For instance, a beam section tends to be deep in one dimension and relatively narrow in the flanges. Such a member would be weak in terms of its use as a column, and would have to be strongly braced on the weak axis. A temporary column might be made, therefore, by welding channel sections at right angles to the web of a beam to make a cross type member.

ANCHORING

Keep in mind that all forces must be balanced. This includes the force imposed upon a pipeline at a turn by the *pressure* in the line. This pressure must be resisted, either through the anchorage of the pipeline, or by a kicker block at the turn. In pressure pipelines of welded metal pipe, the anchorages generally hold the pipe in place with ample strength available in the piping material to provide the holdback. However, in certain special types of pipe, such as concrete water pipe with a slip-on O-ring-type seal, the corner angle is free to blow off unless the force exerted on the turn by the pressure of the fluid is resisted by a static, fixed kicker block of either sufficient dead weight, or combined between dead weight and earth pressure resistance.

The forces imposed upon a pipeline not only by the pressure within it, but also by differential forces created by expansion and/or contraction must be resisted by appropriate anchorage of the material. In hot lines, or very cold lines, the *differential temperatures* between the piping material and the ambient temperature can cause tremendous forces upon the anchorage. For this reason, expansion loops are generally built into the design. The inspection team must be certain that the field force does not "help out" the designer by adding or subtracting anchorages. In certain situations, the installation of too many anchorages will result in a fracture of the piping.

MOTION

Newton's second law of motion is that an unbalanced force acting in a body causes the body to accelerate in the direction of the force, and this *acceleration* is directly proportional to the unbalanced force and inversely proportional to the mass of the body. Mass is a function of the weight of a body, or indirectly its gravitational pull, and force is considered to act through the center of gravity of the mass. For uniform bodies, the center of gravity is at the geometric center of the body. In an irregular shape, the center of gravity must be determined by experiment.

One result of the laws of motion is the fact that two colliding bodies exert equal and opposite forces on each other, so that the mass times velocity of one colliding object must equal the mass times velocity of the other, if the two moving bodies come to stable immobility. However, if motion continues, the resulting vector will be the sum of the momentum vectors of the colliding bodies.

Materials in stable equilibrium may go into harmonic *vibration* depending upon their frequency of vibration. This occurs, for instance, in smoke stacks when wind velocity, although substantially less than the bending strength of the physical configuration of the stack could restrain, may set up vibratory or harmonic motions. This, in some instances, results in the embarrassing situation of a stack waving in a 15 mile per hour breeze when it has been designed to stand 100 mile per hour hurricanes. A caution, therefore, to the inspection staff is to be certain to install all bracing called for without assuming that a stack or vertical member is stiff enough to maintain stability.

Another consideration is wind loading, so that during erection the structure is not made to withstand wind forces for which it was not designed. A typical example of this is the huge vertical assembly building for the Saturn rocket at Launch Complex 39 in Cape Kennedy. This structure was carefully wind tested before erection through the use of wind tunnels. However, a critical consideration was installation of the doors prior to hurricane season. CPM scheduling was utilized to be sure that the doors were in place before the structure could be loaded with hurricane level winds.

PRESSURE

Structures subjected to fluid pressure must be calculated with Pascal's principle in mind. This principle states that an increase in pressure on any part of a confined liquid causes an equal increase throughout the liquid. This principle is utilized in hydraulic presses and hydraulic machinery.

Failure to recognize the liquid pressure principles, as well as the effects of *buoyancy*, result in some situations where box-type foundations may float. In the installation of swimming pools, concern must be maintained for the effect of buoyancy, and the inspection team must be concerned with buoyancy during construction prior to filling.

SUMMARY

Structures are often subject to stresses during erection or construction that exceed design strengths. The inspection force must be alert to these critical potential failure conditions.

8. SAFETY

The construction industry has one of the worst injury experience records according to records kept by the National Safety Council, and has had for the almost 40 years in which these records have been maintained by the Council. While safety affects the contractor's pocketbook directly in terms of his insurance premiums, the owner is also directly involved because accidents can cause delays on the job. Further, in the case of serious accidents or fatalities, the conscientious owner should feel responsibility, usually implied, but under today's liability laws, often legal and real.

Depending on the conditions of the project, it may be appropriate for the owner to insist on certain specific approaches to safety on the part of the contractors. One such approach is the insistence of a safety section, or safety engineer. Naturally, the owner recognizes that these additional requirements may cost additional money. A natural, but slow evolution of safety consciousness has occurred in the construction industry, principally among those owners who build many facilities, such as government or large industrial organizations. This natural evolution will probably plateau now that the Department of Labor *Occupational Safety and Health Act (OSHA)* is being implemented so strongly in the construction field. Regardless of the owner's specific intent in regard to safety, the inspection team has a broad responsibility which must be pursued constantly. The inspection team cannot assume that the contractor will follow the safety precepts specified—they must insist on it. Safety is something which must be worked at by all parties to the project, and when the inspection team observes unsafe practices, they should take action immediately commensurate with the degree of hazard.

Many of the accidents involve equipment which performs below expected loads, particularly cranes. The inspection team should insist on appropriate testing of crane rigs, using either a spring load or lifting of dead weights to simulate difficult lifts.

Other common accident potentials occur in false work, rigging, scaffold, and temporary structures.

The inspection team can be knowledgeable in regard to potential hazards which might not be anticipated by the contractor, particularly if full advantage is taken of the knowledge of the design group. A typical example of an accident which would not have nominally been expected was fire in sprayed fireproofing. In hindsight, the hazard could easily have been noticed by the simple use of common sense and the sense of smell—since the fireproofing was suspended in a volatile liquid with a low flash point. Welding in the vicinity of the freshly placed fireproofing triggered a hot fire which fortunately did not injure the personnel, but delayed the job for some time. The inspection team should be alert to problem areas such as this. Often, a system of welding or burning permits is required in areas of even mild hazard.

New materials and techniques require new safety considerations. Typical of these is the handling of prestressed members. The tendons are placed under tremendous tension, and if mishandled or released can whiplash across a large area. Because of the high stresses, accidental arcing or dropping of molten weld metal on strands may reduce the cross-sectional area, overstressing the members and causing immediate failure.

OCCUPATIONAL SAFETY AND HEALTH ACT (OSHA)

OSHA has the effect of law in the form of Part 1518, "Safety and Health Regulations for Construction," published under Title 29 of the Code of Federal Regulations of the Bureau of Labor Standards, Department of Labor.

Appropriate sections of the OSHA regulation are inserted throughout sections of this book. Those which apply generally to safety are summarized in part in this section.

Inspections—Right of Entry

Any authorized representative of the Department of Labor has the right of entry to any site of a contract performance to inspect or investigate for compliance with the Act, and for the purpose of investigating facts which might apply under the Act. For this purpose, other federal agencies and personnel may be utilized.

General Safety and Health Provisions

It is the responsibility of the employer to initiate and maintain programs to comply with the Act. These programs must provide for frequent and regular inspections of the job site, materials, and equipment by competent persons designated by the employers. The use of unsafe machinery, tools, materials or equipment is prohibited and such items must be identified as unsafe by tagging or by locking their controls to render them inoperable. It is the responsibility of the employer to permit only qualified employees to operate equipment or machinery.

Safety Training and Education

In accordance with the Act, the Department of Labor may establish and supervise programs for the education and training of employers and employees in the recognition, avoidance, and prevention of unsafe conditions.

Employers should be aware of potentially hazardous working conditions, such as the handling of chemicals or poisons. These hazards should be described to employees, as well as procedures—including safety equipment—which can reduce the risk of undue exposure.

Similar admonitions apply to potentially hazardous situations such as the use of flammable liquids, gases, or toxic materials. Cautions and safety measures must be described for employees required to work in confined or enclosed spaces, or with equipment which may be potentially dangerous.

First Aid and Medical Attention

Provisions for medical care must be made available by the employer for every employee covered by the regulations. In the absence of infirmaries, clinics, or hospitals in proximity to the work site, properly trained and certified first aid personnel must be available, and first aid supplies must be provided by the employer. Appropriate equipment for transportation of injured personnel to a physician or hospital must be provided for.

Occupational Noise Exposure

When employees are subjected to sound levels exceeding the following list, the levels must be reduced, or personal protective equipment must be provided. The absolute maximum exposure shall not exceed 140 dB peak sound pressure levels. Other permissible exposures are:

Duration Per Day Hours	Sound Level dBα Slow Response
8	90
6	92
4	95
3	97
2	100
1½	102
1	105
½	110
¼ or less	115

Sanitation

At the work site, adequate supply of potable water must be provided, as well as clean drinking water dispensers. Potable water for clean-up must be provided. Where nonpotable water is used for industrial or firefighting purposes, it must be identified by appropriate signs.

Toilets at construction sites must be provided generally on the basis of one toilet seat and one urinal for 50 workers.

Illumination

Construction areas, ramps, runways, corridors, offices, shops, and storage areas shall be lighted according to the following table:

Foot Candles	Area or Operation
5	General construction areas
3	General construction areas, concrete placement, excavation and waste areas, access routes, loading platforms . . .
5	Indoors, warehouses, corridors, hallways
5	Tunnels, shafts, and general underground working areas
10	Tunnel and shaft heading during drilling, mucking, and scaling
10	General construction plant and shops
30	First aid stations, infirmaries, and offices

Ventilation

Whenever harmful dust, fumes, mists, vapors, or gases exist or are produced in the course of construction work, these hazards must be controlled by the application of general ventilation, local exhaust ventilation, or other effective mechanical means. The ventilation equipment must be designed, constructed, maintained, and operated to insure the required protection by maintaining a volume velocity of exhaust air sufficient to gather the noxious materials and convey them from the work point. To preclude dust settling, ventilation equipment should be operated for a time after the work process is stopped.

When ventilation cannot be provided adequately to preclude the exposure of employees to inhalation, ingestion, skin absorption, or contact with materials or substances at a concentration above those specified in the "Threshold Limit Values of Airborne Contaminants for 1970" of the American Conference of Governmental Industrial Hygienists, then special clothing or equipment shall be used to protect the employees.

Personal Protective and Lifesaving Equipment

Head Protection. Employees working in areas where there is a danger of head injury from impact, falling or flying objects, or electrical shock and burns, shall be protected by protective helmets, which shall meet the specifications contained in ANSI Z89.1. In addition, helmets for the head protection of employees exposed in high-voltage electric shock and burns must meet the additional specifications on such helmets in ANSI Z89.1.

Hearing Protection. When it is not feasible to reduce the noise levels to acceptable exposure, ear protective devices must be provided. These devices, where inserted in the ear, should be fitted or determined individually by competent persons. Plain cotton is not an acceptable protective device.

Eye and Face Protection. Eye and face protection equipment required shall meet the requirement of ANSI Z87.1. Practice for Occupational and Educational Eye and Face Protection. Employees whose vision requires the use of corrective lenses shall wear goggles or spectacles which can either be worn over corrective spectacles, or which provide optical correction.

spectacles which can either be worn over corrective spectacles, or which provide optical correction.

Paragraph 15.18.102 of the Act provides detailed information on the eye and face protector selection, as well as required filter lenses (shade numbers) for protection against radiant energy in welding.

Respiratory Protection. Protective devices shall be approved by the U. S. Bureau of Mines or acceptable to the U. S. Department of Labor for the specific contaminant to which the employee is exposed. In selecting the appropriate respirator, the chemical and physical properties of the contaminant, as well as the toxicity and concentration of the hazardous material, must be considered. In most cases, the optimum respirator has a self-contained breathing apparatus, either a combination airline respirator, or a self-contained air supply or air storage receiver. Respiratory protective equipment which has been previously used must be cleaned and disinfected before being issued by the employer to another employee.

Safety Belts, Life Lines and Lanyards. Safety belt equipment shall be used only for employee safeguarding. Where items are subjected to actual loading or load testing, they must be removed from service and destroyed. Life lines must be secured above the point of operation to an anchorage or a structural member capable of supporting a minimum dead load of 5400 lb. Life lines on rock-scaling operations or in areas where the life line may be subjected to abrasion, must be a minimum of $\frac{7}{8}''$ wire core manila rope. For other applications, a minimum of $\frac{3}{4}''$ manila or equivalent with a minimum breaking strength of 5400 lb shall be used. Safety belt lanyards shall be a minimum of $\frac{1}{2}''$ nylon or equivalent for the maximum length to provide for falls no greater than 6'. This rope must have a nominal breaking strength of 5400 lb or more.

Safety Nets. Safety nets shall be provided when work places are more than 25' above the ground or water surface, or other surfaces where the use of ladders, scaffolds, catch platforms, temporary floors, safety lines, or safety belts is impractical. Nets shall extend 8' beyond the edge of the work surface where employees are exposed, and shall be of a maximum mesh size of nets 6" x 6" of $\frac{3}{8}''$ diameter #1 grade pure manila, $\frac{1}{4}''$ nylon, or $\frac{5}{16}''$ polypropolene rope.

Working on or Near Water. Where danger of drowning exists, employees shall be provided with U. S. Coast Guard approved life jacket or buoyant work vests. Ring buoys with at least 90′ of line shall be provided and available for emergency rescue operations and spacing between ring buoys shall not exceed 200′. At least one lifesaving skiff shall be immediately available at locations where employees are working over or adjacent to water.

Fire Protection and Prevention

The employer is required to provide a trained and equipped firefighting organization (fire brigade). Firefighting equipment is to be maintained in accessible location, conspicuously marked, and maintained in operating condition. A temporary or permanent water supply of sufficient volume, duration, and pressure designed to properly operate the firefighting equipment must be available in direct proportion to the potential for combustion.

Portable firefighting equipment must be available with one extinguisher for each 3000 ft^2 of protected building area (155 gallon open drum of water with 2 fire pails may be substituted for each fire extinguisher). Appropriate protection from freezing must be provided. One hundred feet or less of $1\frac{1}{2}''$ hose with a nozzle capable of discharging water at 25 gallons per minute may be substituted for each fire extinguisher requirement. Other protection systems, such as sprinklers, must be installed and operated in accordance with NFPA recommendations.

Fire Prevention. Electrical wiring equipment for heating, light, or power purposes must be installed in compliance with the requirements of the National Electrical Code, NFPA No. 70, as well as the specific requirements of the OSHA regulations. Internal combustion engine-powered equipment must be located with exhausts well away from combustible materials.

Smoking is to be prohibited in the vicinity of fire hazards, and such areas must be conspicuously posted. Portable battery-powered lighting equipment used in hazardous areas shall be of the approved non-spark type. Care shall be taken properly to ground nozzles, hoses, or steam lines used in hazardous tankage or vessels.

In location of temporary buildings and yard storage, appropriate care shall be taken for proper separation to preclude an accumulation of fire potential. The contractor is responsible to maintain the entire area, but particularly storage areas, free from accumulation of unnecessary combustible materials.

Flammable and Combustible Liquids. Only approved containers and portable tanks may be used for storage and handling of flammable and combustible liquids. Approved metal safety cans should be used for handling and use of flammable liquids in quantities greater than one gallon with certain specific exceptions. Flammable or combustible liquids may not be stored in areas used for exits, stairways, or normally used for the safe passage of people. The indoor storage of flammable and combustible liquids must be limited to no more than 15 gallons. Where larger amounts must be stored, no more than a maximum of 60 gallons of flammable or 120 gallons of combustible liquids may be stored in any one storage cabinet, and such cabinets must be approved. The following limits apply to interior storage:

Fire Protection Provided	Fire Resistance Hours	Maximum Size, ft^2	Maximum Quantity, gal/ft^2
Yes	2	500	10
No	2	500	4
Yes	1	150	5
No	1	150	2

Outside storage of hazardous materials should be limited to container size of less than 60 gallons, and not exceed 1100 gallons in any one pile or area. Piles or groups of containers must be separated by 5' clearance, and none should be closer than 20' to a building.

Portable fire equipment should be located in sensitive areas, such as door openings into any room used for storage of flammable or combustible liquids, and areas in which liquids of this type are dispensed in quantities greater than five gallons at a time. Handling points should observe common sense precautions. Flammable liquids should be kept in closed containers when not actually in use; leakage or spillage of flammable or combustible liquids must be disposed of promptly and safely. Flammable liquids may be used only where there are no open flames or other sources of ignition within 50' and equipment shall be of an approved type, according to NFPA standards.

Liquefied Petroleum Gas (LPG). Liquid gas is important on many construction sites because of its use in temporary heat. Each LPG system must be fully equipped with approved fittings rated at a working pressure of at least 250 psi. Every container and every vaporizer must be provided with one or more approved safety relief valves arranged to afford free vent to the outer air with discharge not less than 5' horizontally away from any opening into a building.

Dispensing of LPG for trucks or motor vehicles shall be performed not less than 10' from the nearest masonry wall building, or not less than 25' from the nearest building of other construction. Equipment utilizing LPG on the job site shall be approved. Storage of LPG within buildings is prohibited.

Temporary Heating Devices. Fresh air must be supplied in sufficient quantities to maintain the health and safety of workmen in the vicinity of the combustion chamber for temporary heaters. Temporary heating devices must be installed to provide clearance to combustible material. Care should be taken to check out the temporary heaters to determine whether they are suitable for sitting on combustible materials or not. Heaters used in the vicinity of combustible tarpaulins, canvas, or similar covering shall be located at least 10' from the covering. Covering should be securely fastened to prevent ignition or upsetting of the heater due to wind action. This is particularly important for temporary enclosures to shield welders, splicers, or in the case of protection for concrete pouring.

Solid fuel heaters and oil-fired salamanders are prohibited in buildings and on scaffolds. Liquid-fired heaters must be equipped with a primary safety control to stop the flow of fuels in the event of flame failure.

Hand and Power Tools

Hand and power tools must be maintained in a safe condition, whether furnished by the employer or the employee. When power-operated tools are designed to accommodate guards, they must be equipped with appropriate guards when in use. Belts, gears, shafts, pulleys, sprockets, spindles, drums, flywheels, chains and other moving parts of equipment must be guarded if the parts are exposed to contact by employees. Guarding requirements must be in accordance with ANSI B15.1.

All hand-held power tools must be equipped with a constant pressure switch which shuts off when the pressure is released. Electric power operated tools shall be of the approved double insulated type, or grounded in accordance with good electrical practice. Pneumatic power tools must be secured to the hose or whip by positive means. Safety clips or retainers must be maintained on pneumatic impact (percussion) tools to prevent attachments from being accidentally expelled. Pneumatically-driven nailers, staplers, and similar equipment provided with automatic fastener feed which operated at more than 100 psi pressure at the tool must have its safety device on the muzzle to prevent the tool from ejecting fasteners, unless the muzzle is in direct contact with the work surface.

Hoses shall not be used for hoisting or lowering tools, and hoses exceeding ½″ inside diameter must have a safety shut-off at source of supply to reduce pressure in case of a hose failure.

All fuel-powered tools must be stopped while being refueled, serviced, or maintained.

Only trained employees may be allowed to operate a powder-actuated tool. Such tools must be tested each day before loading to see that the safety devices are in proper working condition, in accordance with manufacturer's recommended test procedure. Tools shall not be loaded until just prior to intended firing time. Neither loaded nor empty tools are to be pointed at any employee and hands shall be kept clear of the open-barreled end. Fasteners shall not be driven into very hard or brittle materials such as cast iron, glass block, face brick, hardened steel, or hollow tile. When driving into materials easily penetrated, appropriate backing must be available to prevent the pin fastener from passing completely through.

Grinding wheels shall be applied in accordance with ANSI B7.1. "Safety Code for the Use, Care and Protection of Abrasive Wheels."

All employees using abrasive wheels must use eye protection, and other tools shall be operated with personnel utilizing proper personnel safety equipment as appropriate to the tools.

Motor Vehicles

Motor equipment left unattended at night near areas where work is in progress must have appropriate lights or reflectors or barricades to identify the location of the equipment. A safety tire rack, cage or equivalent protection must be used when inflating, mounting, or dismounting tires installed on split rims or rims equipped with locking rings. Heavy machinery which is suspended or held aloft by the use of slings, hoists, or jacks must be blocked or cribbed to prevent falling or shifting before employees are permitted to work under them. Bulldozer and scraper blades and similar equipment shall be either fully lowered or blocked when being repaired or when not in use. All controls must be in neutral position with the motor stopped and brakes set, unless work being performed requires otherwise. Parked equipment must be chocked and parking brakes set. All cab glass shall be safety glass. All vehicles must have a service brake system, an emergency brake system, and a parking brake system. Vehicles which require additional light shall have at least two headlights and two taillights, as well as brake lights.

Other standard vehicle equipment such as seat belts, rear view mirrors, and safety latches on operating levers shall be in accordance with standard vehicle codes, and state inspected where appropriate.

Cranes, Derricks, Hoists

Manufacturer's recommendations on operating conditions shall be followed by the employer. Rated load capacities and recommended operating speeds and special hazard warnings or instructions must be conspicuously posted on all equipment visible to the operator while he is at his control station.

A boom angle indicator and a load indicating device in good working order must also be provided for cranes and derricks. Hand signals to crane and derrick operators shall be those prescribed by the applicable ANSI Standard for the type of crane in use. A thorough annual inspection of hoisting machinery must be made by qualified personnel (qualification by the U.S. Department of Labor).

During operations, wire rope must be taken out of service when any of the following conditions exist: Six randomly distributed broken wires in one lay, or three broken wires in one strand in one lay; wear of $\frac{1}{3}$ of the original diameter of outside individual wires; kinking, crushing, birdcaging, or other damage resulting in distortion of the rope structure; evidence of any heat damage; reductions from nominal diameter of more than $\frac{1}{64}''$ for diameters up to and including $\frac{5}{16}''$, $\frac{1}{32}''$ for diameters $\frac{3}{8}''$ to including $\frac{1}{2}''$, and proportionately similar wear. Wire rope safety factors shall be in accordance with ANDI B30.5. All moving parts, gears, shafts, pulleys, and sprockets must be provided with guards if parts are exposed to contact by employees. Accessible areas within the swing radius of the rear of the rotating superstructure of a crane must be barricaded to prevent an employee from being struck or crushed by the crane.

In operating boom equipment, careful clearance shall be given to electrical distribution and transmission lines. For lines rated 50 kV or below, minimum clearance is 10′, while for loads rated over 50 kV, minimum clearance shall be 10′ plus 0.4″ per each kV over 50—or use twice the length of the line insulator—but never less than 10′.

For hammerhead tower cranes, adequate clearance must be maintained between the moving and rotating structures and fixed objects to allow the passage of employees without harm. Employees required to perform duties on a horizontal boom of hammerhead tower cranes must be protected against falling by guard rails or by safety belts and lanyards. Buffers must be provided at both ends of travel of the trolley.

Overhead and gantry cranes must have the rated load of the crane plainly marked on each side, and if the crane has more than one hoisting unit, each must have its rated load marked on the load block in marking clearly legible from the ground or floor. All operations must be as prescribed in ANSI B30.2, "Safety Code for Overhead and Gantry Cranes."

Derricks in use must meet the applicable requirements for design, construction, installation, inspection, testing, maintenance, and operation prescribed in ANSI B30.6, "Safety Code for Derricks."

Rigging Equipment for Material Handling

Rigging equipment must be operated in accordance with recommended safe working loads as prescribed in Tables H-1 through H-20 of the OSHA Safety and Health Regulations for Construction (Paragraph 1518.251).

OSHA Regulations

Other sections of this book have summary extracts from the Safety and Health Regulations for Construction of the Department of Labor. These are, however, extracts and the inspection team should definitely have current OSHA regulations on the job site as one of the standard references. These regulations, if carefully followed and implemented, will provide the basis for a safe project. While most of the emphasis and direction of the OSHA regulation is toward the employer (contractor), the owner is at least indirectly responsible. Accordingly, his inspection team must be very interested in whether or not the OSHA regulations are being complied with in the letter and the spirit.

SAFETY CHECKLIST

Safety equipment and consideration should include but not be limited to:

1. First aid equipment.
2. Arrangements in advance for medical care.
3. Plan for meeting noise limitations.
4. Appropriate elimination according to required levels.
5. Safe ventilation, adequate for airborne contaminants.
6. Utilization of appropriate protective equipment including hard hats, safety shoes where required, and special hearing and face protection.
7. Provision of safety belts and life lines as appropriate.
8. Safety nets where needed.
9. Fire prevention program.
10. Fire prevention equipment.
11. Safe storage of flammable and combustible materials.
12. Safe use of temporary heating devices.
13. Use of hand and power tools.
14. Use and storage of motor vehicles.
15. Operating of cranes, derricks, and hoists, in accordance with manufacturer's requirements and appropriate ANSI regulations.

9. FIELD TESTING

Some materials, particularly those mixed at the job site, require special testing. The contract may require the contractor to conduct the test, with observation by the inspection team—or the inspection team may take samples and conduct tests.

To accomplish job site testing, appropriate *reference documents* must be available, since the specifications usually refer to testing procedure by reference to standards such as the ASTM, ACI, ASME, ANSI, or others. The inspection team should review specifications and be certain that referenced standards are available and have been reviewed before testing is called for. It is often appropriate to conduct a trial or rehearsal test before actual job materials are mixed.

In accordance with test requirements, required *testing equipment* must be available and checked out at the job site.

Often it is appropriate for testing to be done under the auspices of the materials manufacturer, although the obvious vested interest involved must be considered. Architect-engineers often recommend the use of special testing laboratories or groups as an adjunct to the inspection team. This can be both effective and economical, since it limits the amount of specialized testing equipment which must be purchased, and does not require the inspection team to develop special skills which are used only on a limited basis on a project. Conversely, where large amounts of material are used on the job, it may be appropriate for the inspection team to learn the skills for specific tests. For instance, on large excavation or backfill projects, certain special moisture content tests are run so frequently that it is appropriate for the field team to conduct them.

In other cases where materials were mixed off the job site, such as ready-mix concrete, it may be appropriate to have a special test team assigned to the assembly or batching area. Typically, certified laboratory services are retained to perform the tests at the batch plant. This is particularly appropriate for pre-mixed concrete, asphalt, gravel, and aggregate.

Off-site testing is also used for the inspection of materials at the point of manufacture. Usually, this is during quality control testing or where required, such as water pressure testing. Usually the test is conducted using manufacturer's equipment and personnel under the supervision of the inspection team.

The test requirements for construction sites are many and varied. In the following sections will be described some typical procedures and equipment, but the listing is not inclusive. Once again, the specifications are the explicit directions to the inspection team, both for type of test to be run, equipment to be used, and conditions of test.

SOIL TESTING

Native soil may be checked for loadbearing capacity through actual static loading, or by taking under turf samples for tests. Backfill samples

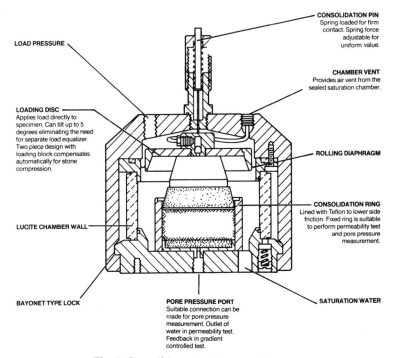

Fig. 1 Consolidometer. *Source: Anteus.*

are checked to determine degree of compaction, which in turn is a measure of the loadbearing capacity of backfill soil.

Figure 1 shows the Anteus consolidometer which is an accurate instrument designed for testing of *soil consolidation.* The unit can also be adapted to test porosity, measuring the ability of the soil tested to resist hydrostatic pressure.

The consolidation chamber is divided into two compartments, with the lower being the sample chamber and the upper the loading head. An extended bayonet-type lock assures an easy and safe locking action between the base and the dome.

More advanced models are available which load the sample automatically and maintain a selected pressure gradient by hydropneumatic force.

One of the most recent devices for measuring proctor values for soil and soil aggregate utilizes a nuclear source and detectors for the nondestructive determination of soil values. This family of instruments is made by Troxler and is known as the 3400D series of surface moisture-density gauges. The devices include solid-state semiconductor components and liquid crystal display readouts. The units are battery powered and contain a microcomputer which holds calibration constants and algorithms necessary to compute and display wet density, moisture, dry density, percent moisture, and percent compaction.

When used to measure soil characteristics, a direct transmission mode can be utilized which places the gamma source into the material by means of a punched access hole. Standard gauges have an 8″ depth capability in 2″ increments. (Other increments and depths are available.) ASTM Standard D-2922-78, "Standard Test Method for Density of Soil and Soil-Aggregate in Place by Nuclear Methods," applies for in-place soil tests.

AGGREGATES

The California State Highways Department has the following suggestions in sampling aggregates:

Samples shall be so chosen as to represent the different materials that are available in the deposit. If the deposit is worked as an open face or pit, the samples shall be taken by channeling the face so that it will represent material that visual inspection indicates may be used. Care shall be observed to eliminate any material that has fallen from the face along the surface. It is necessary, especially in small deposits, to excavate *test holes* some distance back of and parallel to the face to determine the extent of the supply. The number and depth of these test holes depend on the quantity of material that is to be used from the deposit. Samples from open test pits should be obtained by channeling a face of the test pit in the same manner as sampling a face of materials site, described above. Material that will be stripped from the pit as overburden, etc., shall be obtained from the face of the bank and from the test holes and, if visual inspection indicates that there is considerable variation in the material, separate samples shall be obtained at different depths.

Samples from the *stockpile* should be taken at or near the top of the pile, at or near the base of the pile, and at an intermediate point. A board shoved into the pile just above the point of sampling will aid in preventing further segregation during sampling.

Samples from railroad cars should be taken from three or more trenches dug across the car at points which appear on the surface to be representative of the material. The bottom of the trench should be at least one foot below the surface of the aggregate at the sides of the car and approximately one foot wide at the bottom. The bottom of the trench should be practically level. Equal portions should be taken at nine equally spaced points along the bottom of the trench by pushing a shovel downward into the material and not by scraping horizontally. Two of the nine points should be directly against the sides of the car.

Remembering that coarse particles roll to the sides and bottom of a cone-shaped pile, and that fines tend to remain near the center, one of the three trenches across the car should be made at the low point between two cones. Also, it should be observed that if the car has been loaded from a chute which discharges at an angle across the car, that the coarser particles will be next to the far side of the car and finer particles closer to the underside of the chute.

Here again, a crane with a clamshell bucket is very helpful in remixing material in a car before sampling.

Trucks represent the same problem as freight cars, but for a smaller unit. Usually, the problem is to sample the pile as dumped from the truck, and the technique becomes that for sampling a small stockpile.

Belt *conveyors* furnish a handy point for sampling. It is necessary to stop the belt before taking the sample, and a portion of the belt, not less than two feet long, must be scraped clean for the entire width and depth; otherwise, some of the finer sizes and dust will be left out of the sample.

Bucket elevators are difficult to sample and it is not advisable to attempt it. Since they almost invariably empty into a chute or onto a belt, it is recommended that the sample be taken there instead of from the buckets.

Chutes carry their load in surges and, more important, nonuniformly distribute them from one side to the other and from top to bottom. A sample can be caught which will reasonably cross section the flow by passing a box through the column of aggregate at a steady speed so as not to lose any of the particles, large or small, by bouncing out of the box or by overflowing. If the flow is very heavy, a deeper and narrower box should be used, provided it is not less in width than about four times the diameter of the largest particle.

BITUMINOUS PRODUCTS

Representative samples shall be taken of completed asphalt concrete mix at the location where the material is to be spread as soon as possible after delivery of the material.

Under no circumstances should the shovels full of material being obtained for samples be taken from the top or edges of the pile in the truck.

Compressed samples are to be taken from the street at the direction of the engineer.

Since one of the primary purposes of this sample is to measure the relative compaction of the bituminous mixture, extreme care shall be exercised in removing the sample and packing it for shipment in order to prevent the sample from breaking or falling apart. Wherever facilities are available, the density of the sample should be obtained in the field.

An average sample representative of the *asphalt* actually entering into the mix shall be obtained either from a sampling valve which shall be conveniently located in the asphalt line leading from storage tank to mixer, or from the storage tank by means of an oil thief. The following procedure shall be followed in obtaining the samples.

All sampling operations, involving hot asphalt, must be performed with care to avoid burns from spilled material or a possible flash from vapors collecting in storage tanks. There is always the possibility that any asphalt may be delivered to the job site at temperatures sufficiently high to cause the collection of inflammable vapors in the storage tank. These vapors may flash if the correct mixture with air should occur and a spark or other ignition source be present. No smoking shall be permitted on the tops of storage tanks. No sampling shall be done during transfer operations. Gloves shall always be worn and sleeves should be rolled down and fastened at the wrists while sampling.

The recommended method of obtaining the sample is from a valve placed in the line leading from the storage tank to the mixer. The sample shall be drawn from the valve after completion of transfer and sufficient circulation of the contents of the storage tank to assure that the sample represents the contents of the tank. Where different grades of asphalt are being used, at least one-half gallon of asphalt shall be drawn from the valve prior to obtaining the sample.

If no sampling valve or spigot is avilable, an average sample shall be secured by means of an oil thief. A clean oil thief shall be quickly lowered to the bottom of the tank and withdrawn at such a rate that when removed from the asphalt, some unfilled space remains in the thief. In order to prevent contamination of the sample by material remaining in the thief from previous sampling or from traces of solvents used in cleaning, the first sample removed with the thief shall always be poured back into the tank. At least two washings of the thief by the hot asphalt should be carried out before a final sample is obtained, since a very small amount of contaminant can make marked changes in penetration.

The Troxler 3400B can be used in a backscatter mode (i.e, non-destructive) to identify compaction resulting from rolling of bituminous concrete. Typically, maximum density is determined by taking the average of ten random tests. These are normally run on 2800-square-yard sections, and should average 98% of target density, with no test falling below 95%. The maximum obtainable density of asphalt area can be entered into the microcomputer, and the instrument will immediately compute and display the percent compaction of wet density.

Another Troxler device, the asphalt content gauge (Model 2226), provides an accurate reading of percentage of asphalt in an asphalt mix in approximately three minutes. This device is portable and can be utilized at an asphalt plant as well as in a laboratory.

CONCRETE

One of the most consistent testing requirements on construction sites is that of concrete. The following procedures are quoted from the Forney Inc. Test Division notes:

Slump Test. The slump test is the last go-no-go point before the concrete is placed, and a lot rides on the results. But it doesn't take much to affect the test's reliability.

Poor sampling, neglecting to dampen the cone, improper rodding, an unstable base, the wrong type of rod—and the test result bears little relation to actual concrete quality.

Although each step in the test is extremely important, the procedure is relatively simple and straightforward.

The first step—and probably the most important—is sampling. Be sure to take samples from three different parts of the load, directly from the mixer discharge. The total sample should be at least 80 pounds and should be re-mixed in a wheelbarrow.

After sampling, there are four basic operations:

Moisten the inside of the cone and place it on a flat, level, firm surface that extends several inches beyond the base of the cone. Be sure to stand on the foot lugs of the cone to hold it firmly in place as you fill it with concrete.

Fill the cone in three layers, each about $\frac{1}{3}$ of the cone's volume. Rod each layer exactly 25 times with a round spherical nosed iron or steel rod $\frac{5}{8}''$ in diameter. When rodding the second and third layers, penetrate into, but not through, the layer beneath.

Strike off excess concrete with a straight-edge so that the cone is exactly full. Be sure to remove spilled concrete from around the base of the cone. Then carefully lift the cone straight up, slowly and gently. The cone should be removed immediately after the rodding and strike off steps are completed. Be sure to avoid jarring the concrete as you lift the cone.

When the concrete has slumped, measure the distance from the top of the concrete to the height of the metal cone. If the top of the slump is irregular, don't measure the high or low point, but try to get an average.

Concrete Cylinders

This is probably the most critical step in the entire testing operation. After all, what good would it do to follow all standard test procedures if the sample itself is unreliable?

To insure reliable results, take samples from at least three different parts of the load, directly from the mixer discharge, but be sure not to use either the beginning or the end of the load. Either pass a receptacle through the discharge stream, or divert the stream completely so that it discharges into the container.

Before filling the molds, be sure to *remix* individual portions of the sample in either a flat pan, wheelbarrow, or other clean, nonabsorptive surface, but don't over-mix and don't let the sample stand around for more than 15 minutes.

The test should be made in a cylindrical, watertight mold 6″ in diameter by 12″ high. Steel, or paraffin-sprayed paper, is generally recommended. Before filling, place the cylinders on a smooth, firm, level surface. A single test generally consists of three specimens for each age test—7 and 28 days.

Fill molds in three equal layers, and *rod* each layer 25 times with a ⅝″ spherical nose rod. When rodding the second and third layers, penetrate into, but not through, the layer beneath.

The key to effective compaction (the purpose of rodding) is a proper rod. The ⅝″ spherical nose rod slides over aggregates, allowing the concrete to close smoothly after it's withdrawn. A flat-tipped rod forces aggregates downward, leaving voids as it is withdrawn.

After rodding the top layer, tap the cylinder lightly to close any voids that might have formed. Then strike off excess concrete with a trowel, and cover immediately to prevent loss of moisture.

Just as *curing* is the key to high quality concreting, so is it essential for reliable test results.

The cylinders must be kept moist and at temperatures no lower than 60°F or higher than 80°F for the first 24 hours before transporting to the laboratory. If the weather is extremely hot, cover the cylinders with wet burlap and store in a shaded area.

Let the cylinders stand undisturbed from 12 to 24 hours, and then either moist cure at 70°F (according to ASTM specification C-31), or send to a laboratory for curing. When transferring to the cure box or laboratory, be sure to pack the cylinders carefully in sawdust or other cushioning materials. Even the slightest bump can prove fatal to reliable strengths.

Strength Testing

Strength is the accepted measure of concrete quality. However, a low test result does not necessarily mean poor concrete quality. It often indicates poor testing practices.

Because there are so many fingers in the concrete pie, the only way is strict observance of standard methods, such as those established by the American Society for Testing Materials. Without these standards, the construction industry would crumble.

There are three standard tests. Each provides specific information—and each requires specific know-how.

The common 6" x 12" test cylinder measures compression strength. However, results are reliable only if the concrete is sampled, molded, cured, and tested according to ASTM requirements. The following five rules should always be observed.

(1) Take samples from at least three parts of the load; (2) use only nonabsorptive molds (steel or paraffin-sprayed cardboard); (3) fill molds in three equal layers, rod each layer 25 times with a spherical-nose rod; (4) let cylinders stand undisturbed from 12 to 24 hours, with tops covered, at temperatures between 60 and 80°F; and (5) pack cylinders carefully in sawdust and ship to a laboratory for testing. Special precautions: Be sure a qualified person prepares the specimens and supervises the test; be sure the testing laboratory is inspected by the Cement and Concrete Reference Laboratory (CCRL); and be sure the cylinder is damp when tested. Use larger diameter cylinders for aggregate above 2" maximum size or screen out the oversize.

This test measures flexural strength, an important quality of pavements, highways, and airport runways. To be on safe ground, always:

(1) Take samples from at least three parts of the load; (2) use rigid nonabsorptive, water-tight molds at least 6" x 6" in cross section; (3) oil molds lightly before filling, fill molds in 3"-deep layers, and rod each layer once for each 2 square inches of area; (4) spade all sides with a trowel, strike off top with a straight-edge, and finish with a wood float; and (5) cover test beams immediately with a double layer of wet burlap and let specimens stand undisturbed for 24 hours at temperatures between 60° and 80°F.

Special precautions: In addition to the precautions listed for the test cylinder, be sure the maximum aggregate does not exceed one-third the beam thickness, and take special care in handling and curing (because of the shape and larger exposed area).

This is a convenient, though approximate method of testing concrete strength after it has hardened.

Accuracy is usually plus or minus 20%. If the instrument is calibrated very closely, accuracy can be increased to within plus or minus 10%.

The test calls for taking 15 impacts over a one-inch circle, disregarding the highs and lows, and then averaging the remaining results.

(a) **The Schmidt Test Hammer** is available equipped with an automatic recording device which makes a permanent record of each rebound. Type NR is the standard Type N with recorder, and also available is Type LR which is the standard Type L with recorder.

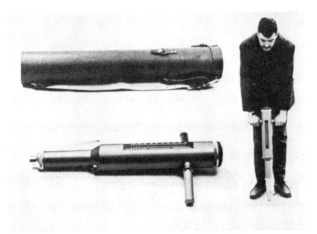

(b) **Schmidt Heavy Duty Test Hammer Type M**—Used extensively in Europe for non-destructive hardness tests on massive structures such as highways, bridges, foundations, dams, and airport runways, the improved Type M hammer has proven to be a dependable method of checking large areas.

Fig. 2 Schmidt concrete test hammer. *Source: Forney.*

Special precautions: Do not use this test on steel-troweled surfaces, surfaces over embedded steel, or structures that are more than 60 days old.

Figure 2 shows a Schmidt concrete test hammer used to test the strength of *in-place concrete.* This unit uses measurement on the theory that soundness is directly proportional to strength, and results in rapid nondestructive testing.

Other special equipment is becoming available for testing of concrete before and after mixing. The James Electronics Company, for instance, has nuclear testing systems for reading accurately moisture and density control of cement/water ratios after mixing.

Several other units are available for testing of concrete after set. Ultrasonic units are used for nondestructive testing of concrete, steel, and other structural units. Sonic resonance testing can be used for concrete, asphalt, and ceramics. The James Electronics Company makes a reinforcement bar testing unit which can determine bar location, depth, and size.

There are other nondestructive devices for the strength testing of concrete in place. Soiltest makes a concrete test hammer (CT-320) which measures strength in terms of hammer rebound. This is a portable device permitting rapid readings. The Windsor probe (Densicon, Inc.) uses a pistollike device to fire a metal probe into concrete to be tested. There are two types of probes, one for standard weight concrete and one for lightweight concrete. The precision power load that powers each probe is the same for standard or low power. The strength of the concrete is measured in terms of the amount of the probe which did *not* penetrate the concrete. For the Windsor probe, the recommendation is that a system of three probes be utilized for an accurate measurement. The company indicates that the test meets ASTM C-803 and ANSI A10.3 precision in accordance with ASTM C-670, and is permissible under ACI 347-78.

The Windsor probe clearly has an advantage for testing concrete which has reflected low cylinder breaks. Densicon also points out that cylinders have a substantial range of accuracy. It is their claim that a standard test cylinder has 10% higher strength in the top third; that cylinders made in steel molds are 10% higher in strength on average than those made in paper molds; and that their core strength can vary to 40% less than that of cast cylinders.

STRUCTURAL WELDS

One basic form of testing welds is the welder test prescribed by AWS. A test tab is made and destructively tested to confirm the welding ability of the welder to be certified. Nondestructive tests of welds in place for structural members and piping are typically made by radiography. These X rays may be made on a sampling basis, or for special situations such as ultimate design or high-pressure pipelines, 100% radiography may be required. An alternative to radiography is the ultrasonic flaw detector. The USK6 unit (Krautkramer-Branson) weighs less than 11 pounds, including batteries. The unit has a CRT screen to provide a video result of the test instantaneously.

SPECIAL TESTING

Special measuring devices are used for setting of equipment. Basically, these are precision millwright tools using dial indicators and similar machine shop equipment to accurately check clearances. Feeler gauges and other manual devices are incorporated in this type of measuring. For measuring speed of rotation, stroboscopic equipment can be used.

For measurements such as ambient noise levels, noise meters can be used on a spot basis, or can be set up to be self-recording. Light meters can measure light delivered in footcandle intensity.

For measurement of moisture in existing or new roofs, the Troxler 3205 surface moisture gauge can be utilized in a nondestructive mode. This device can pinpoint entrapped moisture concentrations within a roof structure, and thus facilitate the cost-effective replacement of only specific areas. The testing device can be used for inspection of roofs in place or as quality assurance on new roofing materials.

EXAMPLE: PENNSYLVANIA

Pennsylvania has specific testing procedures for the inspection team, directing as follows:

1. All materials, equipment, concrete batch plants and their products, fabricators and their products, and erectors and their services must be approved in writing prior to the performance of any tests.

2. If the Contract Documents, laws, ordinances, rules, regulations, or orders of any public authority having jurisdiction require any work to be inspected, tested, or approved, the Contractor must give the State timely notice of its readiness and of the date arranged so the inspection team can observe such inspection, testing or approval.

The Contractor shall bear all costs of such inspections, tests, and approvals unless otherwise provided.

3. The inspectors are authorized to reject work which does not conform to the Contract Documents and to direct any or all Construction Contractors to stop work or any portion thereof, or to require special inspection or testing of the Work as provided in the General Conditions.

4. If such special inspection or testing reveals a failure of the Work to comply (1) with the requirements of the Contract Documents or (2) with respect to the performance of the Work, with laws, ordinances, rules, regulations, or orders of any public authority having jurisdiction, the Contractor shall bear all costs thereof, including additional design services made necessary by such failure.

   If such special inspection or testing shows the Work to be satisfactory, the State shall bear the costs.

5. The contractor prepares or has prepared by a consultant or Testing Laboratory, Form 132 (Fig. 3), for the approval of the concrete mix or mixes to be used on the Project. No concrete is to be incorporated in the Project unless an approved Form 132 has been received.

   The designer or State engineers review the concrete mix computations, as shown on Form 132, for compliance with the Contract Documents.

6. The Resident Inspector prepares Form 129 (Fig. 4) for all materials and equipment samples for testing, in accordance with Contract Specifications. Only one (1) item of material or equipment will be shown on any one Form 129.

   Notify the designer that samples are ready for testing, at the job site, to be delivered to the approved testing laboratory.

7. The Resident Inspector prepares Form 170 (Fig. 5) for all field tests of water, sewer, and steam lines, as well as other major equipment installations.

   Under Para. 6, Comments, a comprehensive report of the results of the test shall be made and conclusions reached by all persons witnessing the test. Include a statement that the item tested meets the requirements of the Contract Documents.

	CONSTRUCTION	Project No. _____
THE GENERAL STATE AUTHORITY	CONCRETE MIX	Contract No. _____
18th and Herr Streets		Project Title _____
Harrisburg, Pennsylvania	COMPUTATIONS	Location _____

Contractor _____ Address _____

Cement	Regular	
Type	Air Entraining	Cement Producer _____

Fine Aggregate Producer _____

Coarse Aggregate Producer _____ (Size: _____)

BATCH WEIGHTS AND VOLUME COMPUTATIONS
ONE CUBIC YARD OF CONCRETE

Material	Quantity	Weight	Volume Computations	Absolute Volume
Cement	Bags	• lbs.		cu. ft.
Fine Aggregate		• lbs.		cu. ft.
Coarse Aggregate		lbs.		cu. ft.
Water **	••• Gals.	lbs.		cu. ft.
Air			Regular Cement - .27 cu. ft. AE Cement - 1.08 cu. ft.	cu. ft.
Total		lbs.		27 cu. ft.

* Saturated, Surface-Dry State
** Use Maximum Allowable Water Content in Computations
*** Mixing Water and Free Water

Remarks:

Specified strength: _____

Computation prepared by: _____

Ready-Mix Supplier _____

_____ _____
Date   Contractor

Approval/Disapproval is recommended.
Remarks:

_____ _____
Date   Designing Architect/Engineer

Approved/Disapproved.
Remarks:

_____ _____
Date   For the Director of Construction

Distribution:	Contractor	District Engineer
	Designing Architect/Engineer	Resident Inspector
	Testing Laboratory	Central File

Fig. 3 Form 132—Approval of concrete mix. *Source: Pennsylvania General State Authority.*

	CONSTRUCTION	Project No. _____
		Contract No. _____
		Project Title _____
18th and Herr Streets	LABORATORY SAMPLE IDENTIFICATION	_____
Harrisburg, Pennsylvania		Location _____
		_____

GENERAL DATA, ALL SAMPLE ITEMS

Contractor _____ _____	Approved Testing Laboratory _____
* Material _____	Item No. _____ App'vd Form 130 dated _____
Place Sampled _____	Sampled By _____	Date Sampled _____
Quantity Represented _____	Ultimate Location In Project _____

Detail of Test(s) Required _____

_____

_____

_____

_____

_____

_____

_____

_____

ADDITIONAL DATA, CONCRETE SAMPLES ONLY

Job Mixed _____	Ready Mixed _____	Specified Strength _____	Supplier If Ready Mixed _____
Consistency (Slump) _____			Manner of Cylinder Curing _____
Weather When Sampled _____		Temperature _____	Cylinders Numbered _____

Batch Weight I Cubic Yard of Concrete

Materials	Type	Size	Weight	Producer
Cement				
Fine Aggregate				
Coarse Aggregate				
Water				
Total				

*lbs. - Dry Aggregate

Remarks:

_____

_____   _____
Date   Resident Inspector

Distribution: Contractor; Architect; Director of Construction; District Engineer; Resident Inspector

Fig. 4 Form 129—Materials and equipment samples for testing. *Source: Pennsylvania.*

FIELD TEST
IDENTIFICATION

Project_____

Contract_____

1. Exact Location of Test on Project_____

2. What is being Tested_____

_____Date of Test_____

3. Paragraph Numbers in Specification and/or reason for making test.

_____

_____

4. Exactly how test was made, Instruments, Gauges, etc. used._____

_____

_____

_____

_____

5. Complete list of the personnel present witnessing test and their title.

_____

_____

6. Comments:_____

_____

_____

_____

Fig. 5 Form 170—Field test identification. *Source: Pennsylvania.*

B.
PROJECT
CHARACTERISTICS

The components of the typical building project are described in this section. Since a truly typical model cannot be established, the components represent an overview, in fact a composite, of a building project. The format is the divisional breakdown of the Construction Specifications Institute (CSI).

The division breakdown is as follows:

Division 1—General Requirements
Division 2—Site Work
Division 3—Concrete
Division 4—Masonry
Division 5—Metals
Division 6—Wood and Plastics
Division 7—Thermal and Moisture Protection
Division 8—Doors and Windows
Division 9—Finishes
Division 10—Specialities
Division 11—Equipment
Division 12—Furnishings
Division 13—Special Construction
Division 14—Conveying Systems
Division 15—Mechanical
Division 16—Electrical

The first edition of the CSI format was issued in 1963. In 1966, the AGC (Association of General Contractors) adopted the CSI format as the basis for a uniform system for cost accounting and filing. This system added subsection numbers to the major divisions, and this subnumbering system was utilized in the first edition of "Construction Inspection Handbook."

Likewise, the parallel 16 Division, Specifications Format, first contained in the 1966 publication of the Canadian Building Construction Index under the aegis of the Specification Writers Association of Canada (now Construction Specifications Canada [CSC]) gained acceptance in that country. In 1972, with the publication of the "Uniform Construction Index," these two formats—in their 16 Division and Broadscope Section headings concept—merged into one. Widespread acceptance of the 16 Division format has helped contractors and increased the accuracy of bids while reducing the efforts normally associated with bidding. Contractors have found products easier to control; architects and engineers have greater assurance that specifications are coordinated and complete. Owners naturally have benefited by reduced expenditures through the increased efficiency provided by the standardization.

In 1978, CSI published the "MASTERFORMAT," which provides an organizational structure for groupings of Bidding Requirements, Contract Forms, and Conditions of the Contract, as well as the merged 16 Division listings of Broad- and Narrowscope section titles and numbering systems published by CSI and CSC subsequent to the 1972 edition of the Uniform Construction Index. It is the intent of the MASTERFORMAT to provide the construction industry with the ability to relate all files, costs, product literature, references, and project manual documentation to a single unified system.

The basis for the 16 Division or "Broadscope" titles (pages 8 and 9 of the MASTERFORMAT) follow. The MASTERFORMAT (pages 12–29) has a further breakdown into "Narrowscope" titles. These appear at the beginning of each divisional breakdown in this book.

The balance of the MASTERFORMAT includes two sections. The first (pages 30–57) is an explanation of the "Broadscope" descriptions. The second (59–127) is an index of key words and their associated "Broadscope" numbers.

From the "MASTERFORMAT," published by the Construction Specifications Institute, 1978, and used with permission:

SPECIFICATIONS—DIVISIONS 1-16

DIVISION 1—GENERAL REQUIRMENTS

01010   Summary of Work
01020   Allowances
01030   Special Project Procedures
01040   Coordination

01050 Field Engineering
01060 Regulatory Requirments
01070 Abbreviations and Symbols
01080 Identification Systems
01100 Alternates/Alternatives
01150 Measurement and Payment
01200 Project Meetings
01300 Submittals
01400 Quality Control
01500 Construction Facilities and Temporary Controls
01600 Material and Equipment
01650 Starting of Systems
01660 Testing, Adjusting, and Balancing of Systems
01700 Contract Closeout
01800 Maintenance Materials

DIVISION 2—SITEWORK

02010 Subsurface Investigation
02050 Demolition
02100 Site Preparation
02150 Underpinning
02200 Earthwork
02300 Tunnelling
02350 Piles, Caissons and Cofferdams
02400 Drainage
02440 Site Improvements
02480 Landscaping
02500 Paving and Surfacing
02580 Bridges
02590 Ponds and Reservoirs
02600 Piped Utility Materials and Methods
02700 Piped Utilities
02800 Power and Communication Utilities
02850 Railroad Work
02880 Marine Work

DIVISION 3—CONCRETE

03010 Concrete Materials
03050 Concreting Procedures
03100 Concrete Formwork
03150 Forms
03180 Form Ties and Accessories
03200 Concrete Reinforcement
03250 Concrete Accessories
03300 Cast-In-Place Concrete
03350 Special Concrete Finishes
03360 Specially Placed Concrete

03370 Concrete Curing
03400 Precast Concrete
03500 Cementitious Decks
03600 Grout
03700 Concrete Restoration and Cleaning

DIVISION 4—MASONRY

04050 Masonry Procedures
04100 Mortar
04150 Masonry Accessories
04200 Unit Masonry
04400 Stone
04500 Masonry Restoration and Cleaning
04550 Refractories
04600 Corrosion Resistant Masonry

DIVISION 5—METALS

05010 Metal Materials and Methods
05050 Metal Fastening
05100 Structural Metal Framing
05200 Metal Joists
05300 Metal Decking
05400 Cold-Formed Metal Framing
05500 Metal Fabrications
05700 Ornamental Metal
05800 Expansion Control
05900 Metal Finishes

DIVISION 6—WOOD AND PLASTICS

06050 Fasteners and Supports
06100 Rough Carpentry
06130 Heavy Timber Construction
06150 Wood-Metal Systems
06170 Prefabricated Structural Wood
06200 Finish Carpentry
06300 Wood Treatment
06400 Architectural Woodwork
06500 Prefabricated Structural Plastics
06600 Plastic Fabrications

DIVISION 7—THERMAL AND MOISTURE PROTECTION

07100 Waterproofing
07150 Dampproofing
07200 Insulation
07250 Fireproofing
07300 Shingles and Roofing Tiles

07400   Preformed Roofing and Siding
07500   Membrane Roofing
07570   Traffic Topping
07600   Flashing and Sheet Metal
07800   Roof Accessories
07900   Sealants

DIVISION 8—DOORS AND WINDOWS

08100   Metal Doors and Frames
08200   Wood and Plastic Doors
08250   Door Opening Assemblies
08300   Special Doors
08400   Entrances and Storefronts
08500   Metal Windows
08600   Wood and Plastic Windows
08650   Special Windows
08700   Hardware
08800   Glazing
08900   Glazed Curtain Walls

DIVISION 9—FINISHES

09100   Metal Support Systems
09200   Lath and Plaster
09230   Aggregate Coatings
09250   Gypsum Wallboard
09300   Tile
09400   Terrazzo
09500   Acoustical Treatment
09550   Wood Flooring
09600   Stone and Brick Flooring
09650   Resilient Flooring
09680   Carpeting
09700   Special Flooring
09760   Floor Treatment
09800   Special Coatings
09900   Painting
09950   Wall Covering

DIVISION 10—SPECIALITIES

10100   Chalkboards and Tackboards
10150   Compartments and Cubicles
10200   Louvers and Vents
10240   Grilles and Screens
10250   Service Wall Systems
10260   Wall and Corner Guards
10270   Access Flooring

10280   Specialty Modules
10290   Pest Control
10300   Fireplaces and Stoves
10340   Prefabricated Steeples, Spires, and Cupolas
10350   Flagpoles
10400   Identifying Devices
10450   Pedestrian Control Devices
10500   Lockers
10520   Fire Extinguishers, Cabinets, and Accessories
10530   Protective Covers
10550   Postal Specialties
10600   Partitions
10650   Scales
10670   Storage Shelving
10700   Exterior Sun Control Devices
10750   Telephone Enclosures
10800   Toilet and Bath Accessories
10900   Wardrobe Specialites

DIVISION 11—EQUIPMENT

11010   Maintenance Equipment
11020   Security and Vault Equipment
11030   Checkroom Equipment
11040   Ecclesiastical Equipment
11050   Library Equipment
11060   Theater and Stage Equipment
11070   Musical Equipment
11080   Registration Equipment
11100   Mercantile Equipment
11110   Commercial Laundry and Dry Cleaning Equipment
11120   Vending Equipment
11130   Audio-Visual Equipment
11140   Service Station Equipment
11150   Parking Equipment
11160   Loading Dock Equipment
11170   Waste Handling Equipment
11190   Detention Equipment
11200   Water Supply and Treatment Equipment
11300   Fluid Waste Disposal and Treatment Equipment
11400   Food Service Equipment
11450   Residential Equipment
11460   Unit Kitchens
11470   Darkroom Equipment
11480   Athletic, Recreational, and Therapeutic Equipment
11500   Industrial and Process Equipment
11600   Laboratory Equipment
11650   Planetarium and Observatory Equipment

11700 Medical Equipment
11780 Mortuary Equipment
11800 Telecommunication Equipment
11850 Navigation Equipment

DIVISION 12—FURNISHINGS

12100 Artwork
12300 Manufactured Cabinets and Casework
12500 Window Treatment
12550 Fabrics
12600 Furniture and Accessories
12670 Rugs and Mats
12700 Multiple Seating
12800 Interior Plants and Plantings

DIVISION 13—SPECIAL CONSTRUCTION

13010 Air Supported Structures
13020 Integrated Assemblies
13030 Audiometric Rooms
13040 Clean Rooms
13050 Hyperbaric Rooms
13060 Insulated Rooms
13070 Integrated Ceilings
13080 Sound, Vibration, and Seismic Control
13090 Radiation Protection
13100 Nuclear Reactors
13110 Observatories
13120 Pre-Engineered Structures
13130 Special Purpose Rooms and Buildings
13140 Vaults
13150 Pools
13160 Ice Rinks
13170 Kennels and Animal Shelters
13200 Seismographic Instrumentation
13210 Stress Recording Instrumentation
13220 Solar and Wind Instrumentation
13410 Liquid and Gas Storage Tanks
13510 Restoration of Underground Pipelines
13520 Filter Underdrains and Media
13530 Digestion Tank Covers and Appurtenances
13540 Oxygenation Systems
13550 Thermal Sludge Conditioning Systems
13560 Site Constructed Incinerators
13600 Utility Control Systems
13700 Industrial and Process Control Systems
13800 Oil and Gas Refining Installations and Control Systems
13900 Transportation Instrumentation

13940 Building Automation Systems
13970 Fire Suppression and Supervisory Systems
13980 Solar Energy Systems
13990 Wind Energy Systems

DIVISION 14—CONVEYING SYSTEMS

14100 Dumbwaiters
14200 Elevators
14300 Hoists and Cranes
14400 Lifts
14500 Material Handling Systems
14600 Turntables
14700 Moving Stairs and Walks
14800 Powered Scaffolding
14900 Transportation Systems

DIVISION 15—MECHANICAL

15050 Basic Materials and Methods
15200 Noise, Vibration, and Seismic Control
15250 Insulation
15300 Special Piping Systems
15400 Plumbing Systems
15450 Plumbing Fixtures and Trim
15500 Fire Protection
15600 Power or Heat Generation
15650 Refrigeration
15700 Liquid Heat Transfer
15800 Air Distribution
15900 Controls and Instrumentation

DIVISION 16—ELECTRICAL

16050 Basic Materials and Methods
16200 Power Generation
16300 Power Transmission
16400 Service and Distribution
16500 Lighting
16600 Special Systems
16700 Communications
16850 Heating and Cooling
16900 Controls and Instrumentation

DIVISION 2: SITE WORK

NUMBER	TITLE
02010	**SUBSURFACE INVESTIGATION**
-11	Borings
-12	Core Drilling
-13	Standard Penetration Tests
-14	Seismic Exploration
02050	**DEMOLITION**
-60	Building Demolition
-70	Selective Demolition
-71	Selective Structural Demolition
-72	Minor Demolition for Remodelling
02080–02099	(Reserved)
02100	**SITE PREPARATION**
-10	Clearing
-11	Tree and Shrub Removal
-12	Tree Pruning
-13	Sod Stripping
-14	Large Tract Tree Clearing
-20	Structure Moving
02130–02149	(Reserved)
02150	**UNDERPINNING**
-51	Shores
-52	Needle Beams
-53	Grillage
-54	Pit Underpinning
-55	Pile Underpinning
-60	Slabjacking
-65	Pressure Injected Footings
02170–02199	(Reserved)

02200	**EARTHWORK**
-10	Site Grading
-11	Rock Removal
-12	Concrete Removal
-20	Structure Excavation and Backfilling
-21	Trenching, Backfilling, and Compacting
-22	Granular Sub-base
-30	Roadway Excavation, Backfill and Compaction
-31	Roadway Re-Grading
-32	Roadway Base
	Subsoil Base
	Soil Cement Base
	Granular Base
-35	Auger Trenching
-40	Soil Stabilization
-41	Soil Stabilization: Lime
-42	Soil Stabilization: Cement
-43	Soil Stabilization: Asphalt
-44	Soil Stabilization: Lime Slurry Injection
-45	Soil Stabilization: Pressure Grouting
-46	Soil Stabilization: Vibro-Flotation
-50	Compaction Control and Testing
-60	Finish Grading
-70	Slope Protection and Erosion Control
-71	Riprap
-72	Gabions
-73	Soil Cement
-74	Slope Paving
-75	Membrane System Slope Protection
-80	Soil Treatment
-81	Termite Control
-82	Vegetation Control
-83	Rodent Control
02290–02299	(Reserved)

02300	**TUNNELING**
-01	Tunnel Ventilation
-02	Tunnel Compression
-10	Tunnel Excavation
-11	Earth Tunneling
-12	Mixed Face Tunneling
-13	Rock Tunneling
-20	Tunnel Liner
-21	Concrete Lining
-22	Prefabricated Steel Liner
-23	Cast Iron Liner
-24	Precast Concrete Liner

-30	Tunnel Grouting
-40	Tunnel Support Systems
-41	Tunnel Rock Bolting
-42	Steel Rings and Lagging

02350   **PILES, CAISSONS, AND COFFERDAMS**

-51	Pile Driving
-52	Pile Performance Specifications
-53	Load Tests
-60	Piles
-61	Wood Friction Piles
-62	Cast-in-Place Concrete Piles (Uncased)
	Bored Friction Concrete Piles
	Bored and Belled Concrete Piles
-63	Concrete-filled Steel Pipe Piles
-64	Concrete-filled Steel Shell Piles
-65	Concrete Displacement Piles
-66	Precast Concrete Piles
-67	Prestressed Concrete Piles
-68	Shaped Steel Section Piles
-69	Sheet Piles
-70	Caissons
-71	Excavated Caissons
-72	Drilled Caissons
-73	Benoto Caissons
-74	Open Caissons
-75	Pneumatic Caissons
-76	Box Caissons
-80	Cofferdams and Excavation Support Systems
-81	Cribbing
-82	Double Wall Cofferdams
-83	Cellular Cofferdams
-85	Piling with Intermediate Lagging
-86	Sheet Piling Cofferdams
-87	Walers
-88	Soil Anchors
-89	Rock Bolting
-90	Ground Freezing
-91	Reinforced Earth
-95	Slurry Wall Construction

02400   **DRAINAGE**

-01	Dewatering
	Sand Drains
	Well Points
	French Drains
	Relief Wells

-10	Subdrainage Systems
-11	Foundation Drainage
-12	Underslab Drainage
-13	Tunnel Drainage
-20	Surface Run-off Collection Systems
-30	Drainage Structures, Pipe and Fittings
-31	Catch Basins, Covers and Frames
-32	Curb Inlets
-33	Drainage Pipe
-34	Culverts

02400   SITE IMPROVEMENTS

-41	Underground Sprinkler Systems
-42	Aboveground Sprinkler Systems
-43	Fountains
-44	Chain Link Fences and Gates
-45	Wire Fences and Gates
-46	Wood Fences and Gates
	Picket Fences
	Stockade Fences
	Rail and Post Fences
-50	Walkway, Roadway, and Parking Appurtenances
-51	Guardrails
-52	Signage
-53	Traffic Signals
-54	Culvert Pipe Underpasses
-55	Parking Barriers
-56	Parking Bumpers
-57	Bicycle Racks
-60	Play Fields and Equipment
-61	Playground Equipment
-62	Recreational Facilities
-63	Play Structures
	Rubble Structures
	Railroad Tie Structures
-70	Site and Street Furnishings
-71	Seating
-72	Tables
-73	Pre-fabricated Shelters
-74	Pre-fabricated Planters
-75	Trash and Litter Receptors

02480   LANDSCAPING

-81	Shrub and Tree Relocation
-82	Plant Materials
-83	Planting Operations
-84	Soil Preparation and Soil Mixes

-85   Lawns and Grass
       Seeding
       Sodding
       Plugging
       Sprigging
       Hydro-mulching
-90   Trees, Plants, and Ground Covers
-91   Trees
-92   Shrubs
-93   Plants and Bulbs
-94   Ground Covers
-95   Aggregate Beds
-96   Wood Chip Beds
-99   Landscape Maintenance

02500   PAVING AND SURFACING

-10   Walkway, Roadway, and Parking Paving
-11   Crushed Stone Paving
-12   Rock Paving
-13   Asphaltic Concrete Paving
-14   Brick Paving
-15   Portland Cement Concrete Paving
-16   Asphalt Block Paving
-17   Stone Paving
-18   Concrete Block Paving
-19   Gravel Surfacing
-20   Shredded Bark Surfacing
-25   Granite Curbs
-26   Precast Concrete Curbs
-27   Asphalt Concrete Curbs
-28   Concrete Curbs
-30   Sports Paving and Surfacing
-31   Asphaltic Concrete Sports Paving
-32   Portland Cement Concrete Sports Paving
-35   Play Surfacing
-40   Synthetic Surfacing
-41   Synthetic Grass
-42   Synthetic Resilient Matting
-43   Synthetic Cinders
-50   Highway Paving
-60   Airfield and Aerodrome Paving
-75   Paving Repair and Resurfacing
-76   Pavement Sealing
-77   Pavement Marking
       Roadway and Parking Marking
       Highway Marking
       Airfield Marking
       Sports Courts Marking

| 02580 | Bridges |
| 02585–02589 | (Reserved) |

02590   PONDS AND RESERVIORS

-91	Stabilization Ponds
-92	Storm Water Holding Ponds
-93	Cooling Water Ponds
-94	Sewage Lagoons
-98	Pond and Reservoir Liners
-99	Pond and Reservoir Covers

02600   PIPED UTILITY MATERIALS AND METHODS

-01	Manholes and Cleanouts
-02	Packaged Lift Stations
-03	Packaged Pump Stations
-10	Pipe and Fittings
-11	Concrete Pipe
-12	Reinforced Concrete Pipe
	Reinforced Concrete Pipe, Non-Pressure Use
	Reinforced Concrete Pressure Pipe
	Reinforced Concrete Cylinder Pipe
-13	Prestressed Concrete Pipe
	Prestressed Concrete Lined Cylinder Pipe
-14	Pretensioned Concrete Cylinder Pipe
-15	Cast-Iron Pipe
-17	Steel Pipe
-18	Corrugated Metal Pole
-19	Asbestos-Cement Pressure Pipe
-20	Asbestos-Cement Non-Pressure Pipe
-21	Vitrified Clay Pipe
-22	Plastic Pipe
-24	Grooved Pipe
-30	Pre-insulated Pipe
-40	Valves, Cocks, and Hydrants
-41	Valves
	Butterfly Valves
	Gate Valves
-42	Corporation Cocks
-43	Curb Stops
-44	Hydrants
02650–02699	(Reserved)

02700   PIPED UTILITIES

-10	Distribution and Transmission Systems
-11	Gas Systems
-12	Oil Systems

-13	Water Systems
-14	Steam Systems
-15	Hot Water Systems
-16	Chilled Water Systems
-17	Fire Water Systems
-20	Collection Systems
-21	Storm Sewerage Systems
-22	Sanitary Sewerage Systems
-23	Combined Waste Water Systems
-30	Water Wells
-31	Well Testing
-32	Test Well Drilling
-33	Well Drilling and Casing
-40	Septic Systems
-41	Sewage Ejectors
-42	Grease Interceptors
-43	Septic Tanks
-44	Siphon Tanks
-45	Distribution Boxes
-46	Leaching Cesspools
-47	Drainage Fields
-48	Sand Filters
02750–02799	(Reserved)

02800	**POWER AND COMMUNICATION UTILITIES**
-01	Towers
-02	Poles
-03	Conduits
-10	Electrical Power Transmission Lines and Distribution Lines
-11	Overhead Power Transmission and Distribution Lines
-12	Underground Power Transmission and Distribution Lines
-20	Communication Lines
-21	Telephone Lines
-22	Police and Fire Signal Lines
-23	Television Lines
-30	Transmission Systems
-31	Radio Transmission Systems
-32	Television Transmission Systems
-33	Shortwave Transmission Systems
-34	Microwave Transmission Systems
02840–02849	(Reserved)

02850	**RAILROAD WORK**
-51	Railroad Ballasting
-52	Railroad Timber Ties
-53	Railroad Precast Concrete Ties
-54	Railroad Track Work

-60   Railroad Service Facilities
-70   Railroad Traffic Control

02880   MARINE WORK

-81   Dredging
-82   Fenders
-83   Seawalls
-84   Groins
-85   Jetties
-90   Docks and Facilities
02900–02999   (Reserved)

02110 CLEARING OF SITE

From an inspection viewpoint, one of the most important facets of clearing is careful identification of the area to be cleared to insure that the *limits* have been established in agreement with the plans and specifications. Further, the area to be cleared should be clearly marked for the contractors, or by the contractors, as required by the specifications.

Proper *land* acquisition or easement rights must have been obtained before the actual clearing. This should be ascertained with appropriate authorities. Generally, it is the inspector's function to review the authority of entry, and not to obtain same.

Safety measures should be in readiness, since the clearing operation generally occurs early in the project before mobilization is complete. Any potential hazards should be identified. Facilities such as fire protection, first aid stations, and similar support functions should be identified, and any liaison established.

Particularly important is the identification of any *utility lines,* above or below ground, including proper protection, warning signs, relocation, or removal as indicated by the specifications.

In wooded areas, the first stage of clearing is swamping, which includes initial removal of underbrush and small undergrowth which could interfere with felling of larger timber. Care should be taken that the spacing of workmen is not too close because of the type of tools utilized, as well as the danger of falling timber.

Tools should be checked for protective devices and proper warning signals. The piling of swamped material should not interfere with operations. Where large timber is involved, safety measures should be reviewed for tree climbing, and work limits of trees to be felled. Equipment such as bulldozers should have overhead protection.

A key role of the inspector is the determination of those trees which are to be left standing, and proper marking and protection of that timber. Trimming of these trees should be carefully performed, and proper measures taken to protect the timber from any tree wounds. If necessary, tree surgeons can be required to stand by.

The grubbing operation involves the removal of stumps and roots. *Disposal* operations must follow appropriate air pollution regulations. Burning permits must be obtained, and burning must not be permitted in periods of high wind. Debris shall be established so that it cannot be carried away into waterways. If debris is to be piled or stored on land, permission must be available and established for the inspection team.

Before any preparation is made for disposal, appropriate local, U.S. Fire Service, County, State and District fire regulations must be reviewed. In particular, factors such as location and size of piles, time of year, safety equipment and similar factors must be checked out.

Air quality regulations, federal, state, and local, increasingly discourage the open burnings of cleared materials.

CLEARING CHECKLIST

1. Check outline control and survey practices to assure that they will be adequate.
2. Ascertain that permission has been received to move onto site, including clear property transfer if required.
3. Determine plan for removal or disposal of debris such as stumps and roots.
4. Obtain burning permits if required.
5. Be sure that topsoil is stripped and stockpiled, not contaminated with subsurface.
6. Review boring logs to determine subsoil conditions expected, and compare with actual.
7. Review existing as-built information in regard to utilities and special conditions.
8. Be sure that excavations are sloped to safe angles, or that required bracing or sheet piling is in place before any personnel enter ditches.
9. Ascertain that required underpinning or sheet piling is in place before excavation proceeds too far below existing structures.
10. Where blasting is required during clearing or excavation, be sure that all safety factors are observed, including licensing of blasters, and curtailment of radio transmissions.

11. If the site is subject to drainage problems, check to be certain that appropriate dewatering equipment and procedures have been installed prior to excavation.

02050 Demolition

Before the initiation of demolition, an inspection should be made to ascertain any existing dangerous structural defects. If, as often occurs, the structure has already been partially wrecked by fire, flood, explosion, or vandalism, walls should be shored or braced to insure the safety of the demolition team.

Demolition is a highly specialized activity, and the contractor should be properly qualified. A work plan should be carefully reviewed with the inspection team before work is initiated. One of the first steps is the safe cut-off of all utilities which enter the demolition area. It is usually good practice to remove glazing prior to demolition, and in populated areas, it may be necessary to close wall openings and to install temporary decking in the lower floors.

The plan for the *removal of material* from the upper sections, such as chutes, shall be carefully reviewed for safety precautions.

Any public *thoroughfares* in the area of the demolition work must be protected and properly identified. Where sidewalk sheds are used, these should have a load limit of more than 150 lb/ft, and the demolition plan should not anticipate any falling material onto the deck area. Where removal of materials is through chutes, material should not be dropped to any point outside the exterior walls of the building, except through enclosed wooden or metal chutes, unless there is no danger to adjacent buildings or thoroughfares. When chutes are at an angle of more than 45° from the horizontal, they should be enclosed on all sides.

Where debris is to be dropped through holes in the floor without chutes, the total hole area should not exceed 25% of the floor area, in any floor below the actual demolition deck.

In removing walls, masonry walls or sections should not be permitted to fall on to lower floors because of the danger of impact loading and premature failure of the building. Care should be taken not to remove structural or load-bearing walls on any intermediate floor until the upper floors have been completely demolished and removed.

In the demolition of chimneys, material which cannot safely be toppled should be dropped down the inside of the chimney.

In the demolition of exterior walls of more than 70' in height, catch platforms should be erected along exterior faces of the walls to prevent injury, and these should not be more than three stories below the story from which material is being removed. These should not be less than 5' in width and should be capable of sustaining a live load of about 150 lb/ft².

In cutting openings in the masonry section between floor beams and girders, care should be taken not to cause structural failure, and shoring shall be used as appropriate.

Where demolition involves *blasting*, special care must be used, and only qualified personnel should handle explosives and/or cause their controlled ignition. The inspection team should specifically approve the qualifications of any blaster, and these should be in writing and on record.

Explosives should be stored in a properly constituted magazine with appropriate security. This storage should protect blasting caps, dynamite, black powder, and other explosives from heat or traumatic impact.

Any blasting operations should be coordinated and planned in accordance with appropriate local codes. For instance, in New York City, the New York City Fire Department Bureau of Combustibles is directly responsible for administering the building code involving "storage, sale, transportation, and use of explosives." The City actually licenses blasters, powder carriers, and watchmen, and issues permits for blasting within the City limits.

Where appropriate, the specifications should require that the blasting be monitored with a seismograph to maintain a control on both amplitude and frequency of the shock. A measurement factor known as energy ratio is often utilized—with a limit of one, where a level of three or less usually will not cause damage. Another measurement dimension is particle velocity, with a particle velocity of 2" per second equal to an energy ratio of one.

Pounds of powder per delay are the controlling factor in the establishment of a safe energy ratio or particle velocity.

DuPont provides a "Blaster's Handbook" which is generally recommended for its information on the handling and use of explosives.

OSHA Regulations. OSHA regulations require that prior to demolition operations an engineering survey be made by competent persons to determine the condition of the structure, with emphasis upon the possibility of unplanned collapse of any portion. The employer is required to maintain written evidence of the survey, and certainly, the inspection team should ascertain that this requirement has been met.

The OSHA regulations require shoring and bracing of any damaged structure, as well as complete cut-off of all utilities.

It is required that a test be made for hazardous chemicals, gases, or flammable materials, and any identified hazard eliminated before demolition starts.

Openings which present a hazard to employees must be protected to a height of about 42″. Dropping of debris is to be only through chutes, as noted previously.

Except for the cutting of holes in floors for chutes and dropping of materials, the demolition of exterior walls and floor construction must begin at the top of the structure and proceed downward. Each story of exterior wall and floor construction is to be removed and dropped into the storage space before commencing the removal of exterior walls and floors below.

Only stairs, passageways, and ladders designated as a means of access—and therefore determined to be safe—are to be used by any persons entering the building under demolition. Other access ways are to be completely closed off at all times. Stairwells being used must be properly illuminated by natural or artificial means.

Proper sheds are to be used to protect entryways outside of the building from falling material. No material is to be dropped to any point outside the exterior walls unless the area is effectively protected. All materials chutes at an angle of more than 45° from the horizontal must be enclosed, and the chutes must have a strong discharge gate to control the discharge of material. Openings cut in the floor for the disposal of materials must be less than 25% of the aggregate of the total floor area, unless lateral supports of the removed flooring remain in place.

Masonry walls and sections are not to be permitted to fall upon the floors of the building in such masses as to exceed the safe carrying capacities of the floors (keeping impact loading in mind).

No wall section more than one story in height may be permitted to stand alone without lateral bracing unless originally designed and constructed to be freestanding.

Employees are not to be permitted to work on top of a wall when weather conditions constitute a hazard.

In mechanical demolition, the weight of the demolition ball must not exceed 50% of the crane's rated load, based on the length of the boom and the maximum angle of operation, nor shall it exceed 25% of the nominal breaking strength of the line by which it is suspended, whichever results in a lesser value. The crane boom and load line should be as short as possible. The ball is to be attached to the load line with a swivel-type connection to prevent twisting of the load line, and shall be attached by positive means so that it cannot become accidentally disconnected.

DEMOLITION CHECKLIST

1. Review engineering survey to determine known structural defects.
2. Establish working plan for demolition.
3. Review utility cut-off plan and implementation.
4. Review bracing and shoring plan, and implementation.
5. Test for hazardous or flammable materials before demolition starts.
6. Designate passages and stairways for access according to plan.
7. Ensure that demolition starts from the top down.
8. Check installation of chutes per plan.
9. Check protection of public thoroughfares adjacent to demolition.
10. Be certain that any cuttings or openings do not introduce structural defects.
11. Be sure that no material is dropped except through approved chutes.
12. Be certain that the demolition does not permit free fall of materials, walls or chimneys, or similar material.
13. Check load of demolition ball (should not be more than 50% of rated load).

Blasting. Only authorized and qualified personnel may handle or use explosives. Explosives must be handled with care, and stored in locked magazines when not in use. Extra precaution should be taken to preclude fire in the vicinity of magazines.

The blaster must take precautions in regard to personnel in the area during any blasting operations.

Transportation of explosives must meet the provisions of Department of Transportation regulations contained in 10 CFR (Code of Federal Regulations) Part 603 Air Transportation, 46 CFR Parts 146–149 Water Carriers, 49 CFR Parts 171–179 Highways and Railways, and 49 CFR

Parts 290–297 Motor Carriers. The OSHA regulations provide detailed instructions in regard to loading of explosives; initiation of explosive charges; use of detonating cords, inspection after blasting; misfires; and storage of explosives. If any substantial amount of blasting has to be done during the project, the inspection team should carefully review Sub-part U of the OSHA regulations, "Blasting and the Use of Explosives."

BLASTING CHECKLIST

1. Check plan for blasting and coordinate with local codes.
2. Check to be sure that blasters are licensed if required by code or by specification.
3. Check the transportation and storage of explosives in accordance with approved regulations.
4. Check safety precautions—particularly for fire in vicinity of magazines.
5. Be certain that vibration testing equipment is installed as required.
6. Post "no radio transmission" signs.

0212 Moving Structures

Moving of structures is a highly specialized operation, and usually assigned to qualified specialized contractors. Nevertheless, the inspection team should insure that appropriate prior preparations for access, permits for moving, and other authorization is available before any actual moving is undertaken. Care should be taken in safety measures where underpinning, jacking, and other pre-moving operations are undertaken. The contractor should carefully explain to the inspection team the moves to be made, and all safety preparations—since the ultimate responsibility still involves the inspection team.

02150 Underpinning

Where excavation is to occur near or under an existing structure, underpinning is often utilized to support the existing building. Since no vertical displacement can be permitted to the existing structure, under-pinning must be carefully preplanned, and must provide—when finished —a solid structure. Also, there must be careful procedures to pick up the load without any displacement or *settlement*. In many cases, because of the settlement which must occur even in well-compacted backfill, the underpinning is designed to remain in place. The nature of the underpinning, where it is major and permanent, would be shown in the drawings. However, in many cases, the underpinning requirement comes under a general specification stipulation that adjoining structures shall not be damaged, allowed to settle, or disturbed in any way. The solution to that problem is left to the contractor, but the inspection team must review the plan for underpinning.

The structural material used to underpin may include many configurations such as concrete piers, piers with beams between, continuous concrete walls, curtain walls, and piling. In all cases, however, the structure should be such that it accepts the existing load of the structure without deflection, which dictates substantial utilization of concrete, and a *final connection* between the concrete structure and the structure which will not shrink or settle. This is often accomplished by grouting the final closure with a nonshrink grout such as Embecco (which has iron oxide aggregate which swells as the normal hydrolysis of the cement causes shrinkage in the volume of mortar).

The sequence of operations usually dictate that a temporary support pick up the load of the structure to be replaced by the permanent underpinning. This is generally oriented to: excavation sufficient to provide working room; replacement of temporary shores or columns; and the threading through of beams—appropriately termed *needle beams*—and jacking up to place load upon this temporary structure, with the structure being sufficient to accept the load. Although the temporary structure would tend to settle, the jacking action can cause a preloading or prestressing which permits no deflection. Since a mistake at any point in the procedure could cause the undesired settlement or disruption, the inspection team must be continually represented during underpinning operations.

0220 EARTHWORK

02210 Site Grading

In some cases, clearing and grubbing is not required before site grading. If so, the inspection team should once again be certain that the local requirements for entry to the job site have been completed, and any necessary paper work is available. Further, a general *control outline*

Fig. 1 Wheel loader clearing rock. Note rock ripper behind. *Source: Caterpillar Tractor Co.*

should have been established by the contractor's survey team, and checked by the owners as required by the specifications and by good practice. In an open site, grade stakes should be set out for guidance of the excavation. In a tight location, such as urban high-rise office buildings, plans for removal of material and appropriate protection of adjacent properties should be established before any equipment starts into the project.

The first stage in site grading is stripping of *topsoil*. There should be a pre-plan for the stockpiling and/or storage of topsoil if it is to be reused in the site work. If not, its removal should be accounted for as separate from the subsoil, and care should be taken that the topsoil is not contaminated by mixture with subsoil. This requires inspection presence to avoid mixture of roots and contamination of materials, as well as too deep a cut in the re-stripping operation. Where topsoil is to be stockpiled, the stockpile should be kept neat, well drained, and in a workable condition.

The *grading* of subsoil material establishes the final grades, or grades to be held during the working phase. The final placement of topsoil occurs after the structure has been established, and usually rather late in the project.

02220 Excavating

Prior to the start of excavation, the inspection team should make preparations including the establishment of cross sections for estimate of material *volume*, if required, and the taking of photographs before and during the excavation process. *Boring* logs should be reviewed to determine water table, rock elevation, and other subsoil conditions

Fig. 2 Caterpillar 992 wheel loader in operation loading heavy duty truck. *Source: Caterpillar Tractor Co.*

Fig. 3 Caterpillar D8H track-type tractor in rock ripping operation. *Source: Caterpillar Tractor Co.*

expected. The purpose is not only to work toward the expected, but to be able to identify the unexpected and to be able to direct the contractor in terms of action to be taken and to issue field change orders as appropriate. This preexcavation check should include a review of utility maps to identify and plan for working around existing utilities, or for disconnection of utilities if they are to be removed.

Experience indicates that even with specific directions, working personnel may dig right through *utilities* in service. Active hands-on inspection is required to avoid the many problems which may evolve from short circuit of electrical lines, burst gas mains, or flooding after water mains are broken. The best move, even though the responsibility is basically the contractor's, is for the inspection team to maintain a presence, and to avoid those problems which can be anticipated. (It is also better to be on hand when the unanticipated is encountered.)

Within the actual location of building foundations or major structure foundations, particular care must be taken in locating for excavation. Also, the boring logs should be reviewed and compared with the material actually removed. Care must be taken that sides of excavations are sloped to *safe angles,* or required sheet piling or bracing is used for purposes of safety to personnel and adjacent property.

The specifications should be reviewed to determine the method of measurement required. Usually, a fixed price contract includes quantity of excavation to a preplanned depth according to the soil borings, but actual conditions may require more or less excavation. This amount is usually measured and used as a *payment* or credit item. It is the inspection team's responsibility to either develop these figures, or to validate their authenticity.

In *rock* excavation, the emphasis is upon safety procedures and proper qualification of personnel.

Fig. 4 Caterpillar D9G track-type tractors in parallel operation. *Source: Caterpillar Tractor Co.*

Fig. 5 Push-pull wheel tractor scraper in operation. *Source: Caterpillar Tractor Co.*

Equipment should be checked, and all appropriate safety regulations followed. Drilling patterns for blasting quantity and fire sequence of explosives should be reviewed for compliance with specifications, and where not specified, should be reviewed with the structural engineers of record. Daily work logs should include records of quantities of explosives, and the results of each blast should be checked against final lines and grades. Continual review for damage to adjacent features, overbreak, and required modification of drilling pattern should be made, if necessary. Generally, rock foundations are required to be cut into rough steps or benches for smooth sloping surfaces, and smooth flat surfaces must be roughed in preparation for foundations.

The *safety* factors including warning notices prior to blasting, curtailment of radio transmissions in the area, appropriate signs and warning systems, and other safety protection is the direct responsibility of the inspection team, even if the contractor is required to carry out the steps. Inspection should be made after every rainstorm because safety hazards are increased thereby. Whenever it is necessary to operate equipment from a level above or near the excavation, the sides should be appropriately braced or sheet piled to resist the extra load. Earth banks should not be undercut, unless they are appropriately shored. Deep excavations, ladders and stairways should have landings about every 20' of vertical interval, and landings should be appropriately equipped with railings and toe boards.

When utility lines have been unearthed, test for proper ventilation and/or leakage. Even the presence of benign gases can cause asphyxiation, and workers should be properly equipped with air masks or air packs when necessary.

It is the responsibility of the inspection team to insure that appropriate safety measures have been taken for protection of passers-by, construction workers, and any other personnel who might be exposed to the hazards inherent in excavation.

02221 Backfill

The preparation for backfilling must be carefully reviewed not only for location, but also to ascertain that the site is prepared for backfill material. Backfilling should not be carried on at a wet site (with the exception of hydraulic fill, and then only when indicated in the specifications). Again, the location of the backfill, whether for purposes of building or for spoiling must be confirmed. Continual review of location and of elevations and quantities should be made to determine compliance with the specifications. Prior to backfilling, all topsoil vegetation and unsuitable material should be removed and the ground surface compacted in accordance with the specifications. The inspection team is responsible for either conduction or review of the compaction soil testing. Since this is a specialized field, either specific training in the *compaction* techniques should be carried out with the inspection team, or the work assigned to an authorized inspection laboratory.

The soil borings will determine whether or not certain hardened overbearing will have to be removed prior to installation of backfill. Old fill areas sometimes develop a hard overcrust which does have a carrying power sufficient to carry equipment, but not sufficient for the ultimate

Fig. 6 Caterpillar standard wheel tractor scraper being push-loaded by D8H track-type tractor. *Source: Caterpillar Tractor Co.*

load-carrying requirements. In such cases, the specifications would require removal and placement.

In areas formerly covered by water, replacement of soft or organic materials is often required.

The filling operation usually consists of placing *layers* from six to twelve inches of depth of appropriate backfill material, with consolidation usually through mechanical methods occurring after each layer is placed. Soil testing for compaction and appropriate moisture density must be accomplished prior to the placement of the next layer. In large backfill areas, vibrating rollers or gravity rollers (often called sheepsfoot rollers) are utilized to achieve the compaction. If properly carried out, compaction often exceeds the density of the natural ground in the area. For small areas, hand-held power tampers can be utilized.

The makeup of approved *stabilized backfill material* is either specified by borrow pit location and/or source, or by qualifications. The material to be used must be approved either through experience or test prior to its utilization. Top soil is not an approved backfill material for placement under structures, and organic silty materials generally are not used. Rock can be a good backfill material, but must have a stabilizing finer rock or gravel interspersed to permit true compaction. Sandy materials or properly graded gravel and sand make up one of the best and most common backfill media.

The proper gradation from rough to fine material is necessary to achieve appropriate compaction of backfill.

In major backfill areas, temporary roads for hauling may be required. The layout of these roads and their construction is subject to the direction of the inspection team. Backfill within the embankment area for haul roads should have the same material placed with the same

Fig. 7 Backhoe in operation. *Source: Bucyrus Erie.*

Fig. 8 Caterpillar Compactor in operation. *Source: Caterpillar Tractor Co.*

moisture-density relationship, since it will become part of the backfill or embankment. In placing construction ramps, do not permit cutting through a previous embankment or backfill. The routing of vehicular traffic should be set up so that compacted and approved areas are not disturbed, and/or so that the heavy backfilling traffic can be utilized as a compacting factor.

In haul roads, dust is often a problem, and in non-backfill areas, a desiccant such as calcium chloride can be utilized.

One item to be checked when rollers are being used is tearing action at the roller turnpoints. If this turnpoint is within the compaction required areas, re-rolling should be required as necessary to obtain the specified density.

In localized backfilling for building foundations, be sure that the material backfilled upon is acceptable. This requires the removal of any wet or unstable material, to be replaced with suitable compacted material. Cold-weather placement requires special consideration, to insure that frozen material is not placed, resulting in unacceptable settlement at a later date.

BACKFILL CHECKLIST

1. Stabilization. If existing ground is muck or marshy, complete replacement may be required. In some cases, slurry mixture of soil cement may be laid at the bottom of the excavation to provide a dry working surface.

2. Check to be certain that backfilling is of appropriate material as described by specifications. Usually this is approved stabilized backfill material from a specified borrow pit or source.

3. In backfilling, be certain that backfill consists of placing layers from 6" to 12" of depth as required by specifications.

4. Be certain that appropriate compaction techniques are utilized in the placement of backfill.

5. Check the compaction testing procedures in accordance with specifications or general practice.

6. Is soil poisoning is required to preclude pests such as termites, ensure that it occurs before backfill starts.

7. If backfill is against a retaining wall or building wall, be certain that the structure is completely cured and/or braced in accordance with design requirements.

8. Check specifications to determine if backfill on opposite sides of a wall or structure is to be placed in a balanced fashion to avoid placing a lateral load upon the wall or structure.

02222 Granular Sub-Base

In most cases, subfoundation drainage is achieved by placement of a gravel layer either across the entire footing area, or at the lowest points.

In some cases, the drainage fill includes drain pipe made of pervious material such as cinder block or inexpensive pipe (such as corrugated or Orangeburg) with holes in the upper half to permit the infiltration of drainage water.

02240 Soil Stabilization

The term stabilization is a relative one, and refers to an improvement in soil operating characteristics. For instance, the replacement of muck or marshy ground with gravel or rock can be referred to as stabilizing the environment. In some cases, stabilization is the development of a better wearing or carrying surface, as in the case of soil cement, where portland cement is mixed with aggregate and water, often by blading of heavy equipment, and then rolled to make a strong, dry carrying surface.

In certain situations, a slurry mixture of soil cement or low-strength cement is laid into the bottom of an excavation to provide a dry working surface, and improve soil bearing capabilities.

In the case of roadways, stabilization is sometimes achieved by mere removal of the topsoil and blading and compaction of the subsurface material.

The inspection team must depend upon the requirements of the specification to determine the type of stabilization required.

02250 Soil Compaction Control

The bearing strength of the soil can be measured by its dry density, with strength being inversely proportional to the volume of voids. In most cases, *consolidation* is achieved by compaction or compression of the soil by physical means. Compaction is usually directly related to the water content, with lower water content being equivalent to higher compaction or strength. This effect is less apparent in granular or rocky soils with limited amounts of fine material.

Compaction by sheepsfoot rollers or high-pressure rubber-tire vehicles is effective to a depth of about 5' with the usual layer of backfill being six to eighteen inches. Other techniques such as the Vibroflotation of sandy soils for compaction are used in specific cases directly related to the characteristics of the soil.

Field testing to determine water content and other consolidation characteristics is carried out by either the Inspection force or an approved laboratory.

02280 Soil Poisoning

Soil poisoning is used to preclude pests such as termites or other ground insects, and also for weed control. Review the requirement in the specification, and check with contractor for material to be used and method of application. During application, check for coverage, quantity, and concentration to be used.

OSHA Regulations—Excavations

Equipment for excavation must meet the requirements of the Society of Automotive Engineers specified in the regulations. In addition, rollover protection structures are required.

The regulations call for walkways to be clear of excavated areas, unless shored and able to carry at least 125 lb/ft^2.

Prior to actual excavation, the contractor is required to locate all utilities, and to either disconnect or support and keep in service. Surface hazards are to be removed or made safe before excavation. The regulations call for the excavations to be inspected by competent management personnel after every rainstorm or other hazard-increasing occurrence, and appropriate protection against slides and cave-ins must be taken.

The regulations call for a careful determination of the angle of repose, and concomitant design of supporting sheeting, shoring, or natural slopes. (The inspection team will normally check carefully for this, and follow the advice of the structural designer.)

The angle of repose should be flattened when an excavation has water conditions, silty materials, loose boulders, deep frost or slide planes. Excavated or other materials should not be stored nearer than 4' from an excavation, and the side slopes and faces of excavations should meet with the structural engineer's approval. When an excavation is in excess of 20' in depth, the support systems must have been planned and designed by a qualified person (presumably a professional engineer). Other precautions as described previously shall be taken in protection of adjacent buildings, avoiding placement of heavy equipment at or above the edge of the excavation, or blasting in close proximity to excavations.

Banks more than 4' high must be shored or sloped to the angle of repose where danger of slides or cave-ins exist. Sides of trenches in unstable or soft material 4' or more in depth must be shored, sheeted, braced, or otherwise supported by means of sufficient strength to protect employees working within them.

02350 PILING

Where the loads imposed by the building or structures are too great for the surface soil to carry, the load can be transferred to deeper, more stable soil by the use of piling. Piling is made of various forms, sizes, shapes, and lengths. The load is transferred from the structure through the pile to the substrata by two means: *skin friction* along the length of the pile, and *end-bearing* between the lower end of the pile and the stratum to which it penetrates.

Most piling is driven into place, although in certain soils water jetting can be utilized.

Types of piling include:

Wood Piling. Timber pilings are perhaps the oldest and at one time the most common type. They are considered to be friction piles, rather than end-bearing, because the relatively soft material cannot be driven through hard layers without danger of brooming (shattering) of top or bottom. Even when driven with a metal shoe, the tip is not generally considered strong enough for end-bearing. Since wood piling must be driven in a single length, the dimensions of a pile are limited in length. Wood piling used in piers is usually pressure-treated to resist the action of salt water, chemicals, and insect life. The most serious area of deterioration is at the water line.

Steel H-Piling. Usually driven as end-bearing piles, since their smooth limited area surface provides little friction. It is possible to drive long lengths of H-piles because of their limited resistance to driving, and the ready ability to splice new lengths. The H-pile, under certain conditions, is subject to lateral displacement, defeating its ability to act as a long column.

Steel Pipe Piles. Can be driven with the end open or closed depending upon the ground and driving conditions. When the end is driven open, the pipe is cleaned out with air or water jet before filling with concrete. Because of their high strength, they can be driven in long lengths without fear of bending as might occur in H-piles, and can be readily spliced. Closed bottom pipe piles cannot be driven as far, since the ultimate load-bearing capacity of pipe piling is based on the material strength of both the tube and the concrete core fill material.

Cast-in-place Concrete Piles. One of the most popular piling types today is the cast-in-place concrete pile, which is driven in a variety of fashions. These tubes are often tapered and are usually driven with a light-gauge metal casing which can be spliced, with an inflated mandrel to maintain stability during driving. The mandrel is then withdrawn and the pile filled with concrete. Some proprietary types include:

Raymond standard concrete pile. Tube of light sheet metal tapered outward with point of tube closed with a steel boot, with tubes made in fixed lengths ranging from 15' to 37'.

Raymond-step-tapered pile. Not driven with collapsible mandrel, but with solid steel core, with sections added to step tapered pile so it can ultimately reach only to 150'.

Monotube pile. Tapered and fluted, ranging in length from 10' to 75'. To extend the length, nontapered sections can be added to the top. The fluting provides sufficient vertical strength to permit driving without using a mandrel.

Armco Helocor pile shell. Uses corrugated metal pipe as shell, which is not tapered but of uniform diameter, driven pneumatically by expandable mandrel.

Pre-Cast and Prestressed Concrete Piles. Prestressing reduces or eliminates tensile stresses placed on the piles during transportation, driving, and service, and because of this strength, prestressed piling has replaced the precast concrete pile in most cases. Pretensioning (stress applied prior to casting the concrete) is used for piling up to 24" in diameter, and post-tensioning (stressing after concrete has been poured and set) used for diameters over 30". Splicing of prestressed piles is difficult and usually avoided.

Piles can be used as anchors, and in special situations to resist lateral forces. In this case, batter piles are driven on a diagonal slant, rather than vertically.

The location of piling depends upon the function to be served. Usually, the purpose is to support vertical loading, and the piles must be located generally below the main columns. The piles are located in clusters or groups, and are capped by a structural slab which carries the concentrated load from the column out to the piles. This footing, or cap, usually is fairly substantial in depth, and heavily reinforced.

In determining the spacing between piles, and the number of piles, a carrying load or capacity must be determined. Because soil is not a homogeneous material, there are design safety factors involved. The formulas used to determine the carrying capacity are subject to substantial interpretation, and are generally derived from a formula postulated by Redtenbacher more than 100 years ago. A simplified version of this formula, which generally suits the nonuniform response of many soils, is known as the *Engineering News* formula as follows:

$$\text{Drop hammer: } P = \frac{2WH}{S+1}$$

$$\text{Single-acting steam hammer: } P = \frac{2WH}{S+0.1}$$

$$\text{Double-acting steam hammer: } P = \frac{2E}{S+0.1}$$

where

P = allowable load (lb);
W = weight of ram (lb);
H = height of hammer stroke or fall (ft);
S = average penetration per blow (last 5 blows or last 20 seconds).
E = rated hammer energy in foot pounds

The formula actually varies with the type of soil encountered, and test factors do indicate that higher coefficients may be used to improve the safety factor, specifically, instead of 0.1 increasing to 0.3 or 0.5.

Safe practice indicates that a field *test* or tests should be conducted to check the calculated value for piling. This is accomplished by driving test piles in each major location, and/or in each type of soil to be encountered. The piles should be in place several days before the test, and in the case of concrete piles, should have achieved design strength. The pile can then be deadloaded using ballast material such as cast iron or other appropriate available loading material. The loading bin should provide an axial load upon the pile.

A more expensive, but more satisfactory method of imposing test loading is to drive three piles in a line, with the center being the test pile. The outboard piles are then connected with a beam, and a jack placed between the pile to be tested, and the beam. Load is applied by jacking and measured by appropriate instrumentation. This method can also be used to test piles for uplift, but jacking is done at the end piles.

In either case, a load should be allowed to remain for approximately 48 hours, with accurate elevation readings taken before and after the placing of the load and at 12 and 24-hour points after the load is in place.

The total settlement (gross settlement less rebound) should be less than 0.1″ per ton of total test load.

The specifications should provide the loading requirements for each pile, and the criteria for determining the allowable load per pile. This should include specifying the proper formula to be used, or other method of ascertaining the pile estimated strength. The specification should also direct the type and number of tests to be conducted. There are cases in which the structural engineer may have sufficient experience in a given area to ascertain that the calculations used are sufficient, and that no test is required. This should be clearly stated in order to relieve the inspection team of the need to validate the assumed carrying capacity.

The type of pile-driving equipment to be used is generally at the option of the contractor, but the inspection team must determine the type of pile driving-rig, the weight of the ram, and the stroke characteristics. Typical ram weights vary between 3000 and 15,000 lb, with the usual stroke being about 3′ for single-acting hammers.

Drop hammers are much slower, and have generally limited use today.

Figure 9 shows three pile-driving rigs in operation on the foundations for the 61-story John Hancock building in Boston. The contractor used two Manitowoc 3900's equipped with pile drivers to set more than 3000 H-beam piles 100′ long. Note the boiler carried at the rear of each pile driving rig to provide steam to the hammer, the vertical leads to guide the pile, and the hammer striking an H-beam.

In driving the pile, it should be placed plumb or vertical in the leads, and set into the ground with several light strokes of the hammer. There is a fairly liberal lateral tolerance for horizontal deflection of either 2% of the length or 3″.

In closely spaced clusters, the driving of piling necessarily displaces a certain volume of earth, and where the earth has a high resistance to compression (such as in clays with a large water volume) there can be upward heave. Often, piles are driven in alternate pattern with half the piles being placed in a forward pass, and then mats placed and the rig backing over the group and placing the second set in between the piles previously driven. In the case of hollow piling which must be filled with concrete, collapse sometimes occurs in the previously driven piles. To avoid this, on the first pass piles should be inspected by dropping a light inside the pile to observe sound condition, and then concrete placed before the second pass is made.

Even with this precaution, or in the case of steel H-piles, driving of tight clusters may cause the first group to rebound or pop out of the

Fig. 9 Steam-driven piledrivers in operation driving cast-in-place piles. *Source: Manitowoc Engineering Co., Division of the Manitowoc Co.*

ground to some extent. Careful elevation checks must be made before and after to reconfirm the cut-off points, so that the pile cap is in the proper vertical location.

An as-built record of the depth of piles should be maintained, and in the case of end-bearing piles, the boring record should be checked to be sure that the pile end is in solid material.

PILING CHECKLIST

1. Location of piling is important; check layout and its agreement with plan requirements. Also check orientation of special piling such as batter piles and anchors.

2. Carry out test piling procedures as prescribed by specifications, if required.

3. Check the type of pile used, and the condition of the pile before driving.

4. Be sure pile-driving rig is load tested before driving starts.

5. Check the guards provided across head block to prevent cable jumping out of the sheaves.

6. Check that leads are provided with ladder or rings for attachments for safety belts for loft workers.

7. Check overhead protection for operator.

8. Check steam hoses for adequate cable connections at joints.

9. For steam drive hammers, quick shut-off valves must be within reach of piledriver operator.

10. Monitor actual driving of pile according to formula for adequate load (either blows per inch for last foot, or in accordance with load tests).

11. In hollow piles to be filled, check pile before filling to ensure that the shell has not collapsed, and that the interior is clean, particularly free of mud and water.

12. For cast-in-place piles, ascertain that reinforcing cage is in place, and not touching casing.

13. For precast concrete piles, check handling to be sure that piling is not overstressed when picking up.

02370 Caissons

In early construction, before the availability of power equipment, the principal method of carrying a foundation through poor soil to rock or better-based foundation conditions was to dig a well-like structure, often lined with rock, to support the structure at significant points. In modern construction, this caisson approach has been continued, and in the early part of the century, was often hand dug or dug with the use of equipment within a large shaft, and then either lined with concrete or filled, as appropriate. The Chicago caisson was an open shaft lined with plank staves braced in turn by steel rings, almost like a water tank within a shaft. The excavation was hand dug in layers equal to the length of the staves and was used frequently in Chicago because it suited the soil situation there.

Today's caissons are generally placed with drilling equipment. In many areas, a well-drilling type of rig is used with a well-shaft driller, which can be opened up with butterfly-like outriggers to broaden the caisson into a bell-like shape (see Fig. 10). These can be drilled rapidly in lengths of 10 to more than 30′. The most popular utilization is in soils with a cohesive nature such as clay, so that the *bell-bottom* caisson maintains its shape when the drilling rig is withdrawn. In sandy or silty soils, the shaft often tends to collapse, and therefore the bell botton is either not used, or the shaft is cased. (The casing need not be structural, and is often a paper board tube).

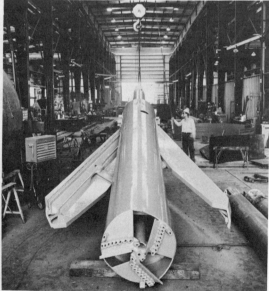

COMBINATION DRILLING & BELLING TOOL FEATURING AN 8″ REVERSE CIRCULATING SYSTEM FOR CUTTINGS REMOVAL. THE TOOL DRILLS 36″ DIA. AND BELLS TO 15′ DIA. AT APPROXIMATELY 60°.

CALWELD,
Div. of Smith Int'l, Inc.
P.O. Box 2875
Santa Fe Springs,
California 90670

Fig. 10 Drilling and belling tool for caissons. *Source: Calweld-Smith-International.*

The advantage of the bell bottom is its rapid installation and low cost. The narrow shaft reduces the amount of concrete required to fill the bell bottom, and the bell, which can extend up to 10' or more in diameter, provides a substantial bearing area at a more favorable depth.

Fig. 11 Caissons being drilled and placed. *Source: Manitowoc Engineering Co.,*

In Fig. 11, larger caisson drilling equipment is shown. In this case, the Caisson Corporation is utilizing Maintowoc 3900 rigs to carry drilling rigs weighing 41 tons, exerting 360,000 ft-lb of torque, drilling holes up to 10'-3" in diameter, and as much as 150' deep.

In inspecting caissons, one of the most important factors is care to insure that earth, rock, and foreign materials have not fallen into the caisson area. In bell bottoms, this may be difficult because of the narrow shaft, while in large caissons, the depth may preclude a visual inspection without special equipment.

When called for by the design, the caisson shaft may be reinforced with steel to produce a resistance to lateral movement in the soils. However, in a great many cases, the caisson shaft is substantial enough so that even the concrete with its low tensile strength can resist this movement.

OSHA Requirements—Piling

Boilers and piping systems used with piledriving equipment must meet the applicable requirements of the American Society of Mechanical Engineers for "Power Boilers" and "Pressure Vessels." When employees are working under the piledriving hammer, a blocking device capable of safely supporting the weight of the hammer must be provided for placement in the leads. *Guards* must be provided across the top of the head block to prevent the cable from jumping out of the sheaves. When the leads are inclined in the driving of batter piles, provisions must be made to stabilize the leads. Fixed leads must be provided with adequate ladder or rings or similar attachments so that the loft worker may engage his safety belt lanyard to the leads. There must be overhead protection for the operator, which at the same time does not obscure his vision. Steam hoses leading to the steam hammer or jet pipe must be securely attached to the hammer with an adequate length of $\frac{1}{4}$"-diameter chain or cable to prevent whipping in the event that the joint is broken at the hammer. Similarly, safety chains must be provided for other hose connections to prevent thrashing. Steam line controls must consist of two shutoff valves, one of which must be quick-acting level-type within reach of the hammer operator.

02380 COFFER DAMS AND EXCAVATION SUPPORT SYSTEMS

This is a general area involving temporary support either vertically or horizontally. It includes the use of timbers to prevent sliding of earth into an excavation, such as the *bracing* of the walls of a deep trench excavation, or support to hold—on an interim basis—a bank which will be later supported by the permanent structure.

In concrete structures, shoring is used to support the finished floors until the concrete develops sufficient strength to be self-supporting, usually at the 14 to 21-day period. On a high rise, the level of shoring becomes progressively less as the floor poured becomes more mature. For instance, when 10 floors have been poured, if the project is on a 3-day pouring cycle, the first two floors would have developed full strength, and all form work would be removed. The floors above would have progressively more shoring, until the floor being poured would have enough flooring and bracing supporting the form work to hold the wet deadload of the concrete—which at the time of pouring would have no structural value at all.

The contractor has broad discretion in the use of shoring and bracing, and it is within this area that many failures occur. Usually, the drawings do not direct specifically the type of shoring or bracing to be used, but leave it to the discretion of the inspection team. Even though these structural episodes or involvements are limited and interim, their failure is easily as dangerous as failure of the ultimate structure, at least to people on the job site. In terms of both life and money, this is an area which must be given the greatest examination.

The inspection team should be certain that areas requiring shoring and bracing are given attention promptly. In some instances, the requirement is general, and is carried out only in response to development of difficult situations. However, the need for shoring and bracing can be anticipated, and the contractor should have a specific written plan including appropriate sketches to describe the size and arrangement of the material to be used. This, in turn, should be approved by the inspection team. Where sufficient engineering knowledge is not available within the team, the inspection team *must* be sure to get formal approval by the engineers of record, or in their absence, through engineers furnished by the owner, designer, or appropriate source.

This is one area where intuition is often used, sometimes with disastrous results.

Sheeting

Sheeting is, as the name indicates, panels or planks of material installed vertically to hold back earth, rock, or other unstabilized material. It is used to hold back material temporarily during construction, in situations such as cofferdams and deep excavations. It can also be left in place permanently in deep highway cuts, railroad excavations, and often as retaining walls where a deep cut has been made with a steep slope ratio.

Sheeting falls into two general categories: passive and active resistors. The *passive* resistor would be wooden planks or concrete planking or sheets installed to resist lateral movement of unstabilized material. While

Fig. 12 Sheet pile coffer dam in place. *Source: Manitowoc Engineering Co., Division of Manitowoc Co.*

each of these may be driven, the principal purpose for the driving is the development of a toe hold so that the bottom of the sheeting will not kick out. In cases of passive resistors, a shoring and bracing system is made up of beams and columns. Columns for this purpose are often termed soldier beams. Where a structural system is provided to resist the load, the sheeting serves only to contain the material.

One of the most popular types of sheeting shown in Fig. 12 is made up of sheet steel with an interlocking flange on each edge. The steel is designed so that it can be driven, and is available in lengths up to about 50′ long. Care must be taken in driving flat sections to be sure that the flange has properly interlocked so that the piece is not overstressed laterally, because buckling will occur.

Figure 13 shows sheet piling in place around the cofferdam, with a clamshell bucket being used to place sand within the cofferdam.

Sheeting should be driven or placed with care to prevent gaps through which material could pass. Steel sheet piling permits only limited water to pass through it, but if not driven to a sufficient imbedment, water can boil beneath the piling, causing *blowouts* where groundwater table is above the bottom of the sheet piling.

Where the sheeting is held in place by a system of soldier beams, beams, braces, or shores, then wedges should be used to develop a close contact between the sheeting and the structural support to prevent movement. Sheeting may be heavy planks known as lagging.

Sheet piling may require additional lateral support; in some cases, this can be achieved by burying *anchors* or "dead men" in a stabilized portion of the material to be retained. These are then connected through cables or stanchions to the sheet piling, giving lateral support.

If sheeting is a temporary measure, and is to be removed, the inspection team must be certain that backfill has been thoroughly consolidated to design requirements before sheeting is removed. The sheeting removal procedure should be approved and carried out carefully so that the consolidated backfill is not disrupted, causing slips or other types of failures.

02400 SITE DRAINAGE

The principal method of planning and achieving site drainage is through the design of a *grading* plan which anticipates the normal and maximum quantities of water and other materials to be handled, and uses the final natural ground grade to carry this water or other fluids to collection

Fig. 13 Sheet piling in place for coffer dam, and clamshell placing sand. *Source: Manitowoc Engineering Co., Division of Manitowoc Co.*

points. These collection points can be manholes, catch basins, or rock drains, as defined by the drawings. The *surface run-off* is collected by a subsurface piping system, where the design conditions have not permitted drainage directly to an existing street surface run-off or storm sewer connection. The proper design, sizing, location, and specifications are the responsibility of the design team. The field inspection group has the responsibility to install the design properly. Normally, this work occurs early in the construction period, after rough grading, and much before the structure area is completed. In fact, it is good practice to plan early installation of the site drainage systems so that the site can be kept dry to enhance working conditions.

Installation conditions have an important effect on the ability of a subsurface piping system to perform its function not only initially, but over a long period of time. Since the subsurface system is backfilled and cannot be easily inspected after installation, on-site inspection presence is required.

Piping systems are either usually reinforced concrete pipe, or corrugated metal pipe. The pipes should be assembled in accordance with manufacturer's instructions, with the pipe being unloaded and handled with care. The pipe is normally installed starting at the low point, with male joints pointing downstream.

Figure 14 shows typical *bedding* conditions and suggests methods for backfilling under different conditions. In the figure, the positive projecting type involves installations where pipe is bedded with its top from 0.9 of the nominal pipe diameter above existing ground to level with it, while negative projecting embankment type involves those installations where the pipe is bedded in a trench with its top below the existing ground surface. The contact between the pipe and the foundation on which it rests is bedding, and the bedding transfers the load on the pipe to the underlying soil. Accordingly, a firm pipe bedding support is required to avoid high deflections of the piping. The foundation material under the pipe should not be too firm, so that rock closer than 12″ to the pipe is usually removed and replaced with a slightly yielding material to allow normal deflections.

One of the most important phases in piping installation, after tight joint connections, is the placing and compaction of *side fill* material around the pipe. The side support material is to be placed in 6″ layers and compacted to the required level of compaction, with care taken not to overstress the pipe during this phase. The backfill material should not contain stones, rocks, or chunks of foreign matter which could place point loading on the pipe.

In some cases, corrugated pipes are elongated vertically in order to carry higher fill loads. In this case, they are installed with the larger axis vertical. The inspection team should check the initial location of pipe, observe its alignment as it is installed, and review the elevation of pipe to ensure that the invert elevations are as required. When multiple lines of

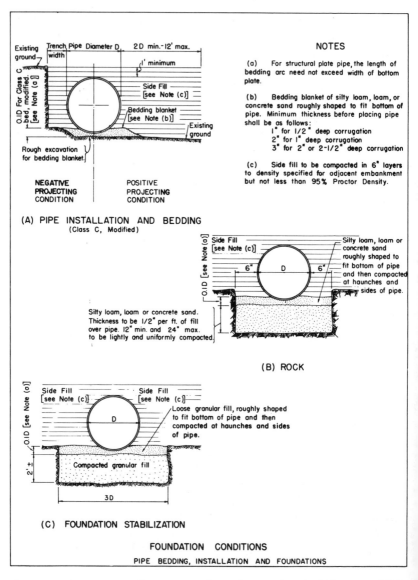

(A) PIPE INSTALLATION AND BEDDING
(Class C, Modified)

(B) ROCK

(C) FOUNDATION STABILIZATION

FOUNDATION CONDITIONS
PIPE BEDDING, INSTALLATION AND FOUNDATIONS

NOTES

(a)   For structural plate pipe, the length of bedding arc need not exceed width of bottom plate.

(b)   Bedding blanket of silty loam, loam, or concrete sand roughly shaped to fit bottom of pipe. Minimum thickness before placing pipe shall be as follows:
    1" for 1/2" deep corrugation
    2" for 1" deep corrugation
    3" for 2" or 2-1/2" deep corrugation

(c)   Side fill to be compacted in 6" layers to density specified for adjacent embankment but not less than 95% Proctor Density.

Fig. 14 Typical type bedding cross sections. *Source: National Corrugated Steel Pipe/U. S. Department of Transportation.*

pipe are installed in the same trench, they should be spaced far enough apart to permit tamping of side fill around each pipe individually, to avoid the occurrence of void spaces in the backfill.

Where temporary crossovers of pipe runs are required, a cover of at least 4' of material shall be placed to protect the piping, and where possible unplanned crossovers should not be utilized. Since site drainage piping is not normally pressure-piped, no special testing is required with the exception of observation by the inspection team of all of the stages to determine joint integrity, alignment, and care in backfilling.

Special care is often required where the site drainage piping crosses over or under utility piping. These special situations may be called out by the drawings or may be observed by site inspection. In either case, special precautions are often required for special situations.

02401 Dewatering

In either excavation or backfill, it is important to keep the work site dry. In many cases, the clearing and grading operation can provide rough *drainage* ditches which can carry away any water on a flat site. In other situations, such as foundations for piers or bridges, or in urban situations, the excavation site is rather deep and may be below the groundwater line. In these cases, special *dewatering* equipment and procedures are required.

Figure 15 shows a Moretrench system, which is a combined pumping and piping system. The piping is made up of pumping points inserted either by driving or jetting several feet below the dry level required, and then connecting the pumping heads with headers connected to heavy duty pumps.

In other situations, direct pumping from a sump made of gravel, pipe, or open may suffice.

Soil conditions may dictate special approaches. For instance, in many sections of the New Jersey Turnpike, a thick impervious clay layer overlaid sandy soil. The effect was to hold water above the impervious layer, and this was corrected by drilling many shafts through the impervious layer and filling them with compacted sand.

The effect was both to drain off surface water, and (in many instances) to lower the groundwater table, causing stabilization of the ground.

Even in periods of dry weather, where a groundwater table or surface water may be no problem, site supervision should be ready for the unexpected.

Fig. 15 Dewatering system in operation for storm sewer. *Source: Moretrench.*

Where underground drainage requires trenches more than 3′ to 4′ in depth, all appropriate precautions for proper cutback of walls, or shoring and bracing must be observed. Underground pipeline trench cave-ins are a common cause of injury and death on construction sites. Observation of the proper safety rules is very much the responsibility of the site inspection team. Similarly, testing for toxic materials, sufficient oxygen supply in deep trenches, and other safety precautions must be enforced by the inspection team.

SUBSURFACE DRAINAGE CHECKLIST

1. Check the location.
2. Be sure that proper safety methods are used for trenches deeper than 3′ or in unstable ground.

Fig. 16 Dewatering ring in operation for deep containment vessel. *Source: Moretrench.*

3. Be sure that proper piping materials per specification are used, also check size.

4. Check bedding conditions before pipe is set.

5. Be sure that tight joint connections are achieved.

6. Check method of backfilling with material placed in compacted layers with equal side stresses so that piping is not offset during backfill.

7. Check utility locations before starting excavation.

8. Drainage lines are usually laid from the deepest point upgrade.

02440 SITE IMPROVEMENTS

Site improvements cover a wide range of items, some of which are described in the following sections. Generally, site improvements are placed when the building structure is well along, so that they are not endangered by the construction traffic, nor vulnerable to vandalism. However, in certain cases, fences may be installed at an early stage for security purposes, and other items may have a temporary utilization during the construction project. When early installation is made, there should be provision for the contractor to rehabilitate the material as part of the final punch-out items, so that—for instance—the "new fencing" does not look old and worn out when surrounding a new building.

02441 Irrigation System

Irrigation systems are placed either in playing fields, golf course greens, or important landscaped lawn areas. The installations are described in the plan, or can be furnished on a performance basis if required by the specifications.

The basic grading should be completed before the installation of the irrigation system. Since the system is operated only in warm weather, it may or may not be below the frost line, depending upon the material and the ability to drain the system completely at the close of the season. Sprinkler heads are of various types, and should be oriented to direct their spray so that sidewalks and pedestrian areas will not be sprayed.

Irrigation systems operate either on a timing system or manual control, and may be total on/off or phased spray. Systems are commonly made from PVC plastic piping, but may be of more standard piping materials if so designated by the specifications.

The principal inspection required is for watertight integrity prior to backfilling, and should include a pressure test. Also, appropriately approved connections should be made with the pressure water system, and there should be no opportunity for a backflow into the potable water system.

02443 Fountains

Fountains are generally for decorative purposes, and may be prefabricated, or fabricated at the site. Most fountains are made of concrete, and the inspection requirements are similar to those of specialized curbing, foundations, and structural concrete. The sub-base must be well prepared, and underground piping in place and inspected prior to pouring of concrete. A drainage system must be provided for fountains, so that they can be drained for maintenance and/or cold weather. The plans should describe the details of the fountain, and where a prefabricated fountain is to be installed, the manufacturer's instructions should be obtained and reviewed by both the contractor and the inspection team.

02444 Fencing

Fence construction is often an intermittent operation, and may extend from anywhere between the start and finish of the contract. The contractor may desire or may be required to install fence as a first order of work, or may delay fence construction to avoid damage or for his convenience.

The inspection team must be concerned with the location of fences for a number of reasons, including the inference that fence lines match boundary lines amongst others. The inspection team should review the ground for proposed access openings to validate the physical feasibility. Good fence construction requires attention to line and grade, and a string line can be used as a guide to maintain alignment and maintain minor variations in the grade at the top of posts as they are being set in their final position. Spacing between posts is center to center, and these distances should be checked. At the same time, the depth of the post holes should be checked. When the total distance in any single run does not check out to even spaces between posts, the end, corner, and brace posts should be set according to plan and any irregular distance divided between line posts, but no spacing shall exceed the spacing specified in the plans or specifications. Careful attention should be paid to connections to existing fences.

Posts are set in the vertical, except in rare cases where terrain does not permit such installation. Concrete and post holes should be poured to a point slightly above the surrounding ground and rounded on top to shed water. Wire and fabric should be taut and spaced as shown on the plans with the required clearance under the fabric to prevent children and small animals from crawling under. To provide this clearance, areas shall not be filled in, but rather high spots between posts should be excavated, assuring a stable surface. Wire and fabric is generally placed on the owner's side of the post. Fences should not be located so as to obstruct streams and flow of drainage areas. Where fences are to be constructed on top of retaining walls or wing walls, careful location of the post hole inserts should be made when that structure is poured.

Aluminum fencing is generally used to enclose a specified area and requires no maintenance. Before installation, check for the approved manufacturer, and make sure that the samples have been approved by the architect. Check the color and inspect for any damage in shipping, handling, and storage. After installation, inspect for good workmanlike operation of gates. Similar inspection should be performed for wire mesh, chain link, and wood fencing.

Wrought iron fencing is usually used as a decorative enclosure, and is prefabricated for assembly at job site. Check shop drawings for details, and after installation, inspect for alignment. The assembly should be firm, plumb, and level, with sleeves solidly filled.

02460 Playing Fields

Playing fields are made up of the components previously discussed: grading, preparation of subgrade, site drainage, and, as appropriate, sidewalks, curbs, and gutters. Playing fields are often surrounded by running track, which also requires good drainage, and may consist of either cinders or specialized running surfaces made out of proprietary bitumastic stabilized cork material. Often, today's playing fields are made of astroturf or similar substances, which are placed on either a rolled clay subgrade, or a lightweight concrete base, well drained and overlaid with about $\frac{3}{8}''$ to $\frac{1}{2}''$ of flexible material to which the carpet-like playing surface is attached.

For baseball fields, the special baseline and the pitcher's mound are of skinned clay, and must be placed in accordance with specifications. Seeding or sodding is used for natural playing surfaces in accordance with landscaping specifications. In natural playing surfaces, it is common to place an irrigation system in the field, with sprinkler heads located to avoid any possible injury to players.

02480 LANDSCAPING

One facet of the design phase is landscape architecture. Presumably, the landscape planters will have explored the site and become familiar with all the natural features—good and bad—and will have moved to determine those items which are really unsuitable, and recognize the uses for which the site is best suited. This should result in a design which fully exploits the potential. The duties of the inspection team are to see that all landscape items are in compliance with the approved samples and placed as detailed in the specifications.

A basic item is the inspection of the rough ground grading to be certain that there is enough slope away from the building structures to allow proper drainage. Further, the inspection during stockpiling of topsoil, and avoiding contamination of the topsoil with subsoil, will be reflected in the successful completion of the landscaping phase. Before the topsoil is spread, the inspection team should be certain that the rough ground has been thoroughly cleaned of any debris. After the topsoil has been spread, it should be checked to be sure that it is spread evenly to the correct depth as called for in the specifications.

02484 Soil Preparation

The inspection team should review the grading of the area to be topsoiled and seeded. Compaction should be checked in fill area, trenches and slopes. A final check should be made to determine that the required depth of top soil can be placed, and if proper drainage can still be maintained. The edges and roads and walkways should be compared with the top soil requirements.

Determine if scarification or tillage has been accomplished to the specified depth in subgrade, and insure that top soil is not applied to untilled or unscarified areas.

In placing top soil, ensure that an even distribution is made, and avoid overcompaction. Keep passing equipment off the top soil.

Check on grades and provide drainage of surface water. Check that watering is within specified limits when top soil with grass is used. Be sure that vegetated or paved flumes are installed, and check matching seed bed below edges of walks and pavements.

Insure that the finished grade is correct, and that light rolling, fine grading, raking, and depth of tillage are as specified.

Check the condition of equipment for application of fertilizer and lime, including emphasis on calibration and check distributor settings for specified amounts. See that the material is not applied more than 24 hours in advance of tillage. Check for even distribution, especially in slopes where distributor should be filled at not less than half its capacity at all times. Check for loss of fertilizer by high winds and postpone applications if necessary. Check the total quantities used against the acreage covered.

02485 Lawns

The time of planting of *seed* is most important, and must be within specified periods. Out-of-season planting is to be permitted only with special written approval. Check for removal of stones and other objects.

When applying seed, check for even distribution and rate of distribution. Because of the cost of seed, calibration tests should be made before the equipment is used. The top soil should not be too firmly compacted, hard, lumpy, or stony. The inspection team should check for loss of seed during windy weather, and if necessary, postpone seeding. Signs and barricades should be installed to avoid traffic over seeded areas. There should be adequate specified equipment on hand for the watering of newly seeded areas. The inspection team should check and record the watering, as well as mulching period. Mulching should be applied in accordance with specifications.

Where *sodding* is required, check the area of coverage, and height of the grass and the sod. Determine that the area to be sodded is relatively free from weeds, undesirable plants, large stones, roots, and anything else which would be detrimental to the development of the grass or the maintenance of the area. Make certain that fertilizer and lime have been applied as required.

Inspect to insure that rectangular sections of sod have uniform widths and do retain native soil on the roots, and that thickness is as specified. Examine the handling, stacking, and care of the sod, and be sure that the time between cutting of sod and placing is within the specification limitations. Check the sod to assure that it is in a moist condition, and be sure that it is placed at a favorable time with respect to weather and season. Sod should be placed in a staggered formation, and tamped or rolled into the prepared sod bed. The cracks between sod should be filled with a good quality screened soil. After placement, be sure that the sod is properly watered. It is usually the contractor's responsibility to maintain seeded areas for a period of 3 grass cuttings. If there are any resulting bare or brown spots, it is the contractor's responsibility to replace these spots with new seed or sod.

02490 Ground Covers and Other Plants

The inspectors should check for certificate of inspection of plants, and check the quality and size against the rules and grading standards included in the American Standard for Nursery Stock, and/or requirements of the drawings or specifications. The inspection team should check the size range of each group, and look at the condition of the root system—rejecting all plants in which the roots have been allowed to dry out or have been damaged. Check foliage for wilt or dryness. In balled or burlap plants, check for a solid ball in conformance with specified dimensions, and reject plants with balls broken, cracked, or pliable. For bare root plants, be sure that plants were dug when fully dormant, and that the root size spread is within the requirements of the American Standard for Nursery Stock. Check to see that the roots contain a mud coating, check for broken roots, and fullness of foliage. Review the method of handling during transit.

Check the dates on which plants should be placed, and check the depth, diameter, and spacing of excavations, as well as proper handling of top soil. When planning, check the grade, soil moisture, and the method of placing and tamping of top soil. Check stability handling and placement of all plants, including the stabilizing stakes or guys. Check wrapping and pruning after planting and insure that the mulching is in accordance with specifications. After planting, review the watering schedule, the pruning, spraying, weeding and appearance.

02491 Trees and Shrubs

Be sure that locations have been properly located for trees and shrubs, and that holes are properly dug as to size and depth. Be sure that there are no broken stones or debris in the bottom of the holes and that the ground is clean, loosely packed, and watered before planting.

The inspector should make sure that all trees and shrubs are planted correctly, set plumb and true, and are properly packed with ground and adequately watered. Plantings may be placed with burlap root cover intact. If trees require anchor wires, be sure that the tie wires are not wrapped directly around the tree trunks, but that there is a rubber wrapping on the tree trunk and the tie wire is around this protective covering.

LANDSCAPING CHECKLIST:

1. Inspect rough grading for proper slope, particularly sloping away from building structures.

2. During rough grading, check stockpiling of topsoil, and be sure that the topsoil is maintained without being contaminated.

3. Before topsoil is spread, check for debris and be certain that it is removed.

4. Before seeding, or sodding, check for proper compaction of any fill area, trenches or slopes.

5. Determine that tillage has been accomplished to specified depth.

6. Topsoil should not be applied to untilled or unscarified areas.

7. In placing topsoil, avoid overcompaction and ensure even distribution.

8. Be certain that watering is done properly.

9. Check for proper fertilizer or lime preparation.

10. The time for planting of seed or transplanting of shrubs and bushes must be within specified calendar limits.

11. Calibration tests should be made before seed is spread.

12. Avoid traffic over seeded areas.

13. For sodding, sod should have uniform width.

14. The handling, stacking, and care of sod should be supervised properly.

15. Sod should be placed in staggered formation.

16. Check for certificate of inspection of plants, comparing quality and size against the specifications.

17. Roots must contain proper coating, or for bare root plants, must be dug when fully dormant.

18. Check location of trees and shrubs.

19. Be sure that new transplants are properly supported.

02500 PAVING AND SURFACING

The installation of roads and walks for a building site is similar in concept and practice to major road building, with the exception that the typical building site does not provide a large enough working area, or a sufficient quantity of work to justify the use of major labor-saving equipment. For instance, excavation for the road bed at a building site will typically be done with a front end loader, while a building site would use scrapers and pans for excavation and for dumping backfill.

02510 Paving

The objective of the inspection team is to be sure that the contractor provides a pavement which satisfies the design criteria, and has uniform characteristics of quality that will furnish a maximum serviceable life with minimum maintenance.

Concrete. The general requirements for concrete production are described in Division 3 and apply to paving as well. Control testing is important, both in design of the mix to meet the requirements of the specification, and maintenance of control to insure that the mix is prepared, delivered, and placed properly.

Bituminous Mix. The contractor must submit for approval a design mix which should be approved by the paving or structural engineer who prepared the specification. Adequacy of the batching or mixing plant should be determined by a joint inspection of the engineer and the inspection team or its representatives. A testing laboratory can be used where appropriate, particularly where large amounts of material are not required. Changes in the mix or plan operation may be requested to modify the bituminous mix hot-bin weights to maintain grading and bitumen content within limits established by the job mix formula or its approved revisions.

Subgrade. Subgrade preparation for paving is very similar to the general preparation, including clearing, grubbing and stripping of the top soil, and placement of a good subgrade, usually gravel, slag, or stone.

Prior to actual commencement of work, the inspection team should ascertain the location accuracy, and review the procedures to be followed by the contractor. Actual soils encountered should be compared with those shown on the contract drawings, and any deviations noted and reviewed with the appropriate engineers. The inspection team should ascertain that suitable material from excavations is being reused to the maximum extent practical, and that all pockets of unsatisfactory material are being removed and replaced with suitable material. Proper drainage and dewatering to maintain the subgrade in suitable condition should be insisted upon. If backfilling is required, appropriate compaction techniques should be employed and appropriate tests made. Subgrade drainage should be installed in accordance with the drawings, and any other subsurface work should be placed before the paving is allowed to begin.

When the subgrade has been prepared, either in totality or in sufficient quantity to permit orderly initiation of the paving, paving work can be allowed to start. The inspection team should be familiar with all factors, including contractor's procedures, maximum and minimum compacted layer thicknesses, base course material types, compaction requirements for the base course, grade and surface control requirements, and any special conditions.

Contractor's equipment should be inspected for both safety and capacity, as well as any special features required in the specifications. When construction starts, the weather and temperature limits must be observed, and batch plant inspections made of paving materials. During the construction, have vehicles vary the path used to deliver materials so that subgrade is compacted, rather than rutted in one area. Review the hauling procedures to determine their compliance with general safety requirements, insisting on items such as backup alarms if required because of operating conditions. Where a base and sub-base are required, insure that the two do not become mixed. If water is being added to the base material, insure that optimum moisture content is being achieved in accordance with specifications. Insure that grade stakes are available, and are being utilized, because once the paving material is in place, it is there to stay.

The contractor should place and compact the specified width of shoulder along with the sub-base and base course, not separately.

Bituminous Pavement. For bituminous paving, be prepared for the retesting requirements, and know the various limitations and specifications involved. Arrange for the appropriate batch plant testing, and check equipment to insure that the distribution equipment conforms to the requirements for proper heating and circulation of bitumen, for control of spreading rate and uniformity of application, as well as any measuring or indicating devices required. Insure that power equipment is available as required for cleaning of surfaces to be primed or tacked. Inspect the *base course* and/or pavement course to be sure that it is clean and free of foreign material or water before paving with bituminous materials is to start. During priming and tacking operations, review the rate of bitumen application, and check to see that the amount of priming applied is completely sealing the surface voids of the base course without causing a surplus to remain on the surface after curing. Also, inspect for uniformity of application and take prompt action to stop or correct unsatisfactory distribution. Check the curing process of the bitumen for compliance with specifications.

The field inspection team can depend upon the laboratory for review of the *bituminous mix* plant for factors such as:

Segregation of aggregate stockpiles

Proper stockpiling methods to avoid contamination or intermixing

Storage tank capacity for bituminous liquids

Insulated pipelines and fittings

Proper heating facilities for storage tanks

Aggregate dryer capacity and moisture control

Operation of hot aggregate screens and sizes

Capacity of cold and hot bins

Accuracy of weigh-in box and control gate

Bitumen weighing devices

Dust collection method

Mixer capacity

Mixer heating provisions

Blade clearances

Capacity of pug mill discharge hopper

Safety features

The field inspection team will be responsible for reviewing the handling of the material in the trucks, and in particular their cleanliness, covers to protect the mix, and ability to maintain mix temperature during delivery.

The *spreader* should be inspected for overall condition and freedom from obvious damage or fault. Inspect the function of the screed heating system and the tamping bar for wear and proper movement and clearance.

Rollers should be inspected for proper weight and rolling surfaces.

Prior to full-scale paving operations, it may be appropriate to require that a quantity of mixture be produced, and a test section constructed to permit the inspection team to inspect all operations and give final approval. This may also be helpful to the contractor to smooth out any operational problems.

In *paving,* the contractor should start at the highest lane in the area, and move in the direction of the main traffic flow. The operation should be laid out to maintain a uniform surface. The lanes should be placed so the joints will have required texture, density and smoothness. String lines should be established to assist in the checking of grade by both the contractor and the inspection team.

In checking temperatures at the field, care should be taken that the mixture is not overheated, as well as not too cold. The minimum acceptable placing temperature for hot mix is usually 235°F for asphaltic mixtures, and 160°F for tar mixtures. Reject mixtures arriving at the spreader with temperatures at less than the specified minimum.

The inspector should check the rate of feed of mixture from the hopper, operation of the distributing devices and screed adjustment, and take corrective measures promptly when irregularities in thickness, surface smoothness, or width of lane are found. Even though rakers can level off any irregular spots, excessive raking is undesirable. Raked-out material should not be cast over fresh surface. The inspector should see that all coarse particles unavoidably raked to the surface are removed.

Preparation and placing of the paving mixture joint should be checked to ensure a well-bonded area and even surfaces after rolling.

The inspection team should maintain a location of truck loads so that identity can be correlated with in-place samples taken and tested, so that if removal is required, the batch can be located.

Often, paving is on a quantity basis and weight tickets should be maintained as a basis for payment. A rough correlation should be maintained between the area paved and the quantities received.

The *rolling* operation should be observed in order to ensure that specified speed and time of rolling is maintained. Rollers should not stand on freshly placed mix, and they should make the required overlap between passes. Rubber-tired rollers are effective only on one mixture, and not when pavement temperature is below 130°. Roll longitudinal joints while the mix is hot to produce a tight, well-bonded joint. Use a straight edge to check for surface smoothness after compliance after the first lift of coverage. The time to correct smoothness in grades is when rolling first begins. To correct depressions, loosen material by raking to a depth of ½″ and add necessary hot material by shoveling and raking. To correct lumps, loosen by raking to a depth below final grade, and remove excess material and rake smoothly.

If proper sub-base conditions have not been established, transverse and longitudinal tracks will probably occur as a result of rolling, and total replacement of both pavement and base will be required.

Safety requirements should be enforced, and unnecessary hazards such as smoking fires, fires, and open flames should be avoided because of the volatile vehicle used as part of the mix.

The requirements for the replacement of the epoxy-asphalt pavement are very similar.

BITUMINOUS PAVING CHECKLIST

1. Check subgrade alignment and grade.
2. Check preparation of subgrade, since final work is no better than the grade it is placed upon.
3. Check contractor's equipment for safety and capacity, as well as any special features required.
4. Arrange for certification of plant mix, or install inspector at batch plant.
5. Check spreader for function, particularly in the screed heating system and the tamping bar.
6. Check rollers for proper weight in rolling surfaces.
7. Start paving at the highest lane moving in the direction of main traffic flow.
8. Mixture should not be overheated nor cold. Minimum placing temperature is 235° for asphalt and 160° for tar.
9. Check rate of feed of mixture from hopper and operation of distributing devices.
10. Check for irregular spots.
11. Avoid excessive raking.
12. Identify truckloads so that any remedial work required can be correlated by area.
13. Observe rolling operation for speed and time.
14. Enforce safety regulations because of volatile materials involved.

Concrete Pavement. The inspection team should review plans and specifications to become familiar with special requirements and conditions of the job.

Methods of reviewing and testing concrete mix, production, and placement are delineated in Division 3.

The base course materials and the location should be carefully checked before paving begins. The base course should be moistened before material is placed, and if paving is carried on during cold weather, the base course should be properly protected before placement of paving. Base materials should be entirely free of frost, including heating of aggregate or material if required.

The *forms* should be free of warps, bends, and/or kinks and should be free of battered top surfaces and distorted faces or bases. The tops and vertical faces of forms should be checked with a straight edge, and dimensions, position, and installation of metal keyway forms should be checked. The base of each form section should have a full bearing for its entire material and width on fully compacted material. The form pins should be of adequate length, and should be properly wedged in the pin pockets. Locking devices between form sections should be free from looseness or play. Forms should be set to required grades, as well as properly aligned or located accurately.

Very often, the paving equipment uses the forms as a track, making form placement even more critical. The contractor should furnish a scratch template to assure that the form placement is exact.

A pre-pouring check of imbedded items should be made. For instance, dowels should be of required diameter and length, and should be clean, straight, and smooth. The expansion joint form work or basket should be according to specification.

Prior to pouring, the usual *batch plant* inspection should be made, usually by a specific and special laboratory testing group. During placing, spreading, and vibration of concrete, the inspection team should slow or stop placement if subsequent operations are lagging too far behind. During placement, vibration in the vicinity of imbedded items must be performed without disturbing the items. Dowel assemblies should be covered by hand-shoveled concrete.

The spreader should be adjusted to strike off the concrete at a proper level, so that when vibrated, the proper amount of concrete will remain for the finishing machines. Vibrations should not be allowed in a single location for more than 20 seconds, and workmen should not walk unnecessarily in the fresh concrete.

Prior to placement of the concrete, *reinforcing steel* must be checked for size and spacing. If reinforcing bars are used, their placement spacing should be reviewed. If mats are used, they should be properly lapped as they are placed.

One end of each dowel should be painted and greased.

If the transverse *finishing* machine produces a slurry ahead of the screed after the first pass, the mix water should be slightly reduced. The transverse finish machine should leave the concrete surface at the proper grade, and essentially the proper smoothness. Hand finishing should be held to a minimum, and finishing machines should be used to achieve the required finish.

The final surface finish is usually produced by burlap dragging, and the timing of the drag is most important to produce the required texture. Drag when most of the surface sheen has disappeared, but while the concrete surface is still in a plastic state.

The inspector should not permit the use of soupy mortar to fill out depressions along joints during hand tooling—use fresh concrete. All spillage of grout and concrete on adjacent concrete surfaces should be cleaned immediately.

Filler strip material should be installed exactly as specified, and should be properly aligned, vertical, and set flush with or slightly below the pavement surface. After finishing is successfully completed, check the application of the curing compound which must be applied to a moist surface. A two-coat application should be obtained, and coverage should be no more than 200 ft^2 per gallon for both coats. If paving has been placed during cold weather, check for any special requirements, such as covering with plastic sheets, heating, salt hay, etc.

Fig. 17 Concrete saw used to cut control joint. *Source: Clipper, Norton Co.*

Do not permit removal of forms until the proper minimum time after placement has elapsed (usually one day or more). If *sawed joints* are required, determine that sufficient equipment is on hand, and make provisions to carry out the sawing operation continuously until completed. The alignment and location of the joint to be sawed should be carefully established prior to sawing. Sawing should be performed as soon as the concrete may be cut without excessive tearing or raveling of the concrete, and without undercutting or washing at the side to the cut. After each cut is made, it should be thoroughly flushed with water. If an uncontrolled crack has occurred at a joint location prior to sawing, do not saw. If a crack forms ahead of the cut during sawing, discontinue sawing. If curing coverings are removed to permit sawing, replace immediately after the joint has been sawed.

Joints should be sealed as specified, and within the weather conditions described in the specifications. Sealers should be heated when required. Joints should be filled flush with pavement surface and excess or spilled material should be removed.

Most random cracks occurring in pavements may be repaired, with the repair either sectional replacement, or limited patching depending upon the scope of the damage. Where obviously minor, repairs can be at the discretion of the inspection team. In cases of doubt, however, the structural engineer of record should be consulted and should approve the repairs.

CONCRETE PAVEMENT CHECKLIST

1. Check for special requirements and location.
2. Check base course materials.
3. During cold weather, protect base course before placing paving.
4. Check that forms are free of warps, bends, or other surface distortion.
5. Forms should be properly wedged and held in place.
6. Check reinforcing, dowels, and placement of contraction or expansion joints.
7. Inspect batch plant.
8. During placement, vibration in vicinity of imbedded items should not disturb the imbedded items.
9. Concrete spreader should be adjusted to strike off at proper level.
10. Check final finish surface.

11. Do not permit use of soupy mortar to fill depressions.

12. Curing compound should be applied properly.

13. If sawed joints are required, check equipment and carry out sawing operation continuously until completed.

02528 Curbs

Curbs and gutters provide the surface range for the roadbed. Finished appearance is important, since it is most noticeable to the public. Alignment and grade of curb forms is required to obtain a good end product. The location and alignment of existing pavement or sidewalks should not be depended upon for establishing uniform grade line for curbs.

When location and grade space are set, the timing should be after subgrade material has been established so that the stakes won't be disturbed during compaction. Where the gutters become a part of the traveled shoulder, the subgrade must be compacted to a completely stable condition to offer a necessary support to the concrete.

Although the curb grades should be set with an instrument, a check by eye will disclose any unusual alignments or problems.

Although the progress of curb work is generally slow because of the small forms, the sequence is the same as the larger-scale paving operation, and the quality control is similar also. *Forms* should be checked for condition, cleanliness, rigidity, and smooth surface. Check the form setting to insure full bearing of the form base on the prepared subgrade and the form bracing, so that displacement or release does not occur during pouring, causing poor appearance in the curbs and gutters. The forms should be pre-oiled to facilitate removal, and contraction and expansion joints should be properly placed according to specifications. The concrete mix should allow maximum workability, while still maintaining required strength. The contractor should be cautioned not to prepare too much concrete at a given time, or deliver too much, since curb pouring tends to be a slow operation.

After removal of forms, the exposed surfaces must be finished properly by required methods. The inspection process should include the tops and faces of curbs and surfaces of gutters to insure conformity with surface smoothness and shape requirements. Curing procedures should be the same as for paving, and continuously applied.

After the concrete curbing has properly cured, spaces in back of the curb should be backfilled with compacted material meeting appropriate specifications.

The flow line of gutters can be checked while the concrete is still in the curing stage by the use of water, permitting correction as work progresses.

The curbing areas should be protected from traffic until the concrete has had a chance to reach full strength.

Walks

Walkways are usually of concrete or bituminous mix, usually placed after the curbing where adjacent to roadways. If the walk is for pedestrian purposes only, it may be an independent structure. In all cases, the preparation of subgrade is, as always, important. Alignment and location are also important, but principally so that they do not encroach upon other areas causing interferences.

Placement of the concrete or bituminous mix should follow the same rules as those for roadway paving, and where adjacent to curbing, the new curbing can be used as an alignment device. The subgrade should be wet down before placement of concrete, or an impervious membrane should be applied, whichever is required by the specifications. Where not specified, the inspection team can make the choice. Finishing of the sidewalks should be in accordance with specifications. Where none is specified, a hard-troweled finish is not preferred for safety purposes, and either a broom finish or wood float surface is generally better. Nevertheless, the design specifications prevail.

Road and Parking Appurtenances

The contractors and inspection team should cooperate to provide temporary roadways or drive-arounds during construction of roads which replace existing facilities. Site parking should be under the control of the inspection team, with the cooperation of the contractor. In many cases, the handling of traffic will depend on the sequence in which the contractor accomplishes the job. Design speed of a detour should not be less than 10 miles per hour lower than the prevailing approach speeds. For multi-lane detours, widths should be 12' per lane, and 15' is adequate for a single one-way traffic lane. Usually, black-top surfacing is used to eliminate dust and maintain safe travel.

Traffic signs for the final roadway and other directional signs shall be in accordance with the specifications and plans, which in turn, should be in conformity with the owner. Parking areas and roadways should be properly striped and painted to indicate pedestrian zones, parking areas, and other specified details. These, again, would be as required by the specification.

In some cases, special equipment will be required within a parking area such as specially activated gates, treadles, and traffic control devices—including aerial lighting. These features will be as specified, and should be inspected with emphasis upon the manufacturer's requirements and special instructions as contained in the specifications.

02700 PIPED UTILITIES

Utilities vary considerably with each site. The scope of utility work depends on the availability of trunk services in the area, as well as the capacity available on the trunk lines. As part of the project, utilities work involves either the introduction of new services to the site, the relocation of existing services, or both. In relocation of existing services, many problems are posed because of inaccurate information on location of utilities.

Utilities in the project area may be publicly or privately owned. New *connections* to services are usually terminated at the boundary or battery limits of the project. In many cases, the contractor puts in the extension of the utilities from the boundary line to the distribution points within the project area. However, there are special situations such as the telephone service, in which the contractor usually places the conduits, but the telephone company places the cable and makes all connections. This is subject to change in remote sites where all installations are placed by the contractors.

The inspection team should be well aware of the utility information presented in the plans, and should make appropriate notation of any changes observed as field excavation proceeds. The inspection team should observe the condition of service of lines to be relocated, keeping in mind the safety implications of potential of flooding, broken gas mains, etc. Where lines are to be relocated or connected, the inspection team should ensure that proper *coordination* has been made with the appropriate utilities, and that any public warning or notification which is appropriate has been made. Since relocation generally does disclose changes in anticipated conditions, the inspection team should be careful with any revisions in the relocation, so as not to cause conflicts with identified or projected lines.

The observation of the existing utilities and possible interference should not be limited to the plans, but should include conferences with local utility representatives (even though this could be presumed to have occurred in the design phase).

The design for underground lines will make available appropriate fixtures for *test points,* air relief valves for testing, and accessibility for valves, hydrants, and manholes. Nevertheless, quite often some of these general features are specified in the specifications and not shown specifically on the drawings. Accordingly, the inspection team should be alert to ensure that all features required are, in fact, included.

As in all things, the location of the underground lines to be put in is important, to avoid interferences on this project and for relocation information in the future.

Quite often utility connections or relocations involve traffic interference, and the inspection team should consider coordination factors to avoid interference with the public vehicle or pedestrian traffic. When there are interferences, appropriate warning signs and obstruction lights should be utilized.

Before underground lines are backfilled, appropriate *pressure tests* should be conducted. Where pressure tests are not necessary, the visual inspection should include every joint, pipe bedding, and backfilling.

Before connections are started in utilities, the inspector should be certain that the contractor has available all material required, so that service will not be interrupted any longer than is necessary. While pipe and valving equipment is durable, care should be taken not to permit the entry of foreign materials into the piping system as a result of dragging or poor handling, nor should there be undue stresses on the equipment as it is being placed.

Steel lines are usually coated with a moisture barrier, and before backfilling, the inspection team should be certain that the coating has been tested in accordance with the specifications. This includes at the minimum visual checks for breaks and abrasions of the pipe coating, and often requires a "jeep" test, which is an electric test to locate breakdowns in coverage.

While alignment and location are important, so is *grade,* in particular for gravity lines.

Where fire hydrants are placed they should be plumb and facing the roadway, and before backfilling the drain plugs should be removed in areas of low groundwater, but left in where high groundwater is encountered. Pressure lines require kicker or reaction blocks at any turn in the line, and inspection includes the assurance that these are in place properly before backfilling.

Water lines must be *sterilized* before utilization, and the inspection team supervises this in accordance with the specifications. Also, bacteria tests may be required.

Underground fuel gas lines should not be located under buildings or in trenches with other utilities. During construction of *gas* lines, safety regulations must be carefully enforced, including current checking with explosion meters where there is an existing gas line. Minimum cover of gas pipes is 2' and additional coverage is preferable.

Sanitary sewer lines should be installed below water lines if the two lines are within 6' horizontally, and the spigot or male end of the pipe should be pointed downstream. A uniform grade must be maintained between manholes, and the bedding should be properly placed so that there is no sag or depression in between manholes. The top elevation of the *manhole* should be flush with the paving grades or higher than finished grade of the existing ground area. Working in sewer lines requires safety precautions, including both explosimeter to preclude inadvertent gas collections, and periodic checking for proper air supply.

Storm sewers are less sensitive than sanitary sewers, but should meet the same tests. Large storm sewer lines are often made out of reinforced concrete with rubber gasket or O-rings, which should be affixed carefully. Visual inspection will determine the adequacy of the connection.

Hot water or steam distribution lines are subject to *expansion* and *contraction,* and the design should provide for movement, through the device of expansion or fabricated pipe loops. These lines must also be anchored at strategic points as indicated by the design. The proper design of hot water/steam lines includes an accurate evaluation of the expansion and contraction anticipated, and the *anchorages* are important in terms of the location, number, and durability. Hot lines are usually insulated, and the insulation should be installed in a dry condition and appropriately waterproofed before backfilling. Welding should be monitored in accordance with the requirements in the specification and applicable codes.

The factors to be observed in pipe joints are basic, including clean joints and proper procedures to suit the joint in question. *Lead joints* should be filled in one continuous pour. The lead should not be displaced to a depth of greater than $\frac{1}{4}''$ during packing. Hot pour joints should have a uniform annular space. In cement mortar joints, proper fillers such as oakum should be used in accordance with the specification, and the mortar curing should meet all the requirements of concrete work.

Where the joint is a *pipe thread,* the thread should be cut with sharp tools and be of proper thread length. All joints should be made up tightly.

For *tubing connections,* the tubing should be cut off square with burrs removed before soldering, silver soldering, or brazing is performed. Where the tubing is joined with flared compression joints, the tools and techniques of the workers should be checked by the inspection team.

For *welded joints,* the welders should be certified, particularly where the work is under ASME code. As much as practical of the joint fabrication should be done at ground level before lowering prefabricated sections into place. The usual precautions for explosive gases should be emphasized in preparing for welding in the trench.

In placing corrugated piping, the laps of all circumferential joints should be on the outside lap on the downstream side of the joint, with the longitudinal laps on the side of the in-place pipe. All markings indicating the top of the pipe should coincide with the specified alignment of the pipe. While the connecting band is being placed, the inspector should ensure that the band is going to fit tightly. If required, be sure that the proper bituminous material is used at the joint after joining.

In *manhole construction,* the location, dimensions, and layout should be checked and confirmed before pouring or installation of precast manholes. Of greatest importance are the invert elevations as well as the flow details of the invert channels in the bottom of the manhole. Manholes should be checked for materials according to the specification, with brick and concrete being the two most common materials used in manhole construction, but some being made out of corrugated pipe.

All of the normal rules for excavating should be followed in trenching for utility lines, and piping should be placed in dry bedding, so that pumps or well point systems should be used as appropriate to dewater trenches. All testing and inspection of the pipes should be made before any backfilling is permitted. After backfilling, the pipes should be checked as appropriate to determine that no broken pipes or settlement in excess of that allowed has occurred.

UNDERGROUND UTILITY CHECKLIST

1. Be sure that lines are properly located in depths as well as plan location.
2. Note any deviations in the location on the as-built plans for coordination with future lines.
3. Be sure that appropriate test fixtures are installed in accordance with specifications.
4. Before underground lines are backfilled, be sure that appropriate pressure tests are conducted.
5. Be sure that lines are properly bedded.
6. Be sure that steel lines are coated with the required anticorrosion coating, and if required the coating is tested for tightness.
7. Be sure that water lines are sterilized, and other special post-installation tests are conducted.
8. For certain types of joints, thrust blocks must be placed at all turns.

02850 RAILROAD WORK

In addition to specifications and shop drawings, the inspection team can utilize the American Railway Engineering Association documents for reference material. In railroad work, location and alignment is of the greatest importance.

When tie-in to active lines is to be made, appropriate arrangements must be made with the adjacent railroad for provision of flagmen, light signals, slow boards, and other requirements of the railroad itself or to meet ICC regulations for protection of equipment and personnel. Any tie-in should be carefully coordinated in terms of schedule. Similarly, local authorities should be coordinated in terms of the detours, barricades, crossing guards, safety lights, and other equipment which may be required.

Rough grading of the *roadbed* is usually prepared under the excavation contract, and the inspection team should have reviewed the angle of repose and condition of the side slopes in the cut-and-fill sections. Drainage should leave the roadbed dry, and culverts should be installed as described in the plans. The size, shape, grading, and compaction of the roadbed should have been inspected carefully when it was done. To avoid the possibility of dispute, the roadbed subgrade should be accepted by the railroad contractor as well as the owner.

Track is divided into types and classified according to degree of curvature, and will be specified on the drawings. Tracks must be laid to the exact gauge as shown on the drawings.

On curves, tracks are usually superelevated so that the outer rail is slightly higher than the inner, and this shall be as specified on the drawings or in the specifications. Maximum value of superelevation is usually about 6″, but again, the specific requirements will prevail. Rail joints should be staggered, and the joint and the running rail should be as near as possible to the center of the running rail on the other side of the track.

The openings between the ends of the rails vary with the temperature of placement and range from zero to $\frac{3}{16}$″, with the larger gap relating to the lower temperatures.

Just prior to installation, *joint bars* should be bent to the curvature of the track and bolted to the track rail to secure a snug fit. Joint bar bolts are applied with their heads alternately inside and outside of the rail, except on guarded curves where the bolts are applied at the joints in the running rail, on the low side with the heads on the outside of the rail, and at joints in the guard rail with heads in the side of guard rail nearest to the center of the track. Spring washers for bolts are required on all rail joints. Nuts are screwed up tight with an automatic nut tightener or track wrench. Heads of the bolts should be tight against the joint bar before tightening the nuts, and as a protection against rust, all joint bars and parts of all rails covered by joint bars should be carefully painted with a protective coating as specified. Joint bars and heads of bolts should be tapped with a light hammer to insure that the top and bottom finishing edges are all the way home.

At key points as specified by the signal design, insulated joints are to be installed. They should be installed when the track has been otherwise completed. Care must be taken to provide at least 1″ clearance between metal parts of the insulated joints and tie plates or abrasion plates which may be installed under the insulated joints and tie plates. All scale dirt, burrs, and rough edges must be removed from the rail and the rail cleaned with a wire brush prior to applying a liberal coat of insulating paint such as Glyptal to all parts of the joint. The joint will be assembled wet. To protect against rust and deterioration, sponges should be wedged between the joint bar and the web of the rail at the ends of the joint bar, and all parts of the insulating joint, including the sponges, bolts, and all metal and fiber parts should be carefully painted with another coat of the insulating paint.

Guard rails are generally installed adjacent to the gauge side or inside rail of curves of less than 2300' radius, including all transitions and extend on the adjacent tangents 50' on the trailing end of all curves, 10' feet on the receiving end of transition curves, and 30' on the receiving end of simple curves unless otherwise shown on the drawing. Guard rails are secured to the running rails by means of adjustable rail braces and separators.

Rails must be handled carefully. In the process of laying, lifting, or surfacing—particular care must be taken to prevent the bending of rails. Rails should be skidded off cars or conveyances by a method approved by the inspection team, and should in no case be dropped. Running rails to be used in curves of less than 1800' radius should be properly curved using approved rail bender. Rails for curves of 1800' and over may be sprung and spiked to the proper curve. Precautions must be taken to obtain a uniform curve throughout the rail. Guard rails for use on curves of less than 1000' should be properly curved using a rail bending machine, and on radius of more than 1000' may be sprung and spiked to the proper curve. When curving rails, desired curvature should be obtained by successive applications of the rail bending machine. Special-length rails necessary to complete work should be cut in the field by being sawed square across the rail, using an approved saw. Burrs should be removed, and ends made smooth. Tracks should be laid out to involve a minimum of special cutting. Tie plates should be used under all rails and all ties, except under emergency rails. The tie plates should be placed with the shoulder tight against the outside base of the rail to insure a full bearing of all the assembled parts. Tie pads should be used under steel tie plates. In ballasted track, no spiking should be done until the ties have been properly spaced and squared across the center line of the track, on tangents and placed in a radial position on curves.

Holes should be drilled in the ties for the spikes. If, for any reason, a spike is drawn, the hole should be filled with a square hardwood plug and a new spike driven in the same location. Screw spikes must be installed normal to the top surface of the tie and in no circumstances should a screw spike be driven.

Ties should be handled carefully, and not thrown from cars or conveyances. No picks, shovels, spike mawls or other tools should be used for pulling or bucking ties. Holes drilled in ties for spikes should be accurately drilled normal to the top surface. After holes are drilled, the cut surfaces should be thoroughly saturated with hot creosote oil.

Ties should be placed square across the center line of the track on tangents and in a radial position on curves. In spacing and squaring the ties before spiking, rails should be raised and ties shifted to the proper position without striking or pounding them with spike, mawls, picks, or other implements. Any ties that are split or damaged should be removed and new ties substituted. The heart side of all ties should be turned down. The ends of the ties should be lined evenly on the side of the track specified under various types of tracks. Where a third rail is used for power, long ties for the support of the power rails should be installed, as shown on the drawings, in a manner not to obtain a greater spacing than 10′ center to center.

For ballasted-type track construction, *ballast* should be thoroughly tamped under and on both sides of the ties from the ends of the ties to a point at least 15″ inside the rails. Approved automatic tampers should be used for tamping of stone ballast, and ballast must be leveled even with the top of the tie and neatly trimmed. Rails should be carefully raised in at least two lifts to the established grade. Rails should be originally set below the established grade.

No work trains should be operated over the rails until all of the ties and tie plates have been properly placed, the rails set in their position and spiked, and ballast installed and tamped sufficiently to prevent damage to the track system.

Bolts for switch rods should be installed with the nuts on top and provided with cotter pins. Particular care should be used in the loading and unloading and handling of switching material, particularly switch points, frogs, and other parts. A preliminary test of each weld should be made to determine that the joint possesses sufficient mechanical strength to meet any requirements. Joints should be raised at least 5″ from the level of the insulator cap surface and repeated to make sure if required to be certain that the joint is satisfactory. If any defect develops, the weld should be cut out and rewelded (this applies to contact rail). Standard rail gauge is 4′-8½″.

RAILROAD WORK CHECKLIST

1. Check angle of repose and side slopes in cut and fill sections.
2. Drainage should leave prepared roadbed dry.
3. Culverts should be checked and should be installed as in plans.
4. Compaction of roadbed should be inspected.

5. Tracks should be laid as specified on drawings; check, for instance, that tracks are superelevated on curves with the outer rail higher than inner.

6. Rail joints should be staggered.

7. Openings between the ends of rails should gap slightly for temperature variations.

8. Joint bars should be bent just prior to installation.

9. Check for insulated sections for signaling.

10. Check for guard rails on inside rail of curves.

11. Rails must be handled carefully.

12. Check spike installation for conformance with specifications.

13. Ties should be placed square across center line on tangents and in radial position on curves.

14. Ballast should be thoroughly tamped.

15. Work trains should not be operated over tracks until all ties and tie plates have been placed.

02880 MARINE WORK

Marine facilities are generally quite specialized, but often involve piling, both individual and sheet piling. Typically, the environment is in a high water table, and when in an ocean environment, exposed to salt water corrosion. Marine organisms present a problem in wood piling, as does deterioration at the water line level.

0297 Protective Marine Structures

Protective structures involve stabilization of piers or beachfronts, often using driven sheet piling. In other cases, sheet piling may be used to reestablish a given area, with backfill occurring within the sheet piling resulting in a reestablished jetty or bulkhead.

The inspection team must be concerned not only with the normal problems involved in piling, excavation, and related work, but must also work in hydraulic conditions, often working with soils which are saturated with water or made up of water-borne sands or silts.

The plans and specifications should include special instructions from the designer in regard to the latitude which the inspection team has in using existing material, or directing the inspection team to be sure that certain materials are removed and replaced with more stable elements.

02881 Dredging

Dredging is usually done to deepen or maintain a channel. In some cases, the purpose of dredging may be to get the sandy material from a bay or riverbed for use in backfill or as aggregate.

Dredging can be done in its most elemental method by reaching from land or barge into the channel using a dragline or clamshell. For this type of equipment, other operations such as rigging and lifting can be performed with the same equipment. In dredges dedicated to dredging only, a massive suction line is used, connected to water-borne pumps. The U.S. Army Corps of Engineers operates a number of dredges which can pick up an entire shipload of dredged material, carrying it out to sea and dumping it. A more usual dredge is basically a floating pumping station which pulls material from the channel, and through a barge-mounted floating discharge line, places it within cofferdam areas or high areas along the river or water bank.

The inspection team must be concerned with the rights to discharge material on the backfill or spoil sites, and must be certain that the contractor has properly aligned his working area. In this case, marine navigational aids such as buoys and range markers can be utilized to establish position rather readily.

DIVISION 3: CONCRETE

NUMBER	TITLE
03010	CONCRETE MATERIALS (Data Filing Only)
03050	CONCRETE PROCEDURES
−51	Cold Weather Concreting Procedures
−52	Hot Weather Concreting Procedures
03100	CONCRETE FORMWORK
−10	Structural Cast-in-Place Concrete Formwork
−20	Architectural Cast-in-Place Concrete Formwork
−30	Structural Precast Concrete Formwork
−40	Architectural Precast Concrete Formwork
03150	FORMS (Data Filing Only)
−51	Formliners and Coatings
−52	Wood Forms
−53	Prefabricated Forms
−54	Panel Forms
−55	Pan Forms
−56	Steel Forms
−57	Plastic Forms
−58	Prefabricated Stair Forms
−60	Permanent Forms
−61	Permanent Steel Forms
−62	Wood Chip Form Blocks
−63	Polystyrene Form Blocks
−64	Void Forms
03170–03179	(Reserved)
03180	FORM TIES AND ACCESSORIES (Data Filing Only)
03190–03199	(Reserved)
03200	CONCRETE REINFORCEMENT
−10	Reinforcing Steel
−20	Welded Wire Fabric
−30	Stressing Tendons
−40	Fibrous Reinforcing

03250	**CONCRETE ACCESSORIES**
−51	Expansion and Contraction Joints
−52	Anchors and Inserts
−53	Waterstops
03260–03299	(Reserved)

03300	**CAST-IN-PLACE CONCRETE**
−01	Concrete Admixtures
−10	Structural Concrete
−11	Normalweight Structural Concrete
−12	Heavyweight Structural Concrete
−13	Lightweight Structural Concrete
−14	Prestressed Structural Concrete
−15	Shrinkage Compensating Concrete
−20	Concrete Topping
−30	Architectural Concrete
−31	Normalweight Architectural Concrete
−32	Lightweight Architectural Concrete
−34	Prestressed Architectural Concrete
−40	Low Density Concrete
−41	Insulating Concrete
−45	Concrete Finishing
	Finishing Vertical Surfaces
	Finishing Horizontal Surfaces

03350	**SPECIAL CONCRETE FINISHES**
−51	Exposed Aggregate Concrete
−52	Tooled Concrete
−53	Blasted Concrete
−54	Grooved Surface Concrete
−55	Colored Concrete
	Integral Coloring
	Surface Coloring

03360	**SPECIALLY PLACED CONCRETE**
−61	Shotcrete
−62	Pumped Concrete

03370	**CONCRETE CURING**
03380–03390	(Reserved)

03400	**PRECAST CONCRETE**
−10	Structural Precast Concrete
−11	Precast Concrete Wall Panels
−12	Precast Concrete Deck
−13	Precast Structural Concrete Sections
−20	Precast Prestressed Concrete Sections
−30	Tilt-up Concrete Construction

-50   Architectural Precast Concrete
-51   Architectural Precast Concrete Wall Panels
-60   Precast Concrete Storage Tanks
03470-03499 (Reserved)

03500   **CEMENTITIOUS DECKS**
-10   Gypsum Concrete Deck
-15   Precast Gypsum Plank and Tile
-30   Cementitious Wood Fiber Systems
-31   Cementitious Wood Fiber Plank
-32   Cementitious Wood Fiber Deck
-33   Cementitious Wood Fiber Tile
-40   Composite Concrete and Insulation Deck
-50   Asphalt and Perlite Concrete Deck
03560-03599 (Reserved)

03600   **GROUT**
-01   Catalyzed Metallic Grout
-02   Nonmetallic Grout
-03   Epoxy Grout
03620-03699 (Reserved)

03700   **CONCRETE RESTORATION AND CLEANING**
-10   Concrete Cleanup
-20   Concrete Resurfacing
-30   Concrete Rehabilitation
03740-03999 (Reserved)

OSHA REGULATIONS—CONCRETE

Equipment and materials used in concrete construction must meet the requirements of ANSI A10.9. "Safety Requirements for Concrete Construction and Masonry Work." This section has miscellaneous comments such as the requirements that employees shall not work above vertically protruding reinforcing steel unless it has been protected to eliminate the hazard of impalement. Vertical reinforcing structures must be guyed and supported to prevent collapse, and wire mesh rolls must be secured at each end to prevent recoiling action; float handles on power floats must be constructed of non-conductive material and similar safety factors.

Form work and shoring is to be designed to support safely all vertical and lateral loads that may be imposed. These shall be shown on shop drawings which should be approved by the design team structural personnel.

Vertical steel slip forms are to be specifically designed for the purpose, and again approved by qualified engineers.

Single post shores must be horizontally braced in both longitudinal and transverse directions, and diagonal bracing should also be installed. Base plates or shore heads must be in firm contact with the form materials.

Jacks for lift slab construction must have a safety device which will support the load in any position in the event of jack malfunction. If lift slabs are automatically controlled, a device must be installed which will stop the operation when the ½″ leveling tolerance is exceeded. When it is necessary to provide a firm foundation, the base of the jack shall be blocked or cribbed.

03100 CONCRETE FORMWORK

Forms, most commonly made of wood, provide the shape of the final concrete configuration. From the viewpoint of inspection two factors are primary: safety and finish. During the placing of the concrete and its development of initial set, safety is of the greatest importance, and deserves total emphasis. However, after the forms are stripped and bracing and shoring is removed, the inspection team must live with the finish which the construction techniques have provided. Mistakes are very permanent. The surface of the forms is directly related to the resulting surface of the hardened concrete. If the forms have rough surface, blemishes, knots, or flaws, these will be reflected in the concrete surface. (Figs. 1, 2, and 3).

The following notes are extracted from the National Forest Products Association booklet, "Design of Wood Form Work for Concrete Structures," which was prepared to assist in attaining better concrete through the use of *wood forms:*

The first consideration in the design of a form for concrete is the load to be supported. This consists of two parts, the weight of the concrete, and the construction live load. Ordinarily, the dead load of the forms is small in relation to the weight of the concrete and may be neglected.

For slabs, floor panels, beams, and girders, the weight of the concrete is often assumed to be 145 or 150 lb/ft^3. If the concrete is assumed to weigh 144 lb/ft^3, the weight of the slab in pounds per square foot can be estimated by multiplying its thickness, in inches, by 12. Based on the

Fig. 1 Wall forms anchored to previously poured wall, with vertical columns supporting modular forms. *Source: Symons Corp.*

same assumption, an estimate of the weight of a beam per linear foot is the product of its breadth and depth in inches.

Variations in the concrete mixture can affect the loads, but such effects are usually unimportant. Special aggregates, however, require special consideration. The heavy magnetite iron ore used in the concrete of nuclear reactors, and vermiculite, used in lightweight concrete, are examples.

Fig. 2 Metal frame wood lined wall form panels in place. *Source: Symons Corp.*

The construction live load, made up of workmen, concrete buggies, and other temporary loads, varies considerably depending on conditions. A load commonly assumed is 75 lb/ft² of floor area.

Loads imposed on wall or column forms are of a somewhat different nature. Fundamentally, they are *hydraulic loads* resulting in a lateral pressure on the forms. The amount of lateral pressure is influenced by the weight of the concrete, the rate of placement, the method of placing—by hand or vibrated—and the temperature. Variations in mixture, the shape of forms, and the amount and location of resource reinforcing also affect lateral pressure. Except for special aggregates, these effects are small and usually are neglected.

When concrete is placed by hand, the effect of hydrostatic head (pressure) increases with the depth until—by compaction or by hardening, or by a combination of the two—the concrete tends to support itself without causing other lateral pressure on the forms. For the slower rates of placement, say one to two feet per hour, the maximum pressure will be

Fig. 3 Metal frame wood lined modular form panels in place for offset pour. Note finish on previous pour. *Source: Symons Corp.*

reached at a depth of two to four feet. Faster rates of placement result in maximum pressures at greater depths.

High frequency *vibration,* commonly used to compact freshly poured concrete, results in higher lateral pressures. The vibration causes the concrete to act as a fluid for its full depth. Accordingly, forms must be designed for the increased lateral pressures that are developed, and they must be tighter because of the greater leakage to be expected. In view of the fact that the use of immersion vibrators is very common, it is suggested that the form design be based on vibrated concrete. The design recommendations given herein do not include a consideration of revibration or external vibration. These practices can destroy well-built forms not specially designed to withstand the excessive stresses thereby induced.

Effect of temperature on lateral pressure at the time of placement is considerable. At low temperatures, the hardening of concrete is somewhat delayed. In consequence, a lesser depth of concrete can be poured before the lower portion sets up enough to become self-supporting. The greater hydrostatic pressure thus developed results in higher lateral pressures. A generally accepted rule-of-thumb is that concrete placed at 50°F will develop *one-third more pressure* than concrete placed at 70°F. The design of forms for winter-poured concrete should be based on a realistic appraisal of the higher lateral pressures to be expected."

At one time or another, the pressure of concrete placed in vertical form has been the subject of much speculation. Most authorities have generally reached similar conclusions, and all agree that pressure in column forms is often higher than in wall forms because the rate of pressure is likely to be more rapid. Figures 4 and 5 show the pressures that may be expected for various rates of placement at 70°F and 50°F, respectively.

Forms will deflect under load. A normal deflection limitation of 1/240 is considered adequate. In some cases, a more stringent requirement of 1/360 is specified for horizontal surfaces. Deflection may be important in architectural formwork.

Another consideration is an optical illusion that makes a perfectly straight beam appear to sag. To offset this, a camber is often built into the forms which actually provides a little positive bow. ¼" for each 10' of length is usually considered to be an appropriate camber.

The selection of *lumber* for concrete form construction depends upon the design requirements and the working stresses of the various sizes, grades, and species. Recommended working stresses for commercial grades and species of lumber are given in the "National Design Specification for Stress Grade Lumber and its Fastenings" and "Wood Structural Design Data," publications of the National Forest Products Association.

The tabulated working stresses of the "National Design Specification" are for normal loading, which means the full maximum design load

p = maximum lateral pressure in psf
R = rate of placement in feet per hour
T = temperature of concrete in the forms
    in degrees F.
(Maximum values: 3000 psf for columns and
    2000 psf for walls)

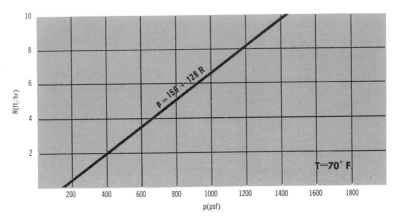

Fig. 4 Lateral pressures for various rates of placement at 70°F. *Source: National Forest Products Association.*

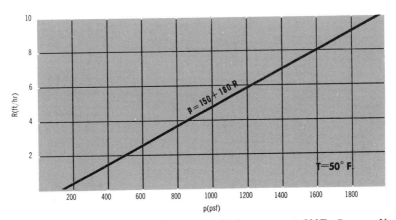

Fig. 5 Lateral pressures for various rates of placement at 50°F. *Source: National Forest Products Association.*

would be applied for not more than 10 years during the anticipated life of the structure.

A unique characteristic of wood is its capacity to absorb large overloads for short periods. Therefore, since the load on a concrete form is rapidly reduced as pouring activity ceases and the concrete hardens around the reinforcing, substantial increases in *working stresses* are justified. Also, loads on a concrete form are distributed among many members, most of which are stronger than the minimum piece which determines the lumber grade.

In view of these circumstances, it is recommended that the working stresses of the lumber used for concrete forms be increased 25% above the values given in the "National Design Specification." To illustrate, there are a number of grades of the various species that qualify for 1200 psi in bending. In design of concrete forms, it is proper to use these grades at 1500 psi.

Prefabrication of panels for the assembly of forms is common. These are then joined together and supported by whalers and bracing to provide a strong enough container for the concrete. It is good practice to prepare form clean-out access at the bottom of columns and wall forms, so that dirt, chips and foreign material can be washed out prior to the pouring of concrete. (Fig. 6)

Forms should be wet down prior to use to swell them before the concrete is placed. This also has the advantage of preventing the dry forms from pulling water out of the concrete mix, changing its water-cement ratio. Also, a form which has a wet surface will be less likely to honeycomb unvibrated concrete.

In wall forms where the forms can be tied together with special form ties, the amount of bracing and shoring can be reduced because the ties resist the lateral expansion forces of the concrete. These snap ties can then be twisted off after the forms have been removed, and the tie holes are patched with mortar.

Forming is often more expensive than the actual cost of the concrete itself. Accordingly, the contractors usually design forms for reuse. The inspection team should be sure that the forms are not used beyond their ability to provide a smooth well-finished surface.

Many devices have been developed to simplify the form erection, stripping, and/or to reduce costs. Typical of these is the doublehead nail which can be easily withdrawn for stripping. Other devices include adjustable shoring and screwjacks.

Clean–out door on column form.

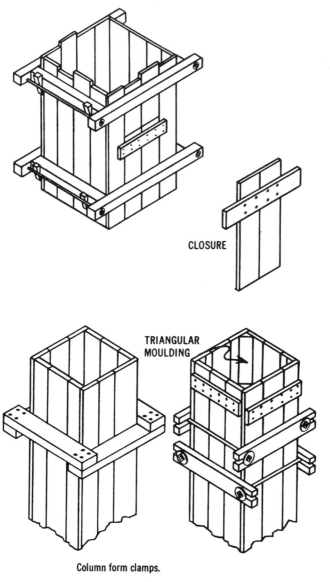

CLOSURE

TRIANGULAR
MOULDING

Column form clamps.

Fig. 6 Various form details. *Source: National Forest Products Association.*

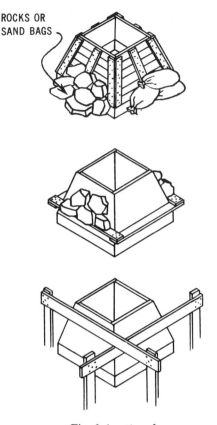

ROCKS OR
SAND BAGS

Fig. 6 (*continued*)

The inspection team must be careful to validate the location of forms, and their conformance in terms of size and configuration with the structural drawings. Mistakes are permanent, and poured in place.

Metal forms should be oiled, rather than wetted, but surplus oil should not be left on the forms. Further, there can be no oil on the steel-reinforcement or other surfaces where a bond of the concrete to the steel is expected.

During placing operations, the forms should be constantly checked for movement, and the pour stopped if any unexpected deflection or shifting of forms occurs.

The economies of form work reutilization often permit the use of modular forms which can be reused many times, and can be reassembled

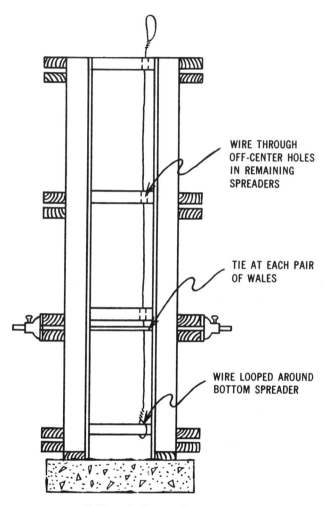

WIRE THROUGH
OFF-CENTER HOLES
IN REMAINING
SPREADERS

TIE AT EACH PAIR
OF WALES

WIRE LOOPED AROUND
BOTTOM SPREADER

Wall form showing spreaders.

Fig. 6 (*continued*)

into various shapes and forms. Several manufacturers specialize in this type of form work, and can be of particular assistance in suggesting proper bracing and shoring to support the formwork for proper strength and safety (Figs. 7 and 8).

A very special type of form used over the past several years is the *slip form*. This form is used on vertical structures such as chimneys, elevator

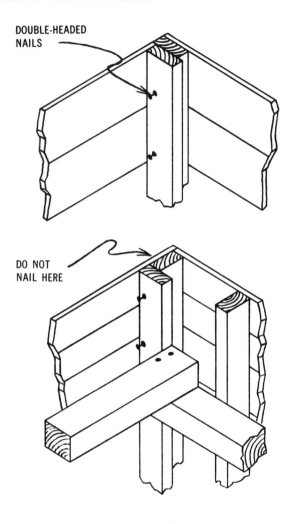

Internal corners designed to facilitate stripping.

Fig. 6 (*continued*)

shafts, and core structures for buildings. The form is jacked vertically, literally pulling itself up by its own bootstraps. The reinforced jacking bars are special dowels placed in the concrete. The rate of vertical rise is timed to a balance between curing of the concrete, and development of sufficient strength to hold the forms, while not being so slow as to

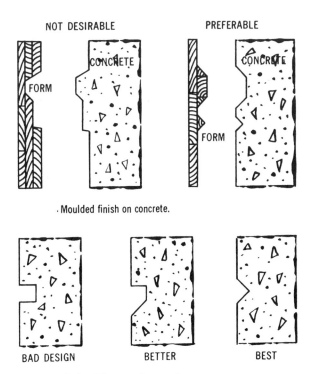

NOT DESIRABLE   PREFERABLE

. Moulded finish on concrete.

BAD DESIGN   BETTER   BEST

Horizontal grooves in concrete.

Fig. 6 (continued)

preclude proper bonding of the new concrete as it is placed on top of the prior material.

Slip forming requires special precautions for structural safety, as well as the usual precautions required for high work.

The AAI Standard 347-68 "Recommended Practice for Concrete Formwork" lists a number of *common problems* encountered in concrete formwork which impacts the quality of the ultimate concrete mixture, including:

1. Diagonal bracing of shores not adequate.

2. Soil under mud sills unstable (resulting in boiling out of concrete).

3. Insufficient connection of formwork, including insufficient nailing.

4. Failure to provide for lateral pressures imposed by the plastic concrete.

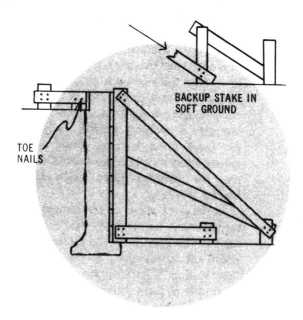

Cellar wall form in firm ground.

Fig. 6 (*continued*)

5. Shoring out of plumb inducing lateral loading.

6. Locking devices on metal shoring not properly installed or missing.

7. Vibration from moving loads or load carriers.

8. Failure to adequately tighten or secure the forms ties or wedges.

9. Loosening of re-shores from floors below.

10. Premature removal of supports, especially under structural sections.

11. Improper positioning of forming from floor to floor, imposing loads not anticipated.

12. Inadequate support, allowing rotation of beam forms, particularly where slabs frame into them only on one side.

13. Inadequate anchorage against uplift, particularly in trough or trench forms.

Fig. 7 Re-use of gang forms. *Source: Manitowoc Engineering Co., Division of Manitowoc Company.*

CONCRETE FORMWORK CHECKLIST

1. Check location of forms.
2. Check interior condition of forms for soundness and proper surface finish.
3. Check form supports for conformance with form drawings and basic soundness.

Fig. 8 Re-use of prefabricated forms lifted into place. *Source: Manitowoc Engineering Co., Division of the Manitowoc Co.*

4. Check shoring and bracing in accordance with the shoring check list.

5. Form work is often made up of reused lumber; check for grade and soundness.

6. Forms should be wet down prior to use to swell them.

7. Forms should be cleaned out before pour.

8. Where form ties are used, check installation and number.

9. Metal forms should be oiled rather than wetted, but surplus oil should not be on formwork.

10. Steel reinforcement to be placed within forms in accordance with rebar drawings. Count number and check sizes.

11. Reinforcing bars should be supported by chairs or, if permitted by specification, concrete blocks to hold them the proper distance away from the outside face of concrete. Where the final surface will be exposed, specifications usually call for plastic, galvanized tipped, or stainless steel chairs or accessories to avoid rust marks.

03200 CONCRETE REINFORCEMENT

The following information on reinforcing steel for reinforced concrete construction is from the very complete book prepared by the Concrete Reinforcing Steel Institute *Placing Reinforcing Bars*.

Reinforcing bars are fabricated from various *grades* of steel according to the grades specified by the American Society for Testing and Materials (ASTM) as follows:

A615—Billet Steel Bars, grades 40, 60 and 75
A616—Rail Steel Bars, grades 50 and 60
A617—Axle Steel Bars, grades 40 and 60

The grades are equivalent to their respective yield strength in thousands of pounds per square inch (i.e., grade 40 is yield strength 40,000 psi, grade 60 is 60,000 psi yield strength.)

In order to improve the bond between the concrete and the reinforcing steel, the bars have *deformations* rolled into them by the mill. Plain bars

ASTM STANDARD REINFORCING BARS				
BAR SIZE	WEIGHT	NOMINAL DIMENSIONS—ROUND SECTIONS		
DESIGNATION	POUNDS PER FOOT	DIAMETER INCHES	CROSS SECTIONAL AREA SQ. INCHES	PERIMETER INCHES
#3	.376	.375	.11	1.178
#4	.668	.500	.20	1.571
#5	1.043	.625	.31	1.963
#6	1.502	.750	.44	2.356
#7	2.044	.875	.60	2.749
#8	2.670	1.000	.79	3.142
#9	3.400	1.128	1.00	3.544
#10	4.303	1.270	1.27	3.990
#11	5.313	1.410	1.56	4.430
#14	7.65	1.693	2.25	5.32
#18	13.60	2.257	4.00	7.09

Fig. 9 Standard rebar sizes. *Source: Concrete Reinforcing Steel Institute.*

used for special purposes are not considered part of the reinforcing steel, but may be used as dowels at expansion joints or jacking bars for slip forms.

Reinforcing bars are designated by number, with the number approximately being the diameter of the bar in eighths of inches; for instance, a #4 bar has a $\frac{1}{2}''$ diameter. Figure 9 provides a table of the standard reinforcing bar dimensions. The bar grade and size is marked as part of the rolling process right on the bar itself.

Generally, the specifications call for the contractor to arrange for *shop drawings* to describe specifically the reinforcing bar arrangements, sizes, and placement. These shop drawings also include bar lists which are used by the fabricator in the shop to fabricate the bars and to identify them for loading and unloading. The bar lists are also used to group work for the fabricating shop. The fabricator also prepares placement drawings which are used by the ironworkers in the field to place the rebars. The shop drawings should be approved by the structural engineer, and can be used by the inspection team to inspect the location of the reinforcing bars after placement, both for location and size, as well as proper material.

The *fabrication* of reinforcing bar is selection of the right material, cutting to specified lengths from longer stock in the shop, and then bending as appropriate. Small radius bends may be made with bending machines either cold or hot. The designer usually specifies typical bends as standardized by the American Concrete Institute, since these can be made using the standard bending machines. Special shapes may be bent using more general tools. Bends of larger radius may be made in the field. In certain instances, there are union regulations which specify all bending to be field bending. The prevailing bending regulations have an effect upon the type of steel which can be used; for instance, some of the high-yield-strength steels are less amenable to handling in the field, and this type of consideration should be included in the design conditions.

In spirally-reinforced columns, piers, and piles a reinforcing cage similar to a coil spring is made. The prefabrication work is more exacting, and these are usually prepared in the shop but shipped flat or collapsed. In the field, the spirals are opened and spacers added according to the shop drawings.

The bars, when delivered to the job site, are usually marked in *bundles.* The bundles are marked for similarity of parts, and often also indicate a column or beam location or section of the building for convenience.

However, bundle tags may be lost or mismarked, so the inspection team must depend on actual measurement. Reinforcing may be delivered to the job site and stored by size where job storage space permits. In other cases, the loads are oriented toward the actual pouring schedule, and the steel is off-loaded in an area adjacent to the area of use. The major impact of the two systems on the inspection team is that the direct shipment offers less notice to the inspection team.

Reinforcing should be lifted from trucks, rather than dropped off. It is permissible, however, to skid the bars using rails or timbers. Dropping reinforcing bars may cause cracking or bending.

Rust and scale, if tight, do not adversely affect the reinforcing capabilities. In fact, they improve the bond between the concrete and the surface of the bar. However, mud, dirt, or oil are detrimental materials, and must be cleaned from rebar before use. Accordingly, the storage area should be either free of mud and deleterious materials, or skids should be provided for storage so that the rebar remains clean. In hoisting the rebar, use slings rather than the wire wrappings used to bundle the bars into groups. The optimum lifting points are at the quarter points for long bars.

Figure 10 shows a typical *placement drawing*. This is a first-floor framing plan for concrete joist construction. The exterior solid and parallel dash lines on the plans show the floor bearing on a wall. Supporting columns are indicated by small rectangles, while beams are shown by dashed parallel lines from column to column. Beams are referenced by number, with a prefix 1B for first floor beams, followed by a serial number. Joist forms are shown in solid lines, which is standard practice. Joists have reference numbers 1J, followed by a serial number. Metal forms are shown tapered near the supports, although they may be square. Holes and openings are indicated by rectangles of heavy lines with light cross diagonal lines to indicate that the space is empty. Midway between the tapered ends of the forms in the larger two spans are double horizontal lines indicating a distribution rib.

The bars to be placed are found on the beam and joist schedule shown in Figs. 11 and 12. In some instances, the bar placing schedule may be shown right on the placement drawing, but usually there is not enough room. The placing drawing always shows typical details and sections to properly locate the bars as shown in Fig. 13.

Bars which cannot readily be scheduled, such as temperature, distribution, rib, and slab bars are shown directly on the plan.

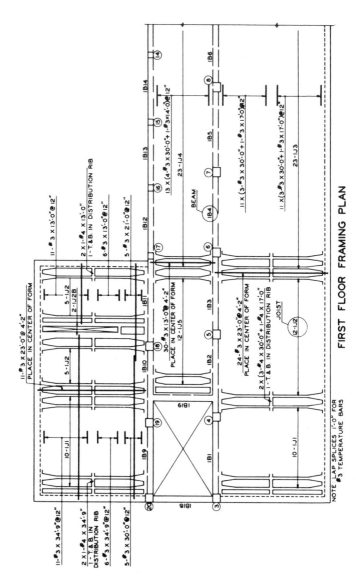

FIRST FLOOR FRAMING PLAN

Fig. 10 Typical rebar placement drawing. *Source: Concrete Reinforcing Steel Institute.*

BEAM SCHEDULE

MARK NO	B	D	Straight No	Straight Size	Straight Lgth	Bent No	Bent Size	Bent Lgth	Bent Mark	Bent Type	Hook	M1	M2	M3	M4	M5	H	O	Stir No	Stir Size	Lgth Mark	Dim	Stir Type	Spacing Each End	Sup No	Sup Size	Sup Lgth	Chairs No	Chairs Type
1B1	12	25	2	#6	16-8	2	#6	24-5	18600	∿	11	2-2	2-5½	9-9	2-5½	6-8	1-9		13	#4	5-10 U400	9/22	⊔	4,2@6,2@10,12 Col-3 / 5,2@8,2@10,12,15 Col-4	2	#4	18-0	4	BB
1B2	12	25	2	#7	16-7	2	#7	27-9	18700	∿		7-6	2-5½	7-10	2-5½	7-6	1-9		7 #4 / 6 #4	#4	5-10 U400 / 5-1 U401	9/22 9/22	⊔	6,2@8,2@10,12,15 Col-4 / 6,2@8,10,12,15 Col-5	2	#4	15-7	4	BB
1B3	12	23	2	#6	16-6	2	#6	27-8	18601	∿		7-8	2-3	7-10	2-3	7-8	1-7		16	#3	4-7 U301	9/20	⊔	4,3@6,2@8,10,12	2	#4	15-6	4	BB
1B4	12	23	2	#6	16-1	2	#6	27-5	18602	∿		7-8	2-3	7-7	2-3	7-8	1-7		16	#3	4-7 U301	9/20	⊔	Do	2	#4	15-1	4	BB
1B5	12	23	2	#6	17-0	2	#6	28-0	18603	∿		7-9	2-3	8-0	2-3	7-9	1-7		16	#3	4-7 U301	9/20	⊔	Do	2	#4	16-0	4	BB
1B6	12	23	2	#7	16-6	2	#7	26-10	18701	∿		7-9	2-3	7-10	2-3	6-9	1-7		16	#3	4-7 U301	9/20	⊔	Do	2	#4	15-6	4	BB
1B7	12	23	2 / 2	#6 / #7	13-9 / 27-1	Bot In Top Extend 7-8					Into beam 18-8								10	#3	4-7 U301	9/20	⊔	5,2@7,10,12				4	BB
*1B8	12	33	3	#8	30-8	3	#8	38-10	18800	∿		8-4	3-2	19-6	3-2	3-5	1-3	2-3	20	#4	6-5 U402	9/30	⊔	6,3@8,2@12,2@15,2@18	2	#4	21-9	4/7	BBU1/BB
1B9	12	25	2	#6	15-4	2	#6	22-10	18604	∿	11	2-1	2-5½	8-9	2-5½	6-2	1-9		12	#4	5-10 U400	9/22	⊔	5,2@7,2@10,13	2	#4	16-7	4	BB
1B10	12	25	2	#7	15-3	2	#7	26-6	18702	∿		6-11	2-5½	7-1	2-5½	7-7	1-7		6 #4 / 5 #4	#4	5-10 U400 / 5-1 U401	9/22 9/22	⊔	5,2@7,2@10,13 Col-19 / 5,2@7,10,13 Col-18	2	#4	14-3	3/5	BBU1/BB
*1B11	12	25	3	#6	18-3	3	#6	28-2	18605	∿		7-9	2-3	8-8	2-3	7-3	1-7		14	#4	5-1 U401	9/22	⊔	4,2@6,2@10,2@14	2	#4	17-3	4	BB
1B12	12	20	2	#6	14-0	1	#6	23-11	18606	∿		7-6	1-10¾	6-6	1-10¾	6-2	1-4		10	#3	4-10 U302	9/17	⊔	4,2@6,2@10	2	#4	15-3	4	BB
1B13	12	20	2	#6	13-7	1	#6	22-4	18607	∿		6-2	1-10½	6-3	1-10½	6-2	1-4		10	#3	4-10 U302	9/17	⊔	4,2@6,2@10	2	#4	14-6	4	BB
1B14	12	20	2	#6	13-8	1	#6	25-3	18608	∿		6-2	1-10½	6-4	1-10½	9-0	1-4		10	#3	4-10 U302	9/17	⊔	4,2@6,2@10	2	#4	15-0	4	BB
*1B15	12	33	3	#8	23-9	2	#8	38-7	18801	∿		8-3	2-11½	11-5	2-11½	13-0	2-1		19	#4	6-5 U402	9/30	⊔	3@4,3@6,8,3@12,4@15 Col-13 / 6,2@10,14,18 Col-14	2	#4	22-9	6/4	BB/BBU1
*1B16	12	33	3	#8	30-9	3	#8	41-8	18802	∿		11-3	3-2	19-7	3-2	3-3	1-3	2-3	12	#4	6-5 U402	9/30	⊔	6,2@10,14,2@18	2	#4	21-9	4/7	BBU1/BB
1B17	12	25	2	#8	23-7	2	#8	28-6	18803	∿	1-2¾	2-9	2-4	15-11	2-4	2-9	12½	24-9	24	#3	5-8 U300	9/22	⊔	4,8@6,2@10,12	2	#4	22-7	6	BB
1B18	12	20	2	#6	13-9	2	#6	17-7	18609	∿	10½	2-0	1-10½	8-1	1-10½	2-0	10½	14-9	4	#3	4-10 U302	9/17	⊔	4,8	2	#4	12-9	6	BB
1B19	12	18	3	#5	13-9	3	#5	16-9	18500	∿	9	1-9	1-8	8-5	1-8	1-9	9	14-3	4	#3	4-6 U304	9/15	⊔	4,8	2	#4	14-9	4	BB

* Place reinforcing steel in 2 layers
All dimensions on bending details are out to out of bar

Fig. 11 Typical beam schedule. Source: *Concrete Reinforcing Steel Institute.*

JOIST SCHEDULE

MARK	No	B	D	No	Size	LGTH	No	Size	LGTH	MARK	TYPE	HOOK						HOOK	H	O	3/4" CHAIRS
1J1	20	5	8+2½	1	#7	24·0	1	#6	26·10	1J600		8	3·7	1·0½	16·6	1·0½	3·4	8	9	24·11	6
1J2	58	5	‖	1	#6	24·0	1	#6	45·7	1J603		8	3·7	1·0½	15·2	1·0½	24·1		9		6
1J2A	2	Var		2	#6	24·0	2	#6	32·2	1J607		8	3·7	1·0½	15·2	1·0½	10·8		9		12
1J2B	2	Var		2	#6	24·0	2	#6	45·7	1J603		8	3·7	1·0½	15·2	1·0½	24·1		9		12
1J2C	1	5		1	#6	20·2	1	#6	22·8	1J606		8	2·6	1·0½	13·8	1·0½	3·1	8	9	20·9	5
1J3	26	5		1	#6	24·0	1	#6	32·2	1J607		8	3·7	1·0½	15·2	1·0½	10·8		9		6
1J4	23	5		1	#6	13·9	1	#6	21·4	1J604		8·4	1·0½	8·4	1·0½	1·11		8	9		4
1J4A	1	5		1	#6	13·9	1	#6	16·4	1J605		8	1·11	1·0½	9·1	1·0½	2·7		9		4
1J4B	5	5	‖	1	#6	13·9	1	#6	21·4	1J601		8·4	1·0½	8·4	1·0½	2·7		9		4
1J5	31	5	8+2½	1	#5	13·9															4

Typical Joist Schedule

Fig. 12 Typical joist schedule. *Source: Concrete Reinforcing Steel Institute.*

Temperature bars are written between two horizontal lines with arrows to show the extent covered. The notation 13 × (4 − #3 × 30′ − 0 + 1 − #3 × 14′ − 0) @12 means 13 lines each consisting of four pieces of #3 bars 30′ long and one piece 14′ long. A note calls for these bars to lap one foot past each other. The distribution rib bars are shown in much the same manner.

The typical beam section shows what would be seen if the concrete beam were cut through at the end, with the details showing the amount of concrete covering, and the stirrup arrangement of the stirrup support bars. Section X-X shows a side view of a typical joist, indicating how far the bottom bars and truss bars are to extend into the wall. Section Y-Y is

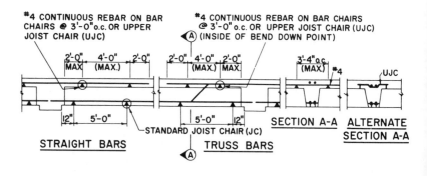

JOISTS (ONE-WAY)

Fig. 13 Typical reinforcing placement section. *Source: Concrete Reinforcing Steel Institute.*

a cut across the exterior wall and several joists. This shows the position of straight and truss bars in the joists, the vertical dotted line being the bent-up portion of the truss bar. The section also shows the clearance under the bars and the location of temperature and tie bars.

The detail puts as much information as possible on the plans, and in the schedules—but conditions will occur which can't be shown in this manner and sectional details are used for clarification in the field.

The *beam, joist,* and *column schedules* combined with the placement drawings provide the information needed for the inspection team. The bar list generally is for the convenience of the fabricator, and would be

Standard Types and Sizes of Wire Bar Supports

Symbol	Bar Support Illustration	Type of Support	Standard Sizes
SB	5"	Slab Bolster	¾, 1, 1½, and 2 inch heights in 5 ft. and 10 ft. lengths
SBR*	5"	Slab Bolster with Runners	Same as SB
BB	2½" 2½" 2½"	Beam Bolster	1", 1½", 2"; over 2" to 5" heights in increments of ¼"; in lengths of 5 ft.
UBB*	2½" 2½" 2½"	Upper Beam Bolster	Same as BB
BC		Individual Bar Chair	¾, 1, 1½, and 1¾" heights
JC		Joist Chair	4, 5, and 6 inch widths and ¾, 1, and 1½ inch heights
HC		Individual High Chair	2 to 15 inch heights in increments of ¼ in
CHC		Continuous High Chair	Same as HC in 5 and 10 foot lengths
UCHC*	8"	Upper Continuous High Chair	Same as CHC

*Available in Class A only except on special order

Fig. 14 Typical types and sizes of wire bar supports. *Source: Concrete Reinforcing Steel Institute.*

principally used to check quantities for unloading, which is really the contractor's responsibility. The shop drawings checked by the structural engineer would be the placement and beam column and joist schedules.

In placing the reinforcing steel, a number of *accessories* are used to support the bars at a proper height away from the finished surface. Since the slab face represents the finished surface, these supports or chairs sit right on the form. In order to protect against rusting, the entire chair is sometimes galvanized or plastic-protected, or in some cases, only those feet touching the form are protected. For special uses, the chair may even be stainless steel, so that the surface can be finished without destroying the rust protection characteristics.

Figure 14 shows typical types and sizes of wire bar supports, while Fig. 15 shows a typical placement plan in section for a one-way slab.

In footings, most specifications require the placement of *chairs* or blocks, such as pre-cast concrete block, under the reinforcing steel to accurately place it at the proper level. Figure 16 shows some of these typical chairs.

The contractor may prefer special support approaches, which can be approved by the inspection team if they do not introduce conditions which could cause the deterioration or corrosion of the reinforcing material. Often, the contractor will tack weld the reinforcing mat to

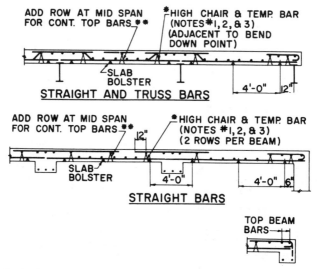

Fig. 15 Typical placement plan for supports. *Source: Concrete Reinforcing Steel Institute.*

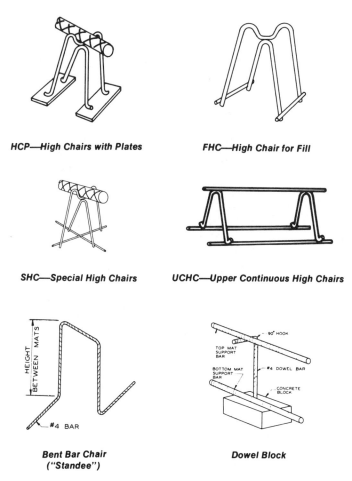

HCP—High Chairs with Plates   FHC—High Chair for Fill

SHC—Special High Chairs   UCHC—Upper Continuous High Chairs

Bent Bar Chair
("Standee")   Dowel Block

Fig. 16 Typical rebar support chairs. *Source: Concrete Reinforcing Steel Institute.*

secure it. In some cases, special supports will be fabricated in the field, and used to support the steel (Fig. 17).

The accurate location of reinforcing bars within the concrete structure is of vital concern. Insufficient coverage can cause corrosion, while placement too far into the shape may cause it to fail.

Where the minimum level of *concrete protection* is not indicated, the following are suggested:

1. Where concrete is cast against earth as in the bottom of footings, 3″.

Fig. 17 Use of special chairs and clips for installation of rebar in slab without tying. *Source: Bethlehem Steel Corp.*

2. Where bar size is larger than number 5, at least 2″ coverage.

3. For beams, girders, and columns, 1-½″ coverage.

4. For the top, bottom and side of joists, ¾″, as well as on the top and bottom of slabs where the surfaces are not exposed directly to ground or weather.

5. The concrete coverage should never be less than a bar diameter where slabs and joists are not exposed to ground or weather, and 2 diameters where they are.

It is important that the placement be accurate; for instance, the raising of the bottom bars or lowering of the top bars in a 6″ slab by as much as ½″ would reduce the load carrying capacity by 20%. The usual tolerance accepted (and assumed in the design) is plus or minus ¼″. The usual important tolerance is in the cross section. The positioning lengthwise of the bars is not critical, but the overlap of bar splices is. Usually, the *splice* is specified, and should be 1' or 20 bar diameters, whichever is larger. In continuous beams, the bottom bars may be specified to extend 6" into supporting columns, and other special splice instructions are usually specified. The location of bends has a tolerance of about 2".

In long runs of slabs or walls, the exact spacing of bars is not critical. For instance, if a drawing calls for 11 bars spaced 6" center to center (10 spaces at 6") the bars should be spread over the 60" space at about 6" spacing. If necessary, a few of the bars could be shifted to avoid an insert or a small opening. However, no space between the bars should be greater than 3 times the thickness of the slab or wall. Bars should not be cut to clear obstructions unless approved by the structural engineer.

Stirrups are placed at right angles to beam reinforcing, and should be placed within a tolerance of about 1".

Bars can be spliced by overlapping, but in some cases there is not enough space to permit double or lapped bars, because there would not be enough room for the concrete to pass through. Also, lap splicing of #14 and #18 bars is not permitted.

Long splices are made by wiring or tying the bars together. Shorter splices can be made by welding—either lap or butt. Welded lap splices are recommended only for bars smaller than #5, and can be accomplished using either spliced bars, a direct lap, or back-up plate.

In precast concrete, *welded splices* are usually necessary. Butt welding requires a special process, such as the Thermit process, where the bar ends are placed in a mold with a gap between, and new material is actually fused to the end of each bar. This process is usually used on large bars, and requires space for the removable jigs to be clamped to the bars. Another method of end coupling is through mechanical devices.

Tie wires are usually #16 gauge annealed wire which are carried in special reels on the ironworker's belt for easy accessibility and use. In wall reinforcement, *snap ties* can be used to secure the form work, and can be used to secure reinforcing bars in position against displacement.

The CRSI cautions in regard to welding of cross reinforcement that the localized heating and arching effect of the spot or tack welding can cause a notching effect which reduces the strength of the material as much as 30–35%.

In tying steel, it is not necessary to tie the bars at every intersection. The tying adds nothing to the strength of the finished structure, and is only for the purpose of holding the reinforcing steel in the proper location. Ties at the fourth or fifth intersection of a mat is usually sufficient to hold in place.

When dowels or reinforcing bars used as dowels are placed in foundations, some adjustment of position is permitted. *Bending* should be with a pipe or hickey, not by sledge or impact blows. A bend of 20° or 3″ per foot with the vertical is about the maximum acceptable. In order to get enough room, it may be necessary to dig or chip out some of the concrete. Often, the design calls for dowels or anchor bolts to be placed within a *sleeve* to permit some movement for final adjustment. These sleeves are then to be grouted full when the equipment or structural steel is set.

A standard sequence for placement of reinforcing steel to suit every condition cannot be anticipated. The sequence should be planned to avoid the necessity for threading bars in and around previously placed bars at intersections. While threading cannot always be avoided, it should be minimized by careful planning. Since girders are deeper than the beams they support, bottom lengthwise bars, truss bars, and stirrups should be placed before beam reinforcement. In some cases, beam reinforcement can be preassembled into a cage and lowered into the beam, and similar techniques are usually used for column reinforcement.

Joist bars are usually placed after beam reinforcement is in place. Form ties, anchors, inserts, and other items are usually placed after the bars have been placed to provide maximum flexibility, in the instance where bars already placed have to be moved or relocated. Distribution ribs, also known as header joists, extend continuously the full length of the joist bay at right angles to the joists, and are used for longer joist spans to brace the joists laterally. There is no fixed rule in regard to the number to be used, but usually one is used at the mid-spans between 18′ and 24′, and at third points (2 ribs) for longer spans.

Temperature reinforcement, being the lightest and on top, is placed after the joist bars and all heavy reinforcement have been placed in the slab.

In inspecting reinforcing steel, the inspection team should first be certain that approved placement drawings are available. Reinforcing bars should be checked for conformance with specification in terms of grade, size, and type.

When placed, the proper amount of spacing for concrete cover should be checked, as well as proper placement of bars in terms of numbers, size, and configuration. The laps and splices should be checked for length and strength. The stability of the reinforcing should be checked prior to placement of concrete, so that the material will not move around when encountering the fluid pressures of the concrete, vibrating equipment, and the workers placing the material.

03300 CAST-IN-PLACE CONCRETE

Concrete is made up of basically two components: aggregates and cement. The *cement* reacts chemically with water to form the binder to tie the aggregates together. With the right mixture of aggregate and cement, an optimum mixture can be designed at which the best strength at the lowest cost can be developed. *Aggregates* fall into two classes: fine and coarse. Fine aggregates are similar to sand, while coarse are larger and reduce the volume of cement and fine aggregate required to complete the mix. Cement paste constitutes 25 to 40% of the usual volume of concrete. With aggregate making up 60 to 80% of the typical concrete mix, the gradation of aggregate size is important. Too much fine aggregate will require too much paste to coat it—too much coarse will similarly require too much paste to fill the voids.

The cementing action is due to a chemical reaction between the cement and water known as *hydration*. The ratio between the water used and the cement is a critical one. More water must be used than is theoretically necessary for hydration in order to insure total availability of moisture, as well as (most importantly to the man in the field) to make the mixture workable enough to handle. However, over-utilization of water to make the concrete mixture more fluid for convenience in the field can drastically lower its ultimate strength.

The ability of concrete to resist alternate cycles of freezing and thawing is also related to the *water/cement ratio*. When the concrete mixture in its plastic form is oversaturated with moisture, the aggregate absorbs some of this moisture, and during the freeze-thaw cycles, displaces the water into the surrounding cement paste. Where severe freeze-thaw cycles are anticipated, air-entrained concrete must be used. This type of concrete uses either an admixture or a special cement to create a homogeneous mixture of fine air bubbles within the cement paste, which continue to exist as fine voids after the hydration. These voids permit expansion space, and also serve to make the concrete lighter. The air entrained into the concrete mixture makes it more workable, so a lower water-cement ratio can be used to place the concrete.

Other special features can be achieved through the use of admixtures,

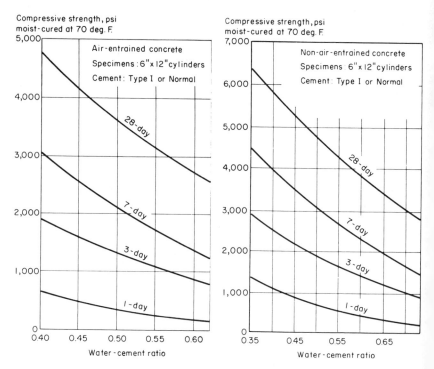

Fig. 18 Strength versus water-cement ratio. *Source: Portland Cement Association.*

although one of the most usual objectives is improved *workability* to permit a lower water-cement ratio.

The hydration reaction is related to the environment—most particularly, moisture and temperature. Figure 18 illustrates the strength of concrete, both air-entrained and standard, under favorable laboratory conditions in terms of curing. The strengths shown are also related to the water-cement ratio. In Fig. 19, the effect of moisture is shown. Air curing robs the cement of moisture required for hydration, and slows down the process. However, Fig. 20 shows the temperature/curing relationship, which is dramatic.

Cement is a manufactured product made up of portions of lime, silica, alumina, and iron components. The raw materials are mixed in the proper propotions, pulverized, then heated in a kiln, at temperatures up to 3000°F. The resulting klinker is cooled and pulverized and a small amount of gypsum is added.

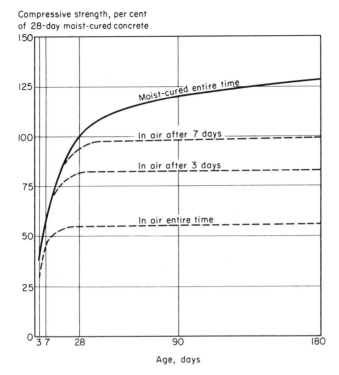

Fig. 19 Effect of moisture on concrete strength. *Source: Portland Cement Association.*

There are five types of *portland cement* classifications by the ASTM, with strength characteristics demonstrated in Fig. 21. The concrete would generally be specified by the structural designer in terms of cement type, water cement ratio, slump, and strength anticipated at 28 days. It is the responsibility of the contractor, usually, to actually order the mix. He may be given flexibility and the opportunity, where appropriate, to use special admixtures or air-entraining cement for workability, or high early-strength cement in order to—for his convenience—strip forms more rapidly, and reduce shoring and bracing sooner.

On major heavy construction projects, cement may be delivered to the site by rail or truck with the dry mixture dispensing aggregate and cement occurring from bins, called batch bins. Usually, however, the batching process occurs at a concrete supplier's facility, where a permanent installation is available. Figure 22 shows a mobile batch plant.

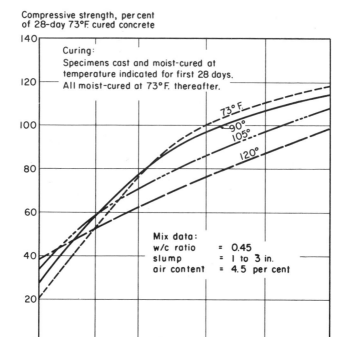

Compressive strength, per cent
of 28-day 73°F. cured concrete

Fig. 20 Effect of curing temperature on strength. *Source: Portland Cement Association.*

APPROXIMATE RELATIVE STRENGTH OF CONCRETE
AS AFFECTED BY TYPE OF CEMENT

Type of Portland Cement		Compressive Strength, per cent of Strength of Type I or Normal Portland Cement Concrete			
ASTM	*CSA*	*1 day*	*7 days*	*28 days*	*3 mos.*
I	Normal	100	100	100	100
II		75	85	90	100
III	High-early-strength	190	120	110	100
IV		55	55	75	100
V	Sulfate-resisting	65	75	85	100

Fig. 21 Types of Portland cement. *Source: Portland Cement Association.*

Fig. 22 Mobile batch plant. *Source: MCP.*

Where the concrete is delivered in bulk to a job site, its storage and the testing of the cement is in the area of responsibility of the inspection team. Usually, the specifications (or the experienced inspection team) direct the contractor to furnish laboratory controlled ready-mixed concrete. In these instances, a certified laboratory conducts the tests at an economical cost.

A crucial role remains for the site inspector. First, the contractor often attempts to add water at the job site in the mixer truck to increase workability. This increases the water-cement ratio, resulting in lower quality concrete. Many contractors are not aware of the real effect of the WC ratio and must be constantly watched or convenience will be traded for strength—possibly with disastrous results.

Another factor which the inspector must check on is the length of time between start of mixing and final placing of concrete. This should not exceed a maximum time period established in ASTM C94. If the time is exceeded, hydration will have proceeded too far and initial set of the mix will have started.

Ready-mix trucks are equipped with water-measuring instrumentation to add accurately more water at the job site, if the maximum water of design was not added at the batching plant. This gives the field inspection force an additional responsibility to monitor the water added.

In *cold weather* concreting, it is often necessary to heat the water, and certain admixtures can be used to improve the placement in cold weather. In ultra-cold weather, even the aggregate may have to be heated before the mixing process—and then protection of the concrete after placement must include covering and introduction of additional heat. Admixtures added as antifreeze agents to lower the freezing point of concrete should not be permitted, since such practices will ruin the strength-producing properties of the mix. However, certain accelerators such as calcium chloride (maximum of 2% by weight of cement) are used legitimately to speed the hardening of the concrete. However, this adds to potential corrosion, and should not be used where aluminum inserts or prestressed concrete are being utilized. Proper curing and frost protection procedures can obviate the need for accelerators. A better approach in cold weather is the use of a high early-strength concrete, which then reduces the amount of time that special care and curing must be given to the concrete by as much as 50% of the curing time.

The *handling* of concrete is critical in any weather. Adding of water to improve the workability of concrete reduces its strength. Dropping the concrete any distance over 3 to 4 feet may lead to separation of the course particles or aggregate. Dropping through reinforcing bars can cause separation, and can be avoided by the use of a drop chute or elephant trunk to contain and preclude free fall of the material (Fig. 23).

Before concrete is placed on grade, the subgrade should be moistened to prevent extraction of the water from the concrete. Such extraction reduces the water-cement ratio, but also provides nonhomogeneous concrete, and may shorten the time available to finish the concrete. It may also cause surface crazing.

The concrete should be placed as near to its final position as practical. It can be moved horizontally a small distance by vibrating, but this should not be overutilized, since this type of movement tends to separate the mixture forcing water and mortar ahead of the aggregate.

In tall narrow forms, such as columns, concrete may be placed through openings or windows. These windows should be of sufficient size to permit a reasonable quantity of concrete to be placed through them. (Fig. 24).

Fig. 23 Placing beam concrete, high elevation, with crane bucket. *Source: Manitowoc Engineering, Division of the Manitowoc Co.*

Concrete should not be placed into water normally, but in certain heavy construction situations, *underwater placement* is required. This requires special techniques and should be approved by the structural engineer. The general method is to place the concrete mixture through a tremie (pipe or tube with a hopper on top) to its final placement and to use the heavier-than-water weight of concrete to displace water. Simply dropping the concrete through water will destroy its usefulness.

Placement of concrete with belt *conveyors* and buggies is essentially a mechanical transfer, and does not require change in the mix. Concrete to be pumped must contain a smaller maximum size aggregate in some cases. Concrete can be pumped up to 1000 feet horizontally, and 300 feet

Fig. 24 Use of crane bucket, hopper and chute to place column concrete in forming tube. *Source: Sonotube Products Co.*

vertically, and even further under special conditions. Other special methods of placing concrete include the pneumatic application, described in Section 03362 (Figs. 25, 26).

One of the major methods of consolidating the concrete in its final position is the use of the *vibrator,* which is usually of the immersion type. These vibrate at speeds between 3600 and 13,000 vibrations per minute, and are available in sizes between 1″ and 7″ in diameter. The vibrator should be lowered vertically into the concrete, and should not be used

Fig. 25 Placement of concrete with conveyor, and tremie trunk in wall forms. *Source: Symons Corp.*

against the formwork. When cement paste begins to accumulate around the vibrator—within 5 to 15 seconds—the vibrator should be removed. Concrete should not be over-vibrated.

Older methods of consolidating concrete included the use of spades or puddling sticks and other tampers. These methods require a more fluid mix, and therefore, a higher water-cement ratio. Wherever possible, vibrators should be used, even where portable generating equipment is required to provide electricity or compressed air to operate them.

Well-placed concrete is finished by first striking off to grade with either a hand *screed,* or a vibrating screed. In this process, the excess concrete is

Fig. 26 Placement of concrete with special crane-supported conveyor. *Source: Manitowoc Engineering Co., Division of Manitowoc Co.*

struck off, and in well-vibrated concrete, this will have a high percentage of paste, and will be readily workable. The hand screed is moved across the concrete with a sawing motion, and then *wood* floats are used immediately after screeding to smooth off the high and low spots. (In screeding, pieces of aggregate sometimes leave holes or voids which are replaced by cement paste during the floating process.) After the initial floating, the mix should be allowed to hydrate sufficiently so that a man's weight is supported with only a slight imprint. The surface is then floated or troweled by hand or machine. This process drives aggregate particles located just below the surface to a deeper position, removes slight imperfections and compacts the concrete at the surface into a more durable, dense working surface.

The best time for floating and *troweling* is established by experience. Too long a delay results in an unworkable surface. When this occurs, the inspection team should not permit the spreading of additional mortar, or water if at all possible. Premature floating or troweling is more common, and can be countered with patience. Dry cement should not be spread on the surface to permit earlier floating or troweling. This will lead to dusting, scaling, or crazing. Hard steel troweling to a smooth, shiny surface is less popular than it once was because of safety considerations. Certain floors, however, must be troweled for good wear resistance. If the surface is primarily for walking purposes, it is often deliberately scored with a broom. Brooming usually follows floating, however, after troweling, the surface may accept a fine broom scoring.

The inspection team should have carefully reviewed the specifications to determine the requirement for special surfaces such as exposed aggregate or special textures for aesthetic purposes.

Forms should not be removed too soon, since early removal tends to dry out the "green" concrete, and the form is part of the support structure for concrete. However, in non-load-bearing concrete or surfaces which must be rubbed, concrete forms are often removed in 12 to 24 hours.

As noted, concrete placed in cold weather requires special precautions for *curing,* particularly when temperatures are below 40 degrees. The concrete must not be allowed to experience freeze and thaw cycles which can destroy the hydration process. Some instances of cold-weather concrete in Alaska and Canada have resulted in frozen concrete which has collapsed with the spring thaw. Conversely, excellent concrete has been placed in below zero weather, but only through the use of warm enclosures, heating, and wind protection. The concrete does, itself, have a heat of hydration which adds to the supplementary heat, but below 40 degrees, this heat of hydration is not sufficient to overcome a cold environment. A good system of insulation over recently-placed concrete can capture and retain the heat of hydration until the concrete has gained sufficient strength. In cold weather, concreting forms should be left in place as long as possible.

The principal problem in *warm weather placement* is loss of moisture and accelerated setting time. This can be offset by admixtures used to retard the time of set, and it may be necessary to cool the ingredients before mixing. Again, the water tends to function as the best and most economical temperature control agent. In some major dams, it has been necessary to include a circulating, cooling system imbedded in the mass concrete, since hydration extends over a long period of time.

Some specifications call for the addition of ice to ready mix to retard flash setting due to high temperature. The water equivalent of the ice is part of the W/C ratio and reduces strength.

In hot weather, forms should be loosened as soon as possible, and water and/or curing compounds applied to exposed concrete surfaces as soon as possible to replace water lost to evaporation, and to preclude evaporation. In hot weather, it is preferable to use a moist curing process (applying water by any appropriate means) for 24 hours, and then apply a curing compound which will slow down evaporation as hydration continues. The materials to be used for curing should be approved by the structural engineer, since certain paraffin-based or plastic-based compounds may react with the concrete, causing discoloration, or even precluding the ultimate use of the concrete—for instance, resisting mastic cement for tile floors.

The inspection team should have available to it the appropriate references from the following:

American Concrete Institute (ACI)
American Society for Testing and Materials (ASTM)
Portland Cement Association (PCA)

Each of these organizations periodically update their standards and *references,* so that a specific list should be obtained from the structural designer. Using this list, the field office should establish a proper library of appropriate standards specified in the contract documents. Since mistakes in concrete are so permanent, the inspection team should have the specific authority to prohibit the start of concrete work until all preparation is satisfactory, and the authority to stop work which is not being done in accordance with the plans of specifications.

As a control on the strength of concrete placed, the inspection team either supervises or actually makes *test cylinders.* Usually, several cylinders are taken for each pour and are tested at the 7 and 28-day interval following placing. In some cases, *cylinders* are divided between laboratory-cured and field-cured, so that the test represents mix control, as well as job environment.

Prior to placing a batch of concrete, the field team has the option of making *slump tests* or penetration tests. Slump tests are the most common, since they require only a test cone, although the test is very subject to variations in technique. A more exact test recently introduced is the penetration test which involves a 6″, 30 lb metal ball in a test described under ASTM C360. In this case, the penetration of the ball in a freshly leveled off concrete surface (using a tub or wheelbarrow) is about half of the usual slump.

Where air entrainment is required, a field test for air content should be carried out using air testing equipment and procedures described in ASTM C231 (ASTM C173 for lightweight concretes).

Where subsequent lifts of concrete are to be placed on top of an earlier placement, this should be done only with the approval of the structural engineer—and the surface of the earlier placement should be cleaned off. Wire brushing or even wet sandblasting can be called for as appropriate. The objective is to remove the laitance which is a weak layer of cement, water, and fine material which has bled or floated to the surface, and is not sound concrete.

There are a number of methods of testing concrete already placed and hardened. An approximate test can be determined by using an impact hammer, a special instrument which measures the rebound of a hard steel ball. This is approximate at best, and where there is real doubt of strength, concrete cores may be drilled from the structure and tested.

03312 Heavyweight Aggregate Concrete

In certain special applications, high-density concrete is required, as in shielding for atomic radiation. In such cases, heavyweight aggregate is used to develop a greater density. Steel punchings, balls, or shot can be used where concrete weight must be more than 300 lb/ft^3. Other aggregates such as barite, ferro phosphorus, hematite, limonite, magnitite, or serpentine may be used. The selection of the heavy aggregate is usually determined by the physical properties required, and when several choices are available, the final selection depends on cost and availability. As in all aggregates, the heavyweight aggregate should be free of foreign coatings or other deleterious material which could affect the bonding of the cement paste. The structural engineer should identify any special design factors in the preparation of the heavyweight concrete. Boron additives can be used to improve the shielding properties of concrete, but may adversely affect the time required for development of early strength, and also the setting time.

While conventional methods of mixing and placing may be used, heavy aggregate is sometimes more amenable to being placed separately, and then grouted in place with a mixture of sand, cement, and water. Through the procedure of preplacing the aggregate, a lower percentage of shrink can be attained.

03313 Lightweight Aggregate Concrete

Lightweight aggregate concrete is used for two general purposes: first, to reduce the weight of a structure and thereby reduce the size of the structural numbers; and second, as an insulating or fill material, or both. The earliest type of lightweight aggregate was produced by kiln drying certain types of shale or clay. Blast furnace slag has proven to be an acceptable lightweight aggregate. The density of shale and clay lightweight aggregate ranges between 30 and 70 lb/ft³, while slag ranges between 75 and 90 lb compared with the normal aggregate weights of 90 and 110 lb/ft³. However, the lightweight aggregates tend to absorb between 5% and 10% more water by weight of dry material than standard aggregates (5% to 15% versus 1% to 2%). This water absorption offsets the advantage of the lightweight aggregate. The resulting lightweight concrete in the plastic state usually weighs between 90 and 120 lb/ft³. Lightweight aggregate tends to require more paste to coat the aggregate because of the gradation including more fines. Air-entrained cement is suggested regardless of application. Normal placing and vibration techniques can be used, with a slump of 3″ or less giving the best results for finishing. Finishing can be started somewhat earlier, and less floating and troweling is required. The development of strength for lightweight concrete is about the same for normal, and equivalent strengths can be produced.

Lightweight concretes used for thermal insulation have lower strength (less than 1000 psi), and a weight range from 15 to 90 lb/ft³. The aggregates used are perlite or vermiculite for the lightest class (group I). Group II utilizes slag, shale, or clay, similar to that used in the lightweight structural concrete (group II), while a third group (group III) achieves a low unit weight by substituting air voids for some or even all of the aggregate. Uses for lightweight insulating concrete include floor fill, fire walls, fire protection, roof construction, and similar nonstructural applications.

Because of the high air content, insulating concrete generally has good workability, and slumps up to 10″ can be used.

03314 Prestressed Structural Concrete

Concrete cracks because of its low tensile strength, cracking in response either to the stresses set up by shrinkage after hydration, or change of temperature, or under load. To improve the resistance to stress, as well as to improve the watertight characteristics of a concrete structure, prestressing can be used. In prestressing, a load is imposed upon the concrete, *preloading* it with a compressive load before it is subjected to the normal tensile loads. In order for the concrete structure to then actually go into tension, the pre-set stress must first be overcome. One of the methods of prestressing is to impose tension on the concrete after it has set. This is known as *post-tensioning,* and is usually done by wires which are grouped together as cables. These are uniformly spaced and kept separate from each other, and have conduits through the concrete so that they do not bond to it. The wires are stressed by drawing up post-tensioning nuts or other special but similar hardware. Only limited special equipment is required for post-tensioning, since the concrete serves to separate the cables, providing the reactive force required to establish the stress. The cables do not have to be straight, but are generally installed from slightly above the center of the beam or joist, drooping to near the low point at the center, and returning to the starting height at the end. This configuration provides a better distribution of stress through the member than does a straight pull (Fig. 27).

The use of post-tensioning also reduces the *sag* or *deflection* of a concrete member, since the clamping force of the tension cables must be

Fig. 27 Tensioning tendon. *Source: Prescon Corp.*

offset before actual sag or deflection can take place. If the member has been supported during post-tensioning, the post-tensioning force will operate to offset the sag caused even by a dead load of the member itself.

Nailable Concrete

Fasteners, including nails, can be driven into almost any concrete with the use of special case-hardened nails and equipment such as Ramset which is explosive operated. The lightweight concretes are more nailable because of their lower density, particularly group III, which has more air voids. However, truly nailable concrete is usually in the category of reinforced *gypsum concrete*, which is generally used only in slabs, poured-in-place, and poured over permanent form boards. Figure 28 shows the nail holding power in gypsum concrete.

03350 Specially-Finished Concrete

White concrete can be cast by using no aggregate or other materials which will discolor the concrete, and utilizing white portland cement as

NAIL HOLDING POWER
(Resistance to direct pull, in pounds, for plain finished nails placed 24 hours after pouring slab)

Type of Nail		6 Penny Cornice	Gypsum Deck Nail (1)	1½" ES NAILTITE (2)
Length of Penetration		1½"	1½"	1¼"
Holding Power in Gypsum Concrete	After 1 day	3 lbs.	3 lbs.	67 lbs.
	After 7 days	18 "	23 "	77 "
	When Dry	152 "	119 "	136 "
Holding Power in Lightweight Gypsum Concrete	After 1 day	8 "	2 "	60 "
	After 7 days	22 "	17 "	67 "
	When Dry	81 "	50 "	141 "
Details				

(1) Manufactured by Simplex Nail and Manufacturing Corporation, Americus, Ga.

(2) ES Products, New Rochelle, New York

Notes: A. Tests were conducted by a gypsum manufacturer's research center. Nails were removed hydraulically from poured gypsum slabs with 2" minimum thickness.

B. Dry density of gypsum concrete was 48 lbs. per cu. ft. Dry density of lightweight gypsum concrete was 38 lbs. per cu. ft. Nail holding power of fill decreases at densities less than those cited.

C. Nails must be driven into wet slabs as soon as possible because most nails depend on rusting to increase holding power.

D. ES NAIL-TITE nail is recommended for smooth coated roofing surfaces since these are more susceptible to damage than aggregated roof surfaces.

Fig. 28 Nail holding power in Gypsum concrete. *Source: Gypsum Association.*

the base. Oil should not be used on the forms, since it may stain the concrete, and particular care is required to prevent rust stains, including the use of properly tipped chairs.

Where *colored concrete* is required, it can be properly tinted by using either color pigments, or colored aggregate. When colored aggregate is used, the specification also calls for exposure of the aggregate at the surface. This can be done by casting against a form which has been treated with retarder. This permits removal of the surface paste by brushing. Aggregate can also be exposed by bush hammering, grinding, terrazzo machine (on slabs), or sandblasting. Surfaces may be acid-washed. In areas where natural aggregates such as granite, quartz, or marble are available, architects may find *exposed aggregate* to be an economical and attractive surface.

Color pigments are usually pure oxides ground very fine, and added to the concrete mixture in amounts less than 10% of the weight of the cement. Where brightness of color is desired, white portland cement should be used.

On horizontal slabs or precast panels, a dry color mixture can be added after the slab has been floated. After the dry material absorbs water from the concrete, it is floated into the surface.

03360 Specially Placed Concrete

Concrete can be *placed pneumatically* by mixing portland cement, fine aggregate, and water, and placing them by means of compressed air. The processes carry trade names such as Gunite or Shotcrete. The maximum aggregate size is usually ¾". The processes use either a dry mix where water is added at the nozzle, with the nozzle velocity of about 400′ per second. Water pressure at the nozzle should be 15 psi higher than the air pressure at the entrance, or about 60 psi.

The nozzle man controls the mixing water, and thereby controls the strength of the mix. An experienced man can put high-strength concrete in place, but a poor nozzle man often results in an oversanded or under-strength product.

Pneumatically applied concrete is particularly useful on repair work of piers, and other structures where limited re-forming is feasible. The technique is also good for free form work, such as swimming pools. This type of placement can be used to convey concrete into small difficult locations, and over a large area where thin sections are required.

CONCRETE PLACING CHECKLIST

1. Check forms and rebar.
2. Be sure that concrete is in accordance with design mix, certified if required.
3. Inspect at batch plant if required.
4. Check time of receipt for ready-mix, amount of water added—if permitted—and mixing time.
5. Be sure trucks do not overstay maximum waiting period before placing concrete, usually less than an hour an on hot days, less than half an hour after water has been added.
6. Be sure that placement is in accordance with specifications, using tremie or other special requirements such as conveyors per approved placement plan.
7. Concrete should be dropped only the allowable distance, and should not bounce across reinforcing bars or other obstructions causing segregation.
8. Be sure that appropriate slump and air content tests are taken as required.
9. Concrete should be vibrated as it is placed, and vibrators should not hit the reinforcing bar or the forms. Concrete should be worked with either a float or troweled finish, as required by the specifications.
10. Curing compound should be placed on finished concrete, unless it is under floor to be tile covered.
11. Concrete should not be placed after initial set has been reached, and should not be remixed after initial set.
12. Screed boards are often utilized to assure proper grade of finished concrete.
13. Be sure that test cylinders are taken and handled carefully. These should be stored under job conditions.

CONCRETE TEST CHECKLIST

1. Check mix versus approved mix design.
2. Check truck time from start of mixing (limit depends upon specifications and temperature).
3. Take slump test if required.

4. Take air test for air-entrained concrete.
5. Take concrete cylinders:
   5.1 Mix and fill cylinders;
   5.2 Rod carefully;
   5.3 Strike off and tap to close voids;
   5.4 Cure under job conditions.
6. Check test results at 3 or 7 days and confirm at 28 days.

03400 PRECAST CONCRETE

In general terms, precast concrete is cast at a location remote from the project, and transported or lifted into its final location. The technique permits erection of concrete without site form work, and can be particularly useful for architectural concrete, since it permits factory control during manufacture. Precast concrete theoretically need not be reinforced, but practically must be, in order to withstand handling stresses. Precast concrete has become much more popular since World War II because of the great interest in cost reduction, and time reduction through the use of industrialized techniques.

While precast concrete is not necessarily prestressed, the great majority of the members and plates which are precast are also prestressed. The following description of prestressing was prepared by Kenneth Braselton of the Prescon Corporation.*

"Prestressing was described by the late Karl Middendorf, founder of the Prescon System of post-tensioning, as the development within a structure of stresses capable of resisting, within preloading of the structure. In layman's language, prestressed concrete can be defined as concrete containing prestressing steel which has been tensioned to create an internal force to counteract external loads. A simple demonstration of this internal force is the act of picking up a row of books by applying hand pressure at each end of the row of books. The pressure applied by your hands creates compression of the row sufficient to counteract the gravitational load on the books. The steel hoops or bands on an old fashioned wooden stave barrel create a similar force.

Both illustrations are external prestressing. The application discussed here, however, is internal prestressing. Prestressing may be done by one

---

* Used with permission, originally presented to the South Texas Section of the American Association of Cost Engineers and printed in the AACE Bulletin, Volume 13.

of two methods: pretensioning or post-tensioning. Pretensioning tensions the steel element before the concrete is placed. Post-tensioning tensions the element after the concrete is placed. The tensioning tendon is a prestressing steel element composed of high tensile wire, strand, cable, or rod.

Prestressed concrete weds two excellent materials: high tensile strength steel, and high compressive strength concrete. Concrete, being a relatively cheap and plastic material, offers inherent advantages to the designer. It has become a dominant building material in the last 50 years. It is weatherproof, fireproof, and almost foolproof.

Concrete itself has virtually no strength in tension; where such forces are encountered, steel is used to absorb the load. Ordinary reinforced concrete must deflect under load in order for the tensile quality of the steel to be utilized. The bottom fibers of the concrete are thus placed in tension, causing it to crack. The bottom reinforcing steel is placed in tension, and the top concrete fiber stresses in compression. Prestressing the concrete tensions the steel prior to the member being loaded, placing the bottom fiber of the concrete in compression and the top fibers in zero or slight tension. Loading the structure will merely shift the compression within the member. This shifting of fibre stresses is predictable.

In the design of a prestressed beam, for instance, forces introduced create a vertical uplift to counteract the gravity loads; this compresses the concrete linearly, so that the beam 'thinks' it is a column. Similarly, a prestressed flat slab with gravity loads actually 'balanced' by the drape of the tendons, 'feels' the compression induced by the tendon end anchorages and 'thinks' it is a wall. Practically speaking, of course, the vertical load must somewhere be directed to the foundation. This is done in a simple span member by supporting the member at each end. In effect, we balance the gravity load along the horizontal member until we come to a support, then we direct it down the column into the foundation. So we divert gravity's path, for a distance or span, from vertical to horizontal, and control deflection as we do so.

In prestressing, we tell the concrete to resist a certain load and it does just what we tell it to do—even if the load isn't there. It 'thinks' it's there and it resists it—so it had better be there or the concrete may fall *up!* It's always 'thinking.' We could say that prestressed concrete is the 'thinking man's concrete.'

The tensile quality of the steel and the compressive quality of concrete provide a unique blend of a small amount of a relatively expensive tensile material and a large amount of a relatively inexpensive compressive material. The result—a smaller amount of each one is required. A normal structure is 50–75% dead load, so if we reduce the dead load—the load of the structure itself—we effect a saving all the way down to and including the foundation. The high cost of labor and material, the requirement for long clear spans, low profile, and demand for efficiency, have made prestressed concrete one of the fastest growing, and certainly *the* most interesting field in the construction industry. A growth rate of 40% per year is normal. Last year Prescon had an 80% increase in sales. Starting in 1950 in the United States at zero, prestressed concrete has grown in 20 years to one-half the size of the 70-year-old structural steel industry. . . .

. . . Linear prestressing in the United States started with post-tensioning. The real beginning dates to the Walnut Lane Bridge in Philadelphia in 1950–51, designed by Magnel and using his system. This was followed by the first contract bridge in the United States, California's Arroyo Seco Bridge in 1951 on which the Prescon Positive End Anchorage System was used. That same year, nine circular structures were built using the Prescon System internally instead of circumferential external wrapping, such as preload used. A new and vital industry was born in 1950. Precasting plants sprang up, often without need or market analysis. Some were practically stillborn. The mortality rate was high, especially among the smaller ones, but today about 300 exist, even though many of the smaller ones have been absorbed by larger organizations.

The cost of a good pretensioning plant may be well over a quarter of a million dollars, and many cost ten times that. Lack of flexibility, problems of erection, connections, camber control, and other factors often limit pretensioning applications and often make post-tensioning the more desirable method.

Both are available to most any markets, and both are quite often used by the precasters. In fact, some designs require both for more efficient prestressing.

The present-day acceptance and widespread use of prestressed flat plates in high-rise construction really owes its start to *lift-slab,* which started about the same time as linear prestressing in the United States. The first prestressed lift-slab was constructed in 1951 at Southwest Research Institute in San Antonio, using positive end anchorage tendons furnished by the Prestressed Concrete Corporation, the forebearer for Prescon.

Fig. 29 Lift-slab construction. Slabs are cast-in-place one upon another, and then jacked into final position. *Source: Prescon Corp.*

In 1953, a few small slabs were prestressed and lifted in California and Nevada, again using the button-head wire. The California State Architect became interested in this construction method, and in 1954, a program got underway which has resulted in over 20 million square feet of prestressed flat slab construction. It has helped launch other applications of post-tensioning as well, so that today, 30- to 40-story structures of prestressed concrete construction have been built. . . ." (Fig. 29).

0341 Precast Concrete Panels

The following material was provided by the Prestressed Concrete Institute.* See Fig. 30.

The designer usually attempts to develop standard modules or sections

---

* Taken from "Production and Design of Architectural Precast Concrete" by C. H. Raths, printed in the *Journal of the PCI*, **12**, 3, June 1967.

Fig. 30 Precast panels being placed. *Source: Lorain.*

so that the precasting can take advantage of reuse of the same forms and connection details. The principal investment both in time and cost in the precast slab manufacturer is the design and fabrication of the form or *mold*. Most of today's forms or molds are made of fiberglass-reinforced plastic, wood, or steel. Fiberglass molds are used where a smooth concrete finish is required, while the increasingly popular exposed aggregate panels are made in wood or steel forms. Under these conditions, a fiberglass mold can be used about 75 times, while a wood mold can be used 50 times.

The *reinforcing steel* is usually made in a jig and lifted into the mold for casting. Since it must be handled as an entity, it must be stable enough not to lose its dimensional characteristics when moved from the jig to the form. The reinforcing is usually hung from the form or mold, rather than supported, so that a minimum of 1" cover is established over all of the reinforcing steel. In placing, vibration is used of all types, varying with the experience of the manufacturer. The curing period is usually one day, with the newly stripped panel having a low strength of between 1500 and 3000 lb/in².

During the stripping, or immediately after it, cracking frequently results because of *thermal shock,* or because of minor handling stress. During *handling* and storage, precast panels are often exposed to higher loads than their actual design configuration intended. In the early years, when manufacturers were less aware of the mechanical strength characteristics of their product, a substantial number of serious failures occurred within the plant in handling and storage. Now, the design in a good production unit includes instructions on where to pick up, where to support during storage, and what characteristics such as edge transport, flat transport, etc., are to be followed.

Typical suggestions include the supporting of the unit at two points only during storage, and storage and shipping in the same manner of support, with emphasis on the two-point support criteria.

In handling, the panels are generally stronger and more resistant to stress if handled vertically, but the height restrictions of bridges and overpasses often dictate shipment flat.

At the job site, the inspection team should inspect the plate on the truck, or in the storage yard—but in all cases, prior to erection. All connections, dimensions, and quality should be reviewed before the unit is placed. In many cases, the panel must be rotated 90° after being lifted from the trailer, and care must be taken not to impose high stresses on the edge acting as the hinge.

The panels are usually cast with the exterior face down, so that at some point in the factory, the unit should have been lifted into working position for inspection, and final work on the exterior face, such as patching, sandblasting (if indicated) dimensional checks, and clean-up.

At this stage, inserts would be cleaned and checked. Basically, there are two types of inserts used for handling—bolted and wire cable. The bolted is most widely used since it requires no cutting or burning after use.

From a safety and field inspection viewpoint, emphasis must be given to the *connections.* They integrate the panel with the structural system of the building and there are four basic types: shear, combined moment-shear, compression, or tension. These connections must be made in the field in accordance with the drawings and specifications in order for the panel to contribute properly to the structural system, and to avoid accidents which could occur if incorrect connections are made. The tension or compression connections keep the panel in a plumb position, and are sometimes referred to as "kicker" connections. In order to achieve a final proper dimension, precast leveling blocks may be grouted in place so that the vertical positioning of the panel is exactly correct. Without this type of preparation, accumulated tolerances could result in substantial misalignment of panels. Most precast concrete panels are not prestressed, but are reinforced. However, prestressing can be achieved through the use of anchorages and/or jigs which can be used to induce stress in the reinforcement prior to the pouring or casting of the concrete material. Prestressing does, of course, reduce the cracking effect.

03410 Precast Structural Concrete

Precast structural members, without prestressing, are made up of standard structural shapes, and a new class—such as the Lin Tee—which have evolved as a combination of design-production-performance experience has been established in the field. The principal reason for the precast structural member is improved control of production, lower cost, and better product through production control. In many cases, the precast members are assembled at the job site in erector-set fashion. The casting yard may be an established one, which permits investment in more permanent casting beds, steam curing facilities, and special handling equipment.

Nevertheless, in certain cases the transportation distance may make it more economically feasible to establish a job site factory or production yard.

The same control requirements exist for precast structural members as for cast-in-place. The inspection team should, therefore, be monitoring the same aspects, but possibly at a different location. An added factor in the precast structural member is assurance of proper handling during transport, and special attention to the connection. Precast structural members are usually joined by welding of reinforcing bars, and then casting in place of high-strength mortar or concrete at connection points. These will be clearly established by the structural drawings, but their installation must be inspected for quality by the inspection team.

03413 Precast Prestressed Concrete

Since prestressing is relatively easy to add to the precasting process, most precast structural members today are prestressed. The comments on inspecting of prestressed concrete which follow were furnished by the Prestressed Concrete Institute and appeared in the Journal of PCI:

Prestressing can be introduced by pretensioning or post-tensioning, or a combination of the two. In all methods, the *tendons* or cables are stressed within close limits, with the stressing measured appropriately by pressure gauge, dynamometer, and/or load cells. If two determinations are used, they should be within a tolerance limit of 5%. The actual measuring equipment should be calibrated to be accurate to within 2%.

The measurement of the stressing force is either a calculation as a result of field measurements, such as the elongation or stressing of the tendon under stress, or direct reading by precalibrated pressure gauges or dynamometers. The measurement system should be approved by the structural engineer, and monitored by the inspection team. The Prestressed Concrete Institute has a procedure for qualification of concrete production facilities, and this procedure can assure the inspection team of qualified operators and a general predisposition to do an effective and thorough job. Conversely, the presence of a qualified production facility does not relieve the inspection team of the responsibility for reviewing the work being done for them.

For instance, a qualified and experienced working team will be attuned to the proper setting of instrumentation, and proper application of stressing loads. The rate of loading is important, as well as the amount—and improper use of the forces involved in prestressing can result in unacceptable damage to the prestressed structural members. Some level of failure is permissible. For instance, if less than 2% of the total area tendons in a member should fail, the structural engineer may, if he is sure that the failure is not indicative of a more extensive problem, permit the member to be utilized.

One of the most important items for inspection is the condition of the tendon. Since the member will ultimately be bonded directly to the tendon, no foreign or deleterious materials must be allowed to coat the tendon which would preclude development of full bonding stress. In the case of post-tensioning, the tendons are grouted into place. In prestress, after stress is applied, the member is cast around the tendons.

The environmental temperature during stressing is important. Because of the high loads, a temperature variation of only 10% will result in a stress variation of 1%. Where stresses of almost 200,000 psi are being used, this amounts to a 2,000-psi variation. When the pouring is done on a cold day, the warming effect of the concrete upon the cold stressed reinforcing will be a relaxation of pressure, while in hot weather, the temperature of the steel could be in the neighborhood of 100 to 120 degrees, and with the concrete temperature about 70 degrees, an increase in stress would result.

One of the most frequent problems encountered during the pouring operation is slippage of the *anchorage devices* at either or both ends of the casting bed. One way of slippage protection is marking with a crayon, or spray paint to form a linear reference line with which slippage can be determined visually.

Inspection of the finished structural member can be initiated after the tendons have been stressed, reinforced in place, hardware and accessories positioned within the forms, and the concrete placed and cured. Detensioning can be done immediately following the curing period, and during this operation, prestressing forces should be kept symmetrical about the vertical axis, and applied in a manner which will prevent shock unloading. Force release can be by jacking with gradual release, or by relaxing the strands by heating one strand at a time.

In post-tensioned members, care must be taken that all anchor devices are set exactly normal to the directions of the axes of the post-tensioned tendons. Accurate measurements must also be made to validate the level of stress applied. *Grouting* is an important phase of the post-tensioning operation with the dual purpose of protection of the tendons, and relieving the anchorages of stress fluctuation. The grout used is fluid, with very accurately measured components.

Generally, steam *curing* is recommended, and holds the temperature of the structural member between 50 and 70° F with almost 100% relative humidity. These are optimum conditions for the proper curing and strength development of the concrete mixture.

In inspection, the factors should include review of the work procedures, and visual inspection to ascertain proper handling. Some cracking is permissible and can be remedied by proper repair. When in doubt, the structural engineer should be consulted.

Concrete which has been steam-cured tends to be whiter in appearance than normal. Accordingly, any surface repairs should be made with a mortar made of white portland cement and light or clean aggregate. Honeycombing and bugholes which have no significance structurally may be patched with application of mortar, hand-rubbed for blending.

Depressions in the lower portion of the pretensioned members after removal of the holding bolts should be plugged with a similar mortar. Before patching, surfaces must be completely clean of oil or grease, and the surface roughened to improve the bond.

Prestressed and precast members require the same concrete breaking techniques as cast-in-place concrete, and this step should not be omitted—even though the production of the concrete has been under controlled methods.

Inspection, of course, continues during shipment and unloading, with emphasis upon handling of the structural members at the job site. This is an area where plant production personnel are no longer in charge, and the inspection team must take over their responsibilities.

CHECKLIST FOR INSPECTION OF PRECAST CONCRETE

1. Identification, examination and acceptance of any plant testing of materials.

2. Inspection and recording of tensioning.

3. Inspection of beds and forms prior to concreting.

4. Checking of dimensions of members, positions of tendons, reinforcing steel, and other incorporated materials, openings, and blockouts.

5. Continual inspection of batching, mixing, conveying, placing, compacting, finishing, and curing of concrete.

6. Preparation of concrete specimens for testing and performing of tests for slump, inner content, cylinder strength, etc.

7. Inspecting operations of detensioning, removal from beds, handling, and storage.

8. General observation of plant, equipment, working conditions, weather and other items affecting the product.

9. Final inspection of finished products prior to shipment.

Source: Prestressed Concrete Institute.

03500 CEMENTITIOUS DECKS

03510 Poured Gypsum Deck

One of the most important materials for low-strength poured decks is gypsum. The following description of poured gypsum roof decks is taken from information furnished by the Gypsum Association:

A gypsum concrete roof deck consists of a reinforced gypsum concrete slab poured on permanent forms which may be supported by subpurlins. These lightweight systems offer low cost, two-hour *fire rating* and acoustical ceiling values. In addition, the material is easily penetrated for vents and other systems openings. Gypsum concrete is quick-setting and usually poured in place over galvanized reinforcing mesh. Where the spacing between roof or floor joists exceeds 36″, subpurlins are used. When completed, the gypsum deck provides a structurally strong, monolithic deck ready for roofing. The typical life span for a properly placed gypsum concrete roof deck is more than 50 years and is ideally suited for use over flat or nearly flat roofs.

Gypsum concrete is a factory-controlled mixture of gypsum and wood chips, shavings, or mineral aggregate. It requires only the addition of water at the job site. It can be poured in place over permanent form boards to an average thickness of not less than 2″.

Two types of *reinforcing fabric* are usually recommended for gypsum concrete: galvanized woven wire mesh, or galvanized welded wire mesh. The wire mesh is laid in the gypsum slab in long continual lengths. The reinforcement mesh provides added strength and shear resistance to the deck. Other types of reinforcing mesh may be used if they provide adequate bond and corrosion resistance, and if their effective cross-sectional area is not less than 0.026 in^2 per foot of slab width. Figure 31 shows typical cross sections of gypsum concrete with related fire resistance ratings.

Gypsum roof slabs can handle normal design loading for wind, uplift, and other live loads, including reasonable human traffic. They cannot handle heavy loads of equipment unless tied directly to the roof structure.

In inspecting the placement of gypsum concrete, the *water-cement ratio* again determines the major factor in the ultimate compressor strength of the gypsum concrete. Naturally, the inspection team must insure that excessive water is not added. Manual mixing or mechanical mixing may be employed, but mechanical is more satisfactory. In placing gypsum concrete, impact on form boards could cause deflection or cracking which should not be permitted. Double pouring (the application of gypsum concrete in more than one layer) is not permissible.

Generally, the minimum thickness of gypsum slab is 2″, but in certain areas where seismic forces are not a factor, and the design load is less that 40 psf, a 1-½″ slab may be permissible.

Form boards should not be placed more than one day ahead of the

One-Hour Deck

The first limiting end point of the two in. deck was reached t 1 hr., 16 min. when cotton waste was ignited on the upper surce. The furnace fire was turned off at 1 hr., 34 min., at which time the 40 lb. per sq. ft. live load was still supported by the gypsum roof deck.

Two-Hour Deck

The first limiting end point of the 2½ in. deck was reached at 2 hr., 19 min. when a single thermocouple reached an end point—limiting temperature on the upper surface. The furnace fire was turned off at 2 hr., 33 min., at which time the 40 lb. per sq. ft. live load was still supported by the deck.

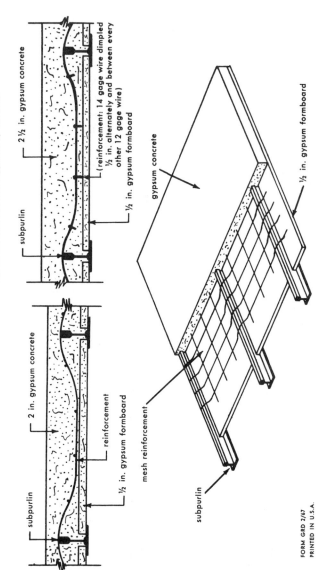

FORM GRD 2/67
PRINTED IN U.S.A.

Fig. 31 Gypsum concrete cross sections. *Source: Gypsum Association.*

gypsum concrete. In placing gypsum concrete, screeds should be utilized just as in standard concrete.

Gypsum concrete should not be mixed or poured when the temperature is 35°F or lower, and accelerators or other antifreeze should not be used with gypsum concrete. (Conversely, gypsum concrete does generate a substantial heat of hydration, and with proper protection, cold weather concreting can be employed.)

In curing gypsum concrete, adequate ventilation should be provided, and water is not added as a curing media. Usually, gypsum concrete is covered by an impermeable membrane, and the application of roof covering should follow as promptly after the pouring of the gypsum concrete as possible, it being necessary to wait only until the top surface of the slab is hard as indicated by the disappearance of the moisture gloss from the surface. The first two plys of built-up roofing felts can be nailed dry to the slab, using five or six penny-cut nails.

One concern of lightweight roofing material is resistance against uplift. This should be considered by the designer, but if not, must be questioned by the field inspection team. Tie wires are not sufficient connections for uplift resistance.

Test specimens should be taken from each slab sample. These are normally 2″ cubes, cut from the slab sample.

0352 Insulating Concrete Roof Deck

The installation of lightweight roof decks and interim floor decks is in accordance with the lightweight aggregate concrete described in Section 03313 or 03341. Usually, this is lightweight vermiculite or perlite concrete, placed with air-entrained concrete. The effect is reduction in weight, and suitable strength for roof or even deck fill.

CHECKLIST FOR GYPSUM CONCRETE

1. Water-cement ratio is important—inspect for amount of water in mix.
2. In placing concrete, avoid impact on form boards.
3. Double pouring (application in more than one layer) is not permitted.
4. Form boards should not be placed more than one day ahead of the gypsum concrete.

5. Screeds should be utilized.

6. Do not allow mixing or pouring when temperature is 35° F or lower.

7. Accelerators may not be used with gypsum concrete.

8. For curing, provide good ventilation. Water should not be added during curing.

DIVISION 4: MASONRY

NUMBER	TITLE
04050	**MASONRY PROCEDURES**
	Cold Weather Masonry Procedures
04100	**MORTAR**
−03	Mortar Coloring Materials
−10	Cement and Lime Mortars
−20	Chemical Resisting Mortars
−30	Premixed Mortars
−40	Epoxy Mortars
−45	Poured Portland Cement Concrete Grouts
04150	**MASONRY ACCESSORIES**
−60	Joint Reinforcement
−70	Anchors and Tie Systems
−80	Control Joints
04200	**UNIT MASONRY**
−10	Brick Masonry
−12	Adobe Masonry
−13	Structural Glazed Tile Masonry
−20	Concrete Unit Masonry
−21	Glazed Concrete Unit Masonry
−22	Exposed Aggregate Concrete Unit Masonry
−23	Split-face Concrete Unit Masonry
−24	Fluted Concrete Unit Masonry
−25	Molded Face Concrete Unit Masonry
−30	Reinforced Unit Masonry
−32	High-Lift Grouted Brick Masonry
−33	High-Lift Grouted Concrete Unit Masonry
−35	Preassembled Masonry Panels
−37	Interlocking Concrete Unit Masonry
−38	Mortarless Concrete Unit Masonry System
−40	Clay Backing Tile
−45	Clay Facing Tile
−50	Ceramic Veneer

−51	Terra Cotta Veneer
−52	Mechanically Supported Masonry Veneer
−70	Glass Unit Masonry
−80	Gypsum Unit Masonry
−85	Sound Absorbing Unit Masonry
−86	Sound Absorbing Structural Glazed Tile

04400	**STONE**
−10	Rough Stone
−20	Cut Stone
−22	Marble
−23	Limestone
−24	Granite
−25	Sandstone
−26	Slate
−30	Simulated Masonry
−35	Cast Stone
−40	Flagstone
−41	Bluestone
−50	Natural Stone Veneer
−51	Marble Veneer

04500	**MASONRY RESTORATION AND CLEANING**
−10	Masonry Cleaning
−20	Masonry Restoration
04530–04549	(Reserved)

04550	**REFRACTORIES**
−51	Flue Liners
−53	Combustion Chambers
−54	Firebrick
−55	Castable Refractory Materials
04560–04599	(Reserved)

04600	**CORROSION RESISTANT MASONRY**
−01	Chemical Resistant Brick
−02	Vitrified Clay Liner Plates
04610–04999	(Reserved)

04100 MORTAR

Mortar is the cementitious binder placed between mortar blocks. It may be a paste of the cement and water, or may include fine aggregate. Generally, mortar is classified by the type of cement which is the primary material. Mortar may be placed in wet condition or dry. Dry mortar is used for dry bedding, and contains enough water to allow proper setting, but not enough to cause it to be sticky.

One of the most common cements is pulverized white lime (hydrated lime), which is the material remaining after the chemical reaction of quick lime and water. It is mixed with portland cement and sand to make mortar.

Masonry cement type II is a premixed lime and cement dry mix, which requires only sand to make mortar.

Where an early set is required, high early portland cement type III can be used as a quick-setting cement mortar.

Portland cement type I is the standard material which mixed with sand and lime is used for joint bedding of brick and concrete blocks.

The aggregate used for mortar is sand meeting ASTM designation C144.

Where it is important not to stain the basic masonry block, a white portland cement Type I can be used with clean aggregate to furnish a mortar which will not stain the final work.

An approved mortar mix is designed by the contractor and reviewed by the structural engineer. The inspection team should check it against laboratory-established proportions, and the field team shall have available accurate volume measuring devices such as one cubic foot boxes for checking mix. The proportioning should be checked on a regular basis, and whenever mortar tenders are changed. Mechanical mixes should be used on all the smallest jobs. The quick lime should be slaked in accordance with manufacturer's instructions, and should be sieved through a # 20 sieve, and allowed to cool before using mortar. Inspection teams should review to determine that mortar is being used within specified time limits.

04200 UNIT MASONRY

Masonry is basically block construction cemented together with mortar. Blocks may be brick, cement block, concrete block, clay tile, glass, stone, and other similar materials. Masonry may be structural and load-bearing, or in other cases used only to fill in either as exterior wall or interior partitions.

The specifications detail the size and make-up of both the mortar and unit block. A useful requirement is the preparation of sample panels which can be used to test as well as monitor progress for materials workmanship and finished appearance.

The inspection team must check *samples* for color or range of colors, texture, grade of material, and tolerances for sizes and defects. If the specifications are general, referring to Federal, ASTM or other materials specifications, the inspection team must maintain these references in their field library, and use them to review the materials. Typical defects to observe are chips, checks, cracks, crawling, crazing, pop-outs and warped or misshapen units.

The inspection team should review *storage* facilities for adequacy. Units should be stored off the ground and covered with waterproof covering such as tarpaulins, polyethylene, or other waterproof material. Covering should be secured in place and resecured whenever rain or snow threatens.

Unit construction may be reinforced or nonreinforced. *Reinforcing* is usually in the form of light wire mesh or bar reinforcement. Where two unit walls are laid together, ties are placed in the one wall for anchorage to the other. Ties can also be used to lock other materials to the masonry units. Anchors can be bent and grouted into voids to hold window and door frames, and other imbedded items. Anything set in masonry must be anchored to that masonry. Inspectors should be careful to ensure that the opening allowed for *imbedded material* is sufficient, including allowances for deflection and caulking when placed. Sleeves and equipment for mechanical and electrical items should be built into the masonry as it is erected, and wherever possible not cut in afterwards. Where there is cutting and fitting to be done, it must be done by masons. Accordingly, it is important that the inspection team be aware of the imbedded items and insert items required.

In erection, the inspection team should be certain that the *ambient temperatures* are at or above the minimums required, usually 35 or 40°F. Where temperatures are below the minimum, there must be an approved method of protecting the masonry against cold weather. Frozen materials cannot be used or built upon. If work is allowed to become frozen after installation, the inspectors must be sure that it is removed and replaced. Mortar must be kept continuously above freezing for at least 48 hours after units are laid. In warm weather, masonry must be protected against rainfall for 48 hours.

In areas which are to be backfilled, the faces of the exterior masonry walls must receive *parging*. This should be troweled to a smooth dense surface at least $\frac{1}{2}''$ thick. Backfill below grade should be carried up evenly on both sides of a wall to avoid overstressing pressures. Where backfill is to occur on one side of the masonry wall, backfill should not start until the top of the wall is laterally supported.

Before erection is started, masonry *dimensions* should be checked with existing foundations and required structural framing. In reinforced concrete buildings, dovetail tie slots should be available—and part of the inspection of the reinforced concrete should ensure the availability of these tie slots.

The inspector should check vertical coursing against the dimensional wall heights. Where a minor change in joint width would eliminate fractional courses, this should be coordinated with the contractor. Horizontal layouts should be checked by template or by tape, again with the object of avoiding fractional length units. Openings should be located so that units are of the same length against both jambs. Check for conflicts between openings and partitions or equipment locations. The inspection team should check control joints for type and location, and should insure that the work is laid in plumb. Control joints should generally not be spaced more than 24' apart, with normal spacing being about 15'. Joint reinforcement should not pass through control joints.

The contractor should use a *story-pole* to establish coursing in leads. String lines can be used for general guidance, but story-poles should be used to check coursing between leads. Masons should use levels to check plumb and straight edge. Block and tile should be cut by masonry saw, and finishing joint tools should be of the same type and shape.

Masons should wait for initial set of *mortar* before tooling the joints. Mortar should be "thumbprint hard" when tooling is done.

If units have to be removed after mortar has taken initial set, they should be replaced using fresh mortar. Excess mortar should be removed from the face of units and joints before it has set up. At sills and copings, as well as at lintels and bond beams, flashing should be installed for waterproofing. Generally, the architectural plans have a typical flashing detail. The inspection team should review the specification for joint reinforcing under sills and over lintels; requirements for bond beams should be reviewed at four levels—under sill, above lintel, at top of wall, or at intermediate locations. The brick pattern should be reviewed for specification for header courses—and whether these are full or dummy. The requirements for weep holes, parging, and other special aspects should also be reviewed *before* the work progresses. It is the responsibility of the inspection team to have foresight, not just belated hindsight.

04210 Brick Masonry

Brick is made of burned clay and comes in a wide variety of sizes and shapes. Common brick is 2-$\frac{1}{4}$" $\times$ 3-$\frac{3}{4}$" $\times$ 8" with a weight of approximately 5 lb, as described in ASTM Specification C62. Grade SW brick is intended where there is a need from high frost; grade MW is used in relatively dry locations where the brick may be exposed to below freezing temperatures (usually used in exterior walls of building); and grade NW brick which is used where brick will not be exposed to any freezing.

Brick walls are usually made in two portions, the inner and outer. The back-up bricks are either common, or on occasion masonry units are used as the back-up wall. The two principal joints in the back-up wall are the bed joint and the head joint.

Exterior wall brick is surface-treated to allow for proper adhesion to the wall and other bricks. This surface treatment makes the brick rougher and gives it a specific color.

The brick exterior wall is tied to the back-up wall forming a structural bond. In solid walls, the tie into the back-up wall is laid up high enough to allow the placing of header brick in correct spacing. This establishes a firm tie between the walls. In cavity walls (where there is a gap between back-up and exterior walls) the connection is provided by the use of metal ties, reaching from the back-up wall to the brick wall. This is necessary in the case of a CMU back-up wall, where the back-up wall dimension is not the same as the exterior wall.

There are six basic *pattern bonds* most usually used in brick work today: running, common, Flemish, English, Dutch, and stack.

The strength of the brick wall is related directly to the strength of the mortar. There must be sufficient water content to allow for a firm bonding between the bricks, but not so much that seepage will occur.

When sample bricks have been furnished, the samples should be weighed for their dry weight, then submerged and after a period of time removed from the water or mortar mixture, and weighed again. The difference between the two weights is the amount of absorption.

One of the problems in brickwork is the *effervescent effect,* represented by a white loose powder forming on the surface of brick or stone walls. Part of the inspection of the brick sample is a laboratory test for effervescence.

In laying brick, the first course should be leveled and laid to line. Coursing sticks give the proper corner elevations. Coursing at windows and doors should be checked to avoid half courses. Three courses of solid brick should be over windows and doors and under windows if headers and sills are to be made of brick. As bricklaying reaches the top of a wall, anchors for the wall plate should be built in and turned down with a washer at the head long enough for a washer nut at the top.

A brick *soldier course* is the course laid standing all on end, used mainly as a design feature. A belt course is another design feature and is a layer of brick carried around at the same level across or around the perimeter.

The specifications should describe the *mortar joint* required. In terms of performance, the best joints are weathered, concave, or vee joints. Raked and set-back joints are used primarily to emphasize patterns that are used in the laying of brick. However, these may require waterproofing of the back of the face units. Where mortar joints are concealed or protected from the weather, a flush joint formed by cutting the excess mortar from the wall face may be used. Where the masonry is exposed, the final finishing operation before cleaning should be the jointing of the surface of mortar joints.

Corbelling involves the installation of courses which protrude beyond the lower courses in order to support structural members above. The maximum protrusion should not exceed 2". Since this involves eccentric loading, corbelling is discouraged in design. In fact, more frequently, the brick load is carried by the main structure of the building, and in order to avoid a high dead load, a shelf angle is attached by bolting to the concrete floor slab every floor, and is used to carry the dead load of the brick between floors.

In *laying brick,* the brick should be placed, not shoved. Joints should be filled solidly as the brick is laid, and the mason can use either the end buttering, or pick and dip method. However, slushing of grout to fill joints after the brick is laid is not acceptable. Where brick facing wall is directly against the back-up masonry, the space should be completely filled with mortar. The inspection team should check header courses and/or metal ties between face and back-up masonry.

04220 Concrete Unit Masonry

For interior partitions and back-up walls, concrete unit masonry is used quite frequently because of its economy in terms of material and labor. (Often the working conditions limit the number of brick or CMU's which can be placed per day, and the advantage is often in terms of the CMU,

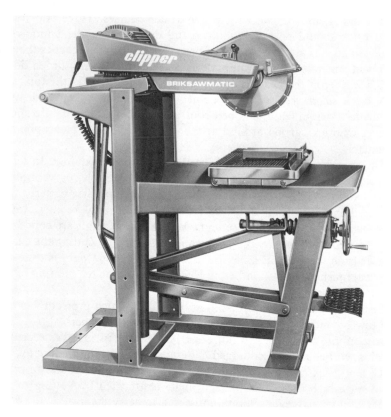

Fig. 1 Brick saw. *Source: Norton Company.*

with concrete masonry units providing more square footage per day than the brick.) The CMU can be hollow or solid construction, but hollow is more common because of its lower weight and better insulating characteristics.

Starting courses and other specified courses should be fully bedded under both the face shell and web. All other courses are face shell bedded. Units should be laid up with a full head joint. Inspection should include rejection of chipped, cracked, or defective units.

Cuts should be made by a dry masonry saw, rather than a mason's hammer (Fig. 1). Where the face of a CMU is exposed, the difference between vertical faces should not exceed $\frac{1}{8}''$. In *control joints,* felt paper should be provided on three sides of the mortar key. Control joint blocks should be available in full and half sizes. (In some cases, the entire control joint is made of half-sized blocks—and the bond pattern must then provide for a half unit, at the control joint, in every course alternating from side to side.)

Bond beams should be constructed entirely of special U-shaped bond beam blocks. Reinforcing should be continuous, including bent corner bars for the bond beam.

CMU's, since they contain a vertical cell, can have vertical cell reinforcing, which is made up of rebars, or masonry tie bars, filled in with mortar or grout. At intersections of partitions, anchors should be installed to tie the walls together.

04230 Reinforced Masonry

Masonry is normally reinforced or combined into structural units by the use of ties or headed courses. It is not unusual to use reinforcing ties even where the masonry is non-load-bearing to maintain integrity of the panel unit over a long life span. Masonry can also be loadbearing, and was once the most popular exterior material for the construction of buildings, including multiple story. Even today, some buildings have recently been completed with masonry loadbearing walls ranging up to about 12 stories. However, these are in the minority, and in most situations, reinforced masonry is limited to two or three story buildings. The reinforcing required is specified in the plans and specifications.

04270 Glass Unit Masonry

For purposes of security or insulation, glass block is sometimes utilized in place of glazing or window glass. These units are hollow, and about 3-⅞" thick x 6", 8", or 12" square. The block is nonbearing, and the area should not be unsupported for more than 144 ft². The block panel should have provision for movement due to temperature changes because of differential nature of the material. Expansion joints are placed at the jams and head, and the block is set on a flexible mastic base.

04280 Gypsum Unit Masonry

Gypsum unit masonry is used because of its desirable installation characteristics, sound attenuation capabilities, and light weight. It is principally used as interior partition material, and dry wall or plaster wall can be placed directly upon it. It is not suitable for direct finishing in itself. The units are laid up with a gypsum mortar.

MASONRY PARAMETERS

The inspection team is dependent upon the design team to properly specify the materials and manner of placement. However, some general awareness of usual practice will alert the inspection team to those areas which are critical:

Anchorage—embedded items must be approved prior to installation and the inspection team should have a tagged approved sample on site for comparison purposes.

Control Joints—a sufficient number of control joints are vital to a serviceable masonry life span. The drawings often depend upon a note or typical detail to indicate these. It is an item easily missed. Control joints are often called for at each column line.

Minimum Dimensions—the following discussion regarding minimum allowable masonry dimensions is extracted from the New York City Building Code:

4.4.1   Slenderness ratio.—

(a) The slenderness ratio (ratio of effective height, h', or length of the wall panel to the effective thickness, t) shall not exceed 30 for walls of solid units, 20 for walls of hollow units, and 25 for walls of filled cell or grouted masonry.

(b) The slenderness ratio (ratio of the effective height, h', to the least effective thickness, t) shall not exceed 25 for columns of solid units, 15 for columns of hollow units, and 20 for columns of filled cell or grouted masonry.

5.1   Method of Support.—Lateral support for masonry walls may be obtained by cross walls, columns, pilasters, or buttresses, where the limiting distance is measured horizontally; or by floors, roofs, spandrel beams, or girts, where the limiting distance is measured vertically. Sufficient bonding or anchorage shall be provided between the walls and the supports to resist the assumed wind or other horizontal forces, acting either inward or outward and shall meet the requirements of section 9.5. All members relied upon for lateral support shall be designed on the basis of allowable stress and shall have sufficient strength to transfer the horizontal force, acting in either direction, to adjacent structural members or to the ground. Where walls are dependent upon floors or roofs for their lateral support, provision shall be made in the building to transfer the lateral forces to the ground.

5.2.   Height and Thickness Limitations.—

5.2.1.   General.—Masonry walls, whether loadbearing or non-loadbearing, shall be provided with lateral support by means of horizontal or vertical members or constructions at intervals not to exceed those specified in section 4.4.1 or, for non-loadbearing walls or for loadbearing walls where it is desired to obviate the need for structural analysis, at intervals not to exceed those specified in this section.

Where a masonry wall containing no openings is supported in both horizontal and vertical spans, the allowable distance between lateral supports as indicated in this section may be increased; but if both horizontal and vertical distances exceed the allowable distance, the sum of the horizontal and vertical spans between supports may be no more than three times the allowable distance permitted for support in only one direction.

5.2.3   Non-loadbearing exterior masonry walls.—In lieu of structural analysis, non-loadbearing exterior masonry walls may be proportioned so that the maximum slenderness ratio does not exceed 20. In the case of a gable, the height of the wall shall be based on the average height. Where the wall panel contains openings having a dimension in excess of 50 per cent of the corresponding dimension of the panel, the wall shall be proportioned by structural analysis.

5.2.4   Interior loadbearing walls.—In lieu of analysis of stresses, interior loadbearing masonry walls may be proportioned so that the maximum slenderness ratio does not exceed 20.

5.2.5   Partitions.—The distance between lateral supports of partitions 3 in. or greater in thickness shall not exceed 48 times the nominal thickness of the partition, excluding plaster, and for partitions less than 3 in. thick, 48 times the actual thickness, including plaster.

Bonding—The following guidelines for bonding requirements is extracted from the New York City Building Code:

7.2   Bonding with Masonry Headers.—Where the facing and backing of solid masonry construction are bonded by means of masonry headers, at least 4 per cent of the wall surface of each face shall be composed of headers extending at least 3 in. into the backing. The distance between adjacent full length headers shall not exceed 24 in. either vertically or horizontally. In walls in which a single header does not extend through the wall, headers from the opposite sides shall overlap at least 3 in., or headers from opposite sides shall be covered with another header course overlapping the header below at least 3 in.

7.3   Bonding with Metal Ties.—The facing and backing (adjacent wythes) of masonry walls shall be bonded with corrosion-resistant $^3/_{16}$ in. diameter ($^1/_8$ in. diameter for veneer), steel ties or metal wire of equivalent stiffness embedded in the horizontal mortar joints. There shall be at least one metal tie for each 2 sq. ft. of wall area. Ties in alternate courses shall be staggered, the maximum vertical distance between ties shall not exceed 24 in., and the maximum horizontal distance shall not exceed 36 in., except that for cavity walls having less than a 4 in. wythe, the maximum vertical distance between ties shall not exceed 16 in. Rods or ties bent to rectangular shape shall be used with hollow masonry units laid with the cells vertical. In other walls, the ends of ties shall be bent to 90 degree angles to provide hooks at least 2 in. long. Additional bonding ties shall be provided at all openings and shall be spaced not more than 3 ft. apart around the perimeter and within 12 in. of the opening.

7.4   Bonding with Prefabricated Joint Reinforcement.—The facing and backing (adjacent wythes) of masonry walls may be bonded with prefabricated joint reinforcement. There shall be at least one cross wire serving as a tie for each 2 sq. ft. of wall area. The vertical spacing of the reinforcement shall not exceed 16 in.

7.5 Bonding Faced or Composite Walls.—Faced or composite walls may be bonded as provided for in sections 7.2, 7.3, and 7.4. Where the facing and backing are bonded by means of masonry headers, such headers shall extend at least 3 in. into a hollow masonry back-up unit specifically designed to receive and provide mortar bedding for the header.

7.6 Bonding Cavity and Masonry Bonded Hollow Walls.—

7.6.1 Cavity walls.—Wythes of cavity walls shall be bonded as required in sections 7.3 or 7.4.

7.6.2 Masonry bonded hollow walls.—Wythes of masonry bonded hollow walls shall be bonded as required in section 7.2.

7.7 Masonry Laid in Stack Bond.—Where unit masonry is laid in stack bond, continuous prefabricated joint reinforcement or other steel bar or wire reinforcement shall be embedded in the horizontal mortar beds at vertical intervals not to exceed 16 in. The longitudinal reinforcement shall be not less than no. 9 steel wire gage. At least one longitudinal bar or wire shall be provided in the prefabricated unit for each 6 in. of wall thickness or fraction thereof.

7.8 Ashlar, Natural or Cast Stone.—In ashlar masonry, bond stones uniformly distributed shall be provided to the extent of at least 10 per cent of the wall area. Such bond stones shall extend at least 4 in. into the backing wall. Rubble stone masonry, 24 in. thick or less, shall have bond stones with a maximum spacing of 3 ft. vertically or horizontally and, if the masonry is thicker than 24 in. shall have one bond stone for each 6 sq. ft. of wall surface on both sides.

7.9 Longitudinal Bond.—In each wythe of masonry loadbearing and nonloadbearing walls, at least 60 per cent of the stretchers in any transverse vertical plane shall lap the units above and below at least 2 in. or $^1/_3$ the height of the unit, whichever is greater, or the masonry walls or partitions shall be reinforced longitudinally as required in section 7.7.

04400 STONE

Once a major loadbearing material for construction, stone today is found principally as a veneer because of its high cost, and the level of skill required in stone masonry.

The inspection team should be certain that samples have been submitted and approved by the designer for stone material. Necessary shop drawings shall also have been approved and should show all details of bedding, bonding, jointing, and anchorage details.

The inspection team will need the designer's description for proper inspection of the stone when it is received at the job site. The inspection team will insure that careful handling of the stone is employed by the contractor to avoid any damage. Also, storage shall be reviewed for prevention of staining and/or exposure to moisture and freezing.

Setting shall be done by competent stone setters, and the contractor should furnish assurance that the craftsmen have been properly qualified. Materials should be free and clean of ice or frost before setting.

Granite facing should not be built up more than two courses above the backing, and no piece having a greater bed width than the one below it should be set until the lower course is backed up.

The following field suggestions are offered by the Indiana Limestone Institute of America in general apply to other stone also:

"If coordination with other trades is required in the fabrication of the stone: were details furnished early enough by other trades to the stone fabricator to enable him to execute the work required?

If an Inspector is present on the job site at the time of stone delivery, a check should be made of the stone to determine whether any damage had occurred during shipment. If damage is noted, a notation should be made on the delivery ticket. If damage is minor, repair work can be approved by the architect, if the repair is expertly done and not readily visible.

Checks should be made of stone stored on the job site, to determine whether the spacers used in piling the stone are located directly over each other. Misplacement of spacers can cause breakage due to overstressing during piling. Has the stone been properly protected by waterproof covering?

Spot checking of stone dimensions is advisable to determine whether the stone is within the allowable tolerances, and if it is as close to square as governed by the tolerances.

Check lifting devices used in setting stone to determine whether they are being used properly. When lewis pins are used with a cable, the cable should travel down from the lifting hook through one ring of one lewis pin, then across and through the other lewis pin, and then return up to the hook, forming an equilateral triangle. When lewis pins are used, check to make sure that the lewis pin is of right size to be inserted into the lewis pin hole.

Do the specifications clearly state that noncorrosive type anchors shall be used?

Check the anchors used in anchoring the stone. Do they meet the specification requirements?

Check to see if anchors are installed properly. Were the required amount of anchors used per stone?

There are two types of expansion bolts: those set into the stone with a sharp blow, and those tightened into the stone by tightening the exterior nut of the assembly. If the expansion bolt that receives the sharp blow appears to be loose in the stone, the bolt should be removed and reset with the aid of an epoxy. Expansion bolts of the nut tightening type should be set into the stone or tightened to a connecting member on the building by the use of a torque wrench. The torque pressure exerted shall not exceed a prespecified figure, which, if exceeded, can cause spalling or fracturing of the stone.

Check relieving angle situations to determine whether clearance has been allowed between the relieving angle and the stone below. If dowel pins are installed through relieving angles, check to make sure that dowel pins are not solidly embedded top and bottom.

Check to see if the mortar specified to be used with Indiana Limestone meets the ASTM C91–70 specification.

Check to see whether the mortar used between stone and any other masonry product is a nonstaining mortar.

Do the specifications include the amount, size, and location of expansion joints in masonry walls?"

04500 MASONRY RESTORATION

New masonry walls should be cleaned after completion, and the same general approach is sometimes required for older walls. Cleaning includes: care and some effort during the laying up of masonry to remove mortar droppings, and to be certain that high-absorption brick has been wetted before laying. Cleaning should start after the mortar has been thoroughly set and cured, and large particles of mortar should be moved by mechanical means before cleaning treatment.

A suitable masonry cleaning agent (but harmful to some stone) is one part hydrochloric *acid* to nine parts water. If acid is used, appropriate safety equipment must be provided. Further, any scaffolding or other suspended equipment must be carefully protected from the acid to avoid accidents.

Soak the area to be cleaned with water before applying acid. Brickwork below the area being cleaned should be kept thoroughly soaked with water. Clean only 10 to 20 ft² at a time, and scrub the brick, not the mortar joints. Wash the wall thoroughly with plenty of water immediately after scrubbing with acid. The acid wash should be coordinated with glazing, and should preferably precede it. Following the acid washing, re-point joints as necessary for water tightness and appearance. Acid wash is never used on concrete masonry. Cleaning of glazed structural tile units can be accomplished with soap or detergent and brushes. Clay tile masonry can be acid washed. Restoration of old masonry includes generally raking out or cleaning out of joints and re-pointing or jointing to restore the waterproof integrity of the wall. Restoration may also include application of parge or stucco, and in this case, any loose material must be removed to establish a good bond between the old and the new material.

MASONRY CHECKLIST

1. Check mortar mix for conformance with design mix.

2. Have accurate volume measuring devices for checking mixes in field available to team.

3. Check samples of masonry for color and conformance with specification.

4. Review storage facilities for adequacy, with masonry units stored off ground covered with waterproofing covering.

5. During installation, check reinforcing for conformance.

6. Check wall anchorages where required.

7. Masonry should not be placed at temperatures 35° F or below.

8. Check masonry dimensions with existing foundations before masonry work initiates.

9. Check vertical coursing for dimension.

10. Be certain that masons wait for initial mortar set before tooling joints.

11. Before work commences, review specifications for any special requirements such as parging.

DIVISION 5: METALS

NUMBER   **TITLE**

05010   **METAL MATERIALS AND METHODS**
−11   Stainless Steel
−12   Bronze
−13   Aluminum

05050   METAL FASTENING
−60   Welding
−70   Bolting
−80   Riveting

05100   **STRUCTURAL METAL FRAMING**
−20   Structural Steel
−21   Architecturally Exposed Structural Steel
−22   Tubular Steel
−30   Structural Aluminum
−31   Architecturally Exposed Structural Aluminum
−50   Steel Wire Rope
−60   Framing Systems
−61   Space Frames
−62   Geodesic Structures
05170–05199   (Reserved)

05200   **METAL JOISTS**
−10   Steel Joists
−11   Standard Steel Joists
−12   Custom Fabricated Steel Joists
−20   Aluminum Joists
05230–05299   (Reserved)

05300   **METAL DECKING**
−10   Metal Roof Deck
−11   Steel Roof Deck
−12   Aluminum Roof Deck
−20   Metal Floor Deck
−21   Steel Floor Deck

−22   Aluminum Floor Deck
05330–05399   (Reserved)

05400   COLD-FORMED METAL FRAMING
−10   Load Bearing Metal Stud System
−20   Cold-Formed Metal Joist System
05430–05499   (Reserved)

05500   METAL FABRICATIONS
−01   Anchor Bolts
−02   Expansion Bolts
−10   Metal Stairs
−15   Ladders
−20   Handrails and Railings
−21   Pipe and Tube Railings
−30   Gratings and Floor Plates
−40   Castings
−50   Custom Enclosures
−51   Heat-Cooling Unit Enclosures
05560–05699   (Reserved)

05700   ORNAMENTAL METAL
−10   Ornamental Stairs
−15   Prefabricted Spiral Stairs
−20   Ornamental Handrails and Railings
−30   Ornamental Sheet Metal
05740–05799   (Reserved)

05800   EXPANSION CONTROL
−01   Interior Expansion Joints
−02   Exterior Expansion Joints
−05   Expansion Joint Covers
−10   Bridge Expansion Control
−11   Bridge Sole Plates
−12   Bridge Bearings
−20   Bridge Expansion Joints
05830–05899   (Reserved)

05900   METAL FINISHES
−01   Anodic Coatings
−02   Enamel Coatings
−03   Acrylic Coatings
−04   Urethane Coatings
−10   Galvanizing
05920–05999   (Reserved)

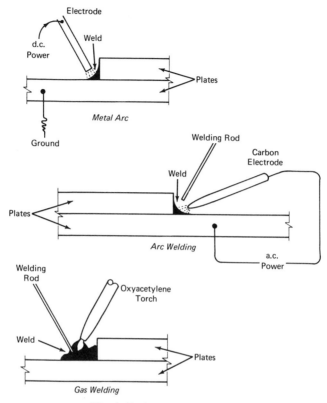

Fig. 1 Various welding modes.

05060 WELDING

Welding employs heat to induce fusion of two like metal surfaces. Usually, a fill-in material is added to the fusion process for increased strength.

Figure 1 shows various welding modes. In *arc welding*, an electrode arcing to ground provides fusion heat, while the welding rod, which may be the electrode or applied externally, furnishes the fill-in material. In the shielded arc electrode, the shielding provides a gaseous shield which excludes oxygen, providing a more complete fusion without oxidation. Arc welding is the most common type utilized in structural steel connections. *Gas welding* is used principally in piping, and applies a flame as the heat source, again with welding rod as the filler material.

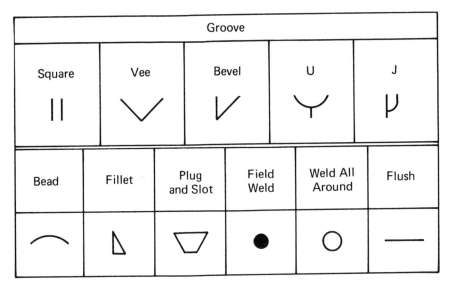

Fig. 2 Typical shop drawing symbols. *Source: American Welding Society.*

A variation of oxyacetylene welding is shielded welding, using proprietary processes such as Heliarc. In *Heliarc welding,* a stream of inert gas is fed around the flame to provide a shield so that the welding process occurs in the absence of oxygen. Heliarc is used in the welding of difficult materials such as stainless steel.

Figure 2 shows typical shop drawing symbols according to the American Welding Society. Figure 3 shows typical welding symbol examples.

Welders should be tested according to an approved welding procedure, with the testing overseen by a qualified testing laboratory.

Welds should be performed in a prearranged pattern which is planned to reduce excessive *distortion* by distribution of heat.

Welds can be inspected in terms of throat dimension, uniformity of weld, absence of cracking, and proper penetration. The inspector should reject coated electrodes which have damaged or wet coating. The coating varies, and can be identified by color code.

The welding qualification is by *position,* so the inspector should be certain that no welder is welding in a position in which he is not qualified. The joint preparation prior to welding should be inspected for parallel beveled surfaces, clean conditions, and proper alignment of

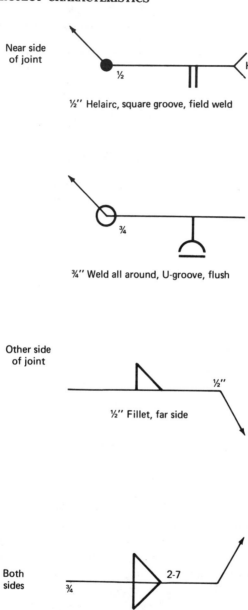

Near side
of joint

½″ Helairc, square groove, field weld

¾″ Weld all around, U-groove, flush

Other side
of joint

½″ Fillet, far side

Both
sides

¾″ Double fillet, 2″ every 7″ of length

Fig. 3 Typical welding symbol examples.

material. Filling of large spaces with welding rod prior to placing weld shall not be permitted. The number of electrodes per welding pass or bead should be checked. All slag or flux should be removed before laying down the next weld bead. Any defects should be ground, chipped or chiseled back to sound metal, before replacing. Welds should be inspected by x-ray, magniflux, or ultrasonic testing means. The percentage of welds to be tested should be in accordance with the specifications. In critical areas, it is not uncommon to have 100% testing. However, in most structural situations, a much lower level of welds can be allowed. In piping welds, it is not unusual to have the welder stamp or identify his work, which is less usual in structural.

OSHA Regulations—Welding

Compressed gas for welding or cutting shall be handled carefully. Valve protection caps must be in place, and secured in a vertical position when moved. The valve protection caps are not to be used for lifting cylinders from one vertical position to another, and cylinders should be moved by tilting and rolling on their bottom edges. Care should be taken not to drop, strike, or apply unusual pressure to the cylinders. Cylinders must be kept far enough away from welding or cutting so that sparks, hot slag, or flame will not reach them. Where this is impractical, fire resistant shields shall be provided for the cylinders. Cylinders, whether full or empty, are not to be used as either rollers or supports.

Fuel gas and oxygen manifolds must bear the name of the substance they contain in 1"-high letters. Manifolds must be in safe, well-ventilated locations, with hose connections such that they cannot be interchanged between the gas and oxygen manifolds.

Torches, regulators and gauges must be kept clean and in good working order, and free of oil or grease.

Arc welding and cutting equipment shall use manual electrode holders which are specifically designed for arc welding and cutting, and of a capacity capable of safely handling the maximum rate of current required by the electrodes used. Current-carrying parts passing through the portion of the holder which the welder grips in his hand must be fully insulated against the maximum voltage and countered to ground. Cables must be maintained in good condition, and splicing, while undesirable, may be done beyond 10' from the cable end to which the electrode holder is connected, using substantial connectors of a capacity at least equivalent to that of the cable used.

Pipelines containing gases or flammable fluids or conduits containing electrical circuits shall not be used as a ground return for welding. Welding on natural gas pipelines must be in accordance with the Department of Transportation Regulation 49 CFR Part 192.

When a structure or pipeline is employed as a ground return circuit, it must be determined that the required electrical contact exists at all joints. Generation of an arc, sparks or heat at any point shall cause rejection of the structure as a ground circuit. All joints shall be bonded and periodic inspection shall be conducted to ensure that no condition of electrolysis or fire hazards exist by virtue of use as a ground return.

The frames of all arc welding and cutting machines must be grounded either through a third wire in the cable or by a separate wire which is grounded at the source of the current.

When electrode holders are to be left unattended, the electrode shall be removed and the holders placed or protected so that they cannot make electrical contact with either employees or conducting objects. Hot electrode holders shall not be dipped in water since to do so will expose the operator to shock. If the operator leaves his equipment for any period of time, the power supply switch to the equipment should be open. Other requirements are as outlined in Article 630 of the National Electric Code.

Welding or cutting generates substantial amounts of heat, and care must be taken that fire hazards are removed and positive means are taken to confine the heat, sparks, and slag. No welding, cutting, or heating may be done where the application of flammable paints or presence of flammable compounds creates a hazard.

Suitable fire watch and fire extinguishing equipment must be immediately available in the work area. When welding, cutting, or heating is performed on walls, floors, or ceilings, since direct penetration of sparks or heat transfer could introduce fire hazard to an adjacent area, a fire watch must be taken on the opposite side from which welding is being performed.

Welding or cutting in enclosed spaces with certain types of rod or electrodes may cause hazardous fumes. In particular, zinc-bearing base or filler materials, lead base materials, cadmium-bearing filler materials, chromium metals, and beryllium base require positive ventilation. Similarly, inert gas metal arc welding generates high ultraviolet concentrations, so chlorinated solvents should be kept at least 200′ from the welding position.

WELDING CHECKLIST

1. Review qualification of welders, including testing where appropriate.
2. Check to see that all welds required by the shop drawings have actually been made, that they are accurately located and are of the specified size.
3. Be certain that the nondestructive shop testing of welds by X-ray, magnaflux, or ultrasonic testing procedures have been performed, and the test locations identified.
4. Be certain that similar field tests required have been made.
5. Make sure that joints to be welded are properly clamped or tack-welded, so that uniform stresses are built up.
6. If stress relieving is required, review the procedure in the field.
7. For certified welds, be certain that welders are placing their stamp and compare unsuccessful welds with individual welders, shifting or disqualifying welders who fail to make a reasonable percentage of their work correctly.
8. Check that proper type and size of welding rod is used, particularly where special strength steel is joined, or for materials that are difficult to weld, such as aluminum.

05100 STRUCTURAL METAL FRAMING

Structural steel construction is becoming increasingly popular, and presents a minimum of problems for the field inspection team because a maximum amount of the work is done under production conditions. In response to the required design prepared by the structural engineer, and approved by the architect, the steel fabricator prepares shop drawings. With these shop drawings, a fabrication or production shop fabricates the individual members from structural shapes provided by heavy steel manufacturers.

The work in the *fabrication shop* is principally cutting to size, and fitting with connection plates, angles, and splices. In the production shop, several members are often preassembled so that connection holes can be aligned, reamed to insure a proper fit—and then disassembled for shipment. In the shop, a prime coat of paint is applied after steel scale and foreign materials have been removed. Usually, the fabricating shop has a sandblasting area and paint shop, which make the operation swift and economical. The members are marked with erection numbers matching erection numbers shown on the erection plan.

When the structural steel members arrive at the job site, they are either placed into storage (if storage is available) or delivered directly to the assembly area. Prearrangement should be made with the fabricating shop for the sequence of delivery. Where *storage* space is available, all steel may be delivered to the job and stored on skids. (Skid storing is preferred to placement on the ground, since it saves cleaning before erection.) However, if site storage space is limited—other arrangements must be made. It is quite common for the higher members in a high or mid-rise structure to be detailed first, and therefore, fabricated first. When this occurs, direct job site delivery is not feasible and interim storage must be arranged.

Care should be taken in *unloading* members, since poor practices such as using connection angles for lifting points, or dragging members from trucks or flat cars from one end only may impose overstressing, resulting in kinks or misalignment. Members should not be dropped or impact loaded.

Many steel fabricators provide erection services, and the resulting coordination between manufacturer and erection can be very useful. Conversely, it requires additional inspection—since there is the opportunity to correct fabricating mistakes in the field. While this is satisfactory if proper corrective action is used, the inspection team must avoid practices such as enlarging of connection holes, caulking of rivets, and hickeying of lighter members.

Erection. Approved *shop drawings* should be available at the job site before the start of erection. The inspection team should check the steel for size and shape against the shop drawing. The type of connectors should be reviewed before erection starts. Members should be free of kinks, bends, and other distortions. No straightening of bent or misaligned members should be allowed in the field unless techniques for straightening are approved by the design team. When in doubt, check with the structural engineer. Unless specific approval is stated to permit use of rehabilitated members, all structural members should be new. If there was shop inspection, it should have included the shop connections. However, where all inspection is in the field, shop connections should also be inspected. Where column ends are to have been milled, check to determine proper milling, and also check whether cap and base plates for columns have been properly welded in the shop.

Check the shop *painting* for painter's holidays, loose scale or rust, or abraded areas, making certain that all defects are satisfactorily corrected immediately. (Painting on the ground, as well as inspection, is much more economical than in the air.) Before erection starts, the contractor should review the erection technique with the inspection team—and where considered necessary, the structural engineer. The inspection team should check the contractor's *equipment*, particularly the cranes for adequate capacity of intended loads, and where appropriate, cranes should actually be boomed and cable tested at the job site. Visual inspection of cables should determine that they are not frayed or worn, and that necessary safety stops are installed.

The contractor should check the *dimensions* of column base plates against actual field bolt dimensions in the foundations—but a confirmation check or field check is in order. Check also to see that there is sufficient length of bolt protruding above the concrete to allow full engagement by the nut, and even double nutting where practical.

Girders and beams are erected in horizontal position with the web vertical, so that the hooks on the wire rope slings should be taken around the girder in opposite directions for stability. Drift pins or the spud end of wrenches are used to align the holes of girders to their supporting members, and then the beams are secured with two to four bolts in each end. Permanently bolted connections should have a bolt in each hole, and heavy girders should be temporarily fastened with at least four bolts, and preferably more (Figs. 4, 5, and 6).

Trusses are triangular modules connected into a structural framework used to span large areas without the necessity of intermediate support. This method requires less material than conventional framing, but requires greater head room. Smaller trusses are shipped to the job in one piece, but the typical truss arrives in sections which must be assembled in the field. Whenever possible, the trusses should be preassembled before erection. *Lateral stability* is a problem with trusses, and if handled improperly, they will have a tendency to twist or bend sideways during erection. The most critical truss is the first one since it has no others to which it can be side-braced. Accordingly, guy ropes are usually attached to the top chord and secured to prevent overturning. The second truss is erected in the same fashion, and then vertical cross bracing is placed between the two truss members. The inspection team should confirm the point at which guy ropes can be removed and the remainder of purlins, sag rods, and bracing can be installed.

Purlins are beams spanning from truss to truss and their spacing may be dependent on the roof or ceiling construction. Roof purlins are

Fig. 4 Tower rig on Manitowoc crane erecting beam on 18th floor of steel frame building. *Source: Manitowoc Engineering, Division of Manitowoc Company.*

usually light enough so that they can be hauled from the uppermost floor by hand lines attached to the connection holes. Despite all precautions, roof trusses may be slightly warped, but within tolerance. Nevertheless, this can make the fit-up of purlins difficult, requiring come-alongs or light tackle to pull them into position, or sledges to force them.

After the rope guys have been removed when the basic vertical *cross bracing* has been placed, the framing system is still susceptible to cracking until the top and bottom chord bracing is in place. The bottom chord bracing is either ceiling purlins or angle sections, either of which is

Fig. 5 Tower rig erecting lower floor of steel frame structure, utilizing sling to lift beam. *Source: Manitowoc Engineering, Division of the Manitowoc Company.*

easily handled by hand labor, and bolting into place should not be troublesome. The diagonal bracing of the top chord strengthens and helps to square the frames. The vertical cross bracing helps to plumb the frames. Top chord bracing usually consists of rods threaded at both ends for a length of about 6″, with each rod provided with two nuts. The inside nuts must be screwed down to the end of the threads and the outside removed so that the rods can be adjusted to the actual field conditions.

In order to align the *base plates* of columns, it is a frequent practice to

Fig. 6 Crane erecting heavy girder section utilizing lifting strong back and tackle. *Source: Manitowoc Engineering, Division of the Manitowoc Company.*

install bearing plates prior to the erection process. This reduces the amount of handling of the column during erection. The bearing plate is set, and firmly anchored down over metal wedges or shims. There should be enough space left (at least $\frac{1}{24}$ of the bearing plate width, or about $1''$ minimum to permit installation of nonshrink grout or mortar.

When sufficient members have been placed to start alignment of the structure, proper instrumentation should be used to determine that beams are level and columns are plumb within *tolerances*. The structure must be viewed in its total sense as well as in tolerances of individual members. That is, the cumulative effect of tolerable errors can be out of the allowable overall. If, however, individual members are within tolerance, the building can be moved and adjusted by releasing and retightening erection bolts. The inspection team should not permit the use of cutting torches for correcting fabrication errors in the field. When the building has been *plumbed*, or levels have been plumbed, erection bolts should be tight and guy lines taut. During the plumbing operation, care should be taken not to overstress back guys, since progressive failure could occur.

The standards of the American Institute of Steel Construction are often incorporated into the specification, and provide excellent guidance in the fabrication, erection, and final connections of structural steel.

OSHA REGULATIONS—STRUCTURAL STEEL ERECTION

As the erection of structural steel members progresses, permanent floors are to be installed, with no more than eight stories between the erection floor and the uppermost permanent floor. The exception to this occurs when the structural integrity is maintained within the design and exceptions may be permitted, but only with the permission of the design structural engineer.

At no time is there to be more than four floors or 48' of unfinished bolting or welding above either the foundation or the uppermost permanently secured floor.

The derrick or erection floor is to be solidly planked over its entire surface, except for access openings. The planking is to be suitable for the working load, but not less than 2" thick and laid tight.

On buildings or structures not adaptable to temporary floors and where scaffolds are not used, safety nets must be installed and maintained whenever the potential fall distance exceeds 2 stories or 25'.

A safety railing of $\frac{1}{2}$" wire rope or equal is to be installed about 42" high around the periphery of all temporary plank or temporary metal deck floors of tier buildings and other multifloor structures during structural steel assembly.

Where erection is being done by means of a crane operating on the ground, a tight and substantial floor must be maintained within 2 stories or 25', whichever is less, directly under that portion of each tier of beams on which bolting, riveting, welding, or painting is being done.

During final placemement of solid web structural members, the load is not to be released from the hoisting line until the members are secured with at least two bolts or the equivalent at each connection and drawn up wrench tight. Open web joists shall not be placed on any structural steel framework until it is solidly bolted or welded.

Where long span joists or trusses 40' or longer are used, a center row of bolted bridging shall be installed to provide lateral stability during construction prior to slackening of the hoisting line.

Containers for storing connectors as well as hand tools shall be carefully secured against accidental displacement aloft. Care shall be taken to avoid falling of any items such as bolts or driftpins.

For plumbing up, turnbuckles must be secured to prevent unwinding while under stress. Guys for plumbing up must be placed so that employees can get at the connection points, and shall be removed only under the supervision of a competent person.

Connections. There are three basic methods of connecting steel members: riveting, high strength bolting, and welding.

For many years, *riveted connections* were the only ones found in structural steel erection. The rivet must be driven through a well-aligned hole, and comes with a head on one end, and is driven while hot. Actually, the rivet fits exactly through the rivet hole, and the driving represents the forming of a new head on the shank end, with the riveter driving against a bucker on the head side. Riveting does not result in a clamping action, so the members must be firmly held together while the riveting is in progress. Any loosening of the fit will result in loose rivets. Even when the members are held tightly together with clamps or shop bolts, poor driving can result in loose rivets. The availability of air-driven tools to improve the productivity in shop riveting and the ability to lay members horizontally provides an opportunity to use equipment hung from overhead cranes. The converse has not been true in the field, and field riveting continues to be an unwieldy process.

Erection bolts are not suitable for permanent connection, since they tend to loosen over a period of time. However, the development of a *high-strength bolted connection* has revolutionized the structural steel erection process. The high-strength bolt is identified by three radial marks on the head and three long indented marks on the nut. The bolt is used with a washer on the side of the element that is turned, and all of the elements are made of high carbon steel. Ordinary washers and nuts must not be used—and this is an element which the inspection team must be aware of. The highest tensile bolt is driven with an air-driven impact wrench which is precalibrated to drive to the right point of refusal. The high-strength bolt is torqued, actually stretching the bolt. The result is not only the resistance to shear which the rivet offered, but also a clamping action which provides movement resistance as well as simple shear and tension.

The high tensile bolt meets ASTM A-325 requirements, and its installation is described in the AISC pamphlet "Structural Joints Using ASTM A-325 bolts."

Even though the high-tension bolts have the ability to clamp, contact services should be clean of mill scale, burns, pits, dirt, oil, and other foreign materials which would prevent solid seating of parts.

Specifications vary as to whether the contact surfaces of high-strength bolted connections should or should not be painted.

The inspection team should review test *calibration* of wrenches used for high-strength bolting—and this should be on a twice-daily basis. The inspecting team should also have a calibrated wrench to check the bolt tension actually achieved, and between 5% and 10% of the bolts should actually be checked.

Another approach to measuring the tension on the high-strength bolt is known as "turn-of-the-nut." After the nut is finger tight, it is turned a specified amount, which can be checked by observing markings on the nut and washer.

The inspectors should also ensure that high-strength bolts are not reused. since they do not develop design strength in this situation.

The introduction of high-tensile bolts in the field revolutionized the erection process, and gradually has moved to the fabrication shop. Installation is easier and faster, and the use of rivets offers no real advantage over high-strength bolting, so most shop connections today are also made with high-strength bolts.

Where the level of strength achieved with high-strength bolting is not required, new bolts have been developed which are the equivalent of rivets called turned bolts and/or ribbed bolts. A325 bolts have a strength at yield of about 81,000 psi and a minimum tensile strength of 120,000 psi. A490 steel bolts are also utilized for high-strength bolting.

The utilization of the joint between members assumes either a friction-type or bearing type connection. The difference is usually discerned principally by use of paint or no paint in accordance with the drawings. Usually, the threaded length of the bolt is kept as short as possible so that the full shank is available to resist shear between the members being connected. Accordingly, selection of the right size bolt is of great importance, with the difference in allowable strength being as much as 50% under AISC specifications.

Where the high-tensile bolt is to be subjected to vibration or dynamic loading, locking devices may be specified. There are a number of special locking features available in proprietary patented devices such as union-nut, powell-nut, and others. Tack-welding the nuts is another method of locking.

Welding is of great importance in the connection of all types of metals in buildings, including structural steel connections. Welding requires careful procedural workmanship—and therefore, careful inspection. Mistakes and errors in welding are often not apparent unless the entire procedure has been reviewed, and even then X-ray testing is often required to be certain that a joint is totally effective. For that reason, welding in the field under difficult conditions may produce less certain results than high strength bolting. Accordingly, the designer considers the efficiency required from the weld specified, and usually over-designs connections on the side of safety factors.

Check finished welds for size, length and standards of workmanship, using appearance of the surface, surface defects, craters, undercutting, cracks, etc., as methods of evaluation.

Weld location is important, and placing welds in the wrong location can be just as serious as omitting them altogether. Check dimensions.

Overwelding, either in size or length of welds is discouraged, since this may result in distortions due to overheating of the connection. Surfaces to be welded should be free of scale slag, rust, grease, and other foreign materials.

STRUCTURAL STEEL ERECTION CHECKLIST

1. Check steel shapes and sizes using approved shop drawings.
2. Check the shop painting for holidays, scale rust or abraded areas, correcting before erection.
3. Check contractor's equipment for condition and adequate lifting capacity.
4. Check column and beam connection plates for matching.
5. Heavy sections should be temporarily fastened with at least four bolts at each connection point.
6. Stability should be maintained at all times during erection.

7. During plumbing operation, take care not to overstress back guys.
8. Review the test calibration of high-strength bolt wrenches on a twice-daily basis.
9. Spot check high strength bolts with a calibrated wrench available to the team.
10. High-strength bolts should not be reused if once removed.
11. Insist that safety nets be rigged or full floor be planked over two floors below erection level, or as required by appropriate safety regulations.

05200 OPEN WEB JOISTS

Open web joists are lightweight steel trusses, used either as roof framing for single-story buildings, or as floor framing within main building structures. Spans up to about 48' are known as short span joists, while those up to 96' are long span joists. The principal advantage to the open web steel joist is its light weight, with respect to strength. A secondary advantage is the ability to place a hung or finished ceiling on the bottom chord, and to thread utilities and services transverse to the joist web. When joists are used for floor framing, it is usual to pour a lightweight or structural concrete slab either on metal pan forms supported from the joists, or as a composite including the upper chord. The depth of open web joists ranges from 8 to 24", with 12 to 24" being the more usual range. The usual construction involves angle iron chords with a steel bar web, but there are variations on the pattern. The Steel Joist Institute provides information on allowable loads and dimensions.

Open-web joists must be well-anchored, and should have lateral bridging because of limited stability to accept transverse loading. Since the joists are light in relation to the load, no cutting or splicing of main tension members is permitted, and no holes should be burned or drilled in the chords.

05300 METAL DECKING

Metal decking is usually used as a form for floor slabs within a steel structure or roof slabs. In some cases, for roof deck, metal deck is topped with insulation and then built up roofing without a poured slab. Metal deck should be clean when put down, and anchored directly to the steel purlins. All joints should be lapped and fastened properly. Decks should be dry, clean, and free of rust or oily material before application of insulation material. The roof deck can be sloped up to 2″ per foot, and should be properly graded to drains. Insulation should be dry when applied, and of a thickness dependent upon the width of the flute openings of the metal deck. No more insulation should be put in place at any one time than can reasonably be protected by roofing pitch and felt to prevent any damage due to sudden weather changes.

In certain types of floor decking, *studs* are welded to the metal pans, making the combined concrete and pans a structural system. This composite installation permits more shallow slab thicknesses. When placing pans for poured slabs or deck, the fit between panels is particularly important to preclude leakage of the fluid concrete mixture.

05400 COLD-FORMED METAL FRAMING

Lightgauge framing is constructed of cold-formed steel sections which are similar in their general configuration to the heavy structural members such as: beams, columns, channels, and zees. However, the cold form shapes are made from a single piece of material, and are therefore uniform in cross section. I- beams are usually made by combining two channels back to back, and other special shapes are made by braking, rolling, or crimping sheet material. Because of the method of fabrication, sharp corners are not achieved, at least not nearly in the degree that they can be hot rolled.

The lightweight members require individual calculations, and tend to be more subject to buckling stresses.

A popular utilization for lightgauge framing is in manufactured or prefabricated buildings, where production techniques can be used to manufacture the members, and the repetitious utilization justifies in-depth calculations to achieve the lightest weight feasible within appropriate safety parameters. Because of their unique characteristics, lightgauge members are made essentially on a custom basis, and therefore, are less susceptible to tabular calculation or reference. Similarly, because of the broad range of shapes available, the lightgauge members do not fall within the ordinary minimum sizes and limitations of the AISC recommendations.

Often, the lightgauge members are made with a groove which permits *nailing*, so that dry wall can be attached directly to the structure.

From the field inspection viewpoint, approval of lightgauge structures by the structural engineer is important, and therefore approved shop drawings are particularly important. During construction, the inspection team should ensure that members are handled and well supported so that buckling does not occur due to inadequate support during erection.

05500 METAL FABRICATION

This section describes a large variety of metal products, the majority of which are constructed of ferrous metals, but also includes many nonferrous metals. Usually, miscellaneous products are fabricated to specific dimensions necessary for the specific utilization, and are often not a standard design.

Typical miscellaneous metal items include:

Access doors and frames	Angles
Balconies	Bearing plates
Bench supports	Bicycle racks
Bleacher seat supports	Bumpers, guards, posts
Cable supports	Canopies, marquees
Catwalks, walkways	Chimney accessories
Chutes	Control joints
Counter accessories	Crosses curb edge bars
Desk accessories	Door guards
Elevator pit and machine room accessories	Expanded metal
Fire escapes	Grating guards
Handrails	Hood support framing
Ladders	Letters and numerals
Manhole rings and covers	Pipe railings

05510 Metal Stairs

The architectural drawings show the conceptual design and dimensional parameters of stairways. The fabricator provides complete details of the structural parts and their connections in his shop drawings, which in turn are approved by the architect. This permits the various manufacturers to utilize standard proprietary details.

The National Association of Architectural Metal Manufacturers (NAAMM) has established a logical classification system for types and kinds of stairs. *Metal stairs* are classified according to types and class, with the type identifying physical configuration or geometry, while the class refers to construction characteristics.

The four *types* of stairs defined are: straight, circular, curved, and spiral. It is not uncommon to find two or more stair types represented in one stair.

According to NAAMM, straight stairs are the most common type representing the bulk of the stair market. A straight stair is one defined in which the stringers are straight members, still permitting a number of arrangements under this type classification. For instance, *straight* stairs can be: straight run, parallel runs, angled, or scissor.

Circular stairs in plan view have an open circular form with a single center of curvature. They may or may not have intermediate platforms between floor levels. *Curved* stairs in plan view have two or more centers of curvature being elliptical or oval in form. Curved stairs may also be a compound curve.

Spiral stairs have triangular-shaped radiating treads, whose smaller ends are attached to and supported by a vertical center post, which may be fabricated from pipe.

The class designation is a key to the type of construction, the quality of materials, the details in finish and relative cost. The four *classes* of stairs, listed in order of decreasing cost, are: architectural, commercial, service, and industrial.

The *architectural* classification includes the more elaborate stairs which are intended to become architectural features in a building. Usually this class has a comparatively low pitch, with relatively low risers and wide treads. Materials and fabrication details vary widely in response to the architect's conceptual design and specifications. In this class, the aesthetic features prevail, while the stair function is mitigated, softened, or blended into an architectural background.

Stairs of the *commercial* class are usually for public use, but of somewhat simpler design than the architectural class. They may be in open locations or can be located in closed stairwells, in low-rise public institutional or commercial buildings. The stringers are usually exposed open channel or plate sections, while treads may be any number of standard types. Railings may be ornamental tube or simple pipe.

The *service* class of stairs is chiefly functional, but not unattractive in appearance. Stringers are usually the same as those used in commercial class, while treads may be standard-type of treadplate or filled. Railings are typically pipe or simple bar.

The *industrial* class of stairs is purely functional and therefore, usually the most economical. Stringers may be open channels or flat plate, while treads and platforms are made of grating or formed floor plate.

Some stairs are designed for preassembly in the plant, and therefore, are easily assembled in the field. Preassembled stairs may be any class, but are generally straight stairs of the commercial and service classes. In multistoried buildings, the use of preassembled units can effect considerable savings and ease of installation.

Metal stairs, regardless of class, are designed to carry a live load of not less than 100 lb/ft^2, and conform with all safety requirements of applicable building codes.

According to NAAMM, metal stairs can usually be installed earlier in a steel frame structure than in one having a concrete frame. Erection tolerances in the steel framing around the stairwell are minimal, so that the stairs can be detailed and shop drawings prepared and the stairs fabricated before the frame is erected. In fact, erection of the stairs in parallel with the erection of the frame can assist in the access for the building trades. In some instances, the stair unit being self-supporting has been erected before the building. In concrete frame structures, the formwork and shoring precludes concurrent erection, so that the stair installation must follow by several floors.

Notes for Metal Stairs. The inspection team should be concerned with the *sequence* of operations in stair installation. This is particularly true where there is a masonry stairwell, and the walls are to be finished with a facing material such as tile, marble, or brick, in which case the stair should be erected before the finished wall material. This permits a better fit between the stair and the wall stringer, and also results in less damage to the finished material. If the sequence of trades is such that this is impractical, the wall stringer should be set out from the wall enough to provide clearance for the finish.

05530 Floor Gratings

The marking system recommended by the NAAMM grating committee includes five characteristics as follows: type of grating; bearing bar spacing; cross bar or rivet spacing; size of bearing bars; and material.

Grating *type* is indicated by letter: W for welded; P for pressure-locked; and R for riveted.

Bearing *bar spacing* is designated by a number which indicates the bearing bar spacing in sixteenths of an inch. For welded or pressure-locked grating, this figure indicates the center-to-center space between bars, while in riveted grating, it is the distance between bearing bar faces.

Cross bar or rivet spacing is designated by a number which indicates the spacing in inches. For pressure-locked or welded grating, this figure is the distance from center to center of cross bars, while in riveted grating, it is the distance center to center of rivets measured along a single bearing bar. The size of bearing bars is expressed in inches of depth and thickness. The metal is designated by names such as aluminum or steel.

05700 ORNAMENTAL METAL

Ornamental metal products are used for functional, architectural, and decorative effects where appearance is of primary consideration. These metal products are often specially designed, lending character and beauty to every type of building construction. They offer the architect an unusual opportunity for imaginative and attractive designs.

Typical ornamental products include: alter rails; bell tower accessories; clock face numerals; counter screens, gates, wickets; decorative screens or panels; fences and gates; foot rails; fountains; furniture and fixtures; grilles; lamps, lanterns, chandeliers; louvre screens; marquees; railings.

DIVISION 6: WOOD AND PLASTIC

NUMBER   TITLE

06050   FASTENERS AND SUPPORTS

06100   ROUGH CARPENTRY
 −10   Framing and Sheathing
 −11   Light Wooden Structures Framing
 −12   Preassembled Components
 −13   Sheathing
 −14   Diaphragms
 −20   Structural Plywood
 −25   Wood Decking
 −26   Fiberboard Decking
 −27   Fiber Underlayment
 −28   Asbestos Cement Panels

06130   HEAVY TIMBER CONSTRUCTION
 −31   Timber Trusses
 −32   Mill-Framed Structures
 −33   Pole Construction
 −34   Trestles

06150   WOOD-METAL SYSTEMS
 −51   Wood Chord Metal Joists

06170   PREFABRICATED STRUCTURAL WOOD
 −80   Glue-Laminated Construction
 −81   Glue-Laminated Structural Units
 −82   Glue-Laminated Decking
 −90   Wood Trusses
 −92   Fabricated Wood Trusses

06200   FINISH CARPENTRY
 −20   Millwork
 −40   Laminated Plastic

06300	WOOD TREATMENT
−10	Pressure Treated Lumber
−11	Preservative Treated Lumber
−12	Fire Retardant Treated Lumber
06350–06399	(Reserved)

06400	ARCHITECTURAL WOODWORK
−10	Cabinetwork
−11	Wood Cabinets: Unfinished
−20	Paneling
−21	Hardwood Plywood Paneling
−22	Softwood Plywood Paneling
−30	Stairwork
−31	Wood Stairs and Railings

06500	PREFABRICATED STRUCTURAL PLASTICS
06550–06559	(Reserved)

06600	PLASTIC FABRICATIONS
06650–06999	(Reserved)

06100 ROUGH CARPENTRY

The American Institute of Timber Construction (AITC) points out many of the advantages of *timber* construction including strength, light weight, and speed. Pound for pound, engineered timber is stronger than other structural materials in most structural properties, including deflection. Conversely, volume for volume and area for area, it is about $\frac{1}{10}$ the strength of steel, and deflects more than steel.

Properly used, wood is an exceptional construction material because it can be readily handled, and can be worked with the simplest of tools. It is also amenable to power tool usage.

The AITC is seeking better treatment of wood, when properly engineered, by planners and fire insurance people in terms of standards. While not acceptable where totally noncombustible materials are required, wood can play a part in fire-resistant construction. Properly installed and protected, wood construction can be economical, safe, and attractive.

Timber and structural construction fall into the rough carpentry category, and usually infer work not open to view. With the advent of glue-laminated wood structural members, the structural framing need not necessarily be hidden, and is not necessarily rough.

06110 Framing and Sheathing

Framing. Framing is usually used for interior partitioning in buildings, but for one- or two-story buildings, may provide the exterior framing wall also. Framing typically in residential structures includes the structure itself—but in commercial or industrial applications, provides the interim structure between the main building structure. One of the purposes of this type of framing is support of the exterior sheathing, or the interior wall board or plaster.

The first item to be reviewed by the inspection team is the type of lumber used in the framing. This should be in accordance with the specifications and is checked according to species and grade. *Grade stamps* are placed in the end of the lumber identifying its type and structural properties.

The Federal Specifications MM-L-00736 and MM-L-00751 provide the appropriate Lumber Association grading rules. These rules give the allowable defects for various grades of lumber, and give guidelines for review and inspection. However, the inspection team can also use common sense, and preclude members which are damaged or have gross defects.

Particularly where the framing is structural in nature, the inspector should check for allowable *moisture content* of lumber, which can be checked with a moisture meter. It also can be checked in the laboratory, and is a factor in determining the proper aging and seasoning of the wood. (Unseasoned lumber will dry out, warping and deforming after installation.) Generally, lumber 1″ thick or less should have a moisture content of less than 12% and over 1″ thick, a moisture content of less than 15%.

Storage at job site is important, and lumber should be stored on blocking off of the ground, and in properly drained areas. Covering should be provided to prevent increase in moisture content if lumber is to be stored for any length of time.

Inspectors should check to be certain that *creosote* has not been used as a preservative for joists, sleepers, or sills when the material will ultimately be in contact with an exposed finished material such as plaster or gypsum board, or when the treated material is to receive a painted finish.

As the framing is assembled, the inspector should check for *accuracy* of line, level, fabrication, and fit. Exterior wall studs should be checked

Allowable load in withdrawal in pounds per inch of penetration into side grain of member holding point.

d = penny weight of nail or spike. G = specific gravity of the wood based on weight and volume when oven-dry.

Specific gravity G	SIZE OF COMMON NAIL (d)										SIZE OF THREADED NAIL (d)*						
	6	8	10	12	16	20	30	40	50	60	30	40	50	60	70	80	90
.75	76	87				127	138	150	162	174	127	127	127	127	150	150	150
.63	60	69	78	78	85	101	109	119	128	138	101	101	101	101	119	119	119
.67	57	66	75	75	82	97	105	114	124	133	97	97	97	97	114	114	114
.66	55	64	72	72	79	94	101	110	119	128	94	94	94	94	110	110	110
.62	47	55	62	62	68	80	86	94	102	110	80	80	80	80	94	94	94
.55	34	39	44	44	49	57	61	67	73	79	57	57	57	57	67	67	67
.54	32	38	42	42	46	54	58	64	70	75	54	54	54	54	64	64	64
.51	29	34	38	38	42	49	53	58	63	68	49	49	49	49	58	58	58
.48	25	29	33	33	36	42	46	50	54	58	42	42	42	42	50	50	50
.47	24	27	31	31	34	40	43	47	51	55	40	40	40	40	47	47	47
.46	22	26	29	29	32	38	41	45	48	52	38	38	38	38	45	45	45
.45	21	25	28	28	30	36	39	42	46	49	36	36	36	36	42	42	42
.44	20	23	26	26	29	34	37	40	43	46	34	34	34	34	40	40	40
.43	19	22	25	25	27	32	35	38	41	44	32	32	32	32	38	38	38
.42	18	21	23	23	25	30	33	35	38	41	30	30	30	30	35	35	35
.41	17	20	22	22	24	29	31	33	36	39	29	29	29	29	33	33	33
.40	16	18	20	20	22	27	28	31	33	35	27	27	27	27	31	31	31
.39	15	17	19	19	21	25	26	30	32	34	25	25	25	25	30	30	30
.38	14	16	18	18	20	24	25	28	30	32	24	24	24	24	38	38	38
.37	13	15	17	17	19	22	24	26	28	30	22	22	22	22	26	26	26
.36	12	14	16	16	17	21	22	24	26	28	21	21	21	21	24	24	24
.34	10	12	14	14	15	18	19	21	23	24	18	18	18	18	21	21	21
.31	9	10	12	12	13	15	16	18	20	21	15	15	15	15	18	18	18

*Loads for threaded, hardened steel nails, in 6d to 20d sizes, are the same as for common nails.

	10	12	16	20	30	40	50	60	5/16″	3/8″
						SIZE OF COMMON SPIKE				
.75	127	127	138	150	162	174	188	188	205	250
.68	101	101	109	119	128	138	149	149	164	198
.67	97	97	105	114	124	133	143	143	158	190
.66	94	94	101	110	119	128	138	138	152	183
.62	80	80	86	94	102	110	118	118	130	156
.55	57	57	61	67	73	79	84	84	93	111
.54	54	54	59	64	70	75	81	81	88	106
.51	49	49	53	58	63	68	73	73	80	96
.48	42	42	46	50	54	58	62	62	69	83
.47	40	40	43	47	51	55	59	59	65	78
.46	38	38	41	45	48	52	56	56	62	74
.45	36	36	39	42	46	49	53	53	59	70
.44	34	34	37	40	43	47	50	50	55	66
.43	32	32	35	38	41	44	48	48	52	63
.42	30	30	33	35	38	41	45	45	49	59
.41	29	29	31	33	36	39	42	42	46	56
.40	27	27	28	31	33	36	39	39	42	51
.39	25	25	26	30	32	34	37	37	40	48
.38	24	24	25	28	30	32	35	35	38	46
.37	22	22	24	26	28	30	33	33	36	43
.36	21	21	22	24	26	28	30	30	34	40
.34	18	18	19	21	23	24	26	26	29	35
.31	15	15	16	18	20	21	22	22	25	30

Fig. 1 Nail sizes and strength. Source: National Forest Products Association.

for plumb and alignment, and this check should be made before any exterior masonry is begun where exterior masonry veneer is utilized. The connection and method of securing members should be checked. Figure 1 is a list of strength of typical nail sizes. Workmanship should be reviewed, and should be of a superior class. The inspector must use judgment in terms of the degree of fit in the final use of the material. If, for instance, the framing is to be covered completely, then the main criteria is structural integrity. Sound joints, and well-placed nails or bolts, are the important factors. Use of the proper size of nails or bolts is of vital importance, as is penetration.

Framing members should not be cut or notched without leaving sufficient strength to carry load. All openings should be framed, and a 2″ space should be maintained between chimney and timber, and 4″ between the fireplace back wall and timber.

Where the framing is anchored to concrete or masonry, *anchorage* should be provided near the end of each section of sill, as well as at the specified spacing. The ends of every fourth joist should be anchored, and joists which parallel masonry walls must be anchored every 8′ at least. Anchors should be extended over (and fastened to) at least 3 joists. Anchors should be provided for window frames and door bucks, and for end studs of partitions abutting the masonry.

Joists should be spaced as detailed or specified, and a minimum of 4″ bearing should be provided. Bridging and blocking provide the cross or lateral support for joists and should be installed as specified. The nailing of the lower ends of cross bridging should remain until the sheathing or subflooring has been placed.

Where joists frame into headers or girders, they should be properly hung, and splices in wood joists, headers, or girders should be lapped with a sufficient length to permit spiking or bolting.

Between joists or sleepers 2″ × 4″ framing should be used to support the cut ends of diagonal subflooring. In some cases, joists bear on the bottom flange of steel beams and should be tied with metal ties across the beam on every fourth joist at least. Where the plans show a partition running parallel with the joists, they should be doubled up to carry the load.

As a guideline, structural members should not be cut, notched or bored more than ¼ of their depth without adequate additional reinforcing which should be approved by the inspection team.

In wall and partition framing, solid *blocking* should be provided for the hanging of fixtures, hand rails, cabinets, baseboard, and other wall hung items. Studs should be doubled at openings, and for exterior framing, diagonal wind bracing may be called for in the specification. For *soundproof* partitioning, split plates may be called for or staggered studs, providing an offset between the opposite wall boards. Insulation may be required to be woven between studs, fastening only to every other stud. Visual inspection should determine that studs are plumb and have minimum bowing where dry wall or sheet paneling is to be used.

The New York City Building Code has the following empirical guidelines for wood framing:

(a) Stud walls and partitions.—
(1) Studs shall be of equivalent or better grade than the minimum grades for the various species as established in reference standard RS.

(2) Corner posts shall be 3-stud members or members of equivalent strength.

(3) Load bearing studs shall be set with the larger cross section dimension perpendicular to the wall or partition. Studs in exterior walls of one story buildings of construction class H-D and H-E shall be at least 2 in. $\times$ 3 in. spaced not more than 16 in. on center, or where studs are 2 in. $\times$ 4 in., spaced not more than 24 in. on center. Studs for other classes of construction shall be at least 2 in. $\times$ 4 in. spaced not more than 16 in. on center.

(4) Stud walls resting on concrete on masonry shall have sills at least 2 in. in nominal thickness. Where such sills bear on concrete, they shall be fastened with minimum $\frac{1}{2}$ in. bolts embedded at least 6 in. Each sill piece shall have at least two anchor bolts, with one bolt located at least 1 in. from each end of the plate, and with intermediate spacing not more than 8 ft. Where such sills bear on masonry, they shall be anchored in accordance with the applicable provisions of reference shall be anchored in accordance.

(5) Stud partitions that rest directly over each other and are not parallel to floor joists or beams may extend down between the joists and rests on the top plate of the partition, partition girder, or foundation below, or may be constructed on sill plates running on top of the beams or joists.

(6) All load bearing stud partitions shall be supported on walls, other partitions, double joists or beams, solid bridging, or on beams at least as wide as the studs. Joists supporting a partition parallel to the joists wherein the joists are spaced apart to permit the passage of piping or duct work shall be provided with solid blocking at intervals of not more than 16 in.

(7) Load bearing partitions perpendicular to joists shall not be offset from supporting girders, walls, or partition by more than the depth of the joists unless the joists are proportioned on the basis of analysis of stress.

(8) In interior walls and bearing partitions, double studs shall be provided at the sides of openings that are greater than 3 ft.-6 in. up to 6 ft. in width, and triple-studs shall be provided at the sides of openings of greater width.

(9) Headers shall be provided over each opening in exterior walls and bearing partitions. Where the opening does not exceed 3 ft., each end of the header shall be

supported on a stud or framing anchor. Where the opening exceeds 3 ft. in width, each end of the header shall be supported on one stud, and where the opening exceeds 6 ft., each end shall be supported on two studs.

(10) All studs in exterior walls and in bearing partitions shall be capped with double top plates installed to provide overlapping at corners and at intersections with other walls and bearing partitions. End joints in double top plates shall be offset at least 24 in. In lieu of double top plates, a continuous header of similar dimensions may be used. For platform frame continuous header of similar dimensions may be used. For platform frame construction, studs shall rest on a single bottom plate.

(b) Bracing of exterior walls.—Exterior stud walls shall be braced by 1 in. × 4 in. continuous diagonal strips let into the face of the studs and into the top and bottom plates at each corner of the building. Bracing may also be provided by one of the following means:

(1) Wood board sheathing of 1 in. nominal thickness, applied diagonally.

(2) For one and two-story dwellings, plywood sheathing at least 4 ft. × 8 ft. (except where cut to fit around openings and for similar purposes) and at least $5/16$ in. thick on studs spaced 16 in. or less on centers and at least $\frac{3}{8}$ in. thick on studs spaced more than 16 in. but not exceeding 24 in. on centers.

(3) For one story dwellings and for the upper story of two story dwellings, fiberboard sheathing applied vertically in panels at least 4 ft. × 8 ft. (except where cut to fit around openings and for similar purposes). Fiberboard sheathing shall be at least ½ in. thick.

(4) For one story dwellings and for the upper story of two story dwellings, gypsum board sheathing applied horizontally in panels at least 2 ft. × 8 ft. (except where cut to fit around openings and for similar purposes). Gypsum boards shall be at least ½ in. thick.

(c) Floor and roof framing.—

(1) Span Tables.—Joists and rafters may be used in accordance with reference standard.

(2) Bridging.—In all floor and roof framing, there shall be at least one line of bridging for each 8 ft. of span. The bridging shall consist of at least 1 in. × 3 in. lumber or equivalent metal bracing. A line of bridging or solid blocking shall also be required at supports unless lateral support is provided by nailing to a beam, header, or to the studs. Midspan bridging is not required for floor or roof framing in one- and two-family dwellings where joist depth does not exceed 12 in. Bridging shall bear securely against and be anchored to the members to be braced.

(3) Notches.—Notches in the ends of joists and rafters shall not exceed one-fourth the depth unless adequate reinforcement is provided or analysis of stress indicates that larger openings are feasible without the necessity for reinforcement. Notches in joists or rafters located in the span shall not exceed one-sixth the depth and shall not be located in the middle third of the span. Bored holes shall not be within 2 in. of the top or bottom of the joists or rafters and the diameter of any such hole shall not exceed one-third the depth. For stair stringers, the minimum effective depth of the wood at any notch shall be 3½ in. unless the stringer is continuously supported on a wall or partition.

(4) Support.—

a. Floor or roof framing may be supported on stud partitions.

b. Tail beams over 12 ft. long and header and trimmer beams over 6 ft. long shall be hung in metal stirrups having anchors, or by other methods providing adequate support. Trimmers and headers shall be doubled where the header is 4 ft. or more in length.

c. Except where supported on a 1 in. × 4 in. ribbon strip and nailed to the adjoining stud, the ends of floor joists shall have at least 1½ in. of bearing on wood or metal, nor less than 4 in. on masonry.

d. Joists framing from opposite sides of and supported on a beam, girder, or partition shall be lapped at least 4 in. and fastened, butted end-to-end and tied by metal straps or dogs, or otherwise tied together in a manner providing adequate support.

e. Joists framing into the side of a wood girder shall be supported by framing anchors, or ledger strips at least 2 in. × 2 in., or by equivalent methods.

(5) Rafters and Ceiling Joists.—
a. Where rafters meet to form a ridge, they shall be placed directly opposite each other and nailed to a ridge board at least 1 in. thick, and not less than the cut end of the rafters in depth.

b. Provisions shall be made to resist the thrust from inclined rafters by connection of collar beams at least 1 in. × 6 in., by connection to joists, or by equivalent means.

c. Where ceiling joists are not parallel to rafters, subflooring or metal straps attached to the ends of the rafters shall be installed in a manner to provide a continuous tie across the building.

d. Ceiling joists shall be continuous, or where they meet over interior partitions, shall be securely joined to provide a continuous tie across the building.

e. Valley rafters shall be double members. Hip rafters may be single members. Valley and hip rafters shall be 2 in. deeper than jack rafters.

(6) Built-up members shall be securely spiked or bolted together and provision shall be made to resist the horizontal shear between laminations.

(d) Nailing schedule.—The size and number of nails for connections shall be in accordance with Table 10-4.

Table 10—4   Nailing Schedule

Building Element	Nail Type	Number and Distribution
Stud to sole plate	Common-toe-nail	4—8d
Stud to cap plate	Common-end-nail	2—16d
Double studs	Common-direct	10d 12 in. o. c. or 16d 30 in. o. c.
Corner studs	Common-direct	16d 30 in. o. c.
Sole plate to joist or blocking	Common	16d 16 in. o. c.
Double cap plate	Common-direct	16d 24 in. o. c.
Cap plate laps	Common-direct	3—16d
Ribbon strip, 6 in. or less	Common-direct	2—10d each bearing
Ribbon strip, over 6 in.	Common-direct	3—10d each bearing
Roof rafter to plate	Common-toe-nail	3—16d

Building Element	Nail Type	Number and Distribution
Roof rafter to ridge	Common-toe-nail	2—16d
Jack rafter to hip	Common-toe-nail	3—10d
Floor joists to studs (no ceiling joists)	Common-direct	5—10d or 3—16d
Floor joists to studs (with ceiling joists)	Common-direct	2—10d
Floor joists to sill or girder	Common-toe-nail	2—16d
Double-joist to joist	Common-direct	10d—staggered at 16 in.
Ledger strip	Common-direct	3—16d at each joist
Ceiling joists to plate	Common-toe-nail	2—16d
Ceiling joists to every rafter	Common-direct	(See table below)
Ceiling joists (laps over partitions)	Common-direct	3—16d
Collar beam	Common-direct	4—10d
Bridging to joists and rafters	Common-direct	2—8d each end
Bridging to studs	Common-direct or toe	2—10d each end
Diagonal brace (to stud and plate)	Common-direct	2—8d each bearing
Tail beams to headers (when nailing permitted)	Common-end	1—20d each 4 sq. ft. floor area
Header beams to trimmers (when nailing permitted)	Common-end	1—20d each 8 sq. ft. floor area
1 in. Subflooring (6 in. or less in width)	Common-direct	2—8d each joist
1 in. Subflooring (over 6 in. in width)	Common-direct	3—8d each joist
2 in. Subflooring	Common-direct	2—20d each joist
1 in. Wall sheathing (8 in. or less in width)	Common-direct	2—8d each stud
1 in. Wall sheathing (over 8 in. in width)	Common-direct	3—8d each stud
Plywood sheathing and subflooring[a]		
1 in. Roof sheathing (6 in. or less in width)	Common-direct	2—8d each rafter
1 in. Roof sheathing (over 6 in. in width)	Common-direct	3—8d each rafter
Shingles, wood	Corrosion-resistive	2—No. 15 B&S each bearing
Weather boarding	Corrosion-resistive	2—8d each bearing
½ in. Fiberboard sheathing	1½ in. galvanized roofing nail 6d common nail 16 ga. galvanized staples, 1⅛ in. long, 7/16 in. crown	3 in. c. c. on all edges and 6 in. c. c. at other bearings

Building Element	Nail Type	Number and Distribution
25/32 in. Fiberboard sheathing	1¾ in. galvanized roofing nail 8d common nail 16 ga. galvanized staples, 1½ in. long, 7/16 in. crown	3 in. c. c. on all edges and 6 in. c. c. at other bearings
½ in. Gypsumboard sheathing	1½ in. galvanized roofing nail 11 ga. ⅜ in. to 7/16 in. head 16 ga. galvanized staples, 1½ in. long, 7/16 in. crown	4 in. c. c. on all edges and 8 in. c. c. at other bearings

[a] For nailing of plywood, see reference standard RS 10-9.

CEILING JOIST NAILING TO EVERY RAFTER
(Number of 16d nails)

Slope of Roof	4/12		5/12		6/12		7/12		9/12		12/12	
Rafter Spacing, o. c. (in.)	16	24	16	24	16	24	16	24	16	24	16	24
Width of building—												
Up to 24 ft.	5	8	4	7	3	5	3	4	3	3	3	3
24 to 30 ft.	7	11	6	9	4	7	3	6	3	4	3	3

Sheathing. Sheathing materials should be checked against specification requirements for type and size. The size and length of nails or fasteners to be used should be checked by the inspector. Check whether diagonal or horizontal application is required for floor planking. During installation, be certain that end joints occur only over framing and that the boards are accurately sawed.

Check the fastening, for deviations such as the use of stapling to secure plywood sheathing or subflooring.

Various *forms* of sheathing can be used including plywood, gypsum, and sheets of special composites which may have bitumastic binders, or plastic.

Gypsum sheathing is ½" thick, and is manufactured in 2′ × 8′, 4′ × 8′, and 4′ × 9′ sheets. The sheathing is manufactured in an envelope of heavy water-repellent paper, and weighs approximately 2 lb/ft². The gypsum core is fire resistant and water repellent. The 2′-wide panels have tongue-in-groove edges, while the 4'-wide sheets are applied with edges over the framing supports, thus providing weather protection. Gypsum sheathing may be applied either with nails or staples. However, the sheathing does not provide a suitable nailing base for materials to be applied to the outside, and exterior finishes must be nailed through to structural members, or nailers must be provided.

06130 Heavy Timber Work

While heavy timber construction is one of the oldest types of building framing in structure used, its utilization declined during the early portion of this century. However, with the advent of laminated members and solid sawn stress-graded lumber, more precise special design procedures can now be applied to heavy timber framing—with a much higher degree of reliability. This has given rise to the use of timber structures for multiple-story buildings such as churches, auditoriums and gymnasiums —as well as for industrial and storage buildings where utility and economy are prime factors.

By placing minimum sizing on wood structural members, and thickness and composition of floors and roofing, by avoiding concealed spaces under floors and roofing and by using approved fasteners, an improved degree of *fire resistance* has been achieved, and in fact is required if the structure is to be termed "heavy timber."

Wood *columns* for heavy timber construction must be at least 8" in any dimension when supporting floor loads, and not less than 6" in width and 8" in depth when supporting roof and ceiling loads only (size is nominal). *Beams* and girders must be at least 6" wide and not less than 10" deep, while arches and framed timber trusses supporting floor loads must have no dimension less than 8". Arches which support only roof loads may have a mimimum dimension of 6" in width, and 8" in depth for the lower half of the height and not less than 6" for any dimension for the upper half. Similarly, arches for roof construction which spring from the top of walls or wall abutments may be 4" in width and not less than 6" in depth.

In heavy timber construction, floors do not have concealed spaces and are of sawed or glued-laminated plank not less than 3" in thickness or of planks not less than 4" nominal, set on edge and well spiked together. The planks are laid so that no continuous line of joints occurs except at points of contact. The planking is covered with a 1" nominal pine groove *flooring* laid crosswise or diagonally, or with ½" plywood. Bearing walls must be of approved noncombustible material, or fire-retardant-treated wood with a resistance rating of not less than 2 hours, except that when a horizontal separation of 3' or less is provided, bearing portions of exterior walls must have a fire resistance rating of not less than 3 hours. For nonbearing portions of exterior walls, similar requirements exist. However, where a horizontal separation of 3' or less is provided—the fire resistance rating must be not less than 3 hours, where the separation is more than 3' but less than 20'—a fire resistance rating of not less than 2 hours is allowed, and where the separation is 20' to 30', nonbearing exterior walls may have a fire resistance rating of not less than 1 hour. If a horizontal separation of 30' or more is provided, nonbearing exterior walls are not required to have a fire resistance rating.

Figure 2 shows a multistory heavy timber construction building with decking on beams and girders. The connection points in heavy timber construction are very important, and must be in accordance with the detailed instructions of the structural engineer, as represented by the drawings and specifications. Metal seats, column caps, brackets, sheer plates, or split rings may be specified. In some cases, special connections such as precast reinforced column caps are called for, but these are not in the majority. Seats for the girders or beams in load-bearing brick or masonry must be carefully inspected for conformance with the specifications and plans.

From the New York City Building Code:

C26-1006.7 Heavy timber construction (construction class II-A).
(a) Minimum sizes of members.—To conform to the fire resistance rating requirements for heavy timber construction (construction class II-A), members shall be solid sawn or solid glue-laminated and of the following minimum dimensions: (Sizes of wood members indicated in this section are nominal sizes).
(1) Columns, Frames and Arches.—
a. Columns shall be at least 8 in. in all dimensions when supporting floor loads, and at least 8 in. deep when supporting roof and ceiling loads only.
b. Beams and girders shall be at least 6 in. wide and 10 in. deep.
c. Frames or arches that spring from grade or the floor line and support floor loads shall be at least 8 in. in all dimensions.
d. Timber trusses supporting floor loads shall have members at least 8 in. in all dimensions.
e. Frames or arches for roof construction that spring from grade or the floor line and do not support floor loads shall have members at least 6 in. wide and 8 in. deep for the lower half of the height, and at least 6 in. deep for the upper half.
f. Frames or arches for roof construction that spring from the top of walls or wall abutments, framed timber trusses, and other roof framing, which do not

support floor loads, shall have members at least 4 in. wide and 6 in. deep. Spaced members may be composed of two or more pieces at least 3 in. thick when blocked solidly through their intervening spaces or when such spaces are tightly closed by a continuous wood cover plate at least 2 in. thick secured to the underside of the members. Splice plates shall be at least 3 in. thick. When protected by approved automatic sprinklers under the roof deck, framing members shall be at least 3 in. wide.

contact surfaces. Mortises shall be true to size for their full depth and tenons shall fit snugly. No shimming in joints, or open joints, shall be permitted.

(b) Erection.—

(1) Assembly.—Joints shall have a tight fit. Fasteners shall be installed in a manner that will not damage the wood. End compression joints shall be brought to full bearing. All framework shall be carried up true and plumb.

(2) Temporary Connections.—As erection progresses, the work shall be bolted, or nailed as necessary, to resist all dead load, wind, and erection stresses.

(3) Alignment.—The structure shall be properly aligned before final tightening of the connections.

(2) Floors.—

a. Planks shall be splined or tongue-and-groove, not less than 3 in. thick, covered with 1 in. tongue-and-groove flooring, laid crosswise or diagonally to the plank, or other surface having equivalent fire resistance; or shall be,

b. At least 4 in. wide, set on edge close together and well spiked, and covered the same as for 3 in. thick plank. The planks shall be laid so that there is no continuous line of end joints except at points of support. Floors shall not extend closer than ½ in. to walls to provide an expansion joint, but the joint shall be covered at top or bottom to avoid flue action.

(3) Roof Decks.—Roof decks shall be splined or tongue-and-groove planks at least 2 in. thick; or tongue-and-groove plywood panels (bonded with exterior glue) at least 1⅛ in. thick, with face grain perpendicular to supports that shall be spaced not more than 48 in. on center; or of planks at least 3 in. wide set on edge close together and laid as required for floors.

06180 GLUE-LAMINATED WOOD

Glue-laminated lumber is made either in custom shapes, such as arches, or in standard shapes. The material is made of properly seasoned lumber which has been usually kiln-dried. The ability to inspect the components of the laminated lumber (that is, inspection of the individual wooden boards/planks which make up the lamination) provides a better quality control, and therefore, a more reliable stress loading can be assigned to laminated members.

The laminated members are made by selecting the proper quality components, gluing them with special high-strength glue, and curing under controlled conditions. Depending then upon the appearance grade required, the laminated member is finished. These grades are described

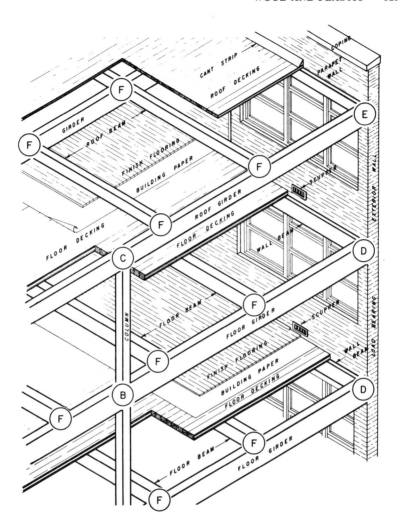

Fig. 2 Multi-story timber structure—decking on beams and girders. *Source: National Forest Products Association.*

in the standards of the American Institute of Timber Construction, and fall into the categories of: industrial, architectural, and premium appearance.

The inspection team must be careful to monitor the handling of laminated members, and their storage. The most obvious problem area is any indication that the laminations are separating.

06200 FINISH CARPENTRY

Finish carpentry is that exposed to view, and installed principally for the purposes of appearance. Inspection, therefore, is oriented principally to conformance with specifications, proper location and use of specified materials, and the highest level of workmanship. Split, damaged, or improperly fastened pieces shall be removed.

Various types of trim and molding will be specifically called for by the architect. The design team must be certain that the architect's intent is carried through. Trim and molding *baseboard* is used to cover the edge of plastered wall where the wall and floor meet, and should be inspected for tight fit and adequate nailing to the wall.

Base shoe is a molding used to close a joint between the baseboard and floor to protect the walls. Again, close fit and adequate nailing are the main criteria in addition to workmanship.

Bead molding is carved to resemble a bead to serve as a decorative feature. Inspect for fit and workmanship, and since the bead molding is so lightweight, check for damaged or broken material.

Chair rail is used around the wall of a room to provide protection from the backs of chairs, and should be carefully checked for position, damaged pieces, and anchorage.

Door trim is the casting around the frame to conceal the break between plaster or other wall cover, and is principally inspected for damage, good fit, and attachment.

Drip cap or beak molding is used on the exterior to cause the rain water to drip on the outside of the structure. This and other drip molding should be approved for configuration, freedom from damage, and installation. Other trim items include jambs and thresholds, and are specified by the architect. Trim is not a standard item, and therefore careful review of the plans and specifications must be carried out to be sure that the inspection team does not allow the omission of any trim items.

06220 Millwork

Millwork is a general term applying to finished work which is precut, and sometimes prefitted in the mill using better equipment than is normally available at the job site. It may be shipped in strips, or prefabricated. In some cases, the term millwork is used to describe built-in casework, cabinets, and other finished fixed items of furniture. Molding and wooden stairways are examples of millwork.

In some cases, millwork items are to be anchored directly to concrete or masonry, and should be provided with appropriate water repellent or preservative treatment. Millwork items are sometimes brought to the job site primed, particularly where the back of the material will be anchored against concrete or masonry. Again this requires careful review of the specifications.

06400 ARCHITECTURAL WOODWORK

06410 Cabinet Work

Upon delivery to the job, each item of cabinet work should be compared with the approved shop drawings and samples for dimensions, gauge, and type of material to be used, types of finishes provided, and provisions for anchoring or fastening.

Check the composition, accuracy, and fit of doors and drawers. Check guides and runners, bearings or rollers for ease and quietness of operation. Check for specified pulls and stops. Check the requirement for adjustable provisions for shelves, and check the shelf material itself.

Treat the cabinet work as delicate material, and monitor its handling and storage. Cabinet work which arrives and is put in storage should be protected from moisture, heat, and any other adverse environmental effects.

Be certain that installation is performed by properly skilled workmen in accordance with the manufacturer's and architect's requirements. Verify the item locations for alignment, level, and plumb. Check the method of fastening, and be certain that it is properly fastened to sound structural blocking or members.

06420 Panel Work

Panels are usually installed on wood framing, or in some cases, metal lath. Paneling is nailable, and should be installed so that proper trim molding can cover cuts or nailing strips. Examination of panel should include the verification of specified quality of panel, and monitoring of care during handling and storage. Panels are not structural, and are inspected principally for workmanship, fit, and appearance.

OSHA REGULATIONS—CARPENTRY

Carpenters in particular make substantial utilization of hand and power equipment, and should follow the cautions noted under the Safety Chapter—in particular the ANSI P15.1, "Safety Code for Mechanical Power Transmission Apparatus."

All fixed power-driven woodworking tools should have disconnect switches that can be locked or tagged in the "off" position. Operating speeds should be permanently marked on all circular saws over 20″ in diameter or operating at over 10,000 peripheral feet per minute. Any saw so marked must not be operated at a speed other than that marked on the blade.

Automatic feeding devices should be installed on machines wherever the nature of the work permits. All portable power-driven circular saws must be equipped with guards above and below the base plate or shoe. When the tool is withdrawn from the work, the lower guard should automatically and instantly return to the covering position.

All woodworking tools and machinery must meet the applicable requirements of ANSI 01.1, "Safety Code for Woodworking Machinery."

DIVISION 7: THERMAL AND MOISTURE PROTECTION

NUMBER   TITLE

07100   **WATERPROOFING**
-10   Membrane Waterproofing
-11   Elastomeric Membrane Waterproofing
-12   Bituminous Membrane Waterproofing
-13   Modified Bituminous Membrane Waterproofing
-20   Fluid Applied Waterproofing
-30   Bentonite Waterproofing
-40   Metal Oxide Waterproofing

07150   **DAMPPROOFING**
-60   Bituminous Dampproofing
-70   Silicone Dampproofing
-75   Water Repellent Coatings
-80   Cementitious Dampproofing
-90   Vapor Barriers/Retardants
-91   Bituminous Vapor Barriers/Retardants
-92   Laminated Vapor Barriers/Retardants
-93   Plastic Vapor Barriers/Retardants

07200   **INSULATION**
-10   Building Insulation
-11   Loose Fill Insulation
-12   Rigid Insulation
-13   Fibrous and Reflective Insulation
-14   Foamed-in-Place Insulation
-15   Spray Applied Insulation
-16   Granular Insulation
-18   High and Low Temperature Insulation
-20   Roof and Deck Insulation
-30   Perimeter and Under-Slab Insulation
-40   Exterior Insulation and Finish Systems

07250	FIREPROOFING
−51	Intumescent Mastic Fireproofing
−52	Magnesium Oxychloride Fireproofing
−53	Mineral Fiber Sprayed-On Fireproofing
−54	Mineral Fiber Board Fireproofing
−55	Cementitious Fireproofing
−60	Thermal Barriers for Plastics
−70	Firestopping
07280–07299	(Reserved)

07300	SHINGLES AND ROOFING TILES
−10	Shingles
−11	Asphalt Shingles
−12	Asbestos-Cement Shingles
−13	Wood Shingles and Shakes
−14	Slate Shingles
−15	Porcelain Enamel Shingles
−16	Metal Shingles
−20	Roofing Tiles
−21	Clay Roofing Tiles
−22	Concrete Roofing Tiles
07300–07399	(Reserved)

07400	PREFORMED ROOFING AND SIDING
−10	Preformed Wall and Roof Panels
−11	Preformed Metal Siding
−12	Preformed Metal Roofing
−20	Composite Building Panels
−40	Preformed Plastic Panels
−60	Cladding/Siding
−61	Wood Siding
−62	Composition Siding
−63	Asbestos-Cement Siding
−64	Plastic Siding
−65	Plywood Siding
−66	Aluminum Siding
07470–07499	(Reserved)

07500	MEMBRANE ROOFING
−10	Built-Up Bituminous Roofing
−20	Prepared Roll Roofing
−30	Elastic Sheet Roofing
−40	Fluid Applied Roofing
−50	Protected Membrane Roofing
−60	Reflective Membrane Roofing

07570	TRAFFIC TOPPING

07600	**FLASHING AND SHEET METAL**
−10	Sheet Metal Roofing
−20	Sheetmetal Flashing and Trim
−21	Galvanized Flashing and Trim
−22	Aluminum Flashing and Trim
−23	Copper Flashing and Trim
−24	Stainless Steel Flashing and Trim
−25	Coated Steel Flashing and Trim
−26	Metal Armored Paper Flashing
−30	Roofing Specialities
−31	Gutters and Downspouts
−60	Gravel Stops
−61	Prefabricated Gravel Stops
07670–07799	(Reserved)

07800	**ROOF ACCESSORIES**
−10	Skylights
−11	Plastic Skylights
−12	Metal-Framed Skylights
−30	Roof Hatches
−40	Gravity Ventilators
−50	Prefabricated Curbs
−60	Prefabricated Expansion Joints

07900	**SEALANTS**
−10	Joint Fillers and Gaskets
−11	Compression Seals
−20	Sealants and Calkings
07930–07999	(Reserved)

07100 WATERPROOFING

Waterproofing is the material, procedure, or combination thereof, which results in the protection of an area, structure, or individual member from the presence of water either in liquid or vapor form. Various materials and coatings can be utilized to achieve this effect, but all require careful quality control during installation on the part of the contractor, and careful inspection, because failure in even a minute section of waterproofing can result in total breakdown of a waterproofing system. Further, when waterproofing systems do break down in practice, it is often quite difficult to determine the point of failure, and therefore, difficult to take remedial action. Where waterproofing is installed, careful and workmanlike installation is well worth the trouble.

07110 Membrane Waterproofing

Membrane waterproofing is achieved by the placement of a moisture-impervious membrane, such as bituminous membrane, polyethylene, or sheet rubber.

Bituminous saturated felt is usually applied in layers with interim sealing by mopped-on asphalt or coal tar. Although three layers of membrane is normal, additional plies are used in sealing against hydrostatic pressures under deep foundations.

Where a membrane is to be applied to a surface, that surface should be clean and free of foreign material, and also dry. Temperature for application of asphaltic material should be above 40°F and the coating should be checked for breaks after each ply has been placed. The plies should be overlapped, so that a common joint does not occur.

In some cases, a rough slab may be poured first, membrane waterproofing placed, and then the finished slab placed. However, if the membrane is to be placed directly upon gravel or soil, *polyethylene* plastic sheeting is often more appropriate than asphaltic coated felt. Alternate sheets of polyethylene may be welded, but overlapping is the usual approach.

Rubber is not usually used below ground as a moisture membrane barrier.

07120 Liquid Waterproofing

Coating materials which have a volatile solvent agent, or react chemically with concrete, masonry, and other surfaces are sprayed on to provide a *vapor barrier*. Care in application must be employed to ascertain a sufficient build up of material on the surface to be waterproofed. Application is on the side toward anticipated water infiltration.

Liquid waterproofing can include pitch, coal tar, and other asphalt preparations. Special solutions of silicone compounds, rubber compounds, oils, waxes, and other solutions have been tested with various success. Most are useful only where the water infiltration is on a limited or cyclic basis. Once again, a clean, dry surface is required for effective application.

07140 Metallic Oxide Waterproofing

Powdered iron, mixed with an oxidizing agent, is sometimes mixed into mortar coatings for concrete or masonry. During hydrolysis, the iron particles oxidize and expand, resulting in a very tight nonporous grout or coating. Expansion of this technique into ferrous cements using iron oxide fibers has resulted in waterproof concrete which can be used to make boats, and even ships.

Application must be on a clean surface, and the preparation must follow carefully the manufacturer's procedures.

07150 DAMPPROOFING

Dampproofing applies principally to degree, and to the decrease in moisture level through capillary action. Also, dampproofing is often applied on inside walls, but can also be applied on the exterior—with the purpose of stopping exposure of the concrete or masonry surface to moisture. Dampproofing, as opposed to waterproofing, should provide an impervious shield, but can accept some limited flaws.

The inspection team should review the scope of dampproofing required according to the plans and specifications. *Temperatures* should be above 40°F for application of dampproofing, and the time of dampproofing should be prior to furring of walls. The material applied for dampproofing should have a complete and unbroken surface. Surface air voids should be sealed in monolithic concrete.

07160 Bituminous Dampproofing

Liquid pitch or asphalt can be used before dampproofing on interior walls. However, where the exterior has not been properly drained, or where waterproofing under hydrostatic conditions is not well installed, the interior dampproofing may blister. The principal use of the bituminous coating is as a vapor barrier, and may be applied before plaster. Bituminous coating for dampproofing also has the problem of condensation of water on the warm side between the plaster coating or paint, and the bituminous dampproofing. Bituminous coatings are usually used for dampproofing inside before the walls are furred, and when an insulation is to be placed between the interior and the dampproofing vapor barrier. The inspector should ensure that the vapor barrier is applied to a clean surface, and that temperature conditions are proper.

07170 Silicone Dampproofing

Silicone dampproofing is water repellent, but not vapor tight. The principal value is in reducing the absorption of moisture into the wall surface by capillary action. The principal value of this type of dampproofing is the control of rate of absorption, particularly for materials such as precast stone or concrete panels. Manufacturer's directions must be followed most carefully in this type of application. Silicones are generally used on the exterior. No trace should be noticeable after application, and the treated areas should be observed for some time to ensure that areas have not been missed or insufficiently coated. This can be observed after the surface has been exposed to moisture.

07180 Cementitious Dampproofing

The application of a light coating of grout, stucco, gunite, or cement paint provides a water-resistant coating. These are applied on the exposed faces of concrete or masonry walls, and combine with the wall to become a water-resistant system.

Again, application must be with a properly prepared mixture, in warm temperature (above 40°F) and on a clean, dry surface.

07190 Preformed Vapor Barrier

Some preformed insulating material is also suitable as a vapor barrier. These boards are made of foamed urethane, pressed fibers, or other materials (which may be combustible or noncombustible) and are placed in a mastic base, principally as an insulating material. Usually, the vapor barrier aspects are secondary. Preformed material, of itself, does not function as a complete vapor barrier because of the joints between sections. The mastic base or cementitious material forms the boundary seal. This material is readily set, and the principal question is the adhesion to the cement.

07200 BUILDING INSULATION

Insulation material is material resistant to the passage of either heat or sound, or both. Usually, these are nonstructural materials, and tend to be relatively lightweight, often utilizing air cells. Typical insulation materials include: foamed glass, foamed plastics, cork, asbestos fibers, or granular materials such as vermiculite or perlite. Often, the insulation material is enclosed in metal foil which reflects heat, even though it is a relatively good conductor.

Insulating materials are often designed to trap dead air space, precluding convection transfer.

Asbestos insulation was, at one time, commonly blown in the form of asbestos wool, but is now more usually placed in bats for convenience in handling, for cleanliness after installation. The inspection team should be certain that the insulation utilized is from an approved manufacturer, and that samples have been approved by the designer. Insulation should be inspected for proper workmanship—principally careful fit and firm attachment, such as stapling.

Most *cork* insulation placed in sheet form has been most usually used for cold box insulation, but is used in other situations also. Before installation, inspection of the cork should be made for damage or moisture. Again, good workmanship is important.

Fiberglass is usually paper covered on one or two sides, and usually at least one of these sides is also foil. The glass should be inspected for approved manufacture and proper thickness. It may also be blown in loose.

Foam glass board is sometimes utilized under concrete floors, or as roof decking. Again, inspection should include check for sound condition prior to installation, and proper fit and workmanship after. Similar inspection is made of insulation boards made of pressed wood pulp, cane fibers, and other manufactured insulation board.

The inspection team should ascertain that storage prior to installation is proper, with emphasis upon protection from moisture.

Insulation on piping and metal surfaces is sometimes sprayed on, most usually for fireproofing. The inspection team should be certain that proper safety procedures are followed, as the asbestos-gypsum mixture often utilized can be obnoxious if ingested. Also, in some cases, the vehicle used to keep the mixture fluid is volatile, and care must be taken to ensure that fires are not started in the insulation material before it has fully set up.

Styrofoam has become popular as insulation, either in board form or foamed in place, and so has *urethane*. Where these materials are foamed in place, care should be taken to be certain that the foaming pressure does not cause damage to forms, walls, or permanent structure.

07300 SHINGLES AND ROOFING TILE

07310 Asphalt Shingles

Asphalt shingles are manufactured from a felt-like material saturated with asphalt with a weight of approximately 2 lb/ft². The shingles have a granule mineral surface imbedded into the material, and are furnished in various sizes, either individual or stripped form.

Before starting shingle roofing, the inspection team should be certain that the roof is clean, dry, and ready for roofing, including patches with the sheet metal plates over knotholes or splits. Also, sheet metal flashing should be available for installation concurrent with roofing, or should already be installed. In better grade roofing, an underlay of roofing felt is applied before the shingles.

Shingle installations begin with a *starter course* at the eaves. This can be a double or triple layer of shingles, or a layer of shingles on top of a strip of shingle-like roll roofing.

Alignment of the layers and rows of shingles is most important, and rows should be started at the center of the roof for spans over 30'. The roofer should use chalklines or guidelines to insure a neat job. At hips and ridges, all nails should be concealed, and each shingle tab should be cemented in place with bituminous cement with a contact surface of at least one square inch. (Some manufacturers provide self-sealing shingles, and the inspection team should check these before they go on to the roof.)

Nails are ordinarily $1\text{-}\frac{1}{4}''$-long zinc-coated roofing nails—although many experts recommend the use of standard roofing nails which are not coated, since they have good holding power. Because many of the problems in roofs occur near the eaves, due either to damming of water or capillary action, good practice is to cement the starting course, rather than nail it.

07312 Asbestos-Cement Shingles

Asbestos shingles are made up of a composite cast material composed of about 20% asbestos fiber and 80% Portland cement. The resulting material is durable and noncombustible, but is subject to breakage because it is brittle. Asbestos cement shingles come in various sizes, with the larger essentially being a double shingle.

07313 Wood Shingles

Wood shingles are either cut by machine or by hand. The widths vary substantially, usually between 5 and 18 inches. The shingles are tapered, so require no initial cant strip. Shingles can be trimmed to fit, so no special sizes are required. A particular caution during inspection is that nails should not be placed in the center of the shingle, and that joints should not be matched in any three courses. The latter precaution is to avoid leakage in the event of a split in any one shingle.

07314 Slate Shingles

Once very popular, slate shingles are used much less today, particularly because the installation requires a craft level of skill. Slate installation starts with an under-eaves course fastened to a batten, which cants the initial course. In nailing slate, coated or aluminum nails should be utilized, and the nail is not driven completely, because the brittle slate cannot take the loading. Caution should be taken in walking on the installed slate, since the material is brittle and can break after placement.

07321 Clay Roofing Tiles

Roofing tiles fall into a number of sizes and shapes. In placing the tiles, guidelines must be used, and should be based on actual size measurements of the material as delivered to the job site. Installation usually starts from the bottom right-hand corner of a roof, proceeding from right to left. The need for particular care in measuring results from the interlocking nature of many tiles, leaving no adjustment.

07400 PREFORMED ROOFING AND SIDING

07411 Preformed Metal Siding

Metal siding is fabricated on the job site from sheet material which may be flat or corrugated, or furnished as. a sandwich panel, built of a prefabricated frame which usually has an insulation core.

In placing panels, field-fabricated siding must be lapped either about 4" or 1-½ corrugations. Sheets should be firmly fastened at splices and intersections to provide a weathertight seal. Specifications may call for caulking, strip sealing of soft material, or may depend on tight fit.

The method of connection again should be in keeping with the requirements of the specification, and the structural designer will have anticipated the effect of temperature changes.

Prefabricated panels should be installed in accordance with the suggestions of the manufacturer. For instance, sealing of the edges of the prefabricated panel may cause internal pressures due to changes in vapor pressure resulting from solar heating. The panel may have surface vents, or more usually is vented by small holes along the edges.

07412 Preformed Metal Roofing

Metal decks are used in slopes up to 2″ per foot. Insulation must be installed before built-up roofing is applied, and the metal deck must be clean-smoothed and properly anchored to the steel purlins with all joints correctly lapped and fastened. Usual design permits no deflection exceeding 1/240 of clear span under all conditions.

Before roofing is placed, the deck must be dry, clean, and free of rust or oil.

07463 Asbestos-Cement Panels

Installation requirements for asbestos-cement siding are similar to that for metal siding, except that greater care in handling of the asbestos-cement sheets is required because of their brittleness. Coated or aluminum fasteners are utilized to avoid rust streaking of material. In some cases, flat sheets may be installed in frames, while in others, joints are covered with battens.

07464 Preformed Plastic Panels

Plastic panels can be installed in flat or corrugated form. Installation should carefully follow manufacturer's recommendations.

07500 MEMBRANE ROOFING

07510 Built-Up Bituminous Roofing

Bituminous built-up roofing consists of alternate layers, usually four or five plys of alternate layers of felt, mopped with bitumen, and topped with gravel or slag, bedded in bitumen on top of the final roof felt.

The roof surface should be dry, smooth, and free of dirt and foreign materials before placement of the first felt. The first layer is the *vapor barrier*, and should be nailed where appropriate, or attached with a mopped bitumen surface. In some cases, a layer of vinyl sheeting is specified as a vapor barrier. Felt should be maintained at a temperature of 50°F for at least 24 hours prior to laying. In most cases, the vapor barrier seals the edge of insulation at gravel stops and openings. However, vapor barriers should not seal if the edge of the insulation is vented. Vapor barrier placed on gypsum roofing should not be mopped. Asphalt saturated felt vapor barrier on concrete should be mopped in place after priming. A vapor barrier is not required under insulation on roof decking. Where vapor barrier is required, inspection should assure that it provides a complete seal over the deck.

Insulation is placed on top of the vapor barrier, and should be applied to the decking in accordance with the specifications. Insulation should be placed immediately preceding placing of the roof felts, and should be kept dry at all times. Joints should be staggered if application is more than one layer. At the close of each day's work, insulation in place should be covered with a layer of felt which has been hot-mopped to the insulation and the deck. During installation, roof drains should be working.

A new approach to insulation is to place it on top of the roofing plies. This adds mechanical protection and—its main purpose—shields the roof plies from external temperature variations. The insulation also restrains loss of the building conditioning, an added benefit. One proprietary inverted insulation system is the IRMA System.

Control joints are very important in roofs of large area. These mechanical joints are calculated by the designer and are tied in to the structural system prior to roofing. The Inspection team should monitor the installation of control joints to be certain that they can move as they are designed to.

The *roofing* application is laid at right angles to the slope of the deck. Use of open flame-producing equipment on the roof during roofing operations is prohibited for safety reasons. Placement should be at an ambient temperature at or above the 40°F minimum. The felt should be placed immediately behind mopping of the pitch, and the felt should be broomed in place so that the layer will be free of air pockets, wrinkles or buckles. There should be two extra plies of felt at eaves. Extra strips should be hot-mopped over the edge of the gravel stop and other metal roof flashings.

The *aggregate* placement on top of the roofing should be dry and free from dust or foreign material. Aggregate should be spread while the bitumen is still hot. It is important that in transporting the aggregate for spreading on the roof, felt and flashing materials not be damaged, and that excessive roof loading not occur. Temporary wood runways can be provided to avoid overloading.

07520 Prepared Roll Roofing

Special plastic sheet applications are developing—and the installation crew must be particularly attuned to the requirements as stated by the manufacturer and the designer. In many cases, the material is placed with epoxy-type adhesives which have instant bond; usually extreme cleanliness is an important factor. In many cases, the use of new materials suggests the use of a sample placement to approve technique and results prior to the installation of a large roof.

07530 Elastic Sheet Roofing

Neopreme sheeting has been used successfully for roofing, and is bonded with fluid adhesive. Again, the manufacturer's recommendations are of greatest importance, as is application to dry, clean surfaces at suitable temperatures.

07540 Elastic Liquid Roofing

Both plastics and synthetic rubbers can be applied to form a membrane, and again the directions of the manufacturer are of great importance. Two of the most successful elastic coatings are neopreme, and hypalon. Cleanliness is important in the application of this type of surface. Application is by spray or brush. Spraying application requires care to avoid spill-over or scatter spray near the edge of the roof.

07600 FLASHING AND SHEET METAL WORK

07610 Sheet Metal Roofing

Various sheet metals are used for roofing. A principal problem in metal roof material is its expansion characteristics, which are usually substantially greater than the material to which it is attached. Most sheet metal is interconnected with a standing-seam, which provides some flexibility for expansion, and is economical to install.

Industrial roofing often utilizes corrugated sheet metal attached either to wood sheathing, purlins, or directly to angles or bar joists. Flexible fasteners, such as lead and neopreme washers are used to provide for the differential movement between frame and metal roofing.

07620 Metal Roof Flashing and Trim

Flashing materials should be detailed on the drawings, and the inspection team should review for compliance. Exposed *edges* of flashing are normally folded back a full ½″ for both stiffness and to avoid sharp cutting edges.

Both in metal roofing and in flashing, *dissimilar metals* should not be in contact with each other. Check the specifications for requirements for fastening—and ensure that the proper fasteners are used.

Gravel stops and fascia should usually have inner flanges 4″ length for application over roofing felt with nailing about 3″ on center. Inner-flange should be set into a bed of plastic cement. Two plies of roofing felt should be mopped over the inner-flange. The nailer used to fasten the flashing to the roof deck should be slotted for venting roof deck where appropriate.

For *metal-based* flashing, installation should be started after the roof felt has been placed in the angle formed by the roof and the vertical surface. Base should usually be at least 8″ up the vertical surface and 4″ on the roof. Inspect for secure anchorage and size spacing and fixing of cleats. The roof flange and cleats should be covered with two plies of felt, similar to the application at gravel stops.

For *cap flashings,* check the fabrication for size and shape. Usually, there should be at least a 2″ space between the top of the metal base flashing and the top of the cap flashing. The cap flashing should lap the base flashing by at least 6″. Flashing should be extended into masonry wall at least 3-½″, or into reglet (prepared slot in wall).

For *through-wall* flashing, check locations and ensure that flashing is being installed in the middle of the mortar joint. Check the design and installation requirement for laps, amount of turn-up required, location of wall, and length of material extended outside the wall. Sill flashing should always extend full depth of sill and at least 4″ beyond the end.

Spandrel beam flashing should be applied to surfaces which have an asphalt primer followed by coating of roofing asphalt. Flashing should be applied before the roofing asphalt sets up.

Other types of flashing such as hip and ridge, valley, and stepped, should be checked for neat installation, proper fabrication and secure anchorage.

07631 Gutters and Downspouts

The inspector should check gutters for type shape, design and layout. A particularly important factor is the hangers, and these should be checked for size, type, location, and spacing, as well as pitch of the gutter to provide drainage to outlets. If the specifications call for them, be certain that basket strainers are provided for gutter openings into downspouts and/or bird screens for downspout heads.

Downspouts are usually factory fabricated and corrugated longitudinally in about 10′ lengths. Check seams and joints to be certain that expansion joints are provided, as well as special requirements such as scuppers, downspout heads, etc. Downspouts should be installed in true plumb, and should be secured away from the wall.

07800 ROOF ACCESSORIES

07810 Skylights

07811 Plastic Skylights

07822 Metal-framed Skylights

07830 Roof Hatchets

07840 Gravity Ventilators

Each of these roof-penetrating accessories should be installed carefully in accordance with any directions furnished by the manufacturer. In turn, it is most important that the inspection team ensure that the frame or mounting for these units is well anchored to the roof, including anchorage to any structural supports required for larger units, and that the flashing around the mounting has been installed properly. Flashing should be preferably installed before roofing starts, and these units should preferably be mounted, or at least their mounting frames should be installed before roofing begins. Roof penetrations of this type are one of the most common causes of roof failure through leakage.

07900 SEALANTS

The purpose of caulking is to obtain a watertight structure. In masonry walls, a space of about $\frac{1}{4}''$ wide and $1\text{-}\frac{1}{2}''$ deep should be allowed around all door and window frames to provide space for adequate caulking. The caulking material is usually specified in the design. Before caulking is applied, the joints should be inspected for cleanliness, and filled with oakum or rockwool to within $1''$ of the face. Open spaces between wood and masonry sill should be caulked, the ends of stone sills or cast stone should be supported on mortar and the rest of the space caulked after the mortar has set. In roofing, it is advisable to caulk and tape all joints. Flashing cement and felt strip about $6''$ wide can be used to minimize the effects of moisture transmission through open joints which are not level, and can prevent possible bitumen seepage on the roofing felt.

Caulking is usually applied with a gun, and inspection should include determination that the proper type of gun is being utilized. Grooves should be solidly filled and appearance should be uniform and relatively smooth.

The configuration of the caulking groove is most important; too large a groove will not provide a solid base for the caulking and ultimately large pieces may fall out, while too small a groove will not provide enough caulking material to establish a good membrane. The presence of foreign material will preclude a good bond of the caulking material to the surface to be sealed.

DIVISION 8: DOORS AND WINDOWS

NUMBER	TITLE
08100	**METAL DOORS AND FRAMES**
−10	Standard Steel Doors
−11	Standard Steel Frames
−12	Custom Steel Doors
−13	Custom Steel Frames
−15	Packaged Steel Doors and Frames
−20	Aluminum Doors and Frames
−30	Stainless Steel Doors and Frames
−40	Bronze Doors and Frames
08200	**WOOD AND PLASTIC DOORS**
−10	Wood Doors
−11	Flush Wood Doors
−12	Panel Wood Doors
−13	Plastic Faced Wood Doors
−14	Steel Faced Wood Doors
−20	Plastic Doors
08250	**DOOR OPENING ASSEMBLIES**
08300	**SPECIAL DOORS**
−05	Access Doors
−10	Sliding Metal Fire Doors
−15	Blast-Resistant Doors
−16	Security Doors
−20	Metal-Clad Doors
−30	Coiling Doors
−31	Overhead Coiling Doors
−32	Side Coiling Doors
−40	Coiling Grilles
−41	Overhead Coiling Grilles
−42	Side Coiling Grilles

−50	Folding Doors
−51	Panel Folding Doors
−53	Accordion Folding Doors
−55	Flexible Doors
−60	Overhead Doors
−61	Sectional Wood Overhead Doors
−62	Sectional Metal Overhead Doors
−63	Sectional Plastic Overhead Doors
−65	Multi-leaf Vertical Lift Metal Overhead Doors
−66	Vertical Lift Wood Doors
−67	Vertical Lift Metal Doors
−70	Sliding Glass Doors
−75	Safety Glass Doors
−80	Sound Retardant Doors
−90	Screen and Storm Doors

08400	**ENTRANCES AND STOREFRONTS**
−10	Aluminum Entrances and Storefronts
−20	Entrance Doors
−25	Automatic Entrance Doors
−50	Revolving Doors

08500	**METAL WINDOWS**
−10	Steel Windows
−11	Roll Formed Steel Windows
−20	Aluminum Windows
−27	Aluminum Jalousie Windows
−29	Aluminum Storm Windows
−30	Stainless Steel Windows
−40	Bronze Windows

08600	**WOOD AND PLASTIC WINDOWS**
−10	Wood Windows
−20	Plastic Windows
−21	Reinforced Plastic Windows
−25	Metal Clad Wood Windows
−30	Plastic Clad Wood Windows
−40	Wood Storm Windows
−45	Plastic Storm Windows

08650	**SPECIAL WINDOWS**
−51	Security Windows
−55	Roof Windows

08700	**HARDWARD**
−10	Finish Hardware
−15	Exit Devices
−20	Operators

−21	Automatic Door Equipment
−25	Window Operators
−30	Weatherstripping and Seals
−40	Thresholds
08750–08799	(Reserved)

08800	**GLAZING**
−10	Glass
−11	Plate and Float Glass
−12	Sheet Glass
−13	Tempered Glass
−14	Wired Glass
−15	Rough and Figured Glass
−16	Bullet Resistant Glass
−17	Spandrel Glass
−20	Processed Glass
−21	Coated Glass
−22	Laminated Glass
−23	Insulating Glass
−30	Mirror Glass
−40	Glazing Plastics
−45	Insulating Plastic Glazing
−50	Glazing Accessories
08860–08699	(Reserved)

08900	**GLAZED CURTAIN WALLS**
−10	Glazed Curtain Walls
−11	Glazed Steel Curtain Walls
−12	Glazed Aluminum Curtain Walls
−13	Glazed Stainless Steel Curtain Walls
−14	Glazed Bronze Curtain Walls
−15	Glazed Wood Curtain Walls
−20	Translucent Wall and Skylight Systems
08930–08999	(Reserved)

08100 METAL DOORS AND FRAMES

Doors are utilized either for internal or external purposes. External doors are often entrances to major or public buildings and the following comments are extracted from the "Entrance Manual" prepared by the National Association of Architectural Metal Manufacturers:

Public Entrances serve principally as ingress and egress, for control and direction of traffic in a safe fashion, to provide security from both weather and people—and in addition, contribute to the aesthetic character of the building.

All of these factors will have been considered by the architect in design of an entrance, and in turn the inspection team should be aware of the general function of the *entry doors.* While the frequency of use of any single entrance varies with the size, location, and purpose of the building, the range may vary from a few hundred to five thousand or more opening and closing cycles per day. Some buildings experience as many as one to two million cycles per year. This dictates the type of performance which entryways must furnish—and therefore, underlines the need for proper installation. The design drawings will indicate the *"hand"* of the door. This is important in proper mounting, and interpretation of shop drawings. Also important is the type of mounting hinge or pivot. These are illustrated in Fig. 1. Correct orientation of door installation is vital to the proper operation of the door hardware. Since entry doors, and doors in general, are fabricated to operate with close tolerances, it is particularly important that the permanent door *frame* be properly manufactured, and correctly installed. The frame must be plumb and square, and firmly secured to the structure.

In masonry openings, the soffit or header must be level, and the jambs plumb. Some shimming can be accomplished in the caulking areas, but provide only a limited flexibility. The floor line at major doors should be level within $\frac{1}{16}''$ tolerance. The outside grade adjacent to the door may have a slight slope away from the door to provide water run-off.

While settlement is undesirable in any structure, it is particularly difficult if it occurs beneath a major entry door or revolving door.

Doors delivered to the job site should be checked against the shop drawings for correct fabrication, although preferably if a large number of doors are to be shipped, the inspection team will have better inspection conditions at the factory, and would also have a better opportunity for any remedial repairs required. The inspection team should be certain that doors received at the job site are carefully handled, and stored under protected conditions. Doors should always be stored inside or under cover.

In some cases, floor closers are used and these must be installed in the concrete floor at an early stage in construction. Any deviation from the vertical will cause problems, making the door difficult to open, and possibly noisy. The inspection team should be certain that the installation is by skilled craftsmen—preferably a representative of the door manufacturer—which will preclude coordination problems should any remedial action be required.

The door frames should be installed only after the finished floor has

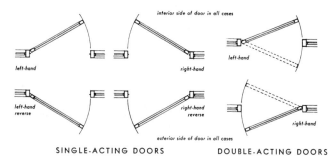

SINGLE-ACTING DOORS   DOUBLE-ACTING DOORS

Types of Mounting

Swing doors may be mounted either on butt hinges or on pivots, as illustrated below:

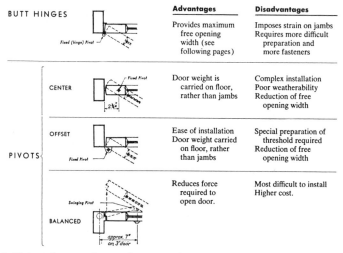

		Advantages	Disadvantages
BUTT HINGES	Fixed (hinge) Pivot	Provides maximum free opening width (see following pages)	Imposes strain on jambs Requires more difficult preparation and more fasteners
PIVOTS — CENTER	Fixed Pivot 2¼"	Door weight is carried on floor, rather than jambs	Complex installation Poor weatherability Reduction of free opening width
OFFSET	Fixed Pivot	Ease of installation Door weight carried on floor, rather than jambs	Special preparation of threshold required Reduction of free opening width
BALANCED	Swinging Pivot approx. 7" on 3' door	Reduces force required to open door.	Most difficult to install Higher cost.

Fig. 1 Type of mounting and nomenclature. *Source: National Association of Architectural Metal Manufacturers.*

been completed, since any premature installation can cause misalignment. When frames are installed, check for accuracy of dimensions in any openings, and sufficiency of anchorage and reinforcement as described by the drawings. When placing door frames in masonry or concrete openings, there should be enough room for expansion of the metal, with alignment being performed by shims. Fitting of a frame into a tight opening may cause later buckling because of heat expansion.

Hollow Metal Doors and Frames

Hollow metal doors can usually be ordered to suit special conditions, although this may require longer delivery times. The custom nature of

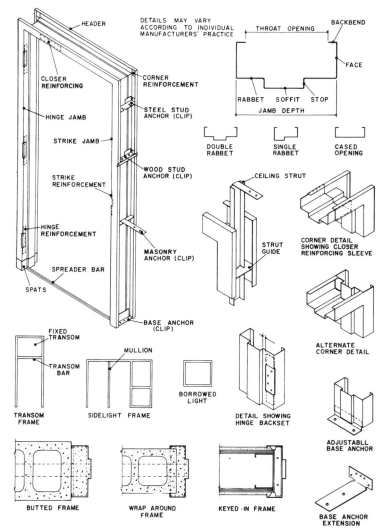

Fig. 2 Nomenclature for steel door frames. *Source: Steel Door Institute.*

the doors underscores the need for the inspection team to check dimensions, or to ensure that the contractor has, so that delays will not occur because of errors in dimensions. Figure 2 shows the nomenclature for steel door frames, and Fig. 3 the nomenclature of *swing* for steel door frames.

Where assembly details preclude ability to inspect hollow metal doors,

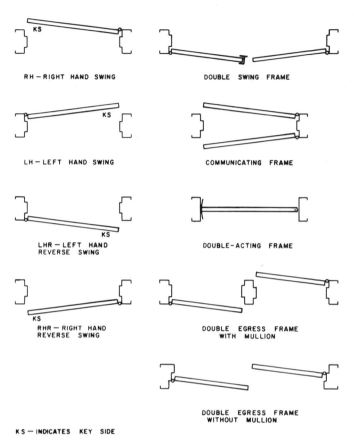

Fig. 3 Nomenclature of swing for steel door frames. *Source: Steel Door Institute.*

they can either be factory-checked, or observations can be made, such as checking for reinforcing or backing plates when doors are drilled for hardware.

08120 Aluminum Doors and Frames

Doors and frames usually come as one unit. Inspect doors for dissimilar material, and where this occurs, make certain that proper protective coating has been applied to surfaces which come in contact with each other to prevent galvanic action.

Other details noted previously apply also to the installation of aluminum doors and frames.

08130 Stainless Steel Doors and Frames

Stainless doors should be checked for proper material, including proper grade of stainless. These doors tend to be heavier, and should be handled with suitable lifting equipment.

08140 Bronze Doors and Frames

Bronze doors are becoming increasingly popular as architectural items. They are heavy, and should be handled with appropriate care.

08150 Metal Storm and Screen Doors

These doors are lightweight, and should be reviewed for specifications. Be certain that the proper grade and make have been used in conformance with approved shop drawings. These doors are generally mounted as a unit, and their alignment is less critical because of the type of mounting.

08200 WOOD DOORS

The handling and storage of wood doors is of the greatest importance, to avoid either damage or marking. Wooden doors should be stored in dry, humidity-controlled space. Most wooden doors utilized are interior doors, and are of the wood flush-type.

Many designs call for doors classified by the National Woodwork Manufacturer's Association. Where used, this NWMA classification is clearly labeled on the appropriate door. This marking indicates the classification and grade of the door, and affirms that the door meets the specifications prepared by the National Woodwork Manufacturer's Association.

Figure 4 contains tables with the standard types of construction for wood flush doors, and standard sizes. The variety of types indicate the importance of the certified label for identification.

Doors can be constructed to be fire-resistant—but must have Underwriter's Laboratories' certification.

08300 SPECIAL DOORS

Special doors may be stock ordered on a custom basis. Other doors such as those separating gymnasiums or performing special functions must be

Table 1 – Kinds and constructions of Wood Flush Doors generally available

KINDS OF DOORS	Solid Cores						Hollow Cores		
	Glued Block	Framed Block-Glued	Framed Block-Non-Glued	Stile and Rail	Wood Compos.	Mineral Compos.	Ladder	Mesh	Im-planted
Exterior Doors: (Entrance, or service)									
Plain	V,H	V,H	V,H	V,H	V,H	V,H	V,H	V,H	V,H
Light Openings	V,H	V,H	V,H	V,H	V,H	V,H	V,H	V,H	V,H
Louver Openings	V,H	V,H	V,H	V,H	V,H	V,H	V,H	V,H	V,H
Interior Doors:									
Dwarf	V,P,H	V,P,H	V,P,H	V,P,H	V,P,H	V,P,H	V,P,H	V,P,H	V,P,H
Fire Doors				V		V,P	--------	--------	--------
Passage or Closet[2]	V,P,H	V,P,H	V,P,H	V,P,H	V,P,H	V,P,H	V,P,H	V,P,H	V,P,H
Sound Resistant	V,P,H	V,P,H	V,P,H	V,P,H	V,P,H	V,P,H	--------	--------	--------
X Ray Resistant	V,P,H				V,P,H	V,P,H	--------	--------	--------

Heading note: Standard Door Construction and Grades[1]

1. V=Hardwood Veneered Faces (Premium, Good, Sound or Specialty Grades), P=Plastic face Grade, and H=Hardboard face Grade. Species of wood faces are described in Table 3.
2. Doors exposed to unusually damp environments (shower rooms) may require Exterior type doors.

Fig. 4 Standard types of construction for wood flush doors, standard sizes.
Source: National Woodwork Manufacturer's Association.

designed and fabricated to meet the requirements of the project. These requirements are carefully spelled out in the plans and specifications, and must guide the inspection team. Keep in mind that large doors are structures in themselves, and must be handled carefully so that failure does not occur when they are in other than their normal support mode.

In large doors, power drives are often included, and tracks are located at the top or bottom of the doors, or both. Tracks must be aligned carefully, and all moving parts should work freely.

08310 Sliding Metal Fire Doors

Check to be certain that the door has the proper Underwriter's classification, and review the door for proper size, hardware, and

Table 2 — Standard Sizes of Wood Flush Doors

Interior Flush Doors Solid-Core
1-3/8" and 1-3/4"

1'6" x 6'6"	2'6" x 6'0"
6'8"	6'6"
2'0" x 6'0"	6'8"
6'6"	7'0"
6'8"	2'8" x 6'0"
7'0"	6'6"
2'4" x 6'0"	6'8"
6'6"	7'0"
6'8"	3'0" x 6'8"
7'0"	7'0"
	3'4" x 6'8"
	7'0"

Interior Flush Doors Hollow-Core
1-3/8" and 1-3/4"

1'6" x 6'6"	2'4" x 6'8"
6'8"	6'10"
6'10"	7'0"
7'0"	2'6" x 6'0"
1'8" x 6'6"	6'6"
6'8"	6'8"
6'10"	6'10"
7'0"	7'0"
1'10" x 6'6"	2'8" x 6'0"
6'8"	6'6"
6'10"	6'8"
7'0"	6'10"
2'0" x 6'0"	7'0"
6'6"	2'10" x 6'0"
6'8"	6'6"
6'10"	6'8"
7'0"	6'10"
2'2" x 6'0"	7'0"
6'6"	3'0" x 6'0"
6'8"	6'6"
6'10"	6'8"
7'0"	6'10"
2'4" x 6'0"	7'0"
6'6"	

Exterior Flush Doors
Hollow-Core
1-3/4"

2'8" x 6'8"	3'0" x 6'8"
6'10"	6'10"
7'0"	7'0"
2'10" x 6'8"	
6'10"	
7'0"	

Exterior Flush Doors
Solid-Core
1-3/4" and 2-1/4"

2'4" x 6'8"	3'0" x 6'8"
7'0"	7'0"
2'6" x 6'8"	3'4" x 6'8"
7'0"	7'0"
2'8" x 6'8"	3'6" x 6'8"
7'0"	7'0"

Flush Dwarf Doors
1-1/8"
(When so specified, doors can be furnished 1-3/8" thick.)

1'6" x 4'0"	2'4" x 4'0"
4'6"	4'6"
5'0"	5'0"
5'6"	5'6"
2'0" x 4'0"	2'6" x 4'0"
4'6"	4'6"
5'0"	5'0"
5'6"	5'6"

Fig. 4 (*continued*)

dimensions. The operating mechanism should be on the inside of the door for exterior doors; check for operation after installation. Check for operating requirements such as actuating devices or safety devices.

08320 Metal-Covered Doors

Where metal-covered doors are intended as fire-resistant doors, check for Underwriter's label and conformance with design.

08330 Coiling Doors and Grills

These are made of metal similar to venetian blinds, with movable slats to allow rolling up and down on a drum. Check for handling and freedom from damage, and after installation, be certain that the door works freely and easily and closes the opening completely.

Each of the following special doors should be reviewed carefully for its compliance with materials, dimensions and condition before installation:

0834 Plastic-faced Doors
0835 Folding Doors
0836 Overhead Doors
0837 Sliding Glass Doors
0838 Tempered Glass Doors
0839 Revolving Doors
0840 Flexible Doors
0841 Hanger Doors

08500 METAL WINDOWS

08510 Steel Windows

The following information is extracted from information furnished by the Steel Window Institute:

Windows should be set as the wall construction progresses, and should be placed without forcing into openings prepared by experienced craftsmen. Windows must be set plumb, level, and in alignment with proper securement to walls and mullions. Metal-to-metal joints shall be caulked at the exterior surface with an approved mastic suggested by the window manufacturer.

Windows may be placed glazed or unglazed.

The inspection team should determine that window units are fabricated in accordance with specifications, and to proper dimensions. Window frame may be set prior to installation of the window unit.

Window cleaner anchors are usually shipped loose and installed in accordance with instructions.

The Steel Window Institute suggests the following factors in field installation:

Avoid unnecessary roughness when unloading windows at the job site.

Stand windows upright on pieces of lumber to keep them off the ground and protect against defects of prolonged storage or unfavorable weather conditions.

Wire or clips holding ventilators closed should remain in place until windows are completely erected and hardware attached. (This protects against accumulations of cement, lime, and other building materials.)

Do not allow workmen to use windows or ventilators as resting points for scaffolds or ladders.

If shop primer paint has become scratched or marred, touch up damaged areas before installation. Apply field coat of paint before glazing.

The window erector should check and adjust all ventilators to close tightly and operate freely before glazing, if units are field glazed.

08520 Aluminum Windows

All of the same suggestions apply for aluminum windows as for steel, with the added suggestion that any dissimilar metals attached to aluminum should be insulated to preclude the effects of galvanic action. Aluminum windows are not usually set directly with mortar, but are isolated from the concrete structure by appropriate caulking.

08530 Stainless Steel Windows

Installation procedure the same as for steel, except that units are not usually painted.

08540 Bronze Windows

Installation procedures are similar to other metal windows, and units are not normally painted.

08600 WOOD WINDOWS

The National Woodwork Manufacturer's Association (NWMA) has published standards for classifications and grades of wood windows. The Standard provides minimum requirements for component materials, as well as for performance of the materials, and assembly into complete window units. The types covered include: awning; casement; double-hung; single-hung; and horizontal sliding. The units are designed and produced by individual manufacturers and vary widely in design, layout, size, and method of operation, and therefore, do not lend themselves to rigid standardization. Wood parts are usually manufactured from ponderosa pine, or other suitable kiln-dried lumber.

Inspection should include freedom of exposed parts from defects, as defined by the NWMA. Basically, all exposed parts shall be free of defects of appearance, and unexposed portions must be free of structural defects. This does not exclude the use of pieces with light stain and light to moderate streaks.

Windows, unless otherwise specified, are glazed with single-strength "B" glass; maximum size glass is 76 combined inches (width plus height), for single-strength glass, and 100 inches for double-strength "B" glass. Where called for by the specifications, insulating glass, double-glazed panels, or screened panels may be furnished in the windows.

Inspection at the job site should indicate that the window unit has been assembled in a workmanlike manner with all joints well-fitted in exposed surfaces, sanded or properly finished. The frame should have been completely assembled and braced in a manner which will maintain dimensions and squareness during shipping.

The window units should be fastened to the jamb, as indicated in the specification. Where there is not an exact specification, the tables prepared by the NWMA can be used as guidance.

08700 FINISH HARDWARE

The following suggestions on hardware are provided by the National Association for Architectural and Metal Manufacturers (NAAMM):

It is advisable that many of the critical hardware items such as pivots, butt hinges, and panic exit devices be factory-installed, and again the follow-through on this is the responsibility of the inspection team. The installation of surface-mounted overhead closers is a critical one, and if done in the field, must be in strict accord with the manufacturer's instructions and templates. While some types of hardware have operational adjustment, all require careful alignment and installation. Some, such as butt hinges, have no vertical adjustment.

After hardware is installed, it must be protected from the various work hazards involved with finishing of other work.

Additional information on hardware is available from the American Society of Architectural Hardware Consultants, and the National

Builder's Hardware Association (1290 Avenue of the Americas, New York, N.Y. 10019).

08730 WEATHER STRIPPING

The installation of weather stripping should be in accordance with the manufacturer's recommendations. The failure to install carefully proper weather stripping in a workmanlike manner can result in poor performance of well-installed floors and windows. Fortunately, much of the modern weather stripping material can be adjusted using a screwdriver adjustment.

Weather stripping is used principally to preclude intrusion or escape of heat or humidity. However, in certain cases, sound attenuation is the required effect.

In weather stripping of doors, one manufacturer indicates that half of the problems occur in the astragal area where double doors meet.

08800 GLASS AND GLAZING

Glazing can be performed either by dry glazing, which utilizes preformed gaskets of small cross-sectional shapes, or wet glazing using bulk sealing compounds. Dry methods are more common for glazing doors, and both methods are used for windows.

Weather tightness depends on several essentials which should be reviewed by the inspection team. The *gasket* material in dry glazing must be in good condition, must be held under compression to exert pressure at all points of the pane, and corner joints must be tight. For wet glazing, standard glazing materials are satisfactory for smaller panes and/or normal environmental conditions. Where wide temperature ranges will be incurred, or large sizes of glass are involved, appropriate elastomeric compounds based on silicones, or other plastics are recommended.

In wet glazing, neatness and good workmanship are particularly important. Labels on the glass identify whether it is single or double strength, and should be reviewed. Where special requirements such as heat tempered, fire rated, tinted, etc., are required, the materials should be validated before glazing is completed.

Where glazing with acrylic plastic sheets is called for, sheets should be carefully protected to avoid scratching, and to keep them clean.

During installation of double insulating glass, the installer should not remove glass tips formed during manufacture, or allow the units to be struck against frames or other objects.

Installation should have specific requirements for use of glazers, points, glazing strips, imbedment of pane, and factors such as priming of wood molding, before installation. Check for a positive seal between the glass and sash. Single panes installed in metal sash being set in continuous vinyl or neoprene channels should fit snugly. Nails or screws used in the beads of exterior units should be corrosion-resistant, aluminum or galvanized.

Masking tape should be removed from acrylic plastic immediately after glazing in spite of the temptation to leave it on to prevent damage. Removal is important because some types of the protective tape when subjected to sunlight become difficult to remove.

The following information on glazing is from the New York City Building Code:

Requirements for Glass Panels Subject to Impact Loads[a,b]

Glass Type	Individual Opening Area	Requirements
Regular plate, sheet or rolled (annealed)	Over 6 sq. ft.	Not less than 3/16 in. thick. Must be protected by a push-bar or protective grille firmly attached on each exposed side, if not divided by a muntin.
Regular plate, sheet or rolled (annealed), surface sandblasted, etched, or otherwise depreciated	Over 6 sq. ft.	Not less than 7/32 in. thick. Must be protected by a push-bar or protective grille firmly attached on each exposed side.
Regular plate, sheet or rolled (annealed), obscure	Over 6 sq. ft.	Not less than 3/16 in. thick. Must be protected by a push-bar or grille firmly attached on each exposed side.
Laminated	Over 6 sq. ft.	Not less than 1/4 in. thick. Shall pass impact test requirements of reference standard RS 10-67.
Fully-tempered	Over 6 sq. ft.	Shall pass impact test requirements of reference standard RS 10-67.
Wired	Over 6 sq. ft.	Not less than 7/32 in. thick. Shall pass impact test requirements of reference standard RS 10-67.
All unframed glass doors (swinging)		Shall be fully-tempered glass and pass impact test requirements of reference standard RS 10-67.

[a] Glass less than single strength (SS) in thickness shall not be used.
[b] If short dimension is larger than 24 in., glass must be double strength (DS) or thicker.

Maximum Area of Glass—Sq. Ft.

Nominal Thickness of Glass-Inches	Elevation Above Grade of Mid-Point of Glass—Ft.			
	0–50	50–300	300–600	600–
Sheet Glass				
Single strength	9	9	8	7
Double strength	13	13	11	10
3/16	25	25	21	19
7/32	32	32	27	24
Plate and Float Glass				
13/64	25	25	21	19
1/4	37	37	32	28
5/16	54	54	46	40
3/8	78	74	63	58
1/2	114	100	86	75
5/8	152	120	103	90
3/4	210	137	118	103
7/8	241	158	135	118
1	312	204	175	153

08900 CURTAINWALL SYSTEM

The following comments on metal curtainwalls are taken principally from the specification prepared by the National Association of Metal Manufacturers:

The use of metal curtainwalls for high-rise structures is just into its second decade. Many of the problems encountered in the initial phase of popularity have been identified and solved. NAAMM notes: "Having successfully outgrown its initial period of over-popular acceptance, when it was often used ineptly and inappropriately, metal curtainwall has become an established, reliable, and ever-modern idiom of architectural design. Its predominant choice for today's most important high-rise commercial structures attests to this fact."

The NAAMM Manual and specification for metal curtainwall is oriented very strongly to manufacturing. The curtainwall offers one of the earliest true building systems approaches, in that most of the work is done in the factory, and the field portion is limited to installation. A key factor in the design is the method of attachment of the curtainwall panel to the structure. The *clips,* or hangers, to receive the curtainwall must be carefully aligned, and should have been inspected by both the contractor and the inspection team prior to the beginning of installation. Often, installation of a sample panel can be meaningful for all parties.

Because the curtainwall is tied to the structure, and is not inter-supporting, its installation can start at any point. Usually, this is from the bottom floor upward, or can be from an interim floor down. Without exception, the curtainwall parts must be erected plumb and true, in proper alignment, or errors will accumulate and serious problems will develop.

The designer sets permissible dimensional *tolerances* in the building frame, and these will have been checked before field erection of curtainwall sections begins. The curtainwall design will have been ready to accommodate these tolerances, with enough closure or overlap to accommodate the tolerances in the structure. The curtainwall is usually fabricated to be true to within $\frac{1}{8}''$ per 12′ of length, or $\frac{1}{2}''$ in any side, with a $\frac{1}{16}''$ maximum offset from true alignment between identical members butting end to end. Where the curtainwalls are to be installed within masonry openings, only the built-in anchor devices will be put in place until the masonry work is completed.

Supporting brackets are usually designed to provide a three-dimensional adjustment in order to accurately locate the wall components. When the wall has been properly positioned, it is then rigidly fixed by positive means such as welding. After the unit is fixed in place, caulking or sealing materials should be applied, if the inspection team is certain that all foreign materials are removed from the surfaces to be contacted.

Quite often, curtainwall units are delivered direct to the job site and erected with only a minimum backlog stored on site. However, the units must be carefully handled, and protected until they are rigged into place.

If the units are to be *pre-glazed,* the manufacturer, curtainwall erector, and inspection team should agree on the method of handling and placing the units. Where the glazing is done in the field, the curtainwall contractor and manufacturer should coordinate their activities with the glazer.

It is common for sections of the wall to be left out for the convenience of moving materials, and this should be anticipated. Further, after the wall is in place, it must be cleaned with approved cleaning agents in order to preclude damage to the sealants.

DIVISION 9: FINISHES

NUMBER	TITLE
09100	**METAL SUPPORT SYSTEMS**
−10	Non-Load-Bearing Wall Framing Systems
−20	Ceiling Suspension Systems
−30	Acoustical Suspension Systems
09200	**LATH AND PLASTER**
−01	Furring and Lathing
−02	Gypsum Lath
−03	Metal Lath
−04	Plaster Accessories
−10	Gypsum Plaster
−15	Veneer Plaster
−20	Portland Cement Plaster
−25	Adobe Finish
09230	**AGGREGATE COATINGS**
09250	**GYPSUM WALLBOARD**
−60	Gypsum Wallboard Systems
−80	Gypsum Wallboard Accessories
09290–09299	(Reserved)
09300	**TILE**
−10	Ceramic Tile
−20	Ceramic Mosaics
−30	Quarry Tile
−31	Chemical Resistant Quarry Tile
−32	Slate Tile
−40	Marble Tile
−50	Glass Mosaics
−60	Plastic Tile
−70	Metal Tile
−80	Conductive Tile

09400	**TERRAZZO**
−10	Portland Cement Terrazzo
−11	Terrazzo Bonded to Concrete
−20	Precast Terrazzo
−21	Terrazzo Tile
−30	Conductive Terrazzo
−40	Plastic Matrix Terrazzo

09500	**ACOUSTICAL TREATMENT**
−10	Acoustical Ceilings
−11	Acoustical Panels
−12	Acoustical Tiles
−13	Metal Ceiling Systems
	Linear
	Leaf
	Pan
−20	Acoustical Wall Treatment
−25	Acoustical Units
−30	Acoustical Insulation and Barriers
09540–09499	(Reserved)

09550	**WOOD FLOORING**
−60	Wood Strip Flooring
−61	Gymnasium Type Hardwood Strip Flooring
−62	Gymnasium Type Steel Splined Flooring System
−70	Wood Parquet Flooring
−80	Plywood Block Flooring
−90	Resilient Wood Flooring System
−95	Wood Block Industrial Flooring

09600	**STONE AND BRICK FLOORING**
−10	Stone Flooring
−11	Flagstone Flooring
−12	Slate Flooring
−13	Marble Flooring
−14	Granite Flooring
−20	Brick Flooring
−22	Industrial Brick Flooring
−23	Reinforced Brick Masonry Slab

09650	**RESILIENT FLOORING**
−51	Cementitious Underlayment
−60	Resilient Tile Flooring
−65	Resilient Sheet Flooring
−70	Fluid Applied Resilient Flooring
−75	Conductive Resilient Flooring

09680	**CARPETING**
−81	Carpet Cushion
−82	Carpet
−83	Bonded Cushion Carpet
−84	Custom Carpet
−90	Carpet Tile

09700	**SPECIAL FLOORING**
−01	Resinous Flooring
−10	Magnesium Oxychloride Floors
−20	Expoxy-Marble Chip Flooring
−21	Seamless Quartz Flooring
−30	Elastomeric Liquid Flooring
−31	Conductive Elastomeric Liquid Flooring
−40	Heavy-Duty Concrete Toppings
−41	Armored Floors
−50	Mastic Fills
−55	Laminated Plastic Flooring

09760	**FLOOR TREATMENT**
−70	Metallic-type Static Disseminating and Spark-resistant Finish

09800	**SPECIAL COATINGS**
−10	Abrasion Resistant Coatings
−11	Chemical Resistant Coatings
−15	High-Build Glazed Coatings
−20	Cementitious Coatings
−30	Elastomeric Coatings
−35	Textured Plastic Coatings
−40	Fire-Resistant Paints
−45	Intumescent Paints
−60	Anti-Graffiti Coatings
−70	Coating Systems
−71	Protective Lining for Concrete Storage Tanks
−72	Interior Coating System for Steel Storage Tanks
−73	Exterior Coating System for Steel Storage Tanks
−74	Linseed Oil Protection for Concrete Surfaces
−75	Coating System for Steel Piping

09900	**PAINTING**
−10	Exterior Painting
−20	Interior Painting
−30	Transparent Finishes

09950	**WALL COVERING**
−51	Vinyl-Coated Fabric Wall Covering
−52	Vinyl Wall Covering
−53	Cork Wall Covering

−54	Wallpaper
−55	Wall Fabrics
−56	Asbestos Wall Covering
−60	Flexible Wood Sheets and Veneers
−70	Prefinished Panels
−90	Adhesives
09955–09999	(Reserved)

09200 LATH AND PLASTER

Lath is the base upon which gypsum plaster is placed. In older construction, wood lath was utilized with multi-coats of plaster placed directly upon it. Many examples of this construction persist, and its life can indeed be substantial. However, today, lath principally means one of two kinds: metal lath or gypsum lath.

Metal lath is most usually expanded metal, welded wire or woven wire, fabricated from steel. The wire lath includes copper and is galvanized. The lath is usually mounted directly to metal studs, or furring. Furring may be attached to structural members of the building, hung from a suspended ceiling, attached to vertical structure of the building or walls. As described in the ANSI Standard Specification for interior lathing and furring, studless solid partitions can also be made up of metal lath and plaster.

The type and configuration of the studs or supports for lath will be specified in the plans. Check the studs for galvanizing, spacing, and anchorage. Usually, studs are doubled up at jambs or openings. Also, channel braces are usually anchored to double studs at the head of each opening, and carried to adjacent studs for lateral support. Studs anchored to walls or columns should be anchored with at least three anchors. Where stud partitions are more than 10′ long or 9′ high, horizontal channel stiffeners should be installed on 6′ centers vertically.

For suspended ceilings, the furring and runner channels should be checked for size and spacing of hangers, and check whether the hanger wires are saddle tied or bolted to the runner channels. Hangers should be suspended plumb. The furring should be in a true level plane surface before attachment of the lath.

The following comments on installation of lath are taken from the Manual of Gypsum Lathing and Plastering by the Gypsum Association:

The most popular base for plaster today is *gypsum lath*, which is usually ⅜″ thick × 16″ wide × 40″ long, either in plain sheets or perforated (other sizes are available). The gypsum lath can be nailed,

GYPSUM LATH

Type	Thickness (Inches)	Width (Inches)	Length (Inches)	Use
Plain	⅜ ½	16 16	48 or 96 48	Application to wood or metal framing, by nails, staples, screws or clips.
Perforated	⅜ ½	16 16	48 or 96 48	Same as above except not used for ceiling attachment where the only attachment is by clips at edges of lath or where insulation is placed on the ceiling.
Insulating	⅜ ⅜ or ½	16 24	48 or 96 as requested to 12 ft.	Same as for plain and where a vapor barrier is required.
Long length	½	24	as requested to 12 ft.	Primarily used in 2 in. solid gypsum lath and plaster partitions.

TYPES AND WEIGHTS OF METAL LATH, WIRE LATH, AND WIRE FABRIC AND SPACING, CENTER TO CENTER OF SUPPORTS.[a]

Type of Lath	Minimum Weight of Lath, lb per sq. yd.	Maximum allowable spacing of supports, in.				
		Vertical Supports			Horizontal Supports	
		Wood	Metal		Wood or Concrete	Metal
			Solid Partitions	Others		
Flat expanded	2.5	16	16	12	0	0
metal lath	3.4	16	16	16	16	13½
Flat rib	2.75	16	16	16	16	12
metal lath	3.4	19	24	19	19	19
⅜-in. rib	3.4	24	..	24	24	19
metal lath[b]	4.0	24	..	24	24	24
Sheet metal lath	4.5	24	24	24	24	24
Wire lath	2.48	16	16	16	13½	13½
V-stiffened wire lath	3.3	24	24	24	19	19
Wire fabric	c	16	0	16	16	16

[a] Lath may be used on any spacings, center to center, up to the maximum shown for each type and weight.

[b] Rod-stiffened or V-stiffened flat expanded metal lath of equal rigidity and weight is permissible on the same spacings as ⅜-in. rib metal lath.

[c] Paper-backed wire fabric, No. 16 gauge wire, 2 by 2-in. mesh, with stiffener

Fig. 1 Allowable spacing of supports for various lath configurations. *Source: Gypsum Association.*

stapled, screwed, or clipped to either wood or metal framing, or furring members. It is economical because of reduced working time, and because it requires only a ½″ coating of plaster to provide a fire-resistant serviceable base. Figure 1 shows the allowable spacing of supports vertically and horizontally for various configurations of metal lath, as well as typical sizes of gypsum lath.

Lath is applied with its long dimension perpendicular to the framing, with end joints staggered between courses. Lath ends must bear on the framing, unless a special clip attachment is used to secure it to the next piece of lath. Lath should be placed with folded or lapped edges toward the framing, and the edges and ends of adjacent pieces of lath should be in moderate contact. If there is more than a ½″ space between laths, it should be reinforced with metal lath strips. Figure 2 shows the use of cornerite to provide a crack-resistant corner in wood or metal stud partitioning. Note that the nailing or stapling of fasteners within

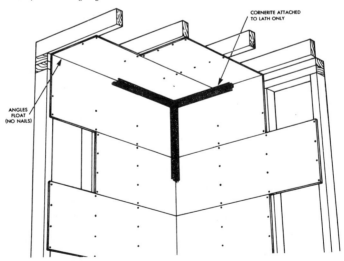

The "Floating Angle" method of application of gypsum lath was developed to reduce plaster cracking caused by stress at intersections of framed walls and ceilings.

It consists of eliminating the last two nails or staples in all interior angles, (both horizontal and vertical).

Conventional nailing or stapling (see other side) of four fasteners per 16″ wide piece of lath on each joist or stud should be used on remainder of the ceiling and wall areas, with the ceiling lath applied first.

Cornerite with flanges of 2 inches or more shall be attached to the lath only (not to the framing members) in all "floating" angles.

CORNERITE ATTACHED
TO LATH ONLY

ANGLES
FLOAT
(NO NAILS)

Fig. 2 Cornerite to provide crack resistant corner. *Source: Gypsum Association.*

approximately 8″ of the corner is omitted so that it has flexibility, and is supported by the cornerite. The cornerite should be secured to the metal lath with staples or tie wire.

Each piece of lath should be securely fastened with four fasteners to each framing member, and these should be driven home to just below the paper surface without breaking the paper. If fasteners are not driven home enough to compress the gypsum core, there is the possibility of the plaster popping over the loose fastener.

All exterior angles should be finished with *cornerbead*. *Grounds* should be provided around all openings and wherever baseboards or moldings are to be used. These provide for a full ½″ thickness of plaster.

Other bases can be used for the application of plaster including gypsum for partition tile, brick or clay tile, concrete block, and concrete. Each has its own special conditions. Gypsum partition tile provides the best block base, since it is a natural bond of gypsum to gypsum. Brick and clay tile provide a good base where they are porous, providing good bonding. However, glazed units do not provide a good bond and should not be used for a plaster base. Concrete block provides an excellent bond for plaster, and is very compatible. Concrete may not be compatible because of its dense, smooth surface, and requires special preparation for plastering. Also, thick plaster should not be placed on concrete; if more than ⅜″ ceilings or ⅝″ plaster walls on concrete is required, metal lath should be secured to the concrete to provide a base for the plaster. Special bonding plaster should be used over concrete (however, where metal lath has been secured to the concrete, normal plaster can be used). Plaster definitely cannot be applied over bituminous compounds which have been placed on masonry or on concrete surfaces for waterproofing. Full furring and/or lath is required to offset the lack of bond furnished by a nonporous surface.

The base for plaster is *gypsum* which is prepared by processing of gypsum rock by crushing, drying, and then calcinating by heating in a rotary kiln. The key to the successful use of plaster is the addition of materials which slow down the setting process. (Pure gypsum sets within a few minutes, and is often known as plaster of paris.) However, the manufacturers of gypsum-based plaster are experts in the addition of proper materials, and, in accordance with specifications, can be depended upon to furnish proper materials for preparation of the plaster material. The first or *base coat* is that portion of the plaster coat applied directly to the lath or masonry base. It fills in whatever additional thickness is required to square off the surface of the partition. The base coat has structural value, and therefore, must resist cracking. There are several distinct base coat plasters, each with a specific use. Neat plaster is gypsum cement plaster. Hard wall is mixed with sand, perlite, or vermiculite aggregate and water. It is available with fibrous materials added or unfibered. Other gypsum plasters come ready-mixed from the mill and require only the addition of water on the job; these are the most commonly used. Gypsum bond plaster is especially formulated plaster for use on interior monolithic concrete and requires only the addition of water at the job site.

The addition of *aggregates*, including wood fibers, sand, perlite, or vermiculite are principally for addition of bulk and coverage of the plaster mass. Sand is the most popular aggregate. Aggregate gradation and proportioning exerts a tremendous influence on the finished plaster performance. Improper gradation of the aggregate, or over-proportioning of aggregate, each operate to reduce the plaster strength of the base coat, which in turn endangers the ability of the plaster to perform without cracking. The Gypsum Association suggests that approximately three-fourths of all plastering employs sand as an aggregate, and about three-fourths of the weight of that plaster mass is sand, indicating that about half of the total mass of plaster utilized is sand. The gradation and composition of the aggregate should either be specified, or should perform as specified by the plans and specifications. Typical mixes based on 100 lb of gypsum plaster are as follows:

1:2—Scratch coat over all lath in a three-coat plaster work.

1:2.5—Over gypsum lath in two-coat plaster work.

1:3—Brown coat over all lath in three-coat plaster work and for all work over masonry.

Even when sand is clean, no greater amount should be used than as noted above, since this will impair plaster strength. Also, any dirt or deleterious materials in the sand, even if used in the above mixes, will impair the strength. The Gypsum Association suggests the use of a measuring box on the job for mixing, since alternate methods of measurement are less accurate.

Perlite aggregate is used most often when a higher fire rating is required, and vermiculite serves a similar purpose.

Vermiculite and perlite plasters are also much lighter. Vermiculite-based plaster has a lower strength rating than either sand or perlite.

Figure 3 shows relationships between strength and aggregate ratios. For inspection of hardness of plaster surfaces, a standard *scratch test*

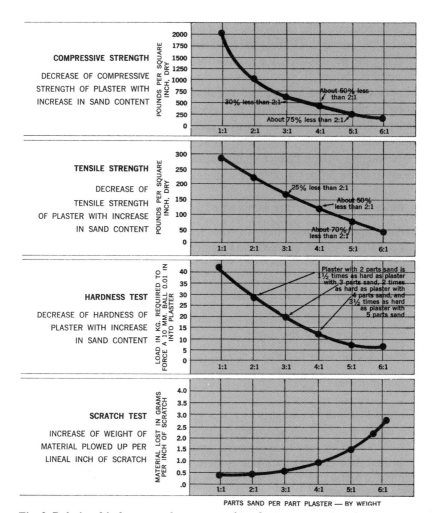

Fig. 3 Relationship between plaster strength and aggregate ratios. *Source: Gypsum Association.*

can be utilized. This consists of a pronged device drawn across the surface of a panel at a uniform rate of speed carrying a fixed weight. The softer the plaster, the deeper the abrasion. While in some cases an experienced inspector can use a nail or similar sharp device for hand scratching the surface of a brown coat to determine its hardness, the test in this fashion is quite inaccurate. Nevertheless, a definitely inferior base coat can readily be detected by this means.

As in concrete, the *water ratio* to gypsum is important, and directly affects the strength. Excess free water also slows down the setting, reducing the productivity for installation.

The fibers added to gypsum plaster do not increase the strength, but in the scratch coat, do serve to fill the openings of the lath to prevent an excess of plaster going through the lath and falling to the floor. For this reason, fibered plaster is always recommended as a scratch coat application.

The mixing of the plaster should always be done with a mechanical mixer, and the proper procedure is generally as follows:

1. Place about 90% of the required water in the mixer.

2. Add approximately half of the required sand, or if perlite or vermiculite are to be used, add all of the aggregate.

3. Add the plaster.

4. Add the remainder of the sand.

5. Mix at least half a minute—but not more than 3 min., depending upon the speed of the mixer.

6. If necessary, add balance of water to obtain proper consistency.

7. Dump the entire batch at once.

In order to avoid too much water, it is good practice to hold back about 10% of the initial water requirement used to bring the plaster to proper consistency. When the mixer is being used continually, this approach will keep the mixer clean, and if intermittent use is being made of it, a small amount of water can be placed in the mixer and used to clean it out prior to the next batch.

The inspection team should be certain that plaster is not used once it has started to set.

A three-coat plaster application is traditional and is described as first a *scratch coat* of plaster, which is cross raked after it has started to set. This scratch coat is allowed to set and partially dry. Next is the application of the *brown coat* of plaster which is surfaced or brought out to the proper grounds and allowed to set and partially dry. Third is the application of the *hard finish coat.* Three-coat plaster is required over metal lath, and over ½″ gypsum lath attached to horizontal supports greater than 16″ on center, or over gypsum lath on ceilings where the lath is attached by clips providing edge support only, and over ¾″ perforated gypsum lath on ceilings.

The two-coat application is similar in every respect, except that the cross raking of the scratch coat is omitted and the brown coat of plaster is applied within a few minutes to the unset scratch coat. This method is generally accepted in applying plaster to masonry and gypsum lath. The three-coat application is preferred because it develops a harder and stronger base coat.

The *thickness* of the base coat plaster is controlled by the use of grounds and screeds. A plaster job should be inspected constantly for base coat thickness, since experience has well documented the proper plaster thickness which will give satisfactory results. For different bases, minumum plaster thicknesses including finish coat are:

1. Gypsum lath and gypsum partition tile—½″.

2. Clay tile and other masonry—⅝″.

3. Metal lath—⅝″ over the face.

Plastering *machines* are used on most large plastering jobs today. The plaster is put in place by the nozzle man, and he is followed by one to four plasterers who darby the wall. Proper drying of plastered surfaces is important to the ultimate performance and strength. Too fast a dry may leave insufficient water for the chemical reaction which produces the set. However, rapid drying after the plaster has set is conducive to high strength and should be encouraged. Slow drying, on the other hand, should be avoided in all cases.

In cold weather, the minimum temperature in the building should be at least 55°F for a period of at least seven days before plastering, during the plastering operation, and then until the plaster has set. If temporary heat is being used, it should be evenly distributed.

In hot, dry weather, rapid evaporative drying should be avoided.

The finish coat will probably be one of six universally available finishes which include:

Gypsum-lime putty troweled finish

Keene's-lime putty troweled finish

Prepared gypsum troweled finish

Keene's cement-lime sand float finish

Gypsum-sand float finish

Acoustical plaster finish

Lime is used in three of the most common finishes, and although not a desirable base for plaster (base coat) does provide plasticity and bulk to the finish coat. The problem with lime-based plaster is in setting, and therefore, it is combined with gypsum to produce a hard, dry finish. Lime also is subject to considerable shrinkage.

The finish plasters are available, ready for use in bag form, and are designated as quick set or slow set. The relative surface hardness can also be established by the ratio of gypsum (Keene's Cement) to the dry, hydrated lime.

The Gypsum Association notes a number of plaster problems and offers solutions:

Plaster delivered to the job has a definite chemical composition which establishes the set, and it is tailor-made for the area and the time of the year. However, this may be somewhat sensitive to unusual conditions and may require job remedial action. For instance, slow set (plaster should set within two to three hours after having been mixed with water, and always within four) can be counteracted by spraying the unset surface with a solution of zinc sulphate and followed by a brisk floating of the surface. The problem of accelerating unmixed plaster is much less difficult, and can be remedied by the addition of accelerators as approved by the specification.

Another problem is quicksetting, which means the premature setting of plaster, sometimes caused by using water in which hardened plaster has been allowed to drop from tools. Since plaster should never be re-tempered, this material should be discarded. Conditions responsible for quicksetting should be investigated and corrected. If atmospheric conditions prove to be the reason, a retarder can be added as approved by the specifications.

If inspection indicates a dry-out which can be identified because the surface is soft and crumbly with a light chalky appearance, water should be added with a soft spray, and openings should be closed to reduce wind through the building to cut down the rate of evaporation.

The reverse of a dry-out is a sweat-out, where the excess water has not evaporated, and walls remain damp for an extended period. This condition is usually due to high humidity and is best remedied by the addition of artificial heat in cold weather, or opening of doors and windows in warmer weather.

When plaster has frozen before set takes place, artificial heat should be added to bring the temperature of the building up to at least 55°F. A single freezing of plaster after the initial set does not injure the plaster, but freeze-thaw cycles should not be allowed to occur.

When plaster has been properly mixed and applied it should not crack or loosen; any such action indicates other problems. Areas to be investigated include deflection of the structural frame, or shrinkage of unseasoned lumber. Settlement of the foundation walls can also cause cracking. Thickness of less than ½″ may make the plaster vulnerable. Also, the ratio of aggregate to gypsum should be reviewed.

Thermal shock can cause cracking either through application of too much heat, or sudden applications of cold. Another cause of cracking can be moving of the lathing due to improper anchorage to the structure.

Loosening of the plaster, when placed on solid base rather than lath, may be due to the use of an inappropriate surface which is too dense, glazed, or made of a material which will not accept bonding of the plaster.

09250 GYPSUM DRY WALLBOARD

09260 Gypsum Dry Wall Systems

The inspection team should ensure that the studs—wood or metal—are properly aligned and present a plane nailing surface for the dry wall installation.

Gypsum board should be stored in dry, well-ventilated and closed-in sheds or buildings, and not brought into the construction area until the building is completely closed in. For cold-weather installation, temperatures should be maintained above freezing at the time of joint cement application.

The board should be installed in a manner which will result in the least number of joints, horizontal or vertical. However, horizontal installation for wall partitions provides greater strength. During the installation, the gypsum board should be held tight against the supports while nailing, and nailing should be started in the center of sheets and carried progressively toward the ends and edges. Figure 4 illustrates sequence of nailing, both single nailing and double nailing. Figure 5 shows the method of attaching dry wall panels, either metal studs or furring channels for a hung ceiling.

SINGLE NAILING

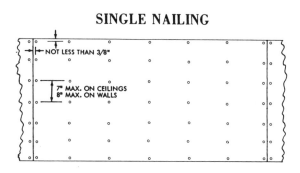

NAILS:

"SINGLE NAILING" SYSTEM PROCEDURE: The wallboard shall be held in firm contact with the nailing member while the nails are being driven. Nailing preferably should proceed from the central portion of each piece of wallboard towards ends and edges. Nails shall not be staggered on adjoining edges or ends. Perimeter nailing at the top and bottom is not required.

DOUBLE NAILING

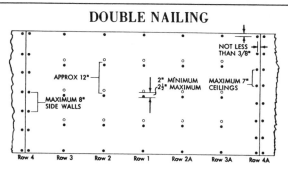

NAILS:

"DOUBLE NAILING" SYSTEM PROCEDURE:

1. Two men are required to nail one ceiling board — one man nails rows 1, 2, 3 and 4; the other nails rows 2A, 3A and 4A.

2. Starting at center of board, nails shown by dot ● are applied in row 1, then rows 2 and 2A, 3 and 3A, 4 and 4A, etc., always nailing from center to edges of sheet.

3. Apply "double" nails shown by circle O in the same manner as first nails, also starting at row 1.

4. As alternate procedure, "double" nails may be applied immediately after nailing of first nails in each row is completed according to step "2" above.

5. Use single nails on perimeter of board.

IMPORTANT: All three field (first) nails in each row should be applied before the "double" (second) nails are applied. It may be necessary to "hit" the first nails in each row after the second nail is applied.

Fig. 4 Sequence of nailing drywall panels. *Source: Gypsum Association.*

EXTERIOR SOFFIT
METAL FRAME CONSTRUCTION

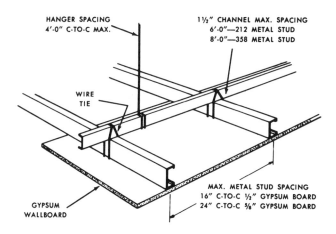

HANGER SPACING
4'-0" C-TO-C MAX.

1½" CHANNEL MAX. SPACING
6'-0"—212 METAL STUD
8'-0"—358 METAL STUD

WIRE
TIE

GYPSUM
WALLBOARD

MAX. METAL STUD SPACING
16" C-TO-C ½" GYPSUM BOARD
24" C-TO-C ⅝" GYPSUM BOARD

METAL STUD DETAILS

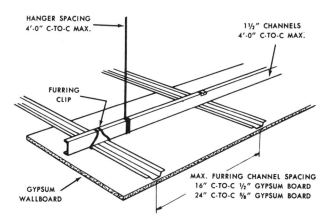

HANGER SPACING
4'-0" C-TO-C MAX.

1½" CHANNELS
4'-0" C-TO-C MAX.

FURRING
CLIP

GYPSUM
WALLBOARD

MAX. FURRING CHANNEL SPACING
16" C-TO-C ½" GYPSUM BOARD
24" C-TO-C ⅝" GYPSUM BOARD

FURRING CHANNEL DETAILS

Fig. 5 Drywall panel attachment to hung ceiling or soffit. *Source: Gypsum Association.*

Gypsum board can be placed single ply or double. Single-ply systems are commonly used in residential and light commercial construction. In double-ply applications, two or more layers of gypsum board are placed, the first layer nailed and subsequent layers glued to provide increased sound isolation.

If paper surfacing is broken around a nailhead, an additional nail should be driven about 2″ away. Sheets should be placed with proper clearance and not driven in place, which will cause bowing. The location of cutouts for electrical boxes and other equipment which will project to the finished surface should be checked so that the sheets will not be damaged when erected.

Check the specifications for special requirements and installation details around bathtubs and showers.

Sheets damaged after erection must be removed and replaced. Patching should be restricted to only minor errors.

Gypsum Dry Wall Finishing

Building temperatures should be held above 55°F for finishing. Paper *tape* should be placed over the first thin coat of finishing *compound*. This should be followed by three coats of cement on the joints and nailing depressions, with a minimum of 24 hours *drying* time between coats. After the final coat of cement is dried, all coated areas should be sanded to provide a smooth, true surface. The final coat should be feathered out 12″ to 16″ on both sides of the joint.

If an occasional isolated bubble appears in taped joints, it can be opened, cement applied under the bubble and tape re-sealed. If tape has to be removed, the joint should be entirely refinished.

Internal and external corners and exposed edges should receive corner bead, and should be finished using the bead as a ground (Fig. 6).

DRY WALL CHECKLIST

1. Before installation, check that studs are properly aligned and present a plane nailing surface.
2. Gypsum boards should be stored in dry area, not brought to the construction area until the building is closed in.
3. Boards should be installed to produce the least number of joints.
4. Wall partition installation is preferably horizontal.

5. Nailing should start in center of sheets carrying progressively to ends and edges.

6. If paper surfacing is broken around a nail head, an additional nail should be driven about 2″ away.

7. Temperature should be above 55° F for finishing joints.

8. There should be a minimum of 24 hours drying time between spackling coats.

9. Final spackling coat should be sanded to provide a smooth true surface, with final coat feathered out 12″ to 16″ on both sides of the joint.

09300 TILE WORK

The following information is extracted from publications by the Tile Council of America, Inc.:

Concrete and masonry walls and ceilings which are to receive ceramic tile must have square corners, and must be plumb and true, with variations limited to plus or minus ¼″ in 8′.

The inspection team should verify that all materials required for the installation of tile have been approved, in particular the size, color, and pattern of the tile.

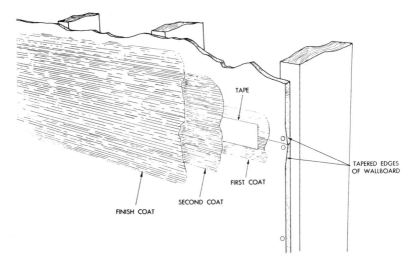

Fig. 6 Drywall finishing sequence. *Source: Gypsum Association.*

The cement should be as called for by the specifications. If Portland cement, it shall be ASTM 150–67 of the color designated, and sand shall be clean and graded in accordance with specifications. All surfaces to be coated with tile shall have clean, dry, firm, and oil-free surfaces. All work to be installed behind the tile such as electrical and mechanical, hangers, anchors, etc., shall be installed prior to start of the tile work.

Tile shall not be installed in areas where the temperature is below 50°F and rapid evaporation of moisture from the mortar bed shall be precluded.

Mortar should be mixed dry before adding water, and water should be carefully worked in. Mix may be reworked from time to time, but mortar should be discarded if it has reached its initial set.

The tile should be installed in a manner conforming with the best current working practices. Cuts shall be held to a minimum, and generally no cuts smaller than half-size of the tile shall be made. Any cut edges should be smoothed with a carborundum stone. Tile should be closely fitted. All joint lines shall be kept straight and of even width. Accessories on the tile work should be placed so that they are evenly spaced, properly centered with tile joints, and are level, plumb, and true to the correct location. Good practice includes *soaking* tile prior to setting, and the tile once placed should be checked for firmness of set, secure installation of accessories, and freedom from damage or defects.

For the *grouting* operation, the edges of tile should be thoroughly wet before application of grout. Grout should be placed in temperatures above 50°F, and should be properly cured with emphasis on avoidance of rapid evaporation.

Expansion of control *joints* should be provided as required by the specifications. Bond-breaker material such as wax paper or foil shall be used to preclude the intrusion of grout into expansion and control joints. The expansion control joints shall be filled with proper sealants as called for by specification.

After completion of setting and grouting of tile, it should be thoroughly cleaned and finally polished with clean dry cloths. Acid or acid cleaners should not be used to clean glazed tile.

09320 Ceramic Mosaics

The inspection team shall check for size, color, pattern, and shape of tile to ascertain that it agrees with the specification. Ceramic mosaic is similar in all respects to ceramic tile, except that it is of smaller size and shape. Preparation of all surfaces shall be clean, dry, and free of oil and firm. Subfloor surfaces must be within $\frac{1}{4}''$ of true per 10'. Wall and ceiling surfaces which are to receive ceramic mosaic tile shall be within $\frac{1}{4}''$ in 8'.

The spaces in which tile is to be set shall be closed to traffic, and installation shall only be carried out with temperatures above 50°F. Prevention of rapid evaporation of moisture from the mortar bed shall be precluded. The inspection team shall ascertain that traffic is routed around newly-tiled floors. If the floors must be crossed, boards or kneeling boards shall be used, at least for a period of three days.

09330 Quarry Tile

Quarry tile is dense and does not have a glazed finish. Conditions of placement and inspection are similar to those of ceramic and mosaic tile surfaces. Surfaces must be true and free of deleterious materials. The inspection team should review the quarry tile for conformity with approved samples for size, shape, and color.

09380 Conductive Ceramic Tile

Conductive tile is used principally in operating rooms to reduce or preclude the development of static electricity. These floors require special installation techniques, and particularly careful inspection. Conductive tile is usually installed in a Portland cement mortar bed, 1 to $1\frac{1}{2}''$ thick, containing carbon (lamp black).

Conductive tile is set generally in accordance with normal practice for tile installation, but after installation and proper curing, the inspection team should check for minimum resistance requirements as called out by the specifications. These tests shall be as described in the specifications, or shall be in accordance with NFPA No. 56. Further curing may improve the ability of the floor to meet requirements, but in no case shall a period of more than 6 months curing and drying be permitted prior to the test.

CERAMIC TILE CHECKLIST

1. Walls or floors which are to receive tile must have square corners, plumb and true.
2. Verify materials to be in accordance with approved samples.
3. Surfaces must be oil free.

4. All anchors inserts etc. must be installed prior to start of tile.
5. Temperature for installation should be 50° F or better.
6. Mortar should be discarded if it has reached initial set.
7. Tile cuts should be held to a minimum.
8. No cuts smaller than half size of the tile should be made.
9. It is preferable to soak tile prior to setting.
10. Check to ensure that control joints for expansion are provided as required by specifications.
11. Check for special tile installation requirements, such as conductive tile for operating rooms.

09400 TERRAZZO

09410 Cast-in-Place Terrazzo

Usually used in hallways, toilet rooms, showers, and corridors where heavy traffic or moisture is evident, terrazzo is cast-in-place floor concrete, in which the aggregates are chips of marble or other decorative stones. Terrazzo is usually ½ to ¾" thick and is placed on a 1-½" bedding of dry mortar. The inspection team should ensure that a vapor barrier is in place before the terrazzo bed is laid in order to separate the terrazzo from existing concrete floor, or bed surface. The surface upon which the terrazzo bed is to be placed should be clean and smooth, and washed down with a mixture of water and neat cement to provide a bonding between the vapor barrier and the floor.

Sections of the terrazzo are separated by *ground strips* made of aluminum or brass which are imbedded in the concrete bedding and in the terrazzo layer. The terrazzo bedding usually contains a mesh for reinforcement of 14-gauge galvanized wire net, or as called for in the specification.

The terrazzo mix should be compacted with heavy rollers after being spread, and the surface troweled, floated, and cured by approved methods. After the mix is hardened, it is *machine* rubbed, ground, and polished. Excess grout should be removed and the surface sealed with sealing compound of a type approved by the specification.

If the specification does not call for a vapor barrier, the surface to receive the terrazzo should be clean of all dirt, oil, and grease, and roughened sufficiently to provide a bond.

A floating terrazzo is used where structural movement might injure the topping as a result of settlement. In this case, the base slab is covered with a sand bed which is then overlaid with the vapor membrane followed by wire mesh reinforcing in the terrazzo bed.

09420 Precast Terrazzo

Terrazzo may be made in precast slabs, usually utilized in staircases, platforms, or similar situations.

09430 Conductive Terrazzo

Conductive terrazzo is used in areas such as operating rooms, to minimize or preclude the development of static electricity charge. The conductivity is developed through the use of carbon in the mortar.

09500 ACOUSTICAL TREATMENT

Acoustical finishes can be made up of sheet, tiles, or a special coating.

Soundproofing acoustical felt is similar to wallpaper, and is applied to wall or ceiling. Installation should be in accordance with manufacturer's recommendations, and adhesive should be applied to clean, dry surface.

Acoustical tiles similarly should be inspected for damage before application. On walls or ceiling, they may be applied with adhesive or may be fitted into hung ceiling grids. The inspection team should be certain that the material installed is the same as the approved samples. The inspectors should also check the manufacturer's label of the adhesive where application is direct to wall surface.

Where the acoustic finish is cement or plaster, the mixture should be approved and in accordance with the specifications. Acoustic plaster is applied in the same fashion as the finish coat of plaster, and may be applied over a scratch coat, or directly to a wall surface. It may be applied with trowel, or sprayed on. The inspector should be certain that the same conditions and preparation are made as for a plaster finish, plus any special requirements as pointed out by the manufacturer's instructions or the specifications.

09550 WOOD FLOORING

09560 Wood Strip Flooring

Wood strip flooring is either hard wood or soft wood, with hard wood being the preference for better wear. The flooring is laid on subflooring which is anchored either to wooden joists or sleepers, which, where appropriate, are also anchored to a concrete subfloor. The anchorage of the sleepers or joists at proper intervals is very important to preclude the actions of expansion and contraction due to the change in moisture content of the wood. Proper installation of fasteners can almost exclude the undesirable aspects of such expansion. Anchors for wood *sleepers* are usually zinc-coated and designed so that they cannot pull out after the concrete is set. Subfloors are used for thicknesses of final flooring of less than 1″ and can be either plywood or 1″ × 4″ or 1″ × 6″ boards laid in diagonal direction over the sleepers. Subflooring is usually made of soft wood. The inspection team should check sleepers for level and plane surface before starting their subflooring. The locations of end joints should be staggered and located over bearing surfaces. The inspection team should approve the condition of the subflooring before the finished floor is laid. Any damaged material or loose boards should be replaced or repaired. Check the floor for cupping, and if observed, add additional fastenings. If this does not correct the cupping, remove and replace subfloor. Ensure that expansion has been provided for around the perimeter of the finished floors, and be certain that the subflooring is clean. Felt paper should be laid down just ahead of the finished flooring. Before finished flooring is brought in, the work should have progressed to the stage where the building is weathertight and any moisture-producing construction, such as plastering, has been completed. The temperature in areas to be floored should be maintained above 50°F for a number of days prior to the installation of the flooring and continued until the building is put into use. The flooring material to be used should be stored in the space to be floored for at least 3 days prior to laying, and the storage temperature should be above 50°F.

In laying the strip flooring, be certain that the proper size and approved nails are being used, preferably 8-penny spiral or screw-type flooring nails. Nails should be spaced no greater than 12″ on centers, and both ends of each strip should be nailed. The joints should be driven up tight, and end joints should be alternated so that there are at least two courses between joints.

09570 Wood Parquet Flooring

The same requirements of dry storage, proper building heat, and weather tightness, as well as storage of the materials in the space to be floored apply for parquet flooring. Subflooring is the same as for strip flooring.

Parquet flooring may be held together with tongue-and-grooved joints and nailed in place, or held with asphalt adhesive.

09580 Plywood Block Flooring

Plywood block flooring is made up of a laminated block of three or more plies of wood glued together. The blocks are usually tongue-and-grooved, and vary in thickness between $\frac{1}{2}$ and $^{13}/_{16}''$. Plywood blocks are usually installed with adhesives.

09650 RESILIENT FLOORING

09660 Resilient Tile Flooring

Resilient tile is laid either on firm base such as concrete or upon wood floors. The inspection team should check the type of tile, color, and quality with approved samples. Asphalt tile is not normally laid on flexible flooring, while vinyl tile is not usually laid below grade. The existing subflooring should be checked for removal of all foreign materials, and preparation. Joints and cracks should be filled with *filler* material as recommended by the flooring manufacturer, and a level surface prepared. In some cases, where resilient flooring is being placed over an existing floor, a latex underlayment is troweled on to provide the suitable smooth level surface required.

For wood subflooring, a rigid fastening should be placed, if the floor is not suitably firm. In some cases, additional underlayment is required. This will be called for by the specifications. Where tile is to be placed over wood subfloor, felt lining or underlayment is usually utilized in between the tile and the subfloor.

Flooring materials should be stored at their specified temperature (above 50°F) for several days, and the building should be maintained at that temperature or warmer. Installation of the flooring should be deferred as long as possible in terms of completion of other work so that the finished floor will not be damaged by the working conditions imposed.

During installation, check the laying pattern to make certain that the opposite borders are of equal width, and that tiles are cut only at the base, at outlets in the floor and at any border where a diagonal pattern is used.

Check the *adhesive* operation to determine if the cement is being applied in a proper manner and at a proper rate. Cement should have a proper dryness and tackiness before application of the floor covering, but cement applications should not be too far in advance of the tile placement. During placement, look for tight joints, straight lines, and smooth level surfaces.

Temporary protective covering should be provided over the finished floor and kept intact until the building is complete. When appropriate, the tile floor should be waxed in accordance with manufacturer's directions.

09665 Resilient Sheet Flooring

Application of resilient sheet flooring is similar in all respects to application of individual tiles, except that the entire floor must be prepared and adhesive ready at the proper level of dryness when installation is started. The sheet flooring should be rolled with a 150-lb roller, starting at the center of the floor and working toward the edges to remove all air bubbles and wrinkles.

09675 Conductive Resilient Floors

A conductive resilient floor is achieved by the dispersal of acetylene black (carbon) in the material during manufacture. The purpose of this is the reduction of electrical static charge in the vinyl floor. In the placement of resilient tiles, copper foil is sometimes placed between the adhesive and the tile for electrical inter-coupling.

09700 SPECIAL FLOORING

09720 Epoxy-Marble-Chip Flooring

A terrazzo-like floor can be prepared using epoxy as a matrix holder, and embedding various marble chips in a fashion similar to the manner in which marble chips are embedded in concrete to make terrazzo.

09730 Elastomeric Liquid Floors

Elastomeric materials are used in accordance with specifications to form a membrane over a fixed floor surface which, after curing, results in a durable, special floor covering. These floors must be installed in careful accordance with both manufacturer's requirements and the specifications.

09740 Heavy-Duty Concrete Toppings

Latex-based concrete has been used to provide hardwearing surfaces to new concrete floors, as well to repair existing floors. These are quick-drying and must be carefully mixed in accordance with manufacturer's directives. Only a limited amount of the material should be mixed before placement is started because of the rapid drying characteristics. Also, the material should be properly cured after placement. Only approved curing compounds should be used. In all cases, foreign and deleterious material should be cleaned from the surface before placement of the special concrete toppings.

09800 SPECIAL COATINGS

09820 Cementitious Coatings

Special applications such as stucco, parging, and gunnite are placed often to prevent water flow or for decorative purposes. They must be applied to a properly prepared surface which is rough enough to achieve a suitable bond.

09830 Elastomeric Coatings

Elastomeric coatings such as hypalon and neoprene can be sprayed or rolled on a surface to provide an elastomeric coating which results after the solvent material has evaporated. The result is a monolithic seamless sheet. Coating can be over almost any properly prepared surface which must be free of oil, dirt, and other foreign materials. Also, the coating must have properly prepared anchorages or boundaries, or it may peel back after placement.

One of the problems in placement of elastomeric coverings is the hazard of spraying.

09840 Fire-Resistant Coatings

Sprayed-on fireproofing is often used on structural steel to meet building code requirements for resistance to heat and flame. The coating may be lightweight concrete, or an asbestos-based mixture. The health problem incurred with spraying of asbestos fibers has resulted in the banning of much of the formerly used asbestos mixtures.

Some newer materials are sprayed on with a solvent base, and have the same general problems of handling as the asbestos fibers. Once in place, they provide a thin fire-resistant coating, which expands on being subjected to heat, providing an increased insulating thickness.

Of particular concern to the inspection team is the inflammable nature of this type of fireproofing, while it is curing. It is possible for a serious fire to occur in the fireproofing material before it has cured.

09900 PAINTING

The following information on the inspection of painting was provided by the Steel Structures Painting Council (4400 Fifth Avenue, Pittsburgh, Penna. 15213), but is applicable also to the general inspection of painting:

"The obvious purpose of inspection of painting operations on structural steel is to assure compliance with the specifications. Inspection of paints and painting is frequently overlooked or waived by the buyer of structural steel, despite the fact that it is in this phase of the work that he is likely to receive less than he has specified. Failure to provide inspection is common also in contracts for maintenance painting work. The general need for competent inspection of painting work stems from the many variables involved, each of which may have an overwhelming effect on the quality of the job as a whole. The paint itself, surface preparation, mixing and thinning, weather conditions, thickness of coats, and handling of coated surfaces are a few of the variables which must be carefully controlled if a durable coating is to be obtained. Unfortunately for the buyer, the final appearance of a poor paint job may be about the same as the appearance of one which was accomplished with the maximum of care.

Failure to provide adequate inspection of painting has inevitably been used to their own advantage by some contractors. Unfortunately, such malpractice by an original few may put pressure on even the well-intentioned to cut corners in order to meet competition. The net result is a lowering of the quality of painting for everyone. Adequate inspection of paint work as a general practice should be welcomed by well-meaning contractors inasmuch as it would put all competitors on the same plane and enable each to bid without conjecturing about how closely the painting will be inspected for compliance with the specifications. Contractors and suppliers should have no arguments against improving the quality of painting work, if the expense of the improvement does not rest on them because of varying inspection practices. This is not to say that the buyer will not encounter resistance if he attempts to specify surface preparation methods and paint materials which slow down or interfere with the contractor's normal course of operations. At the same time the buyer must realize that high quality work with rigid requirements will result in an increase in prices over the usual, commonplace practices.

The need for fully qualified personnel for paint inspection work is obvious if one considers the complexity and multiplicity of modern protective coatings and the considerable number of operations involved in obtaining a satisfactory paint job. Technicians employed to test paints for compliance with specifications should have considerable experience in the field in order to interpret test results after they are obtained. Inspectors of surface preparation and paint application not only need rather broad experience in the actual operations involved, but also should have some background knowledge of paints and paint materials. Those concerned with inspection of painted surfaces, with a view toward making maintenance painting recommendations, should likewise have a thorough background of experience in all phases of protective coating work, including costs involved in the various operations.

The contrast between the desired and actual qualifications of painting inspectors is usually rather painful. It is not at all unusual to find men inspecting the painting of large-scale structures who believe 'red lead' to be a mixed paint rather than a pigment which can be formulated into many different types of paint. Nor is it unusual to find inspection personnel who know of no solvent or thinner except turpentine and who classify anything except a linseed oil paint as a 'synthetic.' When one passes into the realm of surface preparation, where experience and judgment are particularly important, the competence of most inspectors is even more questionable. This situation does not stem from the fact that those who perform inspection of painting operations are peculiarly lacking in ability. Rather it springs, in part at least, from the rather general disinclination of supervisors and management to attach much importance to painting—and this despite the fact that painting is probably the greatest single item of maintenance cost with respect to most steel structures. Hence, management does not make much of an effort to secure or train competent paint inspectors.

Perhaps one reason for the fact that paint inspectors are so often poorly qualified for their work is that in the past very few generally accepted rules and quantitative guides applicable to the painting of structural steel were in operation. Even among those who have diligently attempted to determine the best materials and methods to employ in the cleaning and painting of steel structures, the claims and counterclaims far overshadow the areas of agreement.

Failures of paint on steel structures have been numerous, repetitious and expensive. These failures have occasioned the imperative need of a study of all of the phenomena involved. There are four fundamental factors involved, namely:

1. Quality of the steel itself;

2. Surface preparation prior to painting, whether this preparation is mechanical or chemical;

3. The conditions and methods of application of paint; and

4. The quality of the paint itself.

The immediate concern of this portion of the Steel Structures Painting Manual has to do with the quality control of paints. Prior to the preparation of this section on quality control, contact had been made with representatives of the large purchasers as well as manufacturers of paints. Three surveys had been made of more than 2000 painted steel structures. The owners of these structures have been unbiased in expressing their opinions of the results of the painting of these structures. Only in very exceptional cases did the paints used survive after a few years' exposure. The story of failure is such that curative procedures and materials must be found.

Two typical statements which have been received may be repeated here with telling effect. First, paints secured several years ago and formulated under definite specifications had been found satisfactory, but later purchases of supposedly the same paint were found unsatisfactory. One conclusion is possible—that inferior materials were substituted, or that old and contaminated materials were used. Substitution of inferior materials is sometimes done in an effort to meet competition or to supply a paint product at a cheap price, despite the fact that a quality product may have a high first cost but an ultimate low cost.

Although certain paint products were used very satisfactorily years ago, they may no longer be mixed in accordance with original formulation. Wherever there is a change in procedures of formulation with a lowering in the quality of ingredients, the manufacturer has doubtless endeavored to lower the cost of production or to increase the production volume. An inferior processed product may be the result.

Many times there are other factors, incident to paint failure, which are important. Purchasers of steel structures may prefer to have all painting done by local painters and with locally produced paint products, or the seller of the structures may let the subcontracts locally. The seller of the structures, if he has guaranteed paint life, is in a hazardous position unless he has proved the quality of the paint products to be used and provides competent supervisory inspection. Failure of high-quality proven paint products may occur due to improper conditions of storage, mixing, and thinning of paints. Too great a responsibility is sometimes left to the men in immediate charge of storing, mixing, and thinning of paints both in the plant and in the field. Quality control procedures will not eliminate failures due to these incidental factors.

Paint quality may be maintained at a high level providing the manufacturer is required to meet definite specifications for all *ingredients* used; for quantities of materials; for times; temperatures, and viscosities in the formulation of vehicle; and for fineness of grind and other requirements of the finished paint. This information will be the basis of adequate quality control. Most manufacturers of paints will gladly cooperate in providing this information, except in the case of some proprietary products where the exact formulation will not be divulged, but other qualitative and quantitative requirements may be established. In some cases, quality control procedures will serve to assure the purchaser that he is being supplied with the same paint that he originally selected."

Some general inspection criteria follow, but great emphasis should be placed on the requirements as described in the specifications:

Check paint containers, and reject any paint in damaged or broken containers which show evidence of having the seal broken allowing either contamination or evaporation. Also reject any paint in containers that are not plainly marked with the name, formula, specification, batch number, color, and date of manufacture. Also, pigmented paint should not be delivered in containers over 5 gallons in size. Usually, paint should be delivered to the site in sufficient time to permit 30 days of testing before the paint is needed, or should be separated at the source of supply to allow the similar testing. *Sampling* at the site should be done immediately after delivery by the contractor in the presence of the inspection team. Sampled material should be placed in one quart friction-top cans filled to 90% capacity and properly labeled for identification. A sample should be obtained from paint cans properly mixed. Testing is usually done by an approved laboratory, and a 3- to 4-week cycle is required. If samples fail the tests, batches represented by the sample should be immediately removed from the site.

Prior to painting, all *surfaces* should be properly prepared, including cleaning of all foreign materials, and proper surface preparation. All work such as welding should be completed before painting starts.

The inspection team should have a painting list which lists color, type of paint, and number of coats required on each surface. It is very important to ascertain that the proper shop coat has been applied to equipment. Incompatibility will cause peeling.

Painting should only be done under dry conditions on dry surfaces. No exterior painting should be done in damp or wet weather.

The inspection should include observation of the mixing and thinning of paints. Paint should be mixed until pigments are evenly distributed, and paint should not be thinned in a cold-temperature area. Paint should be allowed to rise to the *ambient temperature* which should be in excess of 50°F. Thinning materials should be those recommended by the manufacturer, and not in excess of one pint per gallon. Under normal conditions, paint of the required viscosity requires only mixing to be ready for use; therefore, the contractor should demonstrate a need for thinning. Paint in containers should be stirred frequently during the period of application. This is especially true of paints with an aggregate or cement filler, which require constant stirring or remixing to maintain even distribution. The inspection team should be certain that paints of different manufacturers and types should not be mixed together.

The storage of paint materials should reflect the safety precautions for flammable materials. Also, certain types of paint, such as latex paints, should not be allowed to freeze, so that the temperature area must be maintained above freezing.

Personnel operating spray guns should wear approved respirators while spraying, and the inspection team should be certain that the proper precautions have been taken to preclude damage to other parts of the construction and/or personnel. Adequate ventilation should be maintained at all times during spray painting in enclosed areas. Where there is doubt, full air supply should be provided to the painters. Ventilation of buildings should be checked after spray painting operations to be certain that there is no danger of explosion or fire.

Where special field tests are required, lab assistance may be required for the inspection team. Temperature requirements vary with different types of paints, and in some cases, the 50°F minimum is not high enough. Be certain that touch-up of abrasions and field connections, as well as shop markings on shop-coated metals are carried out before field coats are applied. Check the number of specified coats and be certain that adequate paint is actually required under proper conditions. Check for the requirement of tinting undercoats for positive identification of the number of coats. If a dry film thickness is specified, rather than or in addition to the number of coats, checking with a thickness gauge will be required.

Check for condensation on surfaces to be painted, especially if warm air heating is used (with the exception of latex-based paints). The *temperature* of the painted surface and the surrounding air must be maintained either above the specified minimum until the paint is thoroughly dried. The ultimate test for dryness can be finger-touch; when paint film is not tacky and moderate rubbing with a finger does not mar the surface, it can be considered dry.

When absorbent materials are to be painted, such as wood and plaster, moisture meter measurements may be required to be certain that moisture content is within allowable limits before painting starts.

Even when paint has been approved as a result of tests, the following conditions can be causes for field rejection:

A. Heavy *skinning*. Identified as a thick rubbery membrane of solidified or congealed paint over the surface of the paint in the container.

B. Livering or *jelling*. Formation of semi-hardened lumps which cannot be readily broken down and remixed with the rest of the paint.

C. *Gassing*. Chemical reaction between paint constituents determined by bubbles of gas rising to the surface of the paint, and sometimes by odor. In latex paints, this may indicate exposure of paint to several cycles of freeze-thaw.

D. Excessive *settlement* of pigment to bottom of container, solidified to the point where normal mixing procedures do not disperse it.

During painting operations, be certain that previously finished work is protected, painted as well as tiles, floor coverings, etc. All finished floor surfaces should be protected by drop cloths, and the application of paint on surfaces should be checked to determine acceptability before continuing with additional coats.

Plaster surfaces should be allowed to dry at least 30 days before painting. Inspect the prime coat for fading caused by hot spots (i.e., incomplete mixing of hydrated lime) or thin spots caused by inadequate troweling.

PAINTING CHECKLIST

1. Observe painted surfaces for paint runs, drops, waves, laps, brush marks, and for variations in color, texture, or finish.

2. Inspect the application of each cost for uniformity of thickness and coverage.

3. See that all masonry surface voids, pores, and cracks are filled with cement filler and cement latex filler when the painting of masonry is specified.

4. Determine between-coat drying requirements.

5. Be certain that the method of application is allowed—for instance, spraying is sometimes disallowed, as may be rolling.

6. Be certain that difficult to reach places, such as edges, corners, crevices, welds, and bolts obtain treatment and coverage equivalent to that of adjacent surfaces.

7. Check to see that both surfaces of wood doors receive the first coat at essentially the same time.

8. Be certain that latex paint is not used over caulking compound.

9. Check to see that exterior poured concrete surfaces are not painted unless specifically authorized.

10. Be certain that wood surfaces are properly prepared with putty, in nail holes, cracks, and other depressions.

DIVISION 10: SPECIALTIES

NUMBER	TITLE
10100	**CHALKBOARDS AND TACKBOARDS**
−10	Chalkboards
−20	Tackboards
10150	**COMPARTMENTS AND CUBICLES**
−51	Hospital Cubicles
−52	Hospital Cubicle Track
−60	Toilet Partitions and Urinal Screens
−61	Laminated Plastic Toilet Partitions and Urinal Screens
−62	Metal Toilet Partitions and Urinal Screens
−63	Stone Partitions
−70	Shower and Dressing Compartments
10200	**LOUVERS AND VENTS**
−01	Metal Wall and Door Louvers
10240	**GRILLES AND SCREENS**
10250	**SERVICE WALL SYSTEMS**
10260	**WALL AND CORNER GUARDS**
10270	**ACCESS FLOORING**
10280	**SPECIALTY MODULES**
10290	**PEST CONTROL**
10300	**FIREPLACES AND STOVES**
−01	Prefabricated Fireplaces
−02	Prefabricated Fireplace Forms
−03	Fireplace Components
−10	Fireplace Accessories
−11	Fireplace Water Heaters
−20	Stoves

10340	**PREFABRICATED STEEPLES, SPIRES, AND CUPOLAS**
10350	**FLAGPOLES**
10400	**IDENTIFYING DEVICES**
−10	Directories and Bulletin Boards
−11	Directories
−15	Bulletin Boards
−20	Plaques
−30	Illuminated Signage
−40	Signs
10450	**PEDESTRIAN CONTROL DEVICES**
10500	**LOCKERS**
−01	Wardrobe Lockers
−02	Box Lockers
−03	Basket Lockers
−05	Coin Lockers
10520	**FIRE EXTINGUISHERS, CABINETS, AND ACCESSORIES**
−21	Portable Fire Extinguishers
−22	Fire Extinguisher Cabinets
−23	Fire Blankets
10530	**PROTECTIVE COVERS**
−31	Walkway Covers
−32	Car Shelters
−35	Awnings
10540−10549	(Reserved)
10550	**POSTAL SPECIALTIES**
−51	Mail Chutes
−52	Mail Boxes
10560−10599	(Reserved)
10600	**PARTITIONS**
−01	Mesh Partitions
−10	Demountable Partitions
−11	Post and Panel Demountable Partitions
−12	Panel, Clip, and Batten Demountable Partitions
−13	Stud Type Demountable Partitions
−16	Movable Gypsum Partitions
−17	Movable Metal Partitions
−20	Folding Partitions
−23	Accordion Folding Partitions
−24	Folding Gates

10650	SCALES
10670	**STORAGE SHELVING**
−71	Metal Storage Shelving
−72	Storage and Shelving Systems
	Manual
	Automated
−73	Wire Shelving
10680–10699	(Reserved)
10700	**SUN CONTROL DEVICES (EXTERIOR)**
10750	**TELEPHONE ENCLOSURES**
−51	Telephone Booths
−52	Telephone Directory Units
−53	Telephone Shelves
10760–10799	(Reserved)
10800	**TOILET AND BATH ACCESSORIES**
−10	Metal Framed Mirrors
10820–10899	(Reserved)
10900	**WARDROBE SPECIALTIES**
10930–10999	(Reserved)

This division covers a wide range of special items. In most cases, they are units manufactured under production conditions.

The following special items are only a small portion of those which could be listed under this Division.

10100 CHALKBOARD AND TACKBOARD

Old types were usually of black slate, while modern chalkboards are manufactured compounds. Special composition boards may also be magnetic, permitting the use of magnetic strips or magnets to hold special displays in place.

Tackboards may be of material such as cork or a composition such as homosote, which is pressed paper.

10150 COMPARTMENTS AND CUBICLES

10151 Hospital Cubicles

Hospital cubicles are created either by curtains, or a combination of curtains and partitions to provide privacy for bed patients in multiple-bed rooms. The cubicle curtain track is a recessed track set in the ceiling plaster or mounted on the surface of ceiling dry wall. Check the operation of the curtain after installation—which should be one of the final steps of finishing.

Office Cubicles

Office cubicles are semi-finished partitions or demountable partitions used to create privacy without a complete room enclosure. Because these are usually less than full partitions, mounting details should be carefully followed, to avoid a flimsy unstable situation.

Toilet and Shower Compartments

Toilet partitions are metal stalls with doors and panels which do not reach the ceiling or the floor, and are used to isolate the area between water closets. These units are fabricated from premanufactured material, but custom fitted to the shop drawings dimensions. Check for hardware and door swings, as well as dimensions.

10300 FIREPLACE EQUIPMENT

Fireplace Accessories

These include spark screen, and broom, shovel, and tongs for handling logs or wood.

Fireplace Dampers

Dampers are installed to isolate fireplaces not in use from the chimney, and should be checked for tight fit.

Prefabricated Fireplaces

These units are sheet metal, and often include a forced-air ventilating system to provide draft and/or to recirculate room air to capture heat from the updraft. After installation, prefabricated fireplaces are usually covered with a brick or stone front for decorative purposes.

10350 FLAGPOLES

Flagpoles may be permanently installed, or may be installed on a hinge which permits folding down for maintenance and replacement of hardware. The standard installation is a single unit placed in the foundation, with a design sturdy enough to permit scaling by a steeplejack to replace any hardware. Design today generally includes installation of top-grade hardware, sometimes able to be replaced from the ground, precluding expensive maintenance.

10400 IDENTIFYING DEVICES

10410 Directory and Bulletin Signs

Sign or bulletin boards are usually hung in the lobby of a building which lists room numbers and the location of different departments or agencies. Check the size frame lock (if any), letter holders, letters, glass, electric fixture if any, and supply and storage place of letters furnished with the board. Be certain that the key is located in the master key file.

Painted Signs

Check painted signs for size and quantity, as well as color coding if used. The use of luminiscent paint for exit signs has become more popular and has to be tested in the absence of light. Check signs for spelling, and for proper direction, as well as quality and type of paint.

Plaques

Plaques may be made from a variety of metals and plastics and should be checked for quantity, accuracy, and installation location.

Three-Dimensional Signs

Quite often, these signs are electrically operated or lighted, and should be operationally checked.

10500 LOCKERS

Lockers are floor-mounted, free-standing metal cabinets with doors used to store clothing and other articles. They may be full height or stacked. The installation should be true in alignment, and the lockers well secured.

10520 FIREFIGHTING DEVICES

This section applies to the portable firefighting equipment such as extinguishers which are essentially portable units, but which may have a fixed cabinet, which often is furnished with a glass or plastic door. The cabinet may be recessed or wall mounted and should be checked for operation. The usual types of extinguisher equipment include carbon dioxide pressure cylinders, dry powder cylinders, foam-type, or soda acid. The foam-type extinguisher is filled with water and washing soda solution, with a small glass bottle with sulphuric acid suspended inside near the hose connection. A foaming agent is added to the water, when the unit is turned upside down. The inspection team should very carefully check the handling of these units, and it may even be appropriate to have a fire marshall check all extinguishers after they are placed. The soda-acid type extinguisher is usually an 8″ diameter cylinder, 30″ high, filled with water and washing soda solution, again with a small glass bottle of sulphuric acid suspended inside near the hose connection.

10550 POSTAL SPECIALTIES

The mail handling room should include location for special equipment such as postage machines and sorting and handling equipment, as specified.

10600 PARTITIONS

10601 Mesh Partitions

Mesh partitions are generally for the purpose of security either to preclude entry, or prevent unauthorized access to areas such as electrical or mechanical equipment. Mesh is usually attached to vertical and horizontal steel channel frames, with center reinforcing bar. Frames are supported by corner posts set into floor sockets and a top capping bar attached to the ceiling. A door of the same wire mesh construction is hinged and set into a frame attached to the wire mesh. Inspection should include checking of the attachment of the mesh, which should be well stretched, and secure firm attachments, with door and locks operating easily.

10610 Demountable Partitions

These partitions may be prefabricated units or built-in-place, may be full partitions or partial, and are built to be taken down and relocated. Relocation depends upon the design of the demountable partition and may be a matter of minutes or hours. Typically, this type of unit is used to create cubicles or classroom space, and does not provide the optimum in sound attenuation.

10620 Folding Partitions

Folding partitions are usually plastic-covered, flexible partitions made of steel and fabric which are used to divide room areas, or in some cases, even large spaces such as gymnasiums. Inspection of the track or rail installation is quite important, both for alignment and anchorage. Unit should work easily and freely, closing without forcing.

10650 SCALES

Scales may be portable or floor recessed. In either case, they are delivered ready to operate, and must be handled with care. After installation, the unit should be checked for calibration to be certain that accuracy has not been disturbed by moving.

10670 STORAGE SHELVING

Shelving may be wood or sheet metal, and installed in a workmanlike fashion, with proper support as required.

10700 SUN CONTROL DEVICES

Sun control devices may be drapes, venetian fans, or tinted glazing. These are special units installed to suit the situation, and may be fixed or movable. Free-standing vertical sun shades may be used on the exterior to provide shade against a low sun. Lightproof shades may also be used to preclude light and reflect some heat radiation.

10750 TELEPHONE BOOTHS

These units are provided either as a special custom-built unit to conform with the architecture of the building, or may be placed on a lease basis by the telephone company. In either case, the unit is usually placed as one of the last items of finishing, and must be protected from damage after installation.

10800 TOILET AND BATH ACCESSORIES

Toilet and bath accessories may be of ceramic material, or hardware. If hardware, they are installed as one of the last finishing items.

10900 WARDROBE SPECIALTIES

Often in dormitories and hospitals, specially designed wardrobe units may be included in the room furnishings. These are installed as one of the last items.

SPECIAL ITEMS CHECKLIST

1. Is the unit the one specified by the plans and specifications?
2. Is it dimensionally correct in accordance with the plans, specifications and the approved catalogue cut or shop drawing?
3. Has it been delivered undamaged?
4. Is it being properly stored and handled on the job site?
5. Is its final location properly prepared?
6. Is is properly installed, mounted, secured, or fastened?

11400 FOOD SERVICE EQUIPMENT

Bar Units

These units are generally prefabricated, but require plumbing connections for water supply and drain.

Cooking Equipment

Includes range and oven, generally similar to home units, with cooking surface on top and oven below. In some cases, ovens may be free-standing, insulated units with doors in front that slide up, and several rotating shelves with trays inside to hold pans.

Cooking equipment may include steamers and deep fryers. Bakery equipment may include a proof cabinet (dough raiser), which is an insulated cabinet with double doors made of stainless steel and equipped with heating equipment used to raise dough in preparation for baking.

In larger kitchens, a steam kettle may be included, which is free-mounted and used for the steam cooking of food.

Dishwashing Equipment

A dishwasher is a large integrated unit, completely enclosed, which houses conveyor, motors, heating elements, water soap tanks and sprayers for washing dishes, glasses, and silverware. Factory supervision of installation and operation is usually required. Larger kitchens may have pot washers with brushes and steam and water jets for washing of pots, and even can washers for washing of refuse cans.

Food Preparation Machines

Food preparation machines such as can openers, blenders, and mixers are essentially portable and free standing. Units such as ingredient tables, which are working tables, are prefab and relatively easy to install. Other food preparation machines such as coffee urns and doughnut fryers are also free standing, and should be installed in accordance with manufacturer's suggestions.

Food Serving Units

Food serving units can include food and milk dispensing units, which are stainless steel cabinets with refrigeration; pass-through food warmers are basically warming ovens which receive food from the back and hold it in one condition; serving counters such as cafeteria counters; water stations which dispense water and store water glasses; and food and ice stations which are stainless steel counters with sink, hot plates, icemaker storage bin dispenser, etc. Most of these units are prepared in accordance with specifications and plans, and approved shop drawings. Inspection team should compare with shop drawing dimensions and details.

Refrigerator Cases

At one time, refrigerators and walk-in coolers were built-in-place. However, most units, except the very large, are now prefabricated, and delivered either as an entire unit, or in sections for erection. The unit should be tested after installation for ability to maintain proper temperatures.

DIVISION 11: EQUIPMENT

NUMBER	TITLE
11010	**MAINTENANCE EQUIPMENT**
−11	Vacuum Cleaning Systems
−12	Window Washing Systems
−13	Scrubbers, Polishers, and Shampooers
11020	**SECURITY AND VALULT EQUIPMENT**
−21	Vault Doors and Day Gates
−22	Service and Teller Window Units
	Bullet Resistant
−23	Package Transfer Units
−24	Security and Emergency Systems
	Electronic Equipment
	Vault Ventilation Systems
−25	Automatic Banking Systems
−26	Depositories
−27	Teller Counter Equipment Systems
−28	Safes
−29	Safe Deposit Boxes
11030	**CHECKROOM EQUIPMENT**
−31	Motorized Checkroom Equipment
11040	**ECCLESIASTICAL EQUIPMENT**
−41	Baptistries
11050	**LIBRARY EQUIPMENT**
−51	Book Theft Protection Equipment
−53	Library Stack Systems
11060	**THEATER AND STAGE EQUIPMENT**
−61	Rigging Systems and Controls
−62	Stage Curtains
−63	Stage Lift Systems
−64	Lighting Systems and Controls
−65	Acoustical Shell Systems

11070	**MUSICAL EQUIPMENT**
−71	Organs
−72	Carillons

| 11080 | **REGISTRATION EQUIPMENT** |
| 11090–11099 | (Reserved) |

11100	**MERCANTILE EQUIPMENT**
−01	Display Cases
	Refrigerated Display
−02	Cash Registers and Checking Equipment
−03	Barber and Beauty Shop Equipment
−04	Food Processing Equipment
	Weighing and Wrapping Equipment

11110	**COMMERCIAL LAUNDRY AND DRY CLEANING**
−11	Cleaning Equipment
−12	Drying and Conditioning Equipment
−13	Ironers and Accessories
−14	Finishing Equipment
−19	(Reserved)

11120	**VENDING EQUIPMENT**
−21	Money Changing Machines
−22	Beverage Vending Machines
−23	Food Vending Machines
−24	Sundry Vending Machines
−25	Cigarette Vending Machines
−26	Candy Vending Machines

11130	**AUDIO-VISUAL EQUIPMENT**
−31	Projection Screens
−32	Projectors
−35	Learning Laboratories

| 11140 | **SERVICE STATION EQUIPMENT** |

11150	**PARKING EQUIPMENT**
−51	Parking Gates
−52	Ticket Dispensers
−53	Key and Card Control Units
−54	Coin Machine Units

11160	**LOADING DOCK EQUIPMENT**
−61	Dock Levelers
−62	Adjustable Dock Ramps
−63	Portable Ramps, Bridges, and Platforms
−64	Seals and Shelters
−65	Dock Bumpers

11170	**WASTE HANDLING EQUIPMENT**
−71	Packaged Incinerators
−72	Waste Compactors
−73	Bins
−74	Pulping Machines and Systems
−75	Chutes and Collectors
11180–11189	(Reserved)

11190	**DETENTION EQUIPMENT**
−91	Cell Doors and Equipment

11200	**WATER SUPPLY AND TREATMENT EQUIPMENT**
−01	Gates
	Sluice Gates
	Slide Gates
−10	Water Pumps
−11	Centrifugal Pumps
−12	Axial Flow Pumps
−13	Mixed Flow Pumps
−14	Vertical Turbine Pumps
−15	Deepwell Turbine Pumps
−20	Mixers and Flocculators
−21	Clarifiers
−30	Water Aeration Equipment
−31	Chemical Feeding Equipment
−32	Coagulant Feed Equipment
−33	Water Softening Equipment
	Lime-Soda Process Equipment
	Base-Exchange or Zeolite Equipment
−34	Disinfectant Feed Equipment
	Chlorination Equipment
−35	pH
−36	Fluoridation Equipment
11240–11299	(Reserved)

11300	**FLUID WASTE DISPOSAL AND TREATMENT EQUIPMENT**
−01	Oil/Water Separators
−02	Sewage Ejectors
−03	Packaged Pump Stations
−10	Sewage Pumps
−11	Non-Clogging Centrifugal Pumps
−12	Centrifugal Pumps
−13	Positive Displacement Pumps
−14	Torque Flow Pumps
−15	Progressive Cavity Pumps
−20	Grit Collecting Equipment
−30	Screening and Grinding Equipment

−31   Bar Racks
            Mechanically Cleaned Racks
−32   Fine Screens
            Rotating Disc or Drum Screens
            Plate Screens
            Vibratory Screens
−33   Grinders
−34   Comminutors
−35   Barminutors
−40   Skimming Tank Equipment
−50   Sedimentation Tank Equipment
−51   Flocculators
−52   Scum Removal Equipment
−53   Chemical Mixing Equipment
−54   Chemical Feed Equipment
−60   Sludge Handling and Treatment Equipment
−61   Sludge Collection Equipment
            Rectangular Sludge Collection Equipment
            Circular Sludge Collection Equipment
−62   Sludge Thickening Equipment
−63   Sludge Mixing Equipment
−64   Sludge Dewatering Equipment
            Vacuum Filters
            Centrifuges
−70   Filter Press Equipment
−71   Trickling Filter Equipment
            Standard Filter Equipment
            High-Rate Filter Equipment
            Controlled Filter Equipment
−72   Compressors
−73   Blowers
−74   Aeration Equipment
            Surface Aerators
            Turbine Aerators
            Air Diffusers
−80   Sludge Digestion Equipment
−81   Sludge Gas Equipment
−82   Digester Mixing Equipment
−90   Packaged Sewage Treatment Plants
11395−11399 (Reserved)

11400   **FOOD SERVICE EQUIPMENT**
−01   Custom Fabricated Equipment
−10   Cooking Equipment
            Gas Cooking Equipment
            Electric Cooking Equipment
            Microwave Cooking Equipment
            Steam Cooking Equipment

−11	Warewashing Equipment
−12	Garbage Disposers
−13	Food Preparation Machines
−14	Food Preparation Tables
−15	Self-contained Refrigeration Equipment
−16	Serving Line Equipment and Units
−20	Exhaust Systems
−30	Bar and Soda Fountain Equipment
11440–11449	(Reserved)

11450	**RESIDENTIAL EQUIPMENT**
−51	Kitchen Equipment
	Gas Ranges, Cooktops, and Ovens
	Electric Ranges, Cooktops, and Ovens
	Microwave Ovens
	Refrigerators
	Ice Makers
	Compactors
	Dishwashers
	Disposers
	Built-In Appliances
	Hoods and Exhausts
−52	Laundry Equipment
	Clothes Washers
	Electric Dryers
	Gas Dryers
	Built-In Ironing Cabinets
−54	Disappearing Stairs

11460	**UNIT KITCHENS AND CABINETS**

11470	**DARKROOM EQUIPMENT**
−71	Revolving Darkroom Doors
−72	Transfer Cabinets
−73	Safe Lights
−74	Darkroom Processing Equipment
	Photographic Processing Equipment
	Radiographic Processing Equipment

11480	**ATHLETIC, RECREATIONAL, AND THERAPEUTIC EQUIPMENT**
−81	Scoreboards
−82	Backstops
−83	Gym Dividers
−84	Bowling Alleys
−85	Shooting Ranges
−86	Gymnasium Equipment
−87	Exercise Equipment
−90	Therapy Equipment
−91	Hydrotherapy Equipment
−99	(Reserved)

11500	**INDUSTRIAL AND PROCESS EQUIPMENT**
−01	Paint Spray Booths
−10	Shop Equipment
11600	**LABORATORY EQUIPMENT**
−03	Temperature Controlled Laboratory Equipment
−04	Safety Equipment
−10	Fume Hoods
−11	Laboratory Sterilizers
−12	Laboratory Water Still
−13	Laboratory Water Processing Equipment
−14	Laboratory Washing Equipment
−15	Laboratory Controlled Temperature Rooms
11630–11649	(Reserved)
11650	**PLANETARIUM AND OBSERVATORY EQUIPMENT**
11700	**MEDICAL EQUIPMENT**
−01	Solution Warming Cabinets
−09	Automatic Washing and Sterilizing Equipment
−10	Medical Sterilizing Equipment
−11	Gas Sterilizing Equipment
−12	Hospital Washing Equipment
−13	Instrument Washers and Sterilizers
−14	Medical and Surgical Lighting Fixtures
−15	Surgical Tables
−16	Film Illuminators
−17	X-Ray and Fluoroscopy Equipment
−18	Teletherapy Equipment
−20	Nursing Floor Equipment
−30	Patient Care Equipment
−31	Patient Wall Systems
−32	Intensive Care Monitor Modules
−40	Dental Equipment
−50	Optical Equipment
11760–11779	(Reserved)
11780	**MORTUARY EQUIPMENT**
−81	Autopsy Tables
−82	Dissecting Tables
−83	Mortuary Refrigerators
−84	Cadaver Lifts
11790–11799	(Reserved)
11800	**TELECOMMUNICATION EQUIPMENT**
11850	**NAVIGATION EQUIPMENT**
11900–11999	(Reserved)

This division describes special equipment, usually completely fabricated at the point of manufacture. Main concerns in the field are handling, proper location, and secure anchorage.

11050 LIBRARY EQUIPMENT

Bookshelving

Bookshelving can be wood or steel. Steel floor shelving is made of steel uprights with shelves bolted to them. Wall shelving can be double or single faced, with backs used as readily accessible storage for books in the library. Shelves are installed around room perimeter or in rows inside of room. Check for firm anchorage.

Book Stacks

Stacks are usually basically steel shelving, located in nonaccessible floors for book storage.

11060 STAGE EQUIPMENT

Stage equipment includes curtains and backdrop, which are usually made of fireproof material. These are operated in the vertical by counterweight and rigging, or in the horizontal by a standard drape-type of opening. Floor lights and floodlights are operated by rheostat controls from backstage, while house lights may also be similarly operated. The degree of sophistication depends upon the design and the purpose of the installation.

Audio-visual equipment, such as projection equipment, may be operated from a projection booth, or may have a less expensive mounting in the assembly area. Some stages may have an oil-hydraulic system for the orchestra pit, or even the stage.

The stage may have special items such as tormentor towers which are steel frames with fastened plywood and rollers at the bottom, used to reduce the stage opening when the full stage size is not required. These special considerations will be specified—and the inspection team can review them on the basis of the approved shop drawings.

11150 PARKING EQUIPMENT

Parking equipment principally includes toll gates and counters for exterior parking areas, and personnel elevators (see Division 14) for inside units.

11190 PRISON EQUIPMENT

Most prison equipment relates to security, and is described in specific areas—such as glazing—requiring special treatment. Prison equipment is designed for security, and careful installation is important so that the design features can operate successfully without interference. Special prison equipment can include cell equipment, cell block equipment, guard tower installations, and other features unique to prisons.

11450 RESIDENTIAL EQUIPMENT

Central Vacuum Cleaner

Central vacuum cleaners work on the same principle as independent stand-up vacuum cleaners, except that they have a central repository to receive waste material, and outlets in various rooms to receive the flexible hose connection. A heavy-duty vacuum pump creates the suitable suction for the unit. Installation, being behind walls, should be carefully inspected for continuity and lack of obstructions.

Kitchen and Laboratory Cabinets

Prefabricated units, installed in properly prepared space.

Residential Kitchen Equipment

Similar to the typical house residential equipment, and should be inspected on the same basis, that is, tested for operation after installation.

Residential Laundry Equipment

Washers and dryers should be checked for proper handling during installation, and if a large number are installed, the contract may require a test operation by the vendor.

Unit Kitchens

Same as the typical residential equipment, but compact.

11600 LABORATORY EQUIPMENT

Laboratory equipment is specialized, and basically completely prefabricated. Equipment can include cabinets and tables, base cabinets, laboratory tables, test equipment, ventilation fume hoods, washing and sanitizing autoclaves, washers, sinks, and special items. In all cases, operational checks should be made after the equipment is installed either by the inspection team, or preferably by the using agency.

11700 MEDICAL EQUIPMENT

Dental Equipment

Dental equipment includes patient chair, drills, case work, and other associated equipment. All of this equipment is free standing, and provided ready to operate.

Examination Room Equipment

Examination room equipment includes patient table, medicine cabinets, instrument storage, sterilizer and other appropriate equipment as required by the specifications. This is free-standing equipment, with the exception of the treating table, which will probably be bolted down.

Hospital Case Work

Case work is semi-fixed cabinetry which is delivered ready for installation, and requires only bolting in place.

Incubators

Incubator equipment may require only a plug-in for lighter units, or permanent mechanical attachments for power.

Manufacturer's representatives should check out the equipment before it is used.

Radiology Equipment

Radiology equipment itself is free standing, but shielding includes lead lining or similar dense material placed within partitions or walls to protect operator against radiation. Tight fit and complete seal for the sheet lead is very important. Inspection team should view the lead lining before it is enclosed by decorative material.

Sterilizers

Sterilizers are free-standing units which require water and electrical connections. Manufacturer's representative should be available to check out equipment if it is heavy duty.

Therapy Equipment

Therapeutic equipment often requires heavy-duty electrical connections, and should be installed in accordance with manufacturer's suggestions.

11780 MORTUARY EQUIPMENT

Mortuary equipment varies with the installation, but for the typical hospital includes mortuary table, which is a unit including sinks and drainboards for post-mortem use. Also, the mortuary should include a cadaver storage unit. The unit is made of large, open shelf brackets supported on four uprights running from floor to ceiling, and is designed to accommodate large trays capable of holding bodies in storage pending or after autopsy. Special cooling equipment may be required either environmentally, or specifically for the CSU.

EQUIPMENT CHECKLIST

1. The building area should be ready for the installation of special equipment.
2. The special equipment should be checked for conformance with specifications and plans and dimensions.
3. Handling and storage for special equipment should be carefully observed by the inspection team.
4. Installation should be in a workmanlike manner.

DIVISION 12: FURNISHINGS

NUMBER	TITLE
12100	**ARTWORK**
−10	Murals
−11	Photo Murals
−20	Wall Hangings
−30	Paintings
−40	Carved or Cast Statuary
−50	Carved or Cast Relief Work
−60	Custom Chancel Fittings
−70	Stained Glass Work
12200–12299	(Reserved)
12300	**MANUFACTURED CABINETS AND CASEWORK**
−01	Metal Casework
−02	Wood Casework
−03	Built-In Tables
−10	Bank Fixtures and Casework
−15	Library Casework
−20	Restaurant and Cafeteria Fixtures and Casework
−25	Educational Cabinets and Casework
−30	Dormitory Casework
−35	Medical and Laboratory Casework
	Nurse Server Cabinets
−40	Pharmacy Casework
−45	Laboratory Casework
−46	Wood Laboratory Casework
−47	Metal Laboratory Casework
−48	Laboratory Tops, Sinks and Accessories
−50	Hospital Casework
−55	Dental Casework
−60	Optical Casework
−65	Veterinary Casework
−70	Hotel and Motel Casework
−75	Ecclesiastical Casework
−80	Display Casework
−90	Residential Casework
−91	Kitchen and Bath Cabinets

412

12500	WINDOW TREATMENT
−01	Drapery and Curtain Hardware
−02	Drapery and Curtain Operators
−10	Blinds and Shades
−11	Vertical Louver Blinds
−12	Horizontal Louver Blinds
−13	Shades
−14	Lightproof Shades
−15	Woven Wood Shades
−20	Drapes and Curtains
−21	Lightproof Drapes
−22	Fabric Drapes
−23	Woven Wood Drapes
−24	Vertical Louver Drapes
−25	Curtains
12530–12549	(Reserved)

12550	FABRICS
−51	Drapery and Upholstery Fabrics
12560–12599	(Reserved)

12600	FURNITURE AND ACCESSORIES
−10	Landscape Partitions and Components
−15	Room Dividers and Screens
−20	Office Furniture
−25	Integrated Office Work Units
−30	Lounge Furniture
−40	Specialized Furniture
−41	Laboratory Furniture
−42	Hospital Furniture
−43	Classroom Furniture
−44	Restaurant and Cafeteria Furniture
−45	Ecclesiastical Furniture
−46	Hotel and Motel Furniture
−47	Dormitory Furniture
−48	Residential Furniture
−50	Accessories
−51	Ash Receptacles
−52	Lamps
−53	Desk Accessories
−54	Waste Receptacles

12670	RUGS AND MATS
−71	Rugs
−72	Foot Grilles
−73	Mat Frames
−75	Floor Mats
−76	Floor Runners

−77   Matting
12680–12699   (Reserved)

12700   **MULTIPLE SEATING**
−01   Stacking Chairs
−02   Portable Folding Chairs
−03   Interlocking Chairs
−10   Auditorium and Theater Seating
−30   Stadium and Arena Seating
−40   Booths and Tables
−50   Multiple-Use Fixed Seating
−51   Pedestal Table Armchairs
                       Single Unit
                       Tandem Mounted
−60   Telescoping Bleachers
−70   Pews and Benches
−61   Telescoping Chair Platforms
12780–12799   (Reserved)

12800   **INTERIOR PLANTS AND PLANTING**
−10   Interior Plants
−15   Planters
12820–12999   (Reserved)

12100 ARTWORK

Many public regulations require a certain amount of artwork in the form
of fine arts. This should definitely be one of the final items installed, and
should be carefully protected in accordance with direction of the
designers and/or artists.

12300 CABINETS AND FIXTURES

Cabinets are manufactured units, and they can be built in, semi-fixed, or
free standing.

12510 BLINDS AND SHADES

Window shades are made of opaque fabric on spring rolls mounted
above windows, for precluding light, while venetian blinds of wood or
metal slats provide a variable capability. Be certain that the blinds fit the
opening properly and that the length is sufficient to cover the entire
window opening. Check for operation.

12520 DRAPERY AND CURTAINS

Check for material, quality, dimensions, and operation. Drapery and curtain track should be well anchored and secure.

12670 CARPETS AND MATS

Carpeting is being utilized more and more as a finished flooring. Selection of the grade depends upon the specification, with various results possible in either wool or synthetics. Placement of the mat is most important, and the inspection team should ascertain that the subfloor is dry before placement. An experienced placement crew is necessary to place the mat and carpeting with proper stretch and freedom from wrinkles. Nailing strips may be used along the floor edge and joints, and/or the mat and rug may be glued in place. For glued installation constant heat in the 50°-70°F range is required.

12700 SEATING

12710 Auditorium Seating

Auditorium seating is usually upholstered-type attached chairs with padding, curved back, and full spring seat. Full hinge enclosement in the seat automatically raises to position. Some in student areas have plywood tablet arm which folds under the seat when occupant rises.

Check for compliance with specifications, and after installation, inspect rows for uniformity and compliance with seating plan.

DIVISION 13: SPECIAL CONSTRUCTION

NUMBER   TITLE

13010   AIR SUPPORTED STRUCTURES

13020   INTEGRATED ASSEMBLIES

13030   AUDIOMETRIC ROOMS

13040   CLEAN ROOMS

13050   HYPERBARIC ROOMS

13060   INSULATED ROOMS
−61   Insulated Cold Storage Rooms

13070   INTEGRATED CEILINGS

13080   SOUND, VIBRATION, AND SEISMIC CONTROL

13090   RADIATION PROTECTION
−91   Lead Radiation Shielding

13100   NUCLEAR REACTORS

13110   OBSERVATIONS

13120   PRE-ENGINEERED STRUCTURES
−21   Pre-Engineered Buildings
−22   Metal Building Systems
−23   Greenhouses
−24   Portable and Mobile Buildings
−25   Grandstands and Bleachers

13130	**SPECIAL PURPOSE ROOMS AND BUILDINGS**
−31	Prefabricated Rooms
	Office Shelters
	Saunas
	Steam Baths
	Athletic Rooms
13140	**VAULTS**
13150	**POOLS**
−51	Swimming Pools
	Swimming Pool Equipment
	Pool Cleaning Systems
	Chemical Treatment Systems
−52	Aquaria
−53	Therapeutic and Massage Pools
13160	**ICE RINKS**
13170	**KENNELS AND ANIMAL SHELTERS**
−71	Dog Play Pens
13180−13199	(Reserved)
13200	**SEISMOGRAPHIC INSTRUMENTATION**
13210	**STRESS RECORDING INSTRUMENTATION**
13220	**SOLAR AND WIND INSTRUMENTATION**
−21	Solar Control Packages
	Solar Differential Thermostats
	Solar Devices and Sensors
	Sun Tracking Devices
13230−13409	(Reserved)
13410	**LIQUID AND GAS STORAGE TANKS**
−11	Ground Storage Tanks
	Welded Steel Tank
	Galvanized Steel Tank
−12	Elevated Storage Tanks
−13	Underground Storage Tanks
−14	Prestressed Concrete Tanks
−15	Fiberglass Reinforced Plastic Tanks
−20	Tank Cleaning Procedures
−30	Tank Lining Systems
−31	Fiberglass Reinforced Plastic Tank Lining System
13440−13509	(Reserved)

13510	**RESTORATION OF UNDERGROUND PIPELINES**
−11	Inspecting of Pipelines
−12	Water Mains
−13	Sewer Mains
−14	Sealing Existing Pipelines
−15	Re-Lining Existing Pipelines

13520	**FILTER UNDERDRAINS AND MEDIA**
−21	Filter Bottoms
−22	Filter Media
	Sand Filter
	Coal Filter
	Mixed Filters
	Charcoal Filter
	Diatomaceous Earth Filter

13530	**DIGESTION TANK COVERS AND APPURTENANCES**
−31	Floating Covers
−32	Fixed Covers
−33	Gasholder Covers

13540	**OXYGENATION SYSTEMS**
−41	Oxygen Generators
−42	Oxygen Storage Facility
−43	Oxygen Dissolution System

13550	**THERMAL SLUDGE CONDITIONING SYSTEMS**

13560	**SITE CONSTRUCTED INCINERATORS**
−61	Solid Waste Incinerators
−62	Sludge Incinerators

13570–13599	(Reserved)

13600	**UTILITY CONTROL SYSTEMS**
−10	Water Supply Plant Operating and Monitoring System
	Metering Devices
	Display Panels
	Control Panels
	Sensing and Communicating Devices
−20	Waste Water Treatment Plant Operating and Monitoring Systems
	Metering Devices
	Display Panels
	Control Panels
	Sensing and Communicating Devices
−30	Power Generating and Transmitting Control Systems
	Display Panels
	Control Panels

	Meters
	Relays
13640–13699	(Reserved)

| **13700** | **INDUSTRIAL AND PROCESS CONTROL SYSTEMS** |
| 13710–13799 | (Reserved) |

13800	**OIL AND GAS REFINING INSTALLATIONS AND**
	CONTROL SYSTEMS
13810–13899	(Reserved)

13900	**TRANSPORTATION INSTRUMENTATION**
−10	Airport Control Instrumentation
−20	Subway Control Instrumentation
−30	Railroad Control Instrumentation

13940	**BUILDING AUTOMATION SYSTEMS**
−41	Energy Monitoring and Control Systems (EMCS)
−42	Engineering Control Systems
13950–13969	(Reserved)

| **13970** | **FIRE SUPPRESSION AND SUPERVISORY SYSTEMS** |

13980	**SOLAR ENERGY SYSTEMS**
−81	Solar Flat Plate Collectors
	Liquid Collectors
	Air Collectors
−82	Solar Concentrating Collectors
−83	Solar Vacuum Tube Collectors
−85	Solar Collector Components
	Solar Absorber Plates or Tubing
	Solar Reflectors
	Solar Glazings
	Solar Coatings and Surface Treatment
	Solar Collector Insulation
	Solar Housing and Framing
−87	Packaged Solar Systems

| **13990** | **WIND ENERGY SYSTEMS** |
| 13995–13999 | (Reserved) |

SPECIAL CONSTRUCTION

Special construction, as its name indicates, is unique—and requires inspection commensurate with the level of uniqueness and the special demands and characteristics of the materials utilized. In many cases, inspection of special construction is a composite of the components making up the special construction. In other cases, manufacturer's or designer's direction must be carefully followed. Wherever possible, those parties involved in the specification and production of the special construction components should be made a party to their inspection. Often, the contractor or vendor provides a special field erection representative for direction and quality control.

DIVISION 14: CONVEYING SYSTEMS

NUMBER	TITLE
14100	**DUMBWAITERS**
−01	Manually Operated Dumbwaiters
−02	Electrically Operated Dumbwaiters
14200	**ELEVATORS**
−10	Passenger Elevators
−11	Electric Passenger Elevators
−12	Hydraulic Passenger Elevators
−20	Freight Elevators
−21	Electric Freight Elevators
−22	Hydraulic Freight Elevators
−30	Inclined Elevators
14300	**HOISTS AND CRANES**
−01	Rails
−02	Lifting Magnets
−10	Hand Operated Hoists
−20	Motor Operated Hoists
−30	Bridge Cranes
−40	Gantry Cranes
−50	Jib Cranes
−60	Mobile Cranes
14400	**LIFTS**
−10	People Lifts
−11	Sidewalk Lifts
−15	Wheelchair Lifts
−20	Aerial Tramways
−30	Powered Platforms
−40	Funiculars
−50	Vehicle Lifts

14500	MATERIAL HANDLING SYSTEMS
−10	Trucks and Fork Lifts
−20	Automatic Transport Systems
−21	Hospital Transport Systems
−30	Postal Conveying Systems
−40	Baggage Conveyors and Dispensers
−50	Conveyors
−51	Selective Vertical Conveyors
−52	Horizontal Screw Conveyors
−53	Vertical Screw Conveyors
−54	Inclined Screw Conveyors
−55	Bucket Conveyors
−56	Pneumatic Conveyors
−57	Roller Conveyors
−58	Oscillating Conveyors
−59	Hopper and Track Conveyors
−60	Chutes
−70	Feeder Equipment
−71	Rotary Airlock Feeders
−72	Vibratory Feeders
−73	Reciprocating Plate Feeders
−74	Rotary Flow Feeders
−75	Apron Feeders
−80	Tube Systems
−81	Pneumatic Tube Systems

14600	TURNTABLES
14610–14699	(Reserved)

14700	MOVING STAIRS AND WALKS
−10	Escalators
−20	Moving Walks
−30	Aircraft Passenger Loading Bridges
14750–14799	(Reserved)

14800	POWERED SCAFFOLDING
14810–14899	(Reserved)

14900	TRANSPORTATION SYSTEMS
14950–14999	(Reserved)

14100 DUMBWAITERS

Originally dumbwaiters were manually operated vertical transportation units for transporting materials. Today, they may be essentially small elevators or may still be manually operated. The powered version is more likely, and can be used for transporting food, office materials, medicines, and other items.

The comments which follow on the inspection of elevators also apply to the installation and inspection of dumbwaiters.

14200 ELEVATORS

The elevator combines many machines and devices for its operation, traveling on vertical *tracks* mounted on the walls of the hoistway. The tracks or guard rails position the car, and also provide a surface for the operation of safety clamps provided to prevent runaway or free fall. Vertical travel motor *power* is through either a hydraulic plunger or by motor-driven cables or wire ropes. The use of the hydraulic lift is usually limited to low-rise buildings of about 50', with the wire rope or cable-driven elevator being most common for high rise and also very common in low rise.

Wire rope units are either traction or winding drum hoist machine types. Both utilize counterweights to reduce the amount of driving power required for lifting or lowering. Machines can be a.c. or d.c. driven, with the d.c. having a wider range of control options. *Controllers* for all elevators perform the same basic functions including closing doors, controlling travel speed to the desired landing, opening the doors and repeating the sequence as required.

The operation of most elevators is carefully controlled by local *building codes* and includes typical factors such as the amount of illumination, operating characteristics, safety features such as dielectric floor matting, capacity placards, test tags, and inspection certificates.

The inspection team should be certain that the characteristics of the *wire cable* including diameter, manufacturer's rated breaking strength, grade of material used, and month and year of installation are in conformance with the specification and properly registered on the tag. Care should be taken to inspect the cables for any evident signs of failures such as side or bottom wear or broken strands. The cables are strung after the hoistway is available, and the cab is often operated in a working mode to carry construction personnel and equipment. In some cases, the elevator manufacturer may install a temporary working cab, but the more usual practice is to install the finished cab, complete with padding to protect it.

The inspection team should have available the manufacturer's representative during installation and initial operation. The inspection team should be certain that the *safety shoes* are installed and operable before any personnel ride the cars, even installation personnel. Further, *counterweight guards* should be installed in the elevator pits to protect any personnel working in the pits; these are of particular importance during the installation phase.

Pit depth and shaft weight clearances vary with rated car speed and feet per minute. A car traveling at 100′ per minute requires a pit of 4′ in depth, while a car traveling at 600′ per minute requires a pit of 7′, even though both require only 2′ of clearance when resting on the fully compressed buffers in the pit. The clear space is to protect a man working in a pit should a car overrun the final limits switch and descend into the pit. Overrun is predicated on car speed and is related to the distance required to bring the car to a gradual stop when the friction safety clamp is applied.

Similarly, shaftway clearance overhead is predicated on car speed and clearance.

Building codes require that there be adequate *access* to the elevator shaft for emergencies or inspection work. A single car must be provided with means of access from all floors. Where two or more elevators are in a common shaft, access may be gained through side exits or from the roof of the operative car. As a minimum, usually three hoisting cables must be used on all traction-type elevators with a minimum cable diameter being ½″. The usual minimum for drum or sheave size is 40 times the diameter of the cable used, indicating a minimum of 20″ in size.

Installation of the elevator should be in accordance with the shop drawings, which, in turn, should have been approved by the appropriate Department of Buildings or Department of Labor and Industry or a similar controlling authority.

The inspection team should have carefully checked the dimensions of the hoistway, and the penthouse or machine room prior to installation of the elevator. The door bucks must be in accordance with elevator manufacturer's dimensions. Installation of the *guide rails,* which are T-shaped steel rails of weight directly proportional to load-bearing strength, must be carefully aligned and plumb. The rails are set and connected with fish plates. Alignment of the rail is by rail clips which attach to inserts in the wall.

Conduit and wire installed in the hoistway must be in conduit, with the exception that Greenfield cable can usually be used for short sections connecting the wall stations and hoistway switches to the conduit. The hall station is a flush station equipped with directional buttons for up and down at intermediate floors, and a single button at terminal floors.

The controls for the elevator are fairly complex, and include interlocks which stop the car whenever the hall door is open, as well as limit switches which shut down the elevator if it goes past certain limits at the terminal ends of the rails.

The stranded steel cables for hoisting are continuous and nonspliced. When the inspection team is operating the elevator, a remote inspector's station can be set up so that the inspector can operate the elevator at slow speed from the car top without interference from the automatic call controls. The inspector switch is often also installed in the elevator pit and is interlocked through the safety line circuit.

Inspection of the elevator operation should include check of the safety edge on the door operator to be certain that the door does not close if there is any physical contact.

ELEVATOR INSTALLATION CHECKLIST

1. Be certain that wire cable characteristics are in accordance with specifications, and that the cable installed is in sound condition.

2. Inspect hoistway for obstructions and check workmanship before stringing cables.

3. Guide rails should be checked for alignment and plumbness.

4. Any conduit or wire in the hoistway must be in conduit, with the exception of the short cable runs connecting wall stations with hoistway stations which can be in flex cable.

5. The stranded steel cables must be continuous, nonspliced. When cars are installed, check safety shoes for proper operation before any personnel, even installation personnel, ride the cars.

6. Counterweight guards should be installed in elevator pits to protect personnel in the pits—installed before any people work in the pits with the car overhead. Carefully check door bucks for proper dimensions before car installation.

7. Check the safety edge on the door operator for proper operation before the cars are used.

8. Have manufacturer's representative on site check out car operation at installation phase, and preoperational.

14300 HOISTS AND CRANES

Hoists or cranes may be utilized for lifting on an intermittent or regular basis. Included in this category are vertical conveyors which may be continuous chain drive, carrying at constant or variable speed. Installation is similar in many ways to elevators, although the drive machinery is different because of the nature of the trays or compartments carried. Vertical hoists may also be utilized for personnel carriers, for instance in multiple-story garages.

Cranes are operated either by a crane operator who has visual control and sight, or remotely by an operator who is in a position to hook up the load. Installation is in accordance with the drawing specifications and manufacturer's suggestions.

DIVISION 15: MECHANICAL

NUMBER	TITLE
15050	**BASIC MATERIALS AND METHODS**
−60	Pipe and Pipe Fittings
−61	Steel Pipe
−62	Cast Iron Pipe
−63	Copper Pipe
−64	Plastic Pipe
−65	Glass Pipe
−66	Stainless Steel Pipe
−67	Aluminum Pipe
−68	Bituminized Fiber Pipe
−69	Clay Pipe
−70	Brass and Bronze Fittings
−71	Prefabricated Insulated Piping
−75	Hose
−80	Piping Specialities
−81	Gaskets and Calking
−82	Swivel Joints
−83	Strainers, Filters, and Driers
−84	Vent Caps
−85	Traps
−86	Vacuum Breakers
−87	Shock Absorbers
−88	Waterhammer Arrestors
−90	Supports, Anchors, and Seals
−91	Anchors
−92	Wall Seals
−93	Flashing and Safing
−94	Hangers and Supports
−100	Valves, Cocks, and Faucets
−101	Gate Valves
−102	Blowdown Valves
−103	Butterfly Valves
−104	Ball Valves
−105	Globe Valves
−106	Refrigerant Valves

−109	Wall Hydrants
−110	Check Valves
−111	Swing Check Valves
−112	Backwater Valves
−113	Vertical Check Valves
−114	Stop and Check Valves
−115	Faucets
−116	Washer Outlets
−120	Self Contained Control Valves
−121	Pressure Regulating Valves
−122	Pressure Relief Valves
−123	Automatic Temperatures and Pressure Relief Valves
−124	Solenoid Valves
−125	Steam Traps
−130	Tempering Controllers
−131	Photo Lab Tempering Controllers
−132	Mixing Stations
−133	Refrigerant Control Valves and Specialties
−134	Feed Water Regulator
−140	Pumps
−141	Centrifugal Pumps
−142	Rotary Pumps
−143	Turbine Pumps
−144	Reciprocating Pumps
−145	Sump Pumps
−146	Submersible Pumps
−147	Pneumatic Ejectors
−150	Compressors
−151	Vacuum Pumps
−152	Air Compressors
−160	Expansion Compensation
−162	Piping Expansion Joints
−164	Flexible Connections
−170	Meters and Gages
−171	Temperature Gages
−172	Pressure Gages
−173	Flow Measuring Devices
−174	Liquid Level Gages
−175	Tanks
−176	Steel Tanks
	Underground Steel Tanks
−177	Plastic Tanks
−178	Cast Iron Tanks
−180	Metering and Related Piping
−181	Water Meters
−182	Gas Meters

15200   **NOISE, VIBRATION, AND SEISMIC CONTROL**

15250	INSULATION
−51	Cold Water Piping Insulation
−52	Chilled Water Piping Insulation
−53	Refrigerant Piping Insulation
−54	Hot Water Piping Insulation
−55	Steam and Condensate Return Piping Insulation
−56	Underground Piping Insulation
−57	Outside Piping Insulation
−58	Duct Insulation
−59	Breeching Insulation
−61	Equipment Insulation
−62	Boiler Plant Insulation

15300	SPECIAL PIPING SYSTEMS
−10	Air and Gas Piping Systms
−11	Compressed Air Piping Systems
−12	Oxygen Piping Systems
−13	Vacuum Piping Systems
−14	Helium Piping Systems
−15	Nitrous Oxide Piping Systems
−16	Instrument Air Piping Systems
−20	Laboratory Gas Piping Systems
−30	Compressed Industrial Gas Piping systems
−40	Industrial Piping Systems
−41	Chemical Waste Drainage Systems
−42	Chemical Distribution Systems
−43	Acid Distribution Systems
−44	Alkaline Distribution Systems
−45	Lubricating Oil Piping Systems
−50	Natural Gas Piping Systems
−60	Liquid Petroleum Gas Piping
−70	Process Piping Systems
15380–15399	(Reserved)

15400	PLUMBING SYSTEMS
−01	Cold Water Systems
−02	Hot Water Systems
−03	Refrigerated Water Piping Systems
−04	Distilled Water Piping Systems
−05	Soil and Waste Piping Systems
−06	Roof Drainage Systems
−20	Plumbing Equipment
−21	Floor and Shower Drains
−22	Roof Drains
−23	Cleanouts and Cleanout Access Covers
−24	Domestic Water Heaters
−25	Aftercoolers and Separators
−27	Anti-Syphon Equipment

−28	Sediment Interceptors
−29	Laundry Utility Units
−30	Package Waste, Vent, or Water Piping Units
−31	Solar Preheat Domestic Water Heaters
−35	Domestic Water Conditioners
−40	Pool and Fountain Equipment
−41	Pool Circulation and Filtration Equipment
−42	Pool Drains, Inlets, and Outlets
−43	Pool Cleaning Equipment
−45	Fountain Piping and Nozzles

15450	**PLUMBING FIXTURES AND TRIM**
−51	Special Fixtures and Trim
−52	Fixture Carriers
−55	Domestic Watercoolers
−56	Washfountains
−57	Showers
−58	Receptors
−59	Tub and Shower Doors/Enclosures

15500	**FIRE PROTECTION**
−01	Wet Automatic Sprinkler Systems
−02	Dry Automatic Sprinkler Systems
−03	Pre-Action Automatic Sprinkler Systems
−04	Deluge Automatic Sprinkler Systems
−05	Outside Protection Systems
−06	Foam Fire Protection Systems
−07	Carbon Dioxide Fire Protection Systems
−08	Halon Fire Protection System
−10	Sprinkler Equipment
−21	Foam Equipment
−22	Carbon Dioxide Equipment
−30	Standpipe and Fire Hose Systems
−31	Fire Hose Connections
−32	Fire Hose Cabinets and Accessories
−33	Fire Hose Reels
−34	Fire Hose
−40	Fire Pumps
−60	Hood and Duct Fire Protection Systems
−70	Non-Electrical Fire Alarm Systems

15600	**POWER OR HEAT GENERATION**
−01	Hot Water Heat Generation Systems
−02	Steam Generation Systems
−05	Fuel Handling Equipment
−06	Oil Storage Tanks, Controls, and Piping
−07	Liquid Petroleum Gas Tanks, Controls, and Piping
−10	Ash Removal Systems

−15	Lined Breechings
−16	Lined Prefabricated Chimneys and Stacks
−17	Exhaust Equipment
−18	Draft Control Equipment
−20	Boilers
−21	Cast Iron Boilers
−22	Firebox Boilers
−23	Scotch Marine Boilers
−24	Water Tube Boilers
−25	Absorption Boilers
−30	Burners and Controls
−35	Stokers
−38	Fuel Preheaters
−39	Boiler Accessories
−40	Boiler Feedwater Equipment
−41	Packaged Boiler Feed Pump Systems
−42	Deaerators

15650	**REFRIGERATION**
−51	Refrigeration Piping Systems
−55	Refrigerant Compressors
−56	Centrifugal Compressors
−57	Rotary Compressors
−58	Reciprocating Compressors
−60	Condensing Units
−61	Air Cooled Condensing Units
−62	Water Cooled Condensing Units
−63	Evaporative Condensing Units
−70	Chillers
−71	Reciprocating Chillers
−72	Air Cooled Chillers
−73	Ethylene Glycol Chillers
−74	Centrifugal Chillers
−75	Absorption Chillers
−76	Rotary Chillers
−80	Propellor Type Cooling Tower
−81	Centrifugal Type Cooling Tower
−90	Evaporators
−91	Unit Coolers
−95	Condensers
−99	Refrigeration Accessories

15700	**LIQUID HEAT TRANSFER**
−01	Hot Water Heating Systems
	One-Pipe Hot Water Heating Systems
	Two-Pipe Direct Return Hot Water Heating Systems
	Two-Pipe Reversed Return Hot Water Heating Systems

Radiant Hot Water Heating Systems
Snow Melting Systems

−02	Chilled Water Piping System
−03	Steam Supply and Return Piping System
−10	Hot Water Specialties
−15	Steam Specialties
−20	Condensate Pump and Receiver Sets
−30	Heat Exchangers
−31	Storage Water Heaters
−32	Converters
−34	Clean Steam Heat Exchangers
−35	Water Heat Reclaim Equipment
−40	Terminal Units
−41	Induction Units
−45	Radiant Panels
−50	Coils
−51	Baseboard Units
−52	Finned Tube
−53	Convectors
−54	Radiators
−60	Unit Heaters
−61	Fan Coil Units
−62	Unit Ventilators
−63	Air Handling Units with Coils
−70	Packaged Heating and Cooling Units
−72	Air to Air Packaged Heat Pumps
−73	Water to Air Packaged Heat Pumps
−74	Water to Water Packaged Heat Pumps
−80	Humidity Control
−81	Humidifiers
−83	Centrifugal Type Humidifiers
−85	Dehumidifiers
−86	Desiccant Dehumidifiers
−90	Process Heating
−95	Storage Cells
−99	Special Devices

15800	**AIR DISTRIBUTION**
−01	Heating and Cooling Systems
−02	Heating Systems
−03	Cooling Systems
−04	Ventilating System
−10	Furnaces
−11	Direct Fired Furnaces
−12	Cast Iron Furnaces
−13	Steel Furnaces
−14	Rooftop Furnaces
−15	Direct Fired Unit Heaters

−16	Direct Fired Duct Heaters and Reheaters
−18	Energy Recovery Units
−20	Air Distribution Equipment
−21	Centrifugal Fans
−24	Propellor Fans
−25	Attic Exhaust Fans
−26	Fly Fans
−27	Axial Flow Fans
−28	Induced Draft Fans
−29	Exhaust Fans
−30	Power Roof Ventilators
−31	Power Wall Ventilators
−32	Roof Ventilators
−34	Air Handling Units
−35	Air Curtains
−40	Ductwork
−41	Low Pressure Steel Ductwork
−42	High Pressure Steel Ductwork
−43	Nonmetallic Ductwork
−44	Special Ductwork
−46	Prefabricated Insulated Ductwork
−47	Flexible Ductwork
−48	Duct Lining
−49	Duct Hangers and Supports
−50	Special Ductwork Systems
−51	Tailpipe Exhaust Equipment
−52	Dust Collection Equipment
−53	Paint Spray Booth System Equipment
−54	Fume Collection System Equipment
−55	Breeching and Smokepipe
−60	Duct Accessories
−61	Manual Dampers
−62	Gravity Backdraft Dampers
−63	Barometric Dampers
−64	Fire Dampers
−65	Smoke Dampers
−66	Turning Vanes
−67	Distribution Devices
−68	Duct Access Panels and Test Holes
−70	Outlets and Inlets
−71	Wall and Floor Diffusers
−72	Ceiling Diffusers
−73	Ceiling Air Distribution Systems
−74	Light Troffer-Diffusers
−75	Warm Air Baseboard
−76	Cabinet Diffusers
−77	Air Floors
−78	Roof Mounted Air Inlets and Outlets

−79	Air Inlet and Outlet Louvers
−80	Air Treatment Equipment
−81	Disposable Filters
−82	Permanent Filters
−83	High Efficiency Filters
−84	Roll Filters
−85	Oil Bath Air Filters
−86	Electronic Air Filters
−87	Air Washers
−88	Dust Collectors
−89	Fume Collectors or Dispensers
−90	Sound Attenuators
−95	Special Devices
15900	**CONTROLS AND INSTRUMENTATION**
−02	Electrical and Interlocks
−06	Identification
−10	Control Piping, Tubing, and Wiring
−15	Control Air Compressors and Dryers
−20	Control Panels
−25	Instrument Panelboards
−30	Primary Control Devices
−31	Thermostats
−32	Humidistats
−34	Aquastats
−35	Relays and Switches
−36	Timers
−37	Control Dampers
−38	Control Valves
−39	Control Motors
−50	Sequence of Operation
−60	Recording Devices
−70	Alarm Devices
15990–15999	(Reserved)

15950 CONTROLS

This category is an important one with a broad scope. The fact that such an important segment of building construction is presented in one section is indicative, perhaps, of the relative dominance of the design and construction industry by nonmechanical disciplines. Here in one of sixteen sections, work representing perhaps 20% to 30% of the typical building is presented. This section emphasizes environmental equipment to control either heating, cooling, and/or humidity. Also included are the piping systems to operate this equipment, and the controls to direct it.

The electrical systems which power the mechanical equipment are included in the following section—Section 16.

Building mechanical equipment is getting increasingly complex, and inspection of the installations on site is of great importance. Often, one or a group within the inspection team are mechanical specialists, and can be called upon by the general inspectors where areas of expertise are required.

Fortunately, the mechanical equipment is usually prepared under manufacturing conditions and shipped to the job site as operating components. Therefore, the on-site work is usually broken into three phases: installation of equipment; connection of piping systems; and operational testing.

This section also includes plumbing with its piping, water treatment, and special systems.

15050 BASIC MATERIALS AND METHODS

15060 Pipe and Pipe Fittings

Piping may be made of many different materials, with steel and iron perhaps the most usual. Pipe *sizes* vary, and are usually specified by the inner diameter. The pipe thickness or wall is designated by its strength, usually designated as pipe schedules. The pipe *schedule* number is the ratio of the internal pressure in psi divided by the allowable fiber stress multiplied by 1000. Typical schedules of iron and steel pipe are schedules 40, 80, and 160.

In refrigeration work, *tubing* is often used, and controls are usually connected by tubing.

Piping sections are connected by welding, brazing, and threaded and special connections. Special connections include flanges, Dresser couplings and flared tubing fittings.

The inspection team should carefully review the specifications for the type of connection, and the material; for instance, in welding, the type of rod and welding machine (electric arc, heliarc, etc.) is specified, and failure to utilize proper material or equipment will usually result in faulty connections.

The type of *test* is also specified. Some connections are checked for integrity by X-ray or ultrasonic nondestructive testing. In important systems, such as radiological systems or those carrying lethal gases, 100% inspection of the welding for potential flaws is usually required. After the individual weld tests, a pressure test is usually applied, either hydrostatic or gas. Typical test pressures are one and one-half times working pressures. In the case of water or *hydrostatic* tests under pressure, visual inspection is utilized, as well as the requirement that the test pressure hold for a required period of time. Because of the noncompressibility of water, changes in the ambient temperature can cause increase or decrease in the pressure under test. The test hold period is usually one to four hours, with a specified range of variation. In calculating the range, changes in ambient temperature should be calculated and allowed for.

In certain high-pressure, lethal and/or radiological systems, special detection tests are utilized after the hydrostatic, to prevent even *minute leakage.* One test utilizes the introduction of Freon into the system, and the use of sophisticated Freon detecting equipment to sniff the air around joints, particularly flanged or screwed, for evidence of any leakage.

The inspection team should be certain that all materials are handled carefully to prevent damage, and should reject any damaged materials. Before installation, the team should review the piping system drawings to determine how the system fits into the total job. Locations and elevations of pipe should be checked, and spot checked or completely checked as the installation proceeds. Inspection of the piping system often starts with determination of the location number and size of inserts for floors and walls during concrete placement. Sleeves must be large enough.

During erection, trash and debris should be kept out of pipes, fittings, and fixtures. When work is suspended at the end of the day, openings should be closed off to prevent animals from entering into the piping system.

Valves

Valves are utilized for shutting off or controlling flow. The two most common types are globe and gate valves. *Globe valves* are used for throttling, and utilize a circular disk closing against a ground fitted valve seat. The shutoff is complete when the valve is new, but continued service often results in minor scratches or cuts in the seat or disk, and a globe valve cannot assure positive shutoff. However, when used for throttling, the globe characteristics result in less damage.

Gate valves are constructed in a manner indicative of their name. A plate or gate with machined seat, usually in a wedge shape, closes to ensure positive shutoff. However, when used for throttling, the gate valve is susceptible to more damage, and thereby leaks more readily if it has been used improperly for throttling.

Other types of valves include slide valves and control valves, which are usually of the globe type, but are actuated automatically. The material utilized in valves should be checked for compliance with the specification. The material is usually specified in accordance with the type of service required. New coatings and linings are often utilized.

Valves are usually connected into the pipeline by either flanged or threaded connections.

Valves are rated according to pressure, and the rating is usually stamped or cast into the valve body. This should be checked to ensure that valves of proper strength, as well as material, are utilized.

15080 Piping Specialties

There are innumerable special fittings and devices involved in piping systems. The more standard of these involve fittings such as concentric and eccentric size reduction sections, used to convert from one size to another of piping. The type of conversion is dependent upon the material handled, and will be specified. The field team must be careful that no inadvertent changes are made.

A wide range of cast or forged elbows, tees, 45° turns, and other basic fittings are used in welded and screwed piping.

Other fittings may include items such as water hammer arresters and instrumentation fittings to measure flow, pressure, and other characteristics.

From the inspection viewpoint, keep in mind that specialties imply special conditions, and that special items must be installed carefully in conformance with both the specifications and the manufacturer's suggestions.

15090 Supports

The support of piping systems is critical to both their successful operation, and the safety of the operating structure. The basic source of stability for piping supports is the *anchorage* for the support. Many supports are hung from the primary structure, either wall, ceiling, or floor above. The optimum approach in many cases is the use of the basic structure through placement of inserts, particularly in poured concrete structures. The problem is accuracy of placement, which in turn is a function of the inspection team's surveillance and dimensional checking procedures.

Even when the inspection team has carefully positioned or checked the position of inserts, design changes or approved field changes may place the final piping systems in locations which cannot utilize the fixed insert. For this reason, many inserts have some *flexibility,* and certain types offer a broad variety of locations. Typical of this is the Unistrut member, a strip of varying length which can accept inserts in the form of spring-loaded nuts. Other proprietary systems are available.

Other hangers may be clamp-type attachments to steel framing members. This is satisfactory in many cases, but is not permitted where the steel is fireproofed (since it would provide a direct source of heat conduction into the primary structure).

Where inserts have not been provided, or are no longer sufficient, *expansion anchors* are often placed in concrete structure. Anchors are placed by drilling into the concrete, then driving a lead shield or special self-seating shield. The anchorage unit is held in place by the expanding force provided by a screw or bolt being threaded in place. An expansion bolt will hold a tension load, but is dependent upon the soundness of the material drilled into. The larger and deeper the shield receptacle, the greater the holding power. In some cases, inserts can be anchored with explosive-actuated tools. There is a limitation in terms of the size which can be held, and potential danger in driving against rebar locations.

In addition to clamping to steel, inserts can be welded to steel members, again with the reservation that fireproofed devices should not have heat-conducting supports welded to them.

Pipe *hangers* are usually of the threaded type, attached to a collar around the pipe. Often universally threaded rods are utilized for hangers. These are continuously threaded of mild steel, usually $\frac{1}{4}''$ to $\frac{1}{2}''$ in diameter. The convenience of these rods is their ability to be cut to length on the job, still providing threaded ends for threading into the insert, and accepting nuts for the hanger attachment.

Hangers are designed to suit the piping system carried; for instance, piping subjected to dynamic loading may have spring-loaded hangers to reduce the effects of harmonic motion. These must be designed and placed according to design in accordance with the frequency of the piping system.

Pipe supports can directly affect the *stresses* to which the pipe is subjected. Anchorages must be located according to the drawings. Positive connections in certain types of piping systems will subject the pipe to greater than allowable stresses. This is particularly true in piping systems exposed to extremes in temperature, either hot or cold.

15200 Vibration Isolation

Piping subjected to a vibration may be isolated through a number of devices. Anchorage to a heavier structure or a mass may be utilized to absorb the vibration influences. This may, however, induce high loads on the piping. The inspection team should take great care to ensure that the field installation of the supports is in accordance with the design and approved shop drawings. Again, inadvertent anchorages must not be allowed. Where piping is to pass through sleeves with free play, movement must be checked before, during, and after installation. In some cases, bellows valves or flexible braided connections may be called for, and should be installed in accordance with the design drawings.

Some designs call for vibration dampening material between the supports and the job itself. The isolation of vibration should not be a decision of the inspection team, but must be the result of design considerations. If vibration results from the installation according to the design, then the inspection team should immediately call upon the design agency or its approved representatives.

15250 Insulation

Heat travels from warm areas toward cold in an effort to achieve equilibrium. The rate of transmission is a function of the difference in temperatures, and the materials between the hot and cold zones. Insulation is made of material which has a low *conductivity,* or heat transmission characteristic. The rate of transmission is inversely proportional to the thickness of the insulating material. Insulation may be used to contain heat in a warm system, or to keep out heat in a cold system. To maintain the stability of the insulation, a *vapor barrier* is usually specified. Also, the use of a shiny material such as aluminum foil may be required to reflect part of the heat load.

The inspection team should be certain that the type of insulation and the installation is in accordance with the specification. In some cases, wrapping is sufficient, while in others, preformed block material may be attached to the pipe with banding of specified material, and then covered with a vapor barrier.

In particular, check the insulation at points where the pipe passes through sleeves in walls, floors, or ceiling. Ensure that insulation is properly applied over fittings, valves, flanges, and unions, and that it is of the same material and thickness as the adjoining pipe insulation. Where a vapor barrier is called for, it should be continuous and without breaks at hangers. Any discontinuities or penetrations should be sealed with an approved vapor barrier coating. At specified intervals, the vapor barrier seal should be complete, so that any breakdown in vapor barrier in service will be limited to that length.

In older installations, asbestos-based insulation was most popular; however, difficulties in handling of the asbestos fibers has resulted in increasing popularity of formed and flexible fiberglass insulation, as well as foamed plastics such as styrofoam and urethane.

15300 SPECIAL PIPING SYSTEMS

15311 Compressed Air Piping

Check the materials and locations with shop drawings. Keep in mind that compressed air piping is under substantial pressure, although the material which escapes is generally benign. Identify valves with appropriate identification tags. Also check *color coding* requirements in the specification. Compressed air lines often require antivibration couplings. Usually, there is a requirement for an aftercooler and moisture separator in the piping between the air compressor and the receiver. Check the specifications or drawings for requirement, and install as required.

15313 Vacuum Piping

Exhaust piping from vacuum pumps should be piped to outside of building and turned down and screened. Check pressure-reducing regulators for proper settings. Joints for vacuum piping should be made with silver brazing alloys, except valves and equipment requiring threaded connections. Threaded connections should not be caulked, but may be coated with approved material such as litharge and glycerin. Antivibration couplings should be provided between the vacuum pump and piping.

Vacuum piping should be properly color-coded, and identification tags should be provided for valves. Careful testing of the system is required before it is used. Tests may be a hydraulic holding test and/or air pressure with soap solution on all joints.

15312 Oxygen Piping

One of the keynotes in oxygen piping is *cleanliness* of installation. Particularly, combustible items such as oil must be carefully removed from the completed piping to avoid spontaneous ignition in the oxygen atmosphere. Installation and testing should conform with the specifications, and generally be in accordance with National Fire Protection Association Standard 565. Only approved materials and equipment should be used in the oxygen system.

Silver solder may be used for making joints, except valves and equipment requiring threaded connections. Any thread coating must be approved for an oxygen environment.

15315 Nitrous Oxide Piping

Installation and testing should be in accordance with NFPA Standard 565. When installed underground, piping should be encased in nonmetallic ducts or casings. Check for cleanliness of piping before use.

Joints should be silver solder, except valves or equipment requiring threaded connections. Specifications usually require a check valve between each cylinder in a bank of cylinders, and between the bank of cylinders and the manifold connection.

Valves should be identified with tags, and the piping appropriately color coded.

15370 Process Piping

Process piping will be made of materials and connections suitable for the type of material handled, and the pressures at which it is handled.

The type of material and level of workmanship required will vary with the process material. In most cases, process piping is pressure piping, and requires testing either hydrostatically or with air pressure and soaping of joints. Hydrostatic testing is the more suitable method, but in some cases water may not be allowed to remain within the pipe, and even drying out with nitrogen may still leave pools of water collected.

For lines carrying lethal gases, a series of tests may be required. One level of test determines the strength and safety, while a second gas test using a sensitive detector such as a Freon sniffer is used to test the tightness of joints.

15360 Liquid Petroleum Gas Piping

Gas piping should not be installed under buildings or in trenches with other utilities. If such installation is suggested, carefully check specifications in design for special requirements or permission. In the installation of gas piping, there should be strict adherence to safety regulations. There should be principal *shutoff valves* accessible for operation in case of emergency.

Gas lines should be installed above other utilities where they cross and slope down to drip valves or capped drips at low points. Gas lines should not be soldered.

The local utility will usually furnish on-site assistance for inspection, and instruction in special requirements. Be certain that plug cocks, gas-pressure regulators with automatic cutoffs, and insulating coverings are installed as required. The line to *relief vents* from the gas pressure regulators should not be located adjacent to air intakes or heating intakes for the building. There should be drip pockets at all low points in the gas main.

The traditional piping for gas lines is threaded black iron pipe. Where threaded pipe is to be placed, be certain that the pipes are squarely cut and the threading is performed properly. Check for thread length and be certain that the threads are not undercut. Threaded piping, particularly for gas lines, should not be caulked. The screw joints should be made up tightly, but not tight enough to strain the fitting.

PIPING SYSTEMS CHECKLIST

1. Check for size and strength of materials, as well as proper materials in accordance with specifications.

2. Check valves and fittings for pressure rating and specification items.

3. Check anchorages to be sure that those required are in place, and also that there are no inadvertent anchorages where a line should be free moving.

4. Check welding, and other workmanship in accordance with specification tests.

5. Check for sufficient supports, and freedom of movement where required. Check for special fittings such as vibration isolators where required.

6. When piping system is completed, carry out strength tests and containment tests in accordance with specifications.

7. Ensure that special requirements such as sterilization of water systems are carried out.

8. Be sure that piping installation permits full access and egress as required by the drawings and specifications (i.e., doorways, aisleways) and that piping is accessible for maintenance, if required.

9. Following strength and containment tests, monitor the installation of insulation for materials, amount, and workmanship.

10. For lines which are to be hydrostatically tested, ensure that the system can be both vented and drained, and that the fittings can be capped off after the test.

11. For special systems such as liquid oxygen, ensure that special precleaning operations have been carried out.

15400 PLUMBING SYSTEMS

The source of the project water supply is determined by the design team. In the majority of cases, it will be connected to a municipal service, which will impose certain inspection requirements upon the inspection team, including the *sterilization* of systems which will handle potable water for human consumption.

The distribution system for the project water supply is usually iron or copper piping, although where building codes permit, plastic pipe is becoming increasingly popular.

15401 Cold Water Systems

The traditional material for water distribution systems has been iron threaded piping with cast or forged iron fittings. The threads are usually cut on the job site by the plumbing contractor. The test pressure is usually modest, about 150 psi. Newer practice has been utilizing copper tubing of type K or L, with tubing fittings, or threaded fittings, depending upon the size.

The requirements for installation of the piping are found in the piping specification, or in this section.

15402 Hot Water Systems

The hot water supply system starts at the hot water generator which may be gas, oil, or electric-fired. The hot water generator may be an integral part of a larger boiler, or may be a stand-alone unit. To have good response time for delivery of hot water, the piping system for hot water may have a recirculating line which keeps hot water flowing through the system.

Often, state or insurance requirements dictate that the hot water generator be an ASME-approved pressure vessel. This will be spelled out in the specifications, and if required, the generator(s) shall be inspected at the point of manufacture, and appropriately stamped on an approved ASME label-plate denoting that the design and testing has been approved.

Even where pressure vessel approval is not required, the hot water generator should meet appropriate standards, and should be carefully scrutinized by the inspection team, since these units receive constant utilization and are a common source of malfunction—and even accidents—during operation.

15403 Refrigerated Water Systems

These systems were quite common at one time in hotels and large office buildings, circulating potable chilled water for drinking purposes. Recent practice has made broad utilization of individual chilling units operating at each drinking location. The inspection requirement for these units is not rigorous, but must ensure that they are installed in accordance with the specifications.

15405 SOIL AND WASTE PIPING SYSTEMS

Soil and waste systems include the interior collection system for waste, exterior collection such as storm sewage site system, and the delivery trunk to the treatment medium. Treatment may be on site, or a connection to available utility.

The nature of the soil and waste system is dependent upon its principal characteristics, with the two principal areas being normal human generated wastes and industrial wastes. The industrial waste problem is a very special one, and special equipment will usually be specified to handle it; this is particularly true since the passage of the National Environmental Protection Act (NEPA) in 1968.

The traditional material for soil collection within a building is cast iron, but in recent years, copper piping has been utilized because of its lighter weight and ease of installation. Joints for the cast iron are usually leaded, while for the copper joints, brazed fittings—often using precoated fittings—are popular. Most collection systems are gravity feed, so that the pressure is virtually zero, and inspection in terms of good workmanship may be sufficient. Pressure *tests* may be called for, but usually are not because the fittings are not set up for pressure resistance. Gas or smoke tests may be utilized in some instances.

Size is important, and since the waste systems are usually gravity flow, sizes are substantially larger than those for the pressurized delivery systems. *Traps* are required by the design, and the National Plumbing Code also.

The building discharge is usually made of cast iron pipe. It may empty into a larger cast iron pipe, or into extra-heavy clay tile pipe. The usual method of connecting flow from two types of pipes like this is through a *manhole.* Soil pipe joints are usually made with lead and oakum, but may be poured of special mastic material where a limited amount of corrosive material is anticipated, such as acid.

The vertical piping system may be copper, cast iron, or galvanized wrought iron. The *stacks* and related branch lines carry the discharge from water closets, urinals, and other plumbing fixtures. The vertical stack system must be *vented* at a point above the highest fixture connection to provide a nonpressurized outlet for the gravity system. The usual material for vent piping is standard galvanized steel pipe, although other materials are often used.

Where codes permit, plastic pipe is becoming increasingly popular for all waste collection and venting. The reasons include: lower first cost of material, lower cost in labor per unit of installation, and greater resistance to the corrosive ingredients in the material handled. Availability of plastic piping, such as polyvinyl chloride, in pressure ratings correlated with ASTM and ASME ratings for other piping materials has made engineering calculations easier and receptivity by code officials more frequent.

15406 Roof Drains

The roof drainage system is made up of approved drains located as required at the low points of the roof. The system may be tested by ponding the roof. The down spouts may be inside or outside the building—and are usually checked by inspection of workmanship. The roof drainage system is usually connected directly into the storm drainage system by code requirement. Down spouts are usually galvanized steel pipe or galvanized wrought iron.

The inspection procedure should include assurance that no plugs or stoppages exist within the down leaders.

15450 PLUMBING FIXTURES AND TRIM

Most plumbing *fixtures* are manufactured units, delivered ready for installation. These include: water closets, urinals, sinks, bathtubs, drinking fountains, laundry tubs, and other vitreous china fixtures. Installation is usually required utilizing chromeplated piping in the vicinity of the fixtures for maintenance and cleaning purposes. Inspection of the units should be limited to conformity with specifications for type and quality, and assurance that the equipment has not been damaged in shipment.

Fixtures are set on roughed-in drainage piping, which includes the required trap. In certain cases, such as units with a high flushing volume, a deep seal trap is required, twice the depth of the common trap.

Fixture trim escutcheons are placed over the openings where piping extends through walls and floors, and may be a split or solid pattern. Installation should be inspected, since these are often allowed to work loose.

Faucets should be inspected for proper installation, and tested after the unit is pressurized. Typical problem is damage or omission of washers.

In many cases, lavatories and urinals are placed on chair carriers which have been inserted in the concrete. This allows off-the-floor installation, enhancing cleaning.

The types of traps used in practice are the siphon, which is of uniform cross section, making it self-cleaning. The nonsiphon has an enlarged body slowing down the velocity of water flowing through it, in the shape like an S or an open P. Bathtub traps are shaped like a drum. A grease trap or interceptor may also be placed in the drain or waste pipe, particularly in kitchens to prevent grease getting into the sewer systems.

For flush-type lavatories, a chrome *vacuum breaker,* somewhat larger than the tail piece between the flush valve and the bowl, is used to break or prevent the formation of vacuum in the water supply lines.

Waste lines may have a *backwater* trap installed, essentially a check valve preventing reverse flow of water back into the building drainage system.

15500 FIRE PROTECTION SYSTEMS

There is increasing emphasis upon built-in capability for fire extinguishment. For years, certain categories of buildings have been sprinklered either because of public building codes or fire insurance requirements.

Today, even more buildings are requiring extinguishers and other fire extinguishing systems as part of the integral structure.

In urban areas, high-rise structures are more recognized as a firefighting problem than they previously had been. In rural and suburban areas, the degree of available firefighting protection is often based upon volunteer service, and greater distances have to be covered, so that site capabilities are needed to reduce the vulnerability to fire.

Automatic Sprinkler

These systems have proven successful in the prevention of fire spread, once an area of heat or fire has been established. Sprinkler systems are made up of horizontal piping placed near the building's ceiling supplied by vertical risers. The sprinkler systems are usually *preset* to open automatically at a certain preset temperature, usually in the range of 150°F, or in a local area where temperatures are high enough to melt babbitt metal plugs. There are several common types, including: wet pipe, dry pipe, deluge, and pre-action. The type will vary with the building code and insurance company. In a *wet pipe* system, the pipes contain water at all times, and are connected to water supply so that water is discharged immediately when the sprinkler is open. In a *dry pipe* system, the pipes have pressurized air which is replaced by water pressure when the sprinkler head is open. Dry pipe systems do not react as quickly, but are better for unheated buildings such as warehouses. *Deluge* system sprinklers are attached to a piping system and water is open from a reserve reservoir, similar to the dry pipe, but providing a greater immediate supply. *Pre-action* systems are similar to dry systems, but the air is not necessarily pressurized.

Sprinkler systems generally must meet the requirements of the National Fire Protection Association. Check layout and materials against shop drawings, particularly for size and workmanship. Where work is being accomplished in occupied buildings, existing fire protection systems must not be made inoperable without giving advance notice to operating personnel or suitable authority.

Valves and equipment should be located beyond flood levels in buildings located in flood plains.

In accordance with NFPA 13, *sprinkler heads* should be installed in upright position at recommended clearance between roof or ceiling. If installed in pendant position, return bends must be used if water is subject to sedimentation. If heads might be subject to mechanical injury, guards should be provided. Heads should be new, and not painted. Sprinkler head temperature ratings must be proper for ambient temperatures anticipated. For instance, near heaters or skylights, temperature settings should be appropriately higher. If sprinkler heads are to be installed in special hazard areas such as electronic shops or computer rooms, confirm that installation is required, and be particularly careful to ensure that accidental discharge cannot occur.

Valves or plugs must be provided to insure *drainability* of entire system, and piping runs, particularly outside building, should be slightly pitched to allow drainage.

In wet pipe systems, the alarm check valve assembly must conform with the connection diagram. Observe the installation of water flow indicators for conformance with connection diagram, and be certain that insulation and painting requirements are met. Check the water flow alarm signal by using the wet pipe type of inspector's test connection.

In dry pipe systems, determine that the piping layout is in accordance with drawings, and that the connections are also in accordance with the connection diagram. Inspect installation of the air compressors. The air supply line should include a flexible connection and orifice plate. Check motor control or operation to ensure that the compressor cuts in automatically. If compressor is equipped with an air storage tank, be sure that the condensate water drain is provided at the bottom of the tank. Examine the locations and operation of condensate chambers, and the water flow alarm signal time and dry valve trip test time.

In deluge systems, confirm that piping connections are in accordance with drawings. Examine the installation of the deluge valve assembly, and inspect and test releasing devices. Check tripping devices, both manually and automatically, and check provisions to ensure against accidental water damage. The contractor must provide portable testing units in good working order.

15507 CO$_2$ Systems

For closed-in areas such as diesel generator stations, or electronics equipment, bottled closed CO$_2$ systems may be used to smother combustion. CO$_2$ systems depend upon the proper capacity of CO$_2$ supply, and tripping devices in good operating condition.

CO$_2$ systems are not usually installed within habitable spaces unless the system is confined to cabinetry or lockers.

Installation should be in accordance with manufacturer's instructions, and all fittings should be tight.

15530 Standpipe and Hose

Connected to the elevated water tank or dynamic pumping system, is a series of vertical standpipes or risers with fire hose stations. Installation should be tested according to specification for ability to hold water. The inspection team should also carefully check the quality, length, and manner of installation of firefighting hoses at various stations along the standpipe. Appropriate signs and markings, as well as instructions for use, should be installed in accordance with the specifications.

15600 POWER OR HEAT GENERATION

15606 Oil Storage Tanks

Oil storage tanks must be checked for location, and in particular, must be installed with the proper anchorages. Empty or partially full tanks can easily pop out of the ground, if not properly anchored. Steel tanks are usually coated, and the coating should be tested. In recent years, fiberglass tanks or fiberglass covering of tanks has been used to preclude corrosion.

Fuel piping is usually screwed black iron piping, which must be installed carefully. Piping should be pitched to allow fuel to drain back into fuel tanks when pumps are shut down. Piping should be wrapped in accordance with specifications to preclude corrosion.

Check tanks for tank heaters where required. Check tank instrumentation, and calibrate gauges before backfilling.

15620 Boilers

Each piece of material and item of equipment should be approved well in advance of the time for its installation. When the items arrive on the job, the inspection team should check them carefully, comparing them with the approved equipment list. Check the *nameplate data* on all equipment. Ascertain that appropriate storage conditions are provided. Any damage to materials should be rejected, and replaced or repaired. Examine pressure boilers for conformance with the ASME code, and in particular, an approved ASME inspection nameplate.

Boilers are generally package units in smaller installations, or assembled on site with a manufacturer's representative in charge. The inspection team should review *workmanship* during installation, making inspections for pressure connections before installation of the fire brick or combustion wall material which would obscure the piping connections. In particular, the inspection team should check connections for tightness, usually under hydrostatic pressure.

Proper *refractory* materials should be used on the firebox side. Expansion joints should be provided in accordance with design drawings, and piping on both sides of expansion joints should be properly guided. Packing should be placed properly to prevent gas or air leakage, or short circuiting.

Any chipped, cracked, or damaged fire brick or tile should be rejected. Plastic refractory material should be placed carefully, with proper consistency and rammed into place. Refractories must be kept dry and after placing must be cured with a wood fire heat (or gas-fired if allowed by the specifications). Check for air circulation under the combustion chamber floors.

Accessory equipment such as feed water controllers, dampers, pressure and draft gauges, flow and pressure recorders, soot blowers, water columns, and boiler blowdown must be carefully installed and calibrated after installation, if appropriate. Where gauges can be checked only under operating pressures, they should be calibrated before installation.

In oil or gas burning boilers, there is only a minimum of *feed* equipment and piping. For coal burning boilers, the feed or stoker equipment must be checked.

For all boilers, *draft fans* should be checked for anchorage, alignment, and rotation of blades.

The dampers should be checked for operation, and bearings checked for smoothness and running without overheating. Check vibration and vibration absorbing mountings.

For final hydrostatic test, *safety valves* must be gagged. During initial start-up after hydrostatic pressure testing, the inspection team with the manufacturer's representative should check the boiler operating characteristics at low pressure, moderate pressure, and then under prescribed operating test.

BOILER INSTALLATION CHECKLIST

1. The inspection team should inspect nameplate data on equipment as it arrives at the job site.

2. Check storage conditions.

3. Reject any damaged materials.

4. For pressure boilers, ascertain that ASME stamp has been placed on the equipment.

5. For major units, manufacturer's representative should be on site during erection.

6. Check workmanship during installation, particularly welding and pressure connections.

7. Check hydrostatic test.

8. Check refractory materials and installation.

9. Check instrumentation calibration.

10. Check draft fans for anchorage alignment and rotation prior to start-up.

11. Check feed or stoker equipment.

12. Before start-up test, assure that all components are operable.

13. Monitor start-up test, again with manufacturer's representative available.

Circulating System

Steam circulation is principally via piping using pressure differential as the primary motivation force. This system should be checked carefully for layout, size, and material.

Since steam distribution is under pressure, careful testing hydrostatically is required before steam is introduced. The testing should be completed before insulation is placed on steam distribution piping masking the welds and joints.

Drain plugs should be provided both for testing, and for draining lines for maintenance.

Terminal Units

Heating units are often a combination of heating and cooling. Steam may be utilized in a heat exchanger to heat a secondary loop of warm water to carry heat to the distribution units. These may be radiators, fan coil units, or thin tube piping. These basic heating units are straightforward, and the main concern is leakage and good workmanship on installation. The controls usually actuate circulating water pumps.

15650 REFRIGERATION

Refrigeration equipment is used principally in buildings today for air conditioning. It is also used for maintenance of food at cool or cold temperatures, and other special uses. Units up to about 50 h.p. are usually packaged units installed in single units, or a remote cooling unit and a separate unit for refrigerant gas compression. Larger cooling units are usually made up of large-sized compressors, separate cooling units such as a cooling tower, and appropriate heat exchangers. In either case, the major components are usually standard, with equipment available from a broad range of reputable manufacturers. The principal concern of the inspection team is that the equipment specified and approved by the design team has been delivered to the job site, is maintained in sound and undamaged condition, and properly connected.

In package units, the most common refrigerants are Freon 11 and Freon 12. Larger units may use ammonia or special gases for the compressible refrigerant which cools by expansion. Compressors are of either the reciprocating or centrifugal type, with the reciprocal type more common in the smaller sizes.

Most heat waste sinks are water or air cooled. Environmental considerations have discouraged the direct in and out use of a river water or well water source as a cooling medium, since it either wastes water supply, or raises the ambient heat of natural water bodies. Cooling towers or direct coils to air for smaller units are more common. Where waste heat in the form of steam or other material is available, an absorption machine can be used in which a chemical reaction is utilized for the cooling effect.

All refrigeration equipment should be inspected to determine that it is being installed strictly in accordance with the safety code for mechanical refrigeration. Check the installed equipment to be certain that the condenser and chiller tubes can be pulled out for maintenance. Confirm that all rotating parts, belts, sheaves, shaft couplings, etc. are covered to protect personnel. Before start-up be certain that all equipment is lubricated according to the manufacturer's recommendations.

Reciprocating Compressors

In the installation of these units, check for oil, suction, and discharge pressures on start-up, as well as shaft alignment for direct-driven machines. Check the operation of the high and low oil pressure switches for proper operating level. Ascertain that the proper control gauges and meters are installed, and that the amount, correct type, and proper dryness level exist in the refrigerant charge. The unit should have pressure-holding capability when pumped down. Check the isolator deflection and compressor for vibration. After start-up, check the suction strainer screen mesh, but before start-up remove start-up felts. Check the unloader action, as well as the compressor speed. Check belt tension and alignment, and motor amperage under maximum load. During operation, check the refrigerant flood back and oil for foaming. Check the cylinder head for overheating. Initially, check the rotation of the unit by bumping it briefly.

The automatic oil heater in the crank case should work during shutdown. Before start-up, check refrigerant piping for loops which would permit oil to be trapped.

Centrifugal Compressors

Check for alignment of compressor drive and gear box before start-up, as well as suction damper or inlet vein operation. Check the purge compressor operation, as well as the safety control circuit, and the float valve operation if there is one. Check the oil pump and cooler for flow, and the entire unit for noise and vibration. Ensure that required gauges and meters are installed.

Absorption Refrigeration Machine

Check cleanliness of all parts during operation, and operation of the purge system. Check operation control especially for high and low limit temperature cut-outs or condenser water pump interlock.

The unit should be fully charged with water and a nontoxic absorber after installation. The inspection team should ensure that a factory representative is available for charging, testing, and start-up of the plant.

Auxiliaries

Condensers should be checked to see that air flow is not obstructed and that wind deflectors are installed if required in air-cooled condensers. Inspect water-cooled condensers for leaks and proper flow. Evaporative condensers should be checked for spray coverage, float valve operation (including no chatter), water level, fan rotation, pump discharge strainer, and other special requirements of the specifications.

Water chillers should be checked for water drains, vents, and proper pass arrangement in direct-expansion chillers. Inspect for freeze protection safety devices, and check the strength of the liquid bleed-off at bottom of flooded chillers. Check the adjustment of the level control. Check tubes and shell in brine chiller for type of material.

Unit coolers should be checked for corrosion-protected pan and casing, and that water defrost units have spray coverage with no carryover. Electric defrost units should be checked for cycle timing in accordance with job conditions. Hot gas defrost for suction pressures and refrigerant charge should be in accordance with manufacturer's recommendations. The drain line should be properly trapped on the warm end.

For spray type air washers, check that all nozzles are discharging air spray, and that no water carries over from the eliminators. Eliminators should not rattle and must be removable for maintenance. The float valve should not chatter on opening or closing. For capillary-type washers, the media should not sag in frames, and should be wetted completely.

For *humidifiers* and dehumidifiers, examine for supported coil and corrosion-protected pan. Check refrigerant-type dehumidifiers for frosting and cooling coil and for water carryover. Absorption-type dehumidifiers should be checked for solution level and temperature controls, as well as no carryover from eliminators. Regenerator duct must be drained of specified material and correctly sealed. Check for damper operation, cycle timing, evidence of dusting of the desiccant and regeneration temperatures.

For *pumps* ensure that the manufacturer's nameplate and serial numbers are in view after installation. Check for anchorage of pump as well as alignment with motor and piping. Make sure that all gauges and meters are provided. On the suction side of a pump, eccentric reducers should be used with the flat side turned up. Check for adequate support of piping around pump, to avoid vibration. The discharge piping should have a check valve. Pump packing should be installed to permit free motion of the pump, but preclude leakage. The pump motor should be explosion proof or weatherproof in accordance with specifications. Check oil sumps after operation.

Refrigeration Machines

Check superheat setting of expansion valves and for bulb and equalizer position in accordance with manufacturer's recommendations. Solenoid valves should be checked for vertical stem, correct direction of refrigerant flow, and manual disengagement of the manual opener. There should be an unobstructed view of the sight glass, and float valves or switches should be mounted level at a height which will ensure correct liquid level in the evaporator.

Before opening of the refrigerant drier cannisters, check for air tightness. The drier, if it is the replaceable type, should have the piping arranged to facilitate replacement, that is, a three-valve bypass.

Direct expansion coils should be installed in accordance with the manufacturer's recommendations, and pans of fan coil units should be protected against corrosion. There should be drain pans under all units to collect condensation.

During operation, check the operation of the operator pressure regulator under light load. On start-up, check the hold-back valve.

CHILLERS, COMPRESSORS, AND REFRIGERATION MACHINE CHECKLIST

1. Check nameplate data for conformance with specification.
2. Check piping system for cleanliness and strength prior to hook-up.
3. In start-up of reciprocating compressors, check oil suction and discharge pressures on start-up.
4. After brief start-up, recheck shaft alignment.
5. Check operation of low and high pressure oil cut-off switches.
6. Check that all control gauges and meters are installed before operational testing.
7. Check amount, type, and dryness level of charge.
8. Pump down and check pressure-holding capability.
9. Check compressor for vibration during operation.
10. After start-up, check suction strainer screen mesh for particles or foreign materials.
11. Check unloader action during operation.
12. Check compressor speed with stroboscope or RPM tachometer.
13. Check refrigerant flood back for foaming.
14. Check crankcase oil for foaming.
15. Check cylinder heads for overheating.
16. Check operation of automatic oil heater and crankcase.

Some of these checks are only appropriate for reciprocal machines, but most apply also to centrifugal.

15700 LIQUID HEAT TRANSFER

15701 Hot Water Heating Systems

Boilers for hot water generation are similar to steam boilers, operating at slightly lower temperatures. Hot water boilers still require ASME or appropriate pressure vessel inspection, and approval stamps.

The boilers should be assembled, if delivered in knock-down state, under the supervision of the manufacturer's representative. In any case, the unit should be boiled out with appropriate chemical compounds to clean all foreign material before going on line. All procedures which apply to steam boilers also apply to hot water boilers.

Circulating Systems

Hot water generators require circulating pumps for delivery of the hot water to terminal units and/or heat exchangers. These units should be placed carefully in accordance with manufacturer's directives. Units should be tested for smooth running, alignment, and cool run of bearings.

Check circulating pumps for lubrication, seals for leaks, packing adjustment and type, pressure retention, and correct rotation of the pump.

Terminal Units

Check radiant heating coils for placement, firm anchorage, and tightness under a hydrostatic test pressure of one and one-half times the operating pressure. These tests must be completed before the radiation is sealed within construction, or with the faceplates off where appropriate.

Be certain that balancing valves or orifices are placed in the return connection of each radiator or heating device. The contractor should balance the heating system as required by the plans or specifications.

Check all safety devices and temperature controls to assure safe and working order. Check safety blow-off valves and drain lines from them to ensure conformance with specifications.

Snow Melting

Hot water circulation systems are used in snow melting systems for crucial locations such as helicopter pads or driveways. Placement must avoid the attachment of dissimilar metals which may cause electrolytic action. Coils must be anchored securely so that they do not float when surrounded with wet concrete. Since the melting coils are encased, they should be carefully checked for material conformance with specification. In most cases, aluminum is not suitable for concrete encasement for this type of service. The usual material is steel pipe coated with plastic vapor barrier.

15800 AIR-DISTRIBUTION SYSTEMS

These systems heat incoming cold air to reduce the temperature differential within the final heat exchange device. Check for tightness of tubes or plates, and that dampers operate freely under all temperature conditions.

15810 Warm Air Furnaces

These furnaces direct the warm air which is forced through it by blowers, delivering the warm air either directly through duct work, or by mixing warm air and cold air in a dual system to provide tempered air. Units are usually packaged, and from the inspection standpoint, should be checked for conformance with specifications, and good workmanship of installation. The key to a successful operation in addition to the furnace itself is the filtration system, controls, and workmanlike installation of ductwork.

15820 Air-Handling Equipment

Fans are the principal media for handling of large air volume in buildings. Most are made of sheet metal casings, with a rotating impeller or turbine-like air moving device. Air pressures are usually measured in inches of water equivalent, and are essentially low pressure.

Special fans and air movers are utilized for exhaust systems, and special air circulation problems. Essentially the same air moving equipment can be used either to push air into a duct system, or to exhaust it.

The inspection team should be certain that all air-moving equipment has been approved well in advance of its requirement on the job, to permit delivery lead time. Check material delivered to the job for conformance with approved shop drawings, and ascertain that necessary nameplate information and test certificates are furnished as required by specifications. Check incoming equipment for damage in handling and require repair of minor damage, or reject heavily damaged units.

In checking the installation of fans and air handling units, check *rotation* of fan before permanent power connection is made.

Check the method of drive, including *alignment* of motor and fan. Be certain that motors are of approved type—for instance, explosion-proof, if required. Check the seals, sleeves and bearings provided. For lubricated-type bearings, be certain that fittings are provided for lubrication without dismantling fan or disconnecting the duct.

Usually, a fire safety switch is required in return air ducts of air circulation systems.

Check for pulley and belt alignment. Also, be certain that adequate guards are provided for the rotating equipment.

After start-up, check to be certain that equipment is operating without excessive noise or *vibration*. Approved vibration isolators and flexible connections should be installed where required.

For power roof ventilators, check the flashing at curbs to ensure watertightness. Discharged air should not be directed toward air intakes.

In gravity ventilators, examine the installation for weather-tightness, and make sure that units are oiled and properly adjusted. Check rotation of blades for freedom of movement.

Outside air *intake* dampers control the amount of air admitted into the system for supply fans. In some cases, exhaust air is allowed to mix with incoming air to temper it, raising its temperature in the winter, or cooling it in the summer.

Preheat coils, cooling coils, humidifier, air filters, and volume control dampers may be part of the ventilating unit. The unit may be installed on the foundation, or in some cases, may be suspended from the structure. Suspended units must be securely installed to avoid vibration, and a flexible connection usually separates the unit from the fan section.

Access doors are installed on the casing and ductwork for access to the equipment. Usually, the National Fire Protection Association Pamphlet 90-A, as well as the Sheet Metal and Air Conditioning Contractors National Association Standards for Ventilation Systems are commonly invoked as part of the specifications.

To control the *volume* of air in the duct system, splitter dampers may be provided. The damper plate pivots on a rod and is controlled by a control rod fastened to the front edge of the plate. Multiblade dampers may be used for multizone air distribution.

CHECKLIST FOR AIR-HANDLING EQUIPMENT

1. Check material upon receipt at job site for conformance with drawings.
2. Check rotation of fan before power connection is made permanent.
3. Check alignment of motor and fan.
4. Be sure motors are of approved type for area (i.e., explosionproof, dustproof, etc.).
5. Check for fire dampers and safety switches if required.
6. Check pulling and belt alignment before start-up.
7. Check for vibration or excessive noise after start-up.
8. Ascertain that appropriate access doors for maintenance are made available.

Air Filtration

In smaller air handling units, filter units of fiberglass or similar media are placed in removable cartridges or plates, so that they can be replaced easily by maintenance personnel.

In larger installations, complete *filter banks* may be installed. These are available in many forms and media, and the inspection team should ascertain that they are installed in accordance with the drawings and specifications and as per the manufacturer's recommendations.

Certain special applications to reduce the pollution to the environment of exhaust air or gases are available in the form of electronic *precipitators* and similar devices. In certain situations, the incoming air supply may be similarly treated to remove contaminants. These units do not actually filter, but remove particles through the use of charged electronic particles. These units are more complex, and should be checked for electrical operation, warning lights, interlocks, and other operating parts. Check ionizers for loose wires, sparking, and for free access for operation and maintenance.

For standard *filters,* check the thickness and method of mounting and supporting. Be certain that a proper amount of adhesive and washing tank has been provided for viscous medium-type filters. Inspect ceiling strips which prevent short circuiting of filter media. During the actual flow testing of the air handling system, ensure that the air stream is distributed with reasonable uniformity over the filter area.

Some filtration units have automatic *sprays* for washing, and these must be checked for coverage and operation. Filters with a traveling screen should be checked for operation, and oil change. On renewable roll media-type filters, check for tracking of the roll, media run-out switch, timer setting, static pressure control, and tension on the media.

In any filtration system, check to ascertain that clean filters have been installed upon completion of final tests. Filtration tests are usually performed on the basis of operating tests, with a measurable drop in air pressure required as a function of the dirt collection on the filters. Check the filter supplies and spares to be provided as part of the contract.

Humidity Control

Humidity control, either adding or removing moisture from the air, may be handled within the air-handling segment. This is particularly true where humidity is to be added. It is often added by injection of steam into the air stream, or where steam is not available by passing the air over a water surface.

The inspection team should check for special requirements for humidity control, and for the installation of the equipment in accordance with the specifications.

Air Distribution Duct Systems

Distribution of air, heated or cool, is through ductwork which may be rectangular, oval, square or circular in cross section. The size is determined by the designer, and is a function of the air supply or exhaust pressure, and the volume and velocity of air required.

In buildings where space is a major factor, *velocity* of the duct system is increased through the use of high-pressure velocity systems. The SMACNA high-velocity duct manual classifies high-velocity systems as follows: medium pressure for static pressure in duct up through 6″ water gauge; and high pressure, static pressure in the duct over 6″ water gauge through 10″ water gauge.

High-velocity duct is normally manufacturered in round cross sections where feasible.

To provide even air flow and reduce turbulence, *turning vanes* are used in elbows where designated. Shop drawings will indicate locations for the turning vanes. They will also indicate locations of *fire* and/or *smoke dampers*. These are vital to the safety of the operating installation, and one of the most important items for the inspection team to confirm. (Fire dampers are closed or exercised by use of a fusible link which usually has a temperature rating of about 50°F above the maximum temperature which would normally be encountered in the system.)

Workmanship and fabrication must be inspected for compliance by the inspection team. Observe *lock seams* and breaks in ductwork for cracking. Inspect all joint connections for proper type, and for neat finish. Duct should be smooth on the inside, and laps should be made in the direction of the flow of air.

The duct should have adequate *bracing* and reinforcement for larger sections. Elbow and curved radii should be checked for size. Slope ratio of transitions should also be checked. Any abrupt turn in the air flow usually requires a turning vane or deflector.

The duct *hangers* are also important, since the system may go into vibration if duct hangers are inadequate. The duct should have access to doors at all fire dampers, coils, heaters, filters, thermostats, or other items requiring servicing. These doors should be airtight, securely fastened and accessible. Gooseneck intakes should be turned away from prevailing wind.

Insulation or acoustic *liners* may be used—and in fact are used in the majority of ducts used for heating or cooling. The sample liner should comply with specifications, and should be attached to the duct with a required adhesive or mechanical fastener. Fire-resistive adhesives should be used inside of the ducts. Where acoustical attentuation is not required, insulation is usually attached to the exterior of the duct for ease of installation, as well as better economics, since exterior insulation does not cut down the size of the effective duct capacity. Exterior insulation requires a vapor barrier. Be particularly careful to insure that materials are fire retardant or noncombustible as required by the specifications. Where the duct passes through walls or floor, check for continuity of insulation. Where insulation may be subject to mechanical damage, ensure that appropriate protection has been provided.

Balancing and testing is the last phase of duct installation. All ducts, plenums, and casings must be thoroughly cleaned of debris and blown free of small particles and dust before supply outlets are installed. Equipment should be cleaned of oil, dust, dirt, and paint spots. All bearings should be lubricated, and belt tensions checked, as well as adjustments of fan pulleys. Have all fans and belt guards in place for testing. Before insulation starts, the duct should be tested for air tightness.

The contractor must provide necessary equipment for air flow measurements and development of coefficients of flow. The primary instrument used to measure pressure is the simple water gauge or manometer, which is essentially a U-loop in which difference of water heights can be measured directly.

The inspection team should review the proposed method for recording test data including a comparison to the design air flows. While the inspection team should closely monitor testing, it is usually the contractor's responsibility to actually conduct the tests. The tests should include testing for air quantities at each outlet, and final air flows should be recorded on the appropriate proved test forms. If actual air flows result in objectionable velocities or distribution, even though within design parameters, the inspection team should refer the problem to the design team.

During testing, all dampers should be checked for proper operation. Visual observations should be made through those access ports which provide a meaningful view of the operating equipment.

The optimum method of testing is through utilization of the actual flow or fan equipment. If, however, section by section testing is required—and the basic prime moving equipment is not ready—a temporary high-velocity testing source can be utilized. The inlet damper of the blower should be opened slowly until the duct pressure reaches about 8″ water gauge for medium-pressure ducts, or 12″ water gauge for high-pressure ducts.

DUCTWORK CHECKLIST

1. Check for size and material thickness (usually on delivery to job site, since ductwork is prefabricated).
2. Check for turning vanes where required.
3. Check for fire damper installation.
4. Check lock seams and breaks for cracking.
5. Check hanger system for stability.
6. Check that insulation or acoustical liners have been used where required.

7. Monitor the balancing and testing of the system.

8. During testing operations, check dampers for proper operation.

Package-Type Air Conditioners

For package units, check the high-pressure cutoff setting. Before start-up, check that the compressor hold-down bolts have been removed. The drip pan should be watertight and connected to an open drain, and check the water regulator valve if water is the cooling media.

Ascertain that air filters and strainers have been installed, and check the operation of the thermostat. Where gauges are available, check the suction and discharge pressures of the refrigerant compressors.

Cooling Towers

Cooling towers are normally package units furnished and erected by a manufacturer. Check the mechanical draft cooling towers for unobstructed air intake, fan rotation and speed, belt tension, stack fill, and weather protection of the motor. Open fan motors should not be allowed when totally enclosed or explosion-proof motors have been specified. Check that a water flow-through outlet does not form a vortex which draws air in with the water. Check operation of the water temperature control and drainage devices.

Check spray ponds for evenness of spray and for water drift. Ensure that provisions for adjustment of constant bleed have been made, and see that mist eliminators are installed when specified. Be sure that overflow and drain piping has been installed before start-up, and see that water is at an adequate level after operation.

15900 CONTROLS

Review all control installations for conformance with approved control shop drawings. Check valve operations for tight closing, and electrical equipment for interlocks and safety devices. Check for proper rotation of motors and flow of materials.

Determine that remote-operated valves are being actuated properly, and that flow control valves and other controllers function effectively.

Evaluate pneumatic systems for airtightness, restrictions caused by flattening of the tubing, and general cleanliness.

Inspect electronics systems for grounded shielded cable, location of amplifiers with respect to magnetic fields such as large transformers, and operational characteristics.

View graphic panels for damaged plastic, dirt between plastic and backplate, harnessing of control wires, and access for service to controls.

Verify control instructions including sequence of operations, and check each function of the controls when conducting final acceptance tests.

Make a record of all tests including who attended, methods and procedures used, and results and conclusions.

Before tests are scheduled, be certain that the contractor has available proper tools, equipment and instruments, particularly gauges which are certified and pretested. Make sure that strainers and gauges are clean immediately prior to tests of piping systems.

Be certain that instructions for operating equipment are posted, and that operating personnel are instructed in the proper operation and control of equipment.

CONTROL SYSTEM CHECKLIST

1. Ensure that all material and controls are in accordance with specifications in shop drawings.

2. Check valve operations for tight closing.

3. Air test systems for leakage.

4. Check manual operation of valves.

5. Check remote operation of valves.

6. Check tubing system for any pinches or restrictions.

7. Check electrically or pneumatically that remote connections operate the proper valve or control.

8. Check the graphic panels for any damaged plastic, dirt, or inaccessibility of controls.

9. Record all tests.

DIVISION 16: ELECTRICAL

LEONARD P. SCHAEFER, P.E., AUTHOR

NUMBER	TITLE
16050	**BASIC MATERIALS AND METHODS**
16051–16109	(Reserved)
−110	Raceways
−111	Conduits
−112	Bus Ducts
−113	Underfloor Ducts
−114	Cable Trays
−120	Wires and Cables
−121	Wire Connections and Devices
−125	Pulling Cables
−130	Boxes
−131	Pull and Junction Boxes
−132	Floor Boxes
−133	Cabinets
−134	Outlet Boxes
−140	Wiring Devices
−150	Motors
−155	Motor Starters
−156	Large Motor Protection
−160	Panelboards
−162	Panelboard Switches
−163	Distribution Panelboards
−164	Branch Circuit Panelboards
−170	Motor and Circuit Disconnects
−180	Overcurrent Protective Devices
−181	Fuses
−182	Circuit Breakers
−190	Supporting Devices
−199	Electronic Devices
16200	**POWER GENERATION**
−01	Starting Equipment
−02	Alarms and Instrumentation
−10	Generators
−11	Hydroelectric Impulse Turbine Generators
−12	Hydroelectric Reaction Turbine Generators

−13	Steam Engine Generators
−14	Steam Turbine Generators
−15	Nuclear Turbogenerators
−16	Internal Combustion Generators
−17	Gas Turbine Generators
−22	Solar Engines
−30	Cooling Equipment
−40	Exhaust Equipment
−50	Automatic Transfer Equipment
16260−16269	(Reserved)
−70	Solar Power and Wind Power Generation
−71	Photovoltaic Power Generation Systems
−72	Wind Power Generation Systems

16300	**POWER TRANSMISSION**
−01	Vaults
−10	Packaged Substations
−15	Power Centers
−20	Transformer and Shunt Reactors
−30	Capacitors
−40	Insulator and Lightning Arrestors
−50	Circuit Breakers
−51	Oil Circuit Breakers
−52	Air Circuit Breakers
−53	Gas Circuit Breakers
−54	Vacuum Circuit Breakers
−60	Disconnect Switches and Fuses
−70	Rectifiers and Frequency Changers
−80	Converters

16400	**SERVICE AND DISTRIBUTION**
−01	Overhead Electric Service
−02	Underground Electric Service
−03	Emergency Electric Service
−10	Power Factor Correction
−20	Service Entrance
−21	Lightning Arrestors
−30	Metering
−31	Relay Protection
−40	Service Disconnect
−41	Primary Load Interrupters
−50	Grounding
−55	Ground Fault Protection
−60	Transformers
−70	Distribution Circuits
−71	Feeder Circuits
−72	Branch Circuits

16500	**LIGHTING**
−01	Lamps
−02	Ballasts and Accessories
−03	Poles and Standards
−10	Interior Building Lighting
−11	Fluorescent Fixtures
−12	Incandescent Fixtures
−13	Mercury Fixtures
−14	High Intensity Fixtures
−15	Luminous Ceilings
−16	Signal Lighting
−20	Exterior Building Lighting
−30	Site Lighting
−31	Lighting Bollards
−40	Stadium Lighting
−50	Highway and Roadway Lighting
−60	Aviation Lighting
16570–16599	(Reserved)
16600	**SPECIAL SYSTEMS**
−01	Lighting Protection Systems
−10	Emergency Light and Power Systems
−11	Uninterruptible Power Supply Systems
−12	Generator Systems
−13	Packaged Battery Systems
−14	Central Battery Systems
−40	Cathodic Protection
−50	Electromagnetic Shielding
16700	**COMMUNICATIONS**
−20	Alarm and Detection Systems
−21	Fire Alarm and Detection Systems
−22	Electrical Current Alarms
−25	Smoke Detector Systems
−26	Medical Gas Alarm Systems
−27	Burglar Alarm Systems
−30	Clock and Program Systems
−40	Telephone Systems
−60	Intercommunication Systems
−61	Nurse Call Systems
−70	Public Address Systems
−80	Television Systems
−81	Master TV Antenna Equipment and Systems
16850	**HEATING AND COOLING**
−58	Electric Snow Melting Cables and Mats
−59	Electric Heating Cables
−60	Electric Heating Coils

−65	Electric Baseboard
−70	Packaged Room Air Conditioners
−80	Radiant Heaters
−81	Duct Heaters
−90	Electric Heaters

16900	**CONTROLS AND INSTRUMENTATION**
−10	Recording and Indicating Devices
−15	Relay Systems
−20	Motor Control Centers
−30	Lighting Control Equipment
−40	Electrical Interlock
−50	Electric Heating Controls
−60	Limit Switches
−61	Sensing Devices
−62	Level Switches
−63	Flow Switches
−64	Proximity Switches
−70	Valve Operators
−80	Multiplexing
−90	Cabinets
−91	Control Cabinets
−92	Instrument Cabinets
−30	Power Generating and Transmission
16950–16999	(Reserved)

Electricity is a form of energy and is useful when it is controlled and transmitted; it changes a project from a static structure to a functioning facility. The voltage is lowered at the consumer's end of the system, where the voltages are usually between 13 kV and 4.16 kV. The subfeeders may be three-phase or single-phase, where the single-phase voltage is the three-phase voltage divided by 1.73. The consumer's substation may contain one or more voltage step-down power transformers, and may be part of the facility or owned by the supplying utility. The standard nominal voltage levels at the substation could be 115 V, 230 V, 227 V, 480 V, or 550 V. Within the facility the equipment includes metering feeder cables, busways and conduit, switchgear, panelboards, circuit breakers, motor control centers, device wiring connections, and all similar equipment for control and safety. See Fig. 16-1.

16025 Codes and Fees

The electrical work must conform to the National Electric Code of the National Fire Protection Association (latest applicable, NFPA 70-1981)

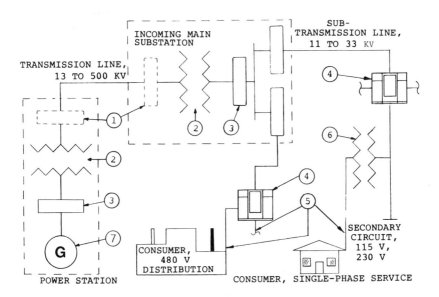

1. OPTIONAL LOCATION OF CIRCUIT BREAKER

2. POWER TRANSFORMER

3. CIRCUIT BREAKER

4. SUBSTATION

5. PRIMARY FEEDERS 2.4 TO 13.8 KV RANGE

6. DISTRIBUTION TRANSFORMER

7. GENERATOR, UP TO 13.8 KV

Fig. 16-1 Basic Electric System

and to any state, county, municipal, or other authority, laws, rules, or regulations having jurisdiction. On completion of the work, a local Department Association of Fire Underwriters Certificate of Inspection is furnished and fees are paid for by the electrical contractor.

16030 Tests

When testing electrical equipment, advise the electrical design engineer and, if possible, have him present for major tests. Unless otherwise specified, testing should be performed by the contractor with the inspector present. For tests of major equipment and high-voltage cables, it may be appropriate to notify the manufacturer to witness tests.

Using agencies or owners should be notified when any unusual testing is to be performed in case they care to be present, particularly if the facility utilizes a common power source with the owner. Precautions should be made to ensure that test voltages are applied only to equipment or circuits under test and that instruments and control circuits, or other associated equipment, are disconnected during the test.

Electrical tests should not be conducted under unsuitable ambient conditions, such as excessively high humidity. Records of tests should be complete, including ambient conditions, equipment tested, extent of test, names and serial numbers of equipment involved, and signatures of those witnessing the tests.

Deenergized operational testing should determine that moving parts are free and unobstructed. Lubrication should also be checked. Operational testing shows that the equipment performs all functions; continuity testing determines that circuits are continuous throughout the equipment.

High-potential testing should determine that the insulation has sufficient insulational strength to withstand surges to which it might be subjected and to ensure that it is free of pinholes or other possible leakage points.

Operational tests of the motor starter should include manually operating the armature or plunger and contact bar to determine that movement is free, contacts are in alignment, contact pressure is adequate, and auxiliary contacts function properly. The starter should be energized from all control points and the operation of all control circuit interlocks checked.

Reduced-voltage starters should be checked for correct sequence and timing of application of incremental and full voltages. Variable and adjustable motor speed controls should be checked to see that operating speeds correspond to the position of the speed control device.

Protective relays should be checked to see that time and current settings have been made as specified. Operational tests of relays should include checking of operation at specified current or voltage and time values.

Special systems for intercommunications, paging, etc., should be given operational tests at all operating points to demonstrate that they will perform all specified functions.

It should be demonstrated that sounding devices are audible under normal ambient sound level conditions in areas for which coverage is specified, and that "off" signals cannot be transmitted over fire alarm systems specified to be noninterfering.

Emergency battery units and power supplies should be tested first manually and then in automatic start.

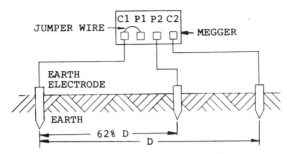

Fig. 16-2 "Fall of Potential" or "Three-Terminal" Earth Resistance Test

1. Grounding tests shall be required for grounding systems for equipment, electrical systems, data handling networks, and earth electrodes for lightning rod systems. Tests shall be made 48 hours or more after a rainfall and shall not exceed 25 Ω in accordance with NEC 250-84, or shall not exceed 5 Ω for major substations. Tests shall be made using the J. G. Biddle "Megger" in the "Three-Terminal" earth resistance test (see Fig. 16-2), where P_2 and C_2 are auxiliary electrodes.

2. The "Hi Pot," or high-potential, testing of cables shall be done in accordance with IPCEA and ASTM procedures for cable type and voltage. The following tests shall be performed and the results recorded: Insulation test with a Megger and Standard High-Voltage dc field test using a Kenotron Test Set, where maximum voltages are determined by the owner.

3. Cathodic Protection (CP) corrosion-prevention systems for underground metallic structures shall be measured to insure proper polarity and adequacy of protection. The measurement is made with a high-resistance dc voltmeter (20,000 Ω/V), where the structure is connected to the negative (−) terminal and a copper–copper sulfate half cell (placed at remote earth) is connected to the positive (+) terminal. If the CP system can be turned off, the change in voltage (off to on) should be −0.3 V. Otherwise, the potential with the CP system working properly should be 0.9 V for a steel structure.

16040 Identification

All electrical equipment such as switches, starters, control centers, distribution panels, junction boxes, substations, etc., shall be properly identified with metallic or plastic nameplates permanently fastened to the equipment. Equipment enclosing high-voltage components should be labeled with permanent red "danger" signs stating the voltage level.

16050 BASIC MATERIALS AND METHODS

Contract Drawings. The electrical contract includes the furnishing of all labor, equipment, material, and services pertaining to the electrical work, but may not include elevators, escalators, etc., and electrical components of the HVAC or plumbing systems other than those involved in supplying feeders and making connections. The following is a typical list of the electrical work to be included:

1. Plot plan
2. Single line
3. Substation
4. Underground conduits and details
5. Above-grade service entrances, conduits and details
6. Equipment layouts and details
7. Lighting layouts and details
8. Power distribution risers and details
9. Offices and details
10. Grounding and lightning protection
11. Corrosion control
12. Public address and telephone systems
13. Fire alarm systems
14. Emergency systems
15. Control and Instrumentation

16110 Raceways

Raceways is a general term applied to all types of enclosures providing space, support, and mechanical protection for electric conductors transmitting power or control between various units of electrical equipment. A function of equal importance is the protection of life and property during normal and electrical-fault conditions.

Fig. 16-3 Raceway System

A raceway system may consist of conduits, tubing, underfloor ducts, cable trays, and bus ducts. The number and size of conductors in a raceway are controlled by the National Electric Code (NEC), article 300-17. The conductor fill and sizes for conduit are covered in the NEC Chapter 9 table section. See Fig. 16-3.

Underfloor ducts are covered in article 354 of NEC, while cellular metal flooring raceways are covered in article 355. Cable trays are utilized in electrical-mechanical equipment rooms and in industrial plants where electrical distribution flexibility is a requirement and article 318 of the National Electric Code covers the allowable cable fill and grounding requirements.

16112 Busways

A busway is an electrical conductor system with its housing or covering built integrally into a compact system. A common version is the plug-in feeder busway designed to transmit power from one point to another and providing tap-off points with switch or circuit breakers. They are available in ratings of 600 A to 3000 A and up to 6500 A, at 600 V.

Busways are large conductors made of solid material such as copper or aluminum, with the capability of carrying large quantities of high-voltage electrical power. The bars are heavy in cross section, and may be insulated or bare. Busways are assembled by bolting, and a variety of special fittings or plug-in switches and taps are available. The inspection team should ascertain that busways are supported at the proper intervals and that sway braces are installed when needed to limit lateral movement. Busway runs should be installed in straight alignment, parallel to floors and walls with sufficient space above, below, or on the side to permit the installation, operating, and servicing of bus plugs. Be sure to check that the types furnished are in accordance with the specification for conductor metal, enclosure type, duct type, wall flanges, and fire stops. Bus duct housing must be grounded in accordance with NEC 250-33. Vertical riser sections up to 6' above the floor must be of the unventilated type (NEC 364-4).

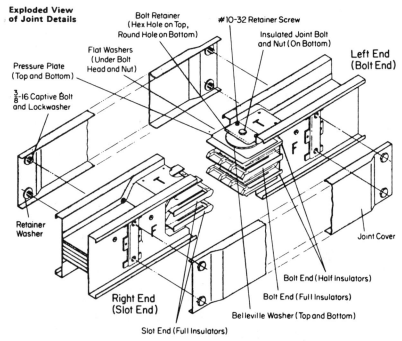

Fig. 16-4 Joint Details of a Busway

Component sections should be legibly marked with their voltage and current ratings. An exploded view of the joint details of a busway are shown in Fig. 16-4.

16120 Wires and Cables

The primary function of wires and cables is to carry energy between the source and the ultimate equipment. When this energy is carried heat is generated; this heat loss must be dissipated, depending on the current rating of conductors and how they are installed.

Conductor selection requires consideration of the source capacity and voltage drops, as well as of fault levels and clearing times. Insulation selection depends on normal operating temperatures, flexibility, and fire resistance. Care must be taken in pulling to avoid stretch and rupture on long pulls and around bends.

The two conductor materials are copper and aluminum. Aluminum should not be substituted for copper without consulting the engineer involved, but should be utilized for sizes #6 AWG (American Wire Gauge) and larger. Aluminum requires special handling and proper terminations. It is softer than copper and has a low yield strength. An aluminum oxide film forms immediately, which is insulating, so a no-oxide flux is required to provide a good conducting surface. The NEC requires all cables in raceways larger than #8 AWG to be stranded. A typical cable for use in conduit systems is shown in Fig. 16-5.

The basic insulating materials are thermosetting, thermoplastic, paper and varnish cloth laminates, and mineral insulation. Heat resistance and heat aging are primary, with ozone, corona, and moisture resistance being the considerations. The paper-insulated, shielded cables can handle the 10-kV to 69-kV voltage class, while polyethylene (natural or cross-linked) can handle the 5-kV to 35-kV voltage class.

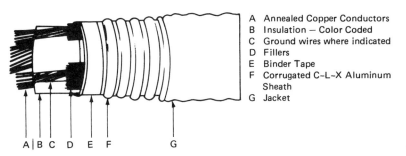

A   Annealed Copper Conductors
B   Insulation — Color Coded
C   Ground wires where indicated
D   Fillers
E   Binder Tape
F   Corrugated C–L–X Aluminum
     Sheath
G   Jacket

A | B C   D   E   F   G

Fig. 16-5 Cable for Conduit Systems

A good "rule of thumb" for branch feeders when considering the voltage drop problems is to increase the wire one size for the following currents and distances:

Amperes	Wire Size #	Increase in Distance (feet)
10	12 to 10	400 to 620
20	10 to 8	325 to 500
40	6 to 4	420 to 620
100	1 to 1/0	430 to 510
150	3/0 to 4/0	410 to 475
230	350 to 400	670 to 700

16121 Wire Connectors and Devices

Wire connections are made using two methods, thermal and pressure. Thermal methods, other than solder or brazing, utilize a thermite-type welding process, commonly called "cad welding." This is suitable for connections to ground rods and ground mats and for attaching connectors to insulated power cables. The pressure connector is either the mechanical type, such as a circuit breaker terminal lug, or the compression type, where a special die and tool indents a surrounding tube by applying crimping to obtain a lasting contact between connector and conductor. See Fig. 16-6.

Connections for aluminum to aluminum and aluminum to copper

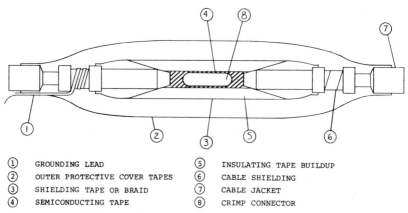

①	GROUNDING LEAD	⑤	INSULATING TAPE BUILDUP
②	OUTER PROTECTIVE COVER TAPES	⑥	CABLE SHIELDING
③	SHIELDING TAPE OR BRAID	⑦	CABLE JACKET
④	SEMICONDUCTING TAPE	⑧	CRIMP CONNECTOR

Fig. 16-6 Typical Taped Splice in Shielded Cable

require special handling. Aluminum forms a light protective oxide which has a relatively high electrical resistance. Aluminum also has a higher thermal coefficient of expansion than copper Corrosion is another problem where the galvanic cell of Al-Cu, with moisture, can cause the anodic aluminum to corrode. Special practices and procedures have been worked out for this, as shown in Fig. 16-7.

Medium- and high-voltage cables require shielding because of the danger from corona or ionization. The shielding provides symmetrical stress distribution, protects the cable from induced voltages, and increases the safety of human life. The shield must be grounded at each splice or termination in cases where stress relief cones are used to relieve the voltage stress at the termination points in switchgears and at splices in junction boxes and manholes. See Fig. 16-8.

16140 Switches and Receptacles

Light-duty, general-use snap switches are for lamp loads and for resistive, inductive, and single-phase appliance and motor loads at 120 V, generally rated 20 A, where the motor loads shall not exceed 80% of the ampere rating (see NEC 380-14).

Duplex outlet receptacles are rated 15 A, 125 V, where normally five outlets are circuited on one branch circuit home-run of 100 feet or less using No. 12 wire on a 20-A breaker (see NEC 410-L).

The National Electric Code, NEC 210-8, requires ground fault circuit interrupters (GFCI) in bathrooms, in garages, and in outdoor receptacles

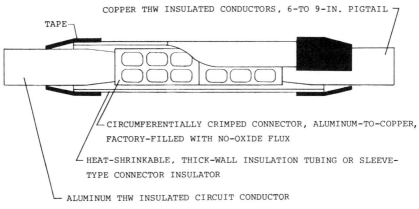

Fig. 16-7 Aluminum-to-Copper Termination Pigtail

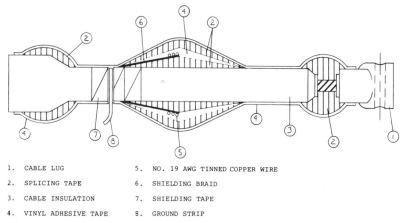

1.	CABLE LUG	5.	NO. 19 AWG TINNED COPPER WIRE
2.	SPLICING TAPE	6.	SHIELDING BRAID
3.	CABLE INSULATION	7.	SHIELDING TAPE
4.	VINYL ADHESIVE TAPE	8.	GROUND STRIP

Fig. 16-8 Insulated Cable Termination with a Stress-Relief Cone

at grade level, near swimming pools, and on construction sites where outlets are not part of the permanent wiring. These types of receptacles are used to save lives, although they also protect equipment and help to prevent fires. The devices are required under these general categories for nursing homes, hospitals, etc. See the receptacle types shown in Fig. 16-9. GFCI circuit breakers are available for one- or two-breaker positions to fit panelboards.

16155 Motor Starters

The specialty equipment to be installed during construction which is electric motor driven will be delivered to the site with the motors as integral parts of the equipment. The motors have been sized and selected for the environmental requirements involved. Also, heating, ventilation, air conditioning, refrigeration, pumping, elevators, and conveyors will involve basic controls which include automatic or programmed devices.

The pilot device control signal causes the motor starter to respond, which in turn energizes the motor and provides the required motor protection. Most motors are induction squirrel-cage, three-phase ac, and low voltage and range from 1 hp for a size 00 starter to 1600 hp for a size 9 starter at 480 V. Horsepower ratings are assigned by NEMA for voltages of 601 V to 5000 V.

Most motors are full-voltage across-the-line starting motors and require only that the starter connect the motor directly to the distribution system.

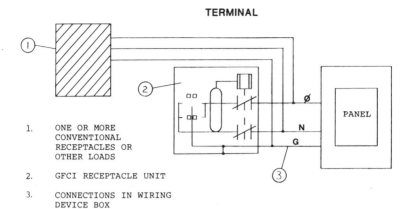

TERMINAL

1. ONE OR MORE
   CONVENTIONAL
   RECEPTACLES OR
   OTHER LOADS

2. GFCI RECEPTACLE UNIT

3. CONNECTIONS IN WIRING
   DEVICE BOX

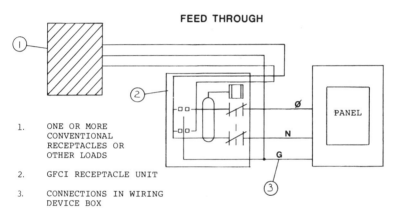

FEED THROUGH

1. ONE OR MORE
   CONVENTIONAL
   RECEPTACLES OR
   OTHER LOADS

2. GFCI RECEPTACLE UNIT

3. CONNECTIONS IN WIRING
   DEVICE BOX

Fig. 16-9 Receptacle Types

The starting current inrush will be 5 to 6 times rates current at a lagging power factor of 35% to 50%.

Large motors for refrigeration and air compressors may provide inrushes of current that result in voltage dips not tolerated by the electric utility, not to mention the building lighting system. This problem, if not resolved at the engineering design stage, may require some type of reduced-voltage starting. The common configurations may be reactor or resistor starting, autotransformer, star-delta controller, part-winding, and series parallel controller.

Short-circuit protection is provided for the motor with a circuit breaker or fused disconnect ahead of the starter; when both are in a common enclosure this is known as a combination starter.

If there is a short circuit in the motor or leads, the interrupt rating of the breaker or fuses must clear the fault. If it does not, high-capacity current-limiting fuses for the system fault current will be required.

Overload protection of the motor is required, which generally consists of thermal elements. When the load current heats the elements to a preset point the overload relay contacts will disconnect the motor from the line. The heater coils are matched to the motor's full load current.

Fire pump controllers must comply with the requirements of Underwriters Laboratories and the National Board of Fire Underwriters. Such controllers may provide either manual operation or manual and automatic operation, the latter of which will put the pumps in operation with a pressure switch or a flow switch. (Note: The starter and circuit breaker are sized to handle locked-rotor current without tripping.)

16156 Large Motor Protection

Reduced Voltage

Star Delta Type

Star-Delta type starters have been applied extensively to industrial air conditioning installations because they are particularly applicable to starting motors driving high inertia loads with resulting long acceleration times. They are not, however, limited to this application. When six or twelve lead delta-connected motors are started star-connected, approximately 58% of full line voltage is applied to each winding and the motor develops 33% of full voltage starting torque and draws 33% of normal locked rotor current from the line. When the motor has accelerated, it is reconnected for normal delta operation.

Class 11-800 and 11-890 starters are suitable for air conditioning application, provided the motors used are open type and horsepower rated.

Autotransformer Type

Autotransformer type starters are the most widely used reduced voltage starter because of their efficiency and flexibility. All power taken from the line, except transformer losses, is transmitted to the motor to accelerate the load. Taps on the transformer allow adjustment of the starting torque and inrush to meet the requirements of most applications.

Closed transition is standard on all sizes assuring a smooth transition from reduced to full voltage. Since the motor is never disconnected from the line there is no interruption of line current which can cause a second inrush during transition.

Part Winding Type

Part winding starting provides convenient, economical one-step acceleration at reduced current where the power company specifies a maximum, or limits the increments of current drawn from the line. These starters can be used with standard dual-voltage motors on the lower voltage and with special part-winding motors designed for any voltage. When used with standard dual-voltage motors, it should be established that the torque produced by the first half-winding will accelerate the load sufficiently so as not to produce a second undesirable inrush when the second half-winding is connected to the line. Most motors will produce a starting torque equal to between $^1/_2$ to $^2/_3$ of NEMA standard values with half of the winding energized and draw about $^2/_3$ of normal line current inrush.

16160 Panelboards

Electric distribution systems include panelboards using fuse or circuit breaker devices or both. They are for lighting/appliance loads and for power-distribution loads. Lighting/appliance panels have more than 10% of their over-current devices rated 30 A or less (see NEC 384-14) where neutral connections are provided; the number of over-current devices is limited to a maximum of 42 in any one box (see NEC 384-15). Power-distribution panelboards handle the heavy power loads, and the 42 over-current device limitation does not apply to them. Ratings are single phase, 2 or 3 wire, three phase, 3 or 4 wire; 120/240 through 600 V; 50-1200 A, and 1200 A maximum branch.

The six-circuit rule, NEC 230-71, provides that a service-disconnecting device for each subset of not more than six main disconnecting devices is provided mounted in a single enclosure, a group of separate enclosures, or a switchboard. Also, be aware of the 30-conductor rule, NEC 362-5, which limits wireways to not more than 30 current-carrying conductors.

16300 POWER TRANSMISSION

16310 Substations

The outdoor structural-type substation is required for incoming voltages of 23 kV and 33 kV and above. The substation may be owned by the utility or the customer, depending on the power contract; the latter will result in a more favorable rate schedule. A substation of any type consists of an enclosure, switching equipment, metering, grounding, lightning equipment, and power transformers. Unit substations will convert a 13.2 kV signal to 4160 V or 2400 V, or the secondary type will transform the above primary range to 600 V or less. A typical substation is shown in Fig. 16-10.

16320 Transformers

The transformer, as the basic unit of the substation, is used to change a voltage level from the utility distribution voltage to a voltage usable within the building. Transformer ratings are as indicated in the following table:

Preferred Kilovolt-Ampere Ratings

Single-Phase	Three-Phase
3	9
5	15
10	30
15	45
25	75
37.5	112.5
50	150
75	225
100	300
167	500
250	750
333	1000
500	1500
833	2000
1250	2500
1667	3750
2500	5000
3333	7500
5000	10000

The transformer cooling medium may be oil, air, or gas. (Note: askarel, a PCB material, is not an acceptable coolant.) Transformers in substations

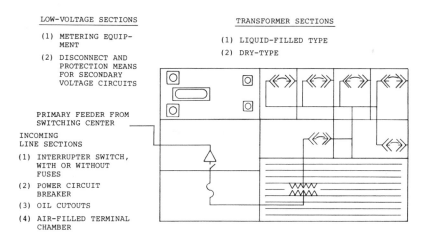

LOW-VOLTAGE SECTIONS

(1) METERING EQUIP-
    MENT

(2) DISCONNECT AND
    PROTECTION MEANS
    FOR SECONDARY
    VOLTAGE CIRCUITS

PRIMARY FEEDER FROM
SWITCHING CENTER

INCOMING
LINE SECTIONS

(1) INTERRUPTER SWITCH,
    WITH OR WITHOUT
    FUSES

(2) POWER CIRCUIT
    BREAKER

(3) OIL CUTOUTS

(4) AIR-FILLED TERMINAL
    CHAMBER

TRANSFORMER SECTIONS

(1) LIQUID-FILLED TYPE

(2) DRY-TYPE

Fig. 16-10 Power Transmission Substation

are used with the secondary connected to a low-voltage switchboard at a voltage range of 208 V to 600 V. The taps, usually at $\pm 2\frac{1}{2}\%$ (two above and two below), are manually adjusted.

Transformers may be pad-mounted outside a building where a conventional substation might be inappropriate. These are tamperproof and do not require fencing. The high-voltage connection is in an air terminal chamber with a disconnect and/or fuses. An indoor distribution transformer would be similar to that shown in the sketch is section 16310; these are normally air cooled.

Several important characteristics should be considered by the field inspector in addition to the basic kVA and voltage ratings. These are the presence of voltage taps and, whether manual or automatic, load changing, voltage insulation class and basic impulse level (BIL), grounding, and sound level.

Oil-insulated transformers installed indoors shall be installed in a vault (see NEC 450-25). Specific construction requirements for vaults, such as ventilation, doors, fire protection, locks, must be met (see NEC 450-41 to 450-48).

16330 Capacitors

The rate structures of most utilities have power-factor clauses which result in a surcharge when the power factor is below a certain level, usually 95%.

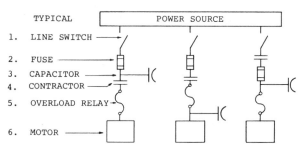

Fig. 16-11 Possible Location of Motor Capacitors for Power Factor Improvement

The poor power factor is usually the result of induction motor equipment; however, other benefits of improved power factor may pertain, such as better voltage regulation and reduction of electric losses. This can result in a gross return of as much as 15% of the investment in capacitors. How many kva are supplied by the capacitors to improve induction motor power factor is usually a design engineering function; however, it may be necessary for the field inspector to be involved in the location of the equipment. A possible location of motor capacitors for power factor improvement is shown in Fig. 16-11. Capacitor ratings, discharge, grounding, and switching are covered in NEC article 460.

16350 Switchgear

Metal-enclosed load interrupter switches are used to supply the transformer substation and may be applied outdoors, in vaults, as well as indoors. They may also be combined with bus connections, instrument transformers, and control wiring.

Standard voltage ratings are 4.16 kV and 13.8 kV, with the main bus rated 100 A, 1200 A, or 2000 A. Power fuses have interrupt ratings to 270 MVA at 4.16 kV.

16400 SERVICE AND DISTRIBUTION

16401 Electric Service

Incoming lines and service laterals should generally be installed, maintained, and owned by the electric utility if qualified personnel are not normally available in industrial/commercial operations. An overhead service lateral terminating at a building should have a ground clearance of 20 ft at a span length of 250 ft or less for voltages of 750 V to 15,000 V. Installation of open-wire lines over buildings should be avoided as poor practice.

16402 Underground Service

Underground cable systems are used for direct burial for lighting, for example, or in conduit where local ordinances may require this in congested areas. The inspector should make sure spare ducts are provided for further expansion or in case of cable trouble, if required.

Conduit shall be laid to pitch and shall drain toward manholes. Away from the building, conduits must be thoroughly cleaned before cables are installed. During construction, and before conduit is installed, open ends must be plugged and spare conduits sealed with tapered wooden plugs.

Electric manholes and handholds are constructed of reinforced concrete, with a cast-iron frame and a cover labeled "Electric." Pulling irons must be provided, as well as cable support racks of noncorrosive material and porcelain insulators. Where more than one set of cables pass through the manhole, all cables shall be adequately fireproofed. Cables shall be tagged for circuit number and phase, and cable sheath grounding shall be provided as required.

16420 Service Entrance

Service-entrance equipment is that part of the system between the customer-owned service equipment and the utilities' supply feeders. This consists of one or more circuit breakers, switches, or fuses, and metering equipment. The billing meters are owned and maintained by the utility, and the metering transformers are usually furnished by the utility and installed by the contractor.

Where service continuity is important, for hospitals for example, multiple or standby service with load-transfer capacity may be required. More than one voltage may be required; tall buildings or building complexes occupying a large area, such as airports, will require multiple services. All of these may require on-site negotiations and reappraisals by the inspection and design engineers.

Several points to be considered and settled by the inspection engineer are 1) awareness by the utility of the complete load characteristics of the building, 2) physical and mechanical requirements of the service entrance, and 3) schedule data: when temporary construction service is required, when service will be required for tests, when full estimated load will be required.

16430 Metering

A single-occupancy building such as a school, a hospital, or a single-tenant office building will be metered by the utility at the service entrance with a watt-hour–demand meter. A multiple-service building may be metered at the several service entrances, with special consideration on watt-hour demands by the utility. Multiple-occupancy apartment and office buildings will have a house meter and separate tenant metering. Also, some localities may permit "submetering," where the owner buys power and legally tenant-meters at a prescribed rate. It is important to provide flexibility in metering, as tenant changes will occur.

Subject to agreement by the utility, meters may be installed indoors, in a separate control house, or in a meter room. Outdoor locations may include pole mounting and wall attachment, generally where accessibility and protection are provided. Where instrument transformer metering is required, usually for large stores, commercial buildings, or industrial plants, the owner may be required to install the instrument transformers. The meter connections, C.T. and P.T. ratios, and meter constants are all required as part of the contractor-furnished information.

16450 Grounding

Electrical equipment is grounded for two main reasons: the system is grounded for operating reasons, and non-current-carrying metal parts are grounded for the safety of personnel. The National Electric Code, article 250, covers the specific requirements. Table 250-94 provides the grounding conductor size for ac systems, and Table 250-95 provides the grounding conductor size for raceways and equipment enclosures. Figure 16-12 illustrates the distinction between the two types of grounding conductors.

The resistance measured at a grounding point, such as a driven rod or rods, a water pipe, or any underground metal system, shall be less than 25 Ω 48 hours after the last rainfall (NEC 250-84; see section 16030 for test procedure). Note that cast-iron water pipe generally have high-resistance joints, which it may be impractical to jumper with cad-welded bonds. A grounding system for a building complex is shown in Fig. 16-13.

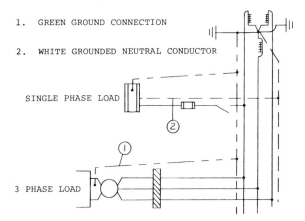

1. GREEN GROUND CONNECTION

2. WHITE GROUNDED NEUTRAL CONDUCTOR

SINGLE PHASE LOAD

3 PHASE LOAD

Fig. 16-12 Two Types of Grounding Conductors

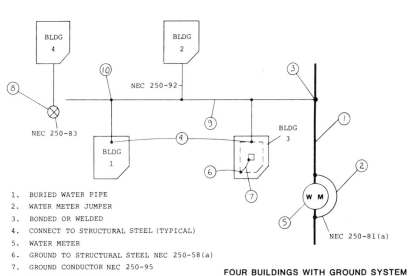

1. BURIED WATER PIPE
2. WATER METER JUMPER
3. BONDED OR WELDED
4. CONNECT TO STRUCTURAL STEEL (TYPICAL)
5. WATER METER
6. GROUND TO STRUCTURAL STEEL NEC 250-58(a)
7. GROUND CONDUCTOR NEC 250-95
8. REMOTE "MADE" GROUND
9. MAIN BURIED GROUND NEC 250-91
10. COPPER - WELDED CONNECTION (TYPICAL)

FOUR BUILDINGS WITH GROUND SYSTEM

Fig. 16-13 Building Ground System

16500 LIGHTING

16510–14 Interior Lighting Fixtures

The successful operation of every commercial building depends on lighting, and in fact this may be the first electrical impact for the owner or tenant. The lighting fixture pattern may supply the right foot-candle level on horizontal surfaces 3 ft above the floor; however, the inspector is alerted here to special tasks or uses which may cause the lighting to appear inadequate. For example, vertical shelving in a high bay warehouse, vertical pigeon holes in a post-office sorting room, or machine tool equipment may require owner-installed special fixture lighting to be supplied later with the furnishings.

Incandescent and fluorescent lamps are used in the normal commercial areas, along with the special-use lighting of the high-intensity-discharge (HID) group, including mercury, metal-halide, and high-pressure sodium lamps. Fluorescent lamps, called rapid start (RS) and slimline, are

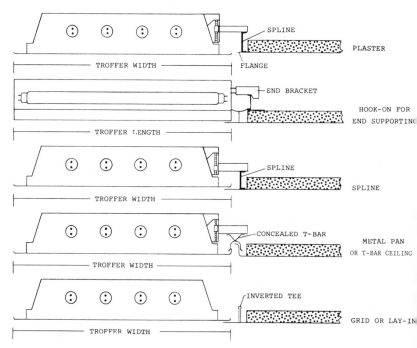

Fig. 16-14 Five Basic Types of Recessed Fluorescent Fixtures

available in the following loadings: normal (430 mA), high-output (800 mA), and extra-high-output (1500 mA).

Lamp operating voltages must be checked. Incandescents are normally 120 V, although high-voltage lamps are available for 230-, 250- and 277-V systems. Fluorescent-lamp ballasts are supplied for 120 V, ±10%, and fluorescent and HID-lamp ballasts are for 277 V (make sure wall switches are approved for 300 V). HID-lamp ballasts are also available at 480 V for these circuits. Check ballast sound levels and make sure ballasts, fixtures, louvers, reflectors, and baffles are all securely mounted to prevent noise amplification.

Mounting types for fluorescent fixtures, incandescent fixtures, and concealed lighting are shown in Figs. 16-14, 16-15, and 16-16, respectively.

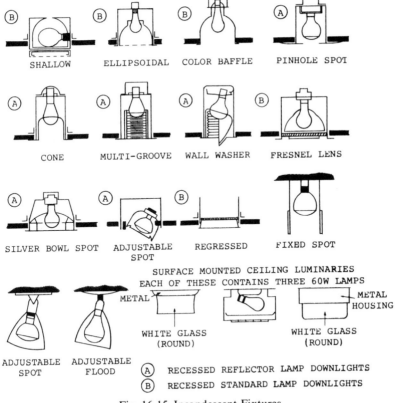

Fig. 16-15 Incandescent Fixtures

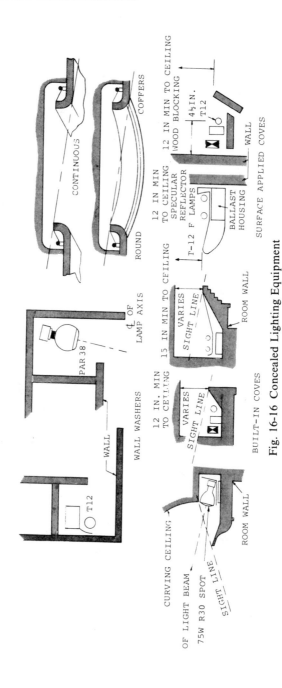

Fig. 16-16 Concealed Lighting Equipment

16530 Exterior Lighting

Exterior lighting for security or architectural esthetics at loading and unloading platforms, streets, parking lots, ballparks, and hazard locations will require fixtures suited to the design. Generally, however, a parking lot will have a 1-2 foot-candle level and a loading or unloading platform a level of 20 foot-candles. The HID lamps, generally mercury vapor or HPS (High Pressure Sodium), are the most suitable for parking lots or building-perimeter security lights. These lamps have about seven times the light output of fluorescent lamps; hence fewer fixtures can be used, and the lamps have a longer life. The fixtures will have integral or remote ballasts. The inspector should determine that the mounting is suitable and that the ballasts are rated for the applicable temperature range and voltage.

16600 SPECIAL SYSTEMS

16601 Lightning Protection

Lightning protection for electrical equipment is a special part of the design function and is in general handled by shielding the overhead line from direct strokes and providing lightning arresters to protect the terminal equipment. When an outdoor substation or a building is to be protected, grounded aerial masts or aerial conductors are provided. The height of the air terminal determines the area of exposed surface protected from direct lightning strokes. The lightning rod's conductor collection grid uses its own ground electrodes and should not be connected to wiring system grounds (NEC 250-86). Also note that electrical equipment should be kept at least 6 ft away from lightning rod conductors (NEC 250-46). A lightning protection system is only as good as the ohmic resistance level of the earth electrode system. The test procedure for this sytem is provided in section 16030, test, part 1.

16620 Emergency Light and Power

Emergency electric services are required for the protection of life and property or against other loss. The extent of the protection required depends on the type of occupancy. Governmental codes define the minimum requirements, and specific provisions are described in NEC section 700-1.

Exit lighting and emergency lights must be sufficient to light the areas where the public may congregate and to permit a safe exit from the

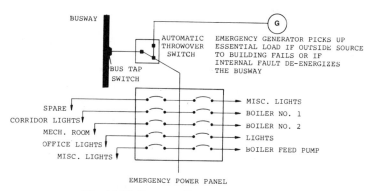

Fig. 16-17 Separate Emergency Service

building. This may be provided from battery packs providing rapid automatic transfer in the event of a power failure.

An emergency source for supplying power loads is required where such loads apply to hospitals, communication systems, alarm systems, and police and fire-fighting facilities. A typical emergency system is shown in Fig. 16-17.

The emergency panel will supply the fire alarm system. Where water requirements cannot be met, emergency power for fire pumps will also be provided. Note that the emergency generator must be sized for five times the full load current of the fire pump's motor.

Specific factory tests are required for the emergency generator, along with the necessary maintenance and instruction manuals, plus a factory technician to make the final run-in of the equipment.

16640 Cathodic Protection

The National Bureau of Standards (NBS), along with Battelle-Columbus Laboratories (BCL), indicates in NBS special publication 511-1 that corrosion costs the United States an estimated $70 billion annually.

Cathodic protection (CP) is an electrical procedure applied to underground and submarine structures for the purpose of halting and controlling corrosion. The Federal Department of Transportation requires cathodic protection on all transmission gas and petroleum pipelines. The electrical inspector will become familiar with these systems in the central urban areas and fuel pipelines and in underground and surface storage tanks.

Corrosion is an electrochemical reaction that occurs when direct current

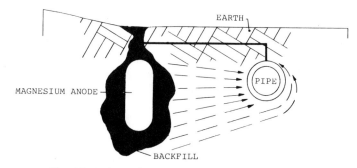

Fig. 16-18 Galvanic Method of Cathodic Protection

is flowing from the corroding metal to the surrounding electrolyte, which may be the earth, fresh water, sea water, cinders, a chemical solution, etc. If the current is prevented from flowing, the reaction stops and no corrosion takes place. Cathodic protection applies to the metal structure a sufficient amount of direct current to back off the electrochemical current. This makes the structure negative or "cathodic" to the surrounding medium.

The two methods of applying cathodic protection are the galvanic method, using magnesium anodes, and the impressed current method, where direct current is forced from a rectifier-driven ground bed through the earth to the metal structure being protected. These two types of systems are shown in Figs. 16-18 and 16-19. Section 16030, Tests, part 3, describes the testing required to determine that the CP system is working properly.

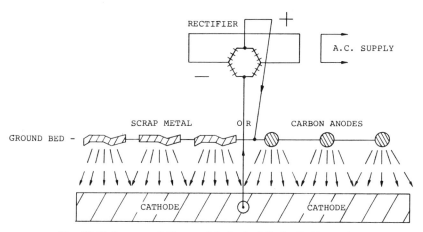

Fig. 16-19 Impressed Current Method of Cathodic Protection

16700 COMMUNICATIONS

These systems are designed for security, fire alarms, and communication, and include peripheral equipment such as clock systems, public address systems, television antenna and cable systems. Details are covered in NEC articles 640, 725, 760, 800, and 810.

16720 Alarm and Detection

In high-vandalism areas, security systems may be used on doors, windows, elevator openings, etc. The sensing devices may be contact switches, foil, photoelectric, TV cameras, or capacity or sound pickup devices. The circuits terminate at a control center at the guard or supervisory headquarters.

16721 Fire Alarm and Detection

Fire-alarm systems most often are the closed-circuit, coded, supervised type. These systems must adhere strictly to city and state ordinance requirements. Small facilities will usually have a noncoded system. Larger facilities have automatic fire detection combined with manual break stations; both are interconnected with a coded system to give a distinguished signal for the protected areas.

The building's sprinkler system can have a number of flow switches, which transmit a signal to the coded fire-alarm system. The flow switches are connected to separate supervised circuits on the fire-alarm control panel. The flow switches operate when a sprinkler head opens or a post indicator valve operates or is left open. See Fig. 16-20.

16725 Smoke Detector

Smoke detectors are used in ventilation and air-conditioning ducts; there are also the familiar ceiling or wall types which give an audible or transmitted signal. Duct detectors are used to close dampers, disconnect blowers, and release interconnecting doors to close to prevent the spreading of fire. A smoke detector may be of the photoelectric or ionized sensor type, or it may be a thermostatic detector set to transmit a signal when the room temperature reaches a preset level. Smoke detectors may or may not be connected to the fire alarm panel—where a trivial signal might bring the fire trucks, this decision must follow local code requirements.

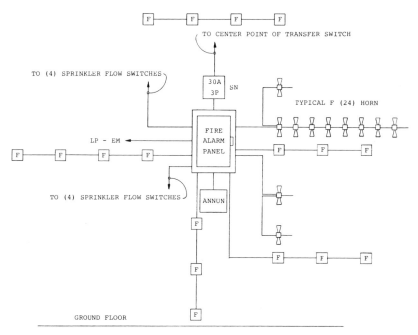

Fig. 16-20 Fire Alarm Riser Diagram

16730 Clock and Program Equipment

A centrally controlled clock system provides a uniform correct time on all clocks, and for clock-operated devices such as bells or horns. The system consists of a master clock that transmits pulses at predetermined intervals. The remote clocks may be hard wired to the master clock, or the pulse may be a carrier current imposed on the building's 120-V branch circuits. These systems permit a number of spinoff controls for outside lighting, tape recorders, time-stamping clocks, etc. See Fig. 16-21.

16740 Telephone

The system for the telephone facility in the building will include provisions for the entrance cable/conduit, the main entrance panel for installation of components, the riser system, the underfloor distribution ducts, and the main terminal room or console. The electrical contractor normally provides the complete duct, sleeves, and wireway system, while the local

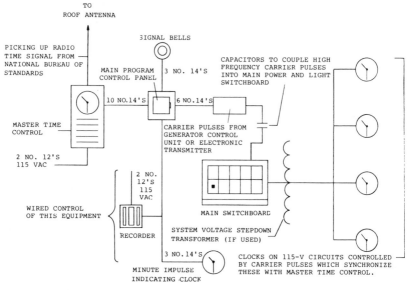

Fig. 16-21 Schematic of Electric Control System for Cables

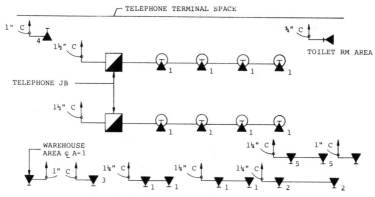

Fig. 16-22 Telephone Riser Diagram

telephone company wires and provides all devices. During construction, the underground service entrance conduit should be sealed to avoid water and gas penetration. A telephone riser diagram is shown in Fig. 16-22.

16770 Public Address Equipment

Sound and radio distribution systems are used in schools, nursing homes, hospitals, airports, warehouse and factory facilities, shopping centers, etc., to supplement security systems. The systems consist primarily of a control center, microphones, and loudspeakers. The additional equipment may be a tape recorder, radio receiver, amplifier, and/or other control devices. If desired, the system can permit distribution to all points, or to selected locations; microphone override; and talk-back features. Typical equipment for sound systems is shown in Fig. 16-23.

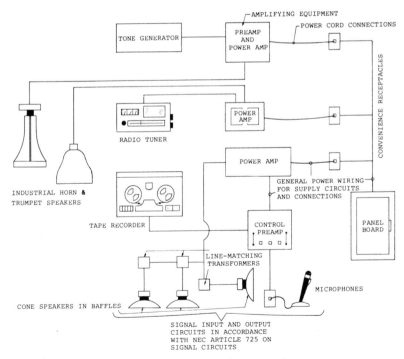

Fig. 16-23 Typical Equipment for Sound Systems

16780 Television Systems

Television facilities for a mid- to high-rise building will require some special considerations by the electrical inspector. If there is to be master TV antenna equipment, this usually has been carefully considered by the design engineer, who will have specified the antenna's location, amplifiers, signal system cables, and circuit devices for the desired channels.

Cable TV service is generally a function of the local cable franchise company. The facility installation is very similar to and requires approximately the same amount of attention as that of the telephone facilities.

16850 HEATING AND COOLING

Planning electric comfort heating involves calculating the heat loss of the structure at the lowest outside temperatures. In facilities where heat pumps are considered for all-electric structures, supplemental duct heaters, electric baseboard heaters, electric heating panels, or possibly compact electric boilers are used.

16858 Snow-Melting Mats

The heating cable buried in concrete for snow-melting must adhere to the design criteria. The inspector must insure that the cable is at least one inch from other metallic bodies, such as pipes and ducts. Leads must be protected by bushings when they leave the floor. The continuity of the cable must be tested with an ohmmeter while the concrete is being poured. The code provisions affecting the installation of heating cable in concrete are shown in Fig. 16-24.

16865 Electric Baseboard Heaters

Electric baseboard heaters are available in lengths up to 10 ft and rated from 100 W to 5000 W. These can be joined to form longer units. The units usually include the thermostat section, although the thermostat may be remotely wall mounted. Baseboards can be equipped with thermal overload protection in case of excessive heat buildup.

16881 Duct Heaters

Generally lumped in this category are cabinet heaters, sill heaters, unit heaters, and unit ventilators. These units are rated from 4 kW to 120 kW

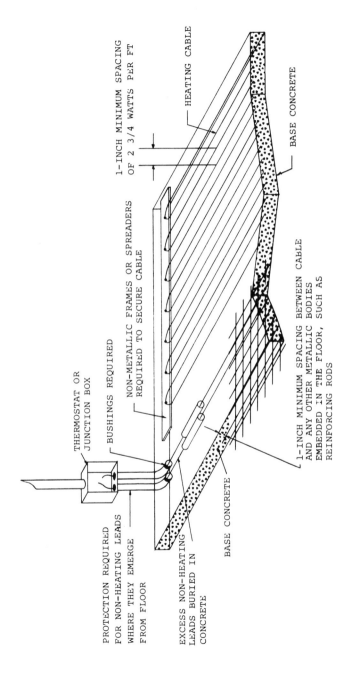

Fig. 16-24 Code Provisions Affecting Installation of Heating Cable in Concrete

and may be adaptable for step or modulating control. The units may also include a chiller to provide ventilation and heating/cooling.

16900 CONTROLS AND INSTRUMENTATION

Controls and instrumentation for an industrial plant may involve a chemical or mechanical process. For a commercial facility they may involve monitoring and limiting electric demand and energy consumption, as well as power for billing purposes. The instruments involved may be of the indicating, recording, or controlling type.

16910 Recording and Indicating Devices

Microprocessors are showing up in designs for buildings more frequently now, and will involve computer-type functions programmed to perform according to a predetermined plan. The electrical inspector needs to determine that the "input," or sensing, device matches the control system and that the "output" circuit load, whether motor, solenoid valve, relay, or data logging, also matches the control system.

In an energy-conscious facility the watt-hour-meter is used to measure the electric energy consumed by the building's load. The input meter for the building is usually installed by the local utility. However, where submetering is involved the electrical inspector must be alert to the following requirements: the meter constant, the instrument transformer ratios, the current, the potential, and the provision of CT's for the number of meter current elements. The inspector must make sure these are all properly connected and the meter multiplier noted on the meter face.

Power companies determine billing based on the peak power demand in kW, with variations based on hourly consumption. The demand function may be integrated; recorded on magnetic, keypunch, or print tape; or totaled from several loads. On a new facility, this type of equipment will require that the inspector coordinate the metering for the various interested parties.

16920 Motor Control Centers

A motor control center is a collection of motor breaker/starters in a centralized freestanding enclosure, permitting a number of motors to be manual/auto-controlled. The wiring and device ratings follow the NEC, NFPA 70-1981, Table 430-150. These equipments are factory preassembled and permit quick shutdown of all motor loads in case of trouble.

BIBLIOGRAPHY

1. L. P. Roe. *Practices and Procedures of Industrial Electrical Design*, New York: McGraw-Hill, 1972.
2. IEEE Recommended Practice for Electrical Power Systems in Commercial Buildings, IEEE Standard 241-1974.
3. J. F. McPartland. *How to Design Electrical Systems.* New York: McGraw-Hill, 1968.
4. Electrical Power Distribution for Industrial Plants, IEEE Standard 141-1969.
5. Electrical Construction and Maintenance, March 1975.
6. Electrical Construction and Maintenance, December 1974.
7. IEEE Recommended Practice for Protection and Coordination of Industrial and Commercial Power Systems, IEEE Standard 242-1975.
8. Electrical Construction and Maintenance, May 1972.
9. National Electric Code, NFPA 20-1981.

C.
PROJECT MANAGEMENT

1. SCHEDULING

Construction men are traditionally good planners, particularly on a day-to-day basis. Methods of recording the day-to-day planning and scheduling have never kept up with the ability of the individual superintendent or a foreman to lay out his work mentally. The traditional method of demonstrating an intended schedule has been the *bar graph* (Fig. 1). In this format, the work items are described along one axis, while the length of time, and in some cases the manpower, to accomplish the work are shown on the horizontal axis, to a time scale. The method is highly effective for demonstration of intent, but breaks down in practice for several reasons. First, for the schedule to be truly effective, it must be in enough detail along the vertical axis to reflect thoughtful planning. For instance, "pour concrete floor" is an insufficient description to use as the baseline for scheduling a 20-story concrete-floored high rise. It leaves no basis for showing interconnecting related activities. However, when the intended pouring sequence for 20 floors is shown, the bar graph tends to become involved in great detail, and therefore difficult to read.

In the 1950's the DuPont Corporation undertook a program to apply computers to construction scheduling. In order to do this, an operations research group addressed itself to a methodology by which construction projects are scheduled—and this of course, mainly centered about the bar graph. The limitation of the bar graph as a planning vehicle was recognized, because of its inability to carry forward the intent of the original planner in terms of interrelationships. By connecting the related activities with vertical lines, planners were able to show the *interdependencies*. Then, on the basis of a mathematician's recommendations, the time scale was dropped, and the logic interconnected diagram or network was drawn in a time flow, but not to an exact scale. This resulted in a more simple drafting operation, cutting down the time for preparation of the arrow diagram.

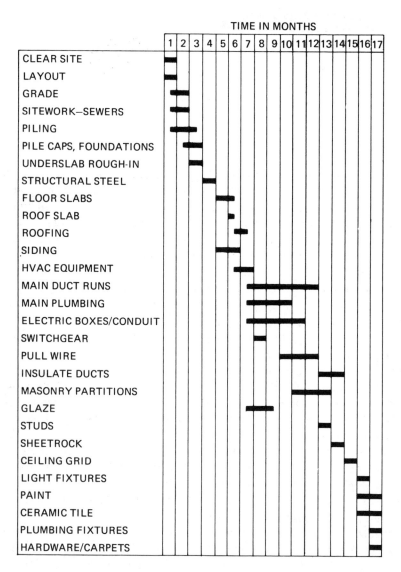

Fig. 1 Bar graph.

Figure 2 demonstrates the method of first interconnecting a bar graph to show interdependencies, and Fig. 3 shows the standard *network diagram*.

Using the network, a project can be planned in terms of its project breakdown and work sequence. This alone is not sufficient to provide the time dimension, so that *estimates* must be added for working time to

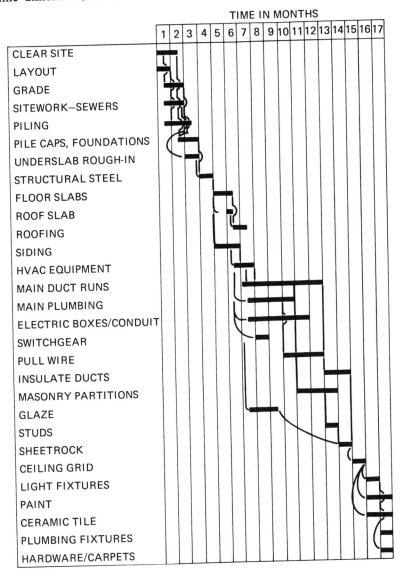

Fig. 2 Connected bar graph.

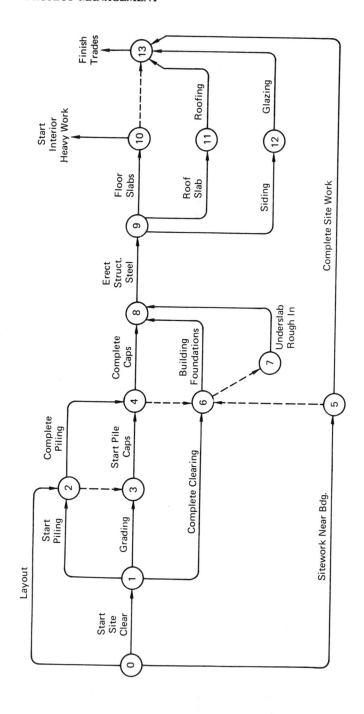

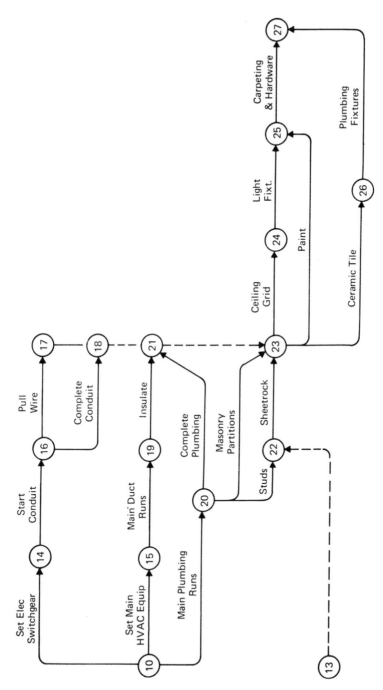

Fig. 3 Basic network diagram.

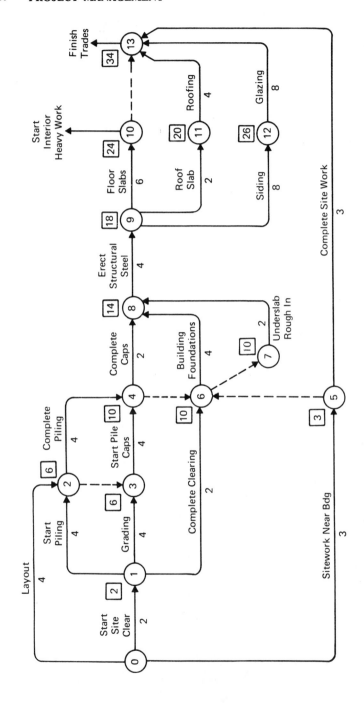

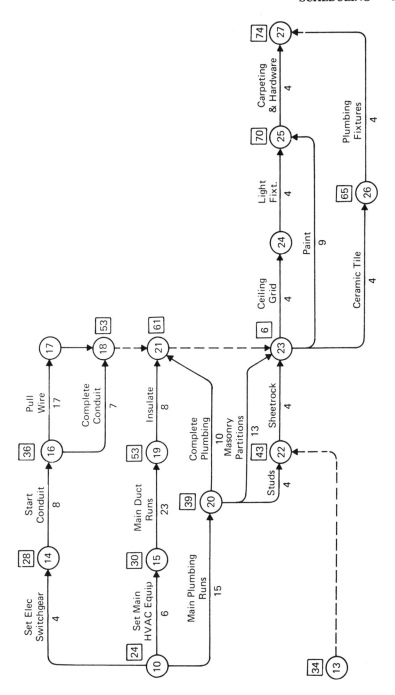

Fig. 4 Early event times.

each activity. These are working day estimates, and not calendar date estimates. However, certain milestones within a project do, of course, have target dates. The planning is, however, initially independent of the dates, and aims at determining feasibility. Because the arrow diagram permits a highly detailed breakdown of the work activities, even to a day-by-day basis, assignment of time estimates is relatively easy, and in the aggregate is also relatively accurate.

The application of network scheduling expanded from construction alone into other fields such as project planning in the aerospace industry.

Once time estimates have been assigned to the network, a calculation is made as shown in Fig. 4, which results in an estimate of the earliest time in which the project, or a segment of it, can be completed. All paths of work must be accomplished, so that the shortest time within which the project can be accomplished is the sum of the longest path through the network. This initial calculation, and the comparison with required results, is planning. Once adjusted to suit specific conditions and the requirements of the user, the plan becomes a schedule. The distinction is significant, since combined planning/scheduling often resulted in a forced schedule. Properly applied, a network calculation will demonstrate problems on paper before they occur in the field, and provide management with a maximum opportunity to respond with remedies such as overtime or a more relaxed schedule if that is the better approach. As one user noted, "It provides the bad news earlier. : . ."

NETWORK CALCULATIONS

The basic premise of the network is completely common sense, as is the determination of the minimum overall time. Intuitively, the *critical path* is that longest path which determines the network time frame. To identify it, however, is perhaps less obvious. In the manual calculation, the identification of the critical path depends upon the establishment of late event times. A late event time is defined as the latest time at which an individual activity can be concluded without delaying the activities which follow it. This is determined by a backwards calculation starting at the end point, and using the overall project time figure as the starting point. This is the reverse of the early time calculation, which is a summation of all paths. In subtracting, the longest path has late event times equal to early event times. While it is possible for an entire network to be totally critical, this requires balancing of all work times, and is in actuality a rarity.

The network *calculation* can be accomplished manually, or with a computer. In the case of a small network, up to 200 to 400 activities, the manual network may be more economical. However, where a network is to be reviewed on a regular basis, a computer output can be anticipated.

Figure 6 shows a summary of event time results taken directly from the example in Fig. 5. Figure 7 is a computer output for the same network. Obviously, the manual and computer outputs are equivalent, but this is often less obvious when a stack of computer output paper is dumped on the inspection team's desk.

The *computer output* can be conveniently arranged into various sorts or edits. Primary among these, for purposes of analysis, is the index sort which arranges the output first in order of the starting event (sometimes called the *i*) and secondarily sorted on the completing or small *j* event. In some cases, this is referred to as the *i-j* sort. It allows the analyst to determine information by addressing the output in terms of specific *i-j* identification of activities.

Another edit which is usual, and particularly useful for the field, is the late start sort. In this case, the work for the next month or two is usually the portion most useful, and the balance can be left off when forwarding to the inspection or construction teams.

Another useful edit is a list of the critical path, since any CPM output should be checked by tracing the critical path. (This precludes inadvertent introduction of errors which can either provide a fictitiously long or accidentally short path, and therefore an incorrect schedule.)

NETWORK FORMS

The first networking approach developed was the DuPont approach, and was named the *Critical Path Method (CPM)*. This was closely followed by the development of PERT. Both techniques have proven effective in construction applications, although CPM is much more common in the industry. In mid-1960's a system called precedence diagramming was developed specifically for the construction industry, and will be encountered on certain projects. The basic approach is similar, although the network may appear simpler, and has the ability to interject overlapping time effects between activities through the device of adding lead time or lag time factors to activities. The same calculated result can be achieved in CPM, but only through additional phased activities, showing the lead or lag by the approach of breaking the activity into multiple parts. (see Fig. 8).

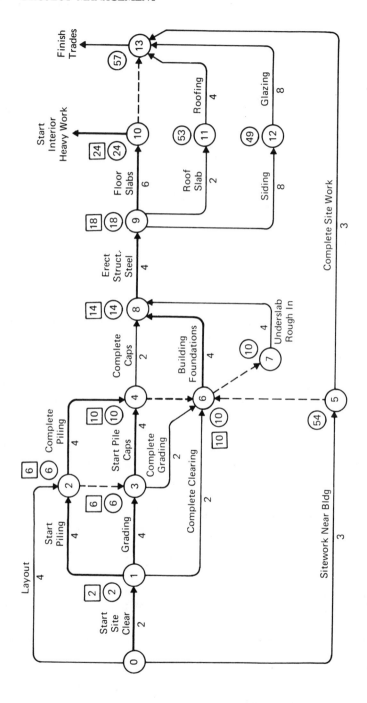

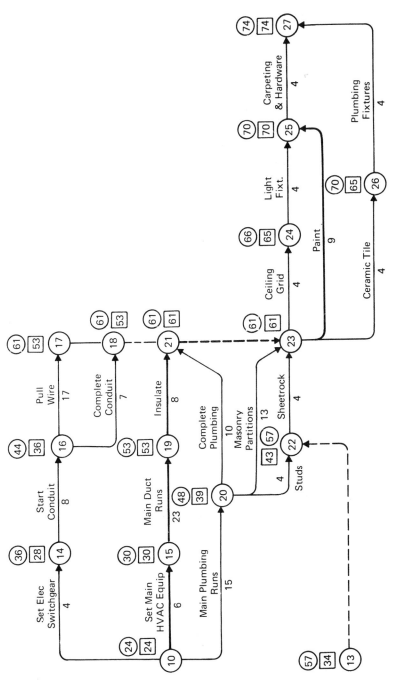

Fig. 5 Late event times.

Event	Follows	Precedes	Early	Late
0	None	All	0	0
1	Clear	Grading–piling	2	2
2	Start piling	Complete piling	6	6
3	Start grading	Start pile caps	6	6
4	Start pile caps	Complete caps	10	10
5	Sitework	Sitework	3	54
6	Clearing	Foundations	10	10
7	Clear–grade	Underslab rough	10	12
8	Pile caps	Erect steel	14	14
9	Steel	Floor slabs	18	18
10	Slabs	Interior	24	24
11	Roof slab	Roofing	20	53
12	Siding	Glazing	26	49
13	Close-in	Interior	34	57
14	Set switchgear	Conduit	28	36
15	HVAC equipment	Ducts	30	30
16	Conduit	Conduit–wire	36	44
17	Wire	Complete conduit	53	61
18	Complete conduit	Complete insulate	53	61
19	Ducts	Insulate	53	53
20	Plumbing	Plumbing	39	48
21	Insulate–plumbing	Ceiling grid	61	61
22	Studs	Sheetrock	43	57
23	Partitions	Ceiling grid	61	61
24	Ceiling grid	Lights	65	66
25	Light paint	Carpentry	70	70
26	Tile	Fixtures	65	20
27	All	None	74	74

Fig. 6 Summary of event times.

CONSTRUCTION SCHEDULING

The inspection team should be prepared to implement network scheduling. In many cases, the owner will call for provision of a network-based schedule by the contractor. In other cases, the owner may provide, through the construction manager, a professional staff service which will prepare a CPM network for the project, directing only the contractor's cooperation in the preparation. However prepared, the CPM plan

I	J	D	Description	ES	EF	LS	LF	F
0	1	2	Start site clear	0	2	0	2	0
0	2	4	Layout	0	4	4	8	4
0	5	3	Sitework—bulldozing	0	3	51	54	51
1	2	4	Start piling	2	6	4	8	2
1	3	4	Start grading	2	6	8	12	6
1	6	2	Complete clearing	2	4	6	8	4
2	3	0	Restraint	6	6	6	6	0
2	4	4	Complete piling	6	10	6	10	0
3	4	4	Start pile caps	6	10	6	10	0
3	6	2	Complete grading	6	10	6	10	0
4	6	0	Restraint	10	10	10	10	0
4	8	2	Complete caps	10	12	12	14	2
5	6	0	Restraint	3	3	10	10	7
5	13	3	Site work	3	6	54	57	51
6	7	0	Restraint	10	10	12	12	2
6	8	4	Foundations	10	14	10	14	0
7	8	2	Ug. rough-in	10	12	12	14	2
8	9	4	Struc. steel	14	18	14	18	0
9	10	6	Floor slabs	18	24	18	24	0
9	11	2	Roof slab	18	20	51	53	33
9	12	8	Siding	18	26	41	49	23
10	13	0	Restraint	24	24	57	57	33
10	14	4	Set switchgear	24	28	32	36	8
10	15	6	Set HVAC equipment	24	30	24	30	0
10	20	15	Main plumbing runs	24			48	
11	13	4	Roofing	20	24	53	57	43
12	13	8	Glazing	26	34	49	57	23
13	22	0	Restraint	34	34	57	57	23
14	16	8	Start conduit	28	36	36	44	8
15	19	23	Main duct runs	30	53	30	53	0
16	17	17	Pull wire	36	53	44	61	8
16	18	7	Complete conduit	36	43	54	61	18
17	18	0	Restraint	53	53	61	61	8
18	21	0	Restraint	53	53	61	61	8
19	21	8	Insulate	53	61	53	61	0
20	21	10	Complete plumbing	39	49	51	61	12
20	22	4	Studs	39	43	53	57	14
20	23	4	Sheetrock	43	47	57	61	14
23	24	4	Ceiling grid	61	65	62	66	1
23	25	9	Paint	61	70	61	70	0
23	26	4	Tile	61	65	66	70	5
24	25	4	Light fixtures	65	69	66	70	1
25	27	4	Carpeting	70	74	70	74	0
26	27	4	Plumbing fixtures	65	69	70	74	5

I	Starting event	ES	Early start
J	Concluding event	EF	Early finish
D	Duration	LS	Late start
F	Float	LF	Late finish

Fig. 7 Computer output, activity times i-j index sort.

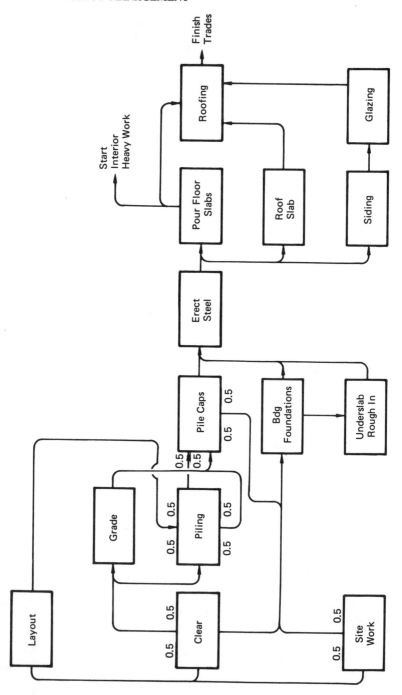

Fig. 8 Precedence diagram method.

reflecting the construction schedule must reflect the manner in which the contractor actually expects to accomplish the project, or it will be worthless exercise.

The method of developing the network is straightforward, and involves discussions between the scheduler and the contractor's key personnel on the project to determine first the general contractor's intended phasing of work, and his estimates for the various work phases. This can be evaluated first, to determine whether or not the overall time frame is acceptable and within schedule parameters. Following this, work of the other prime contractors, mechanical, electrical, plumbing, and key equipment contractors, is placed on the network within the milestones established by the general contractor. It is entirely possible that addition of the work by the other trades will cause an extension of the critical path, which may, for instance, be diverted from completion of structure through installation of major equipment and back into finishing trades. This is, in fact, one of the prime purposes of network scheduling—to communicate the job intent to all parties. Without a positive scheduling base such as this, there is often the tendency on the part of all contractors to expect to have all of the time within the scheduled period available exclusively for their own work. Though impractical, this is a natural tendency.

The inspection team should be well-versed in the network schedule, and can expect to report time and cost progress on the basis of it.

UPDATING

Usually, the network schedule is reviewed on a monthly basis, taking the actual job status and reflecting it in terms of the activities, and then recalculating the network. If the end date stays about on the prior schedule, the job is then considered to be holding its own. If the end date extends outward, the work progress is not sufficient to maintain the pace. If the end date shortens, the indication is that the work is proceeding rapidly enough. The progress of the project depends upon the noncritical items as well as the critical ones—but if the critical items are not on schedule, there is no way in which a project can be maintained on schedule. Often this obvious fact is hidden in a flurry of activity on noncritical work in projects where rigorous network planning is not applied.

In many cases, the specifications will call for a progress payment breakdown by network activities so that progress payments can be directly related to project progress in terms of time. This is cost-effective for the inspection team, since it means that both time and cost can be reviewed concurrently. Further, it maintains a much higher level of interest in the diagram on the part of the contractor, since his progress payments depend upon progress which can be measured on the network.

Networks improve the foresight of projects planners and schedulers because of their interconnected concept. There can be no gaps in thinking permitted, and where uncertainty is involved an activity must be placed spanning . known factors, and the uncertainty identified. However, preparation of a diagram does not include a crystal ball capability, so that certain factors will be forgotten, and others will arise unforeseen. These *new occurrences* are placed into the network in terms of their effects on the work sequence, and with appropriate time impact indicated. The updating calculation will assimilate this difference into the evaluation of the schedule.

Change orders are usually used to describe changed job conditions, and each change order should have a negotiation for time as well as cost. To accomplish an equitable schedule impact, the network calculation can be exercised to demonstrate the effect on the end date. This precludes a traditional negotiating approach in which contractors typically request day-for-day extension on the basis of *any* extra work. Often work occurring in the float or noncritical areas is allowed to substantially extend the job completion. On the other hand, the owner typically doesn't negotiate on a change-by-change basis, and has his designer (or unfortunately his lawyers) negotiate a lump settlement for time. Usually, money negotiations are conducted on a much more businesslike basis. The effect of negotiating individual change orders and utilizing the network as a means of calculating the real impact reduces many arguments, and results in a more equitable solution to time extensions.

SPECIFICATIONS

The requirements for construction scheduling are usually included in the General Conditions. This requirement can be fairly brief; for example (from The Department of General Services, Pennsylvania):

1. A Progress Chart (Fig. 9) is a schedule of the proposed prosecution of work under a project.

2. The General Contractor shall within 14 days from receipt of notice of

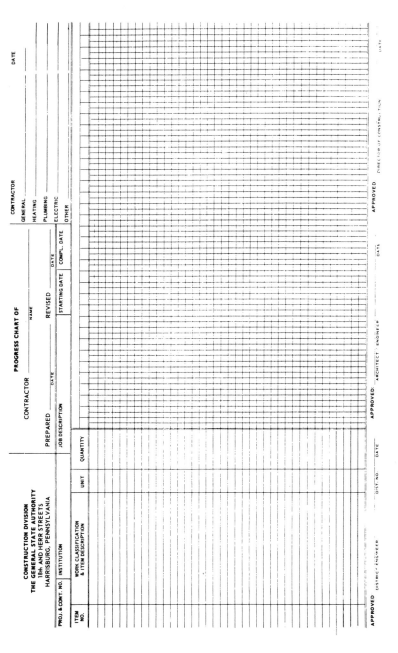

Fig. 9 Progress schedule. *Source: Department of General Services, Pennsylvania.*

the contract award furnish each separate prime contractor a schedule (Fig. 9) of the proposed prosecution of the work under his contract.

3. Each separate prime contractor, upon receipt of the executed work copy of the General Contractor's Schedule, shall prepare his work schedule to coordinate with same. Within 28 days of the issuance of the notice of contract awards, each separate prime contractor will submit his work copy to the General Contractor for coordination.

4. The General Contractor shall submit one complete comprehensive coordinated and final progress chart signed by all prime contractors, to the designer for review and signature. This will then be forwarded to the District Engineer for his review and signature and shall be completed within 42 days of issuance of the notice of contract award.

5. Only one payment can be made on an application for payment until the progress schedule has been submitted in accordance with the contract in the hands of the resident inspector.

The specification includes some more detail, but this is the substance. While network scheduling would meet this requirement, almost any bar chart would also. Pennsylvania has incorporated CPM in many of its schedules in recent years.

A number of organizations, principally in the government sector, require the submission of a network-based schedule. These groups include the Public Building Service of the General Services Administration, the Veteran's Administration, and the Corps of Engineers. Each has variations on the basic requirement, and the inspection team would have to study the individual scheduling requirements.

Preparation of a specification for network scheduling has been relatively well developed by these groups. Unfortunately, it is difficult to establish a specification—even in detail—which can guarantee a useful network. In almost all cases, it is possible for a contractor to evade the intent of the specification, in turn providing a network and computer printout which meets the form and format. The purpose of any network specification is the establishment of a reasonable schedule at the earliest practical point in the contract, a schedule which will point out areas of difficulty or potential difficulty. The *attitude* of the contractor and his interpretation of the potential problems in offering his own scheduling information are both factors. However, even in those cases in which the contractor does not completely meet with the spirit of the specification, experience indicates that the network provides a vehicle for communication in regard to scheduling and scheduling problems, and is a positive factor.

In certain cases, the owner provides consulting help in the preparation of the CPM or PERT network, requiring through the specifications only that the contractor provide his cooperation. This approach has also worked well, and seems to break down the barriers or adversarial roles between the owner and the contractor.

Many of the agencies are utilizing a *prebid construction evaluation* in the form of a CPM network and analysis. This is presented at the prebid conference as one practical way in which the schedule can be met, but is not imposed as the manner or the sequence in which the job must be pursued. Where the prebid network is well done, many contractors utilize a variation for their initial work.

In any event, the inspection team must adapt itself to the specification which was used. It is beyond the scope of the inspection team's authority to impose a change in network scheduling requirements.

Another example of a completed progress schedule on the bar graph approach is shown in Fig. 10 in a sample prepared by the California Department of Highways.

The California Highway specifications call for the contractor usually to start within 15 days after receiving notice. However, many contractors request permission to start work prior to approval, once the low bid has been established. In this way, the contractor can gain time on the operating schedule—but certain problems are presented to the owner. The owner must clearly be responsible and liable for any personal injury or property damage, he must be certain to have the right of access, and the inspection team may not be available to start on the job. Accordingly, a separate agreement must be made with the contractor to agree to the limits of the area to be worked, also to limit the operations to be performed, to obtain his agreement that he will comply with all of the contract plans, specifications, and general conditions, and that he will furnish adequate bond to cover the work contemplated.

The California Highways Department calls for submission of the bar graph type schedule within sixty days. The contractor's proposed schedule for procurement of materials, plant, and equipment should be shown for major items, but smaller equipment and portions may be grouped. The contractor must furnish a practicable and workable schedule showing various salient features. The purpose of the schedule is to furnish a means by which the inspection team can gauge progress. If the contractor fails to properly prosecute the work in accordance with

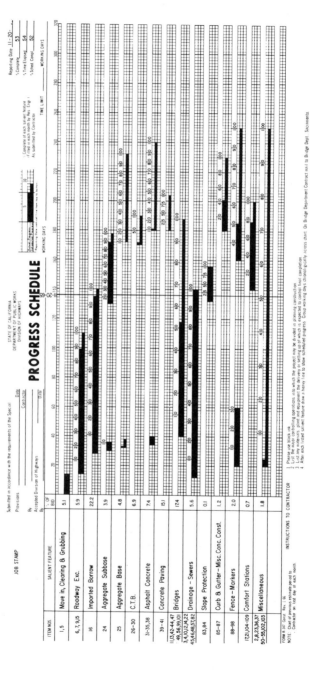

Fig. 10. Progress schedule. *Source: State of California.*

the approved schedule, the inspection team should notify him informally, and then in writing of lack of progress, and request that a revised schedule be submitted showing how the balance of the work is to be completed within the original guidelines.

In the case of suspension of work, notification should be in accordance with the General Conditions, and in writing. If a suspension is ordered due to the failure of the contractor to carry out their orders or to perform a contractual requirement, working days are charged against the contract. If the suspension is ordered for reasons beyond the control of the owner or the contractor, working days are not charged against the contract. When the reason for suspension no longer exists, or it can be expected that favorable conditions will prevail for resumption of work, the contractor should again be notified in writing.

Schedule is important of and within itself, since the time of turnover of a facility is generally of great importance—second only to cost. Monitoring of the approved schedule on a regular basis results in control, and when circumstances occur which preclude that control, then at least forewarning. The inspection team plays a key role, and must recognize that the owner's interest lie not only in the delivery of a facility which meets the quality requirements of the specifications, but also that that delivery be timely.

2. CHANGE ORDERS

Change orders are legal documents describing and approving changes in conditions or scope to the contract. In some cases, they are written with a net no-cost, with the specific purpose of permitting a change in the contractual documents which are legal and binding. In other cases, the change order may result in a credit to either the contractor or the owner. Change orders may be the result of mutual agreement, or may emanate from the request for relief from change in scope or conditions on the part of either party.

TYPES OF CHANGE ORDERS

There are four basic reasons for change orders:

1. *Change in scope.* The owner, either because of special conditions or change in his requirements, elects to change the scope of the contract. He more usually adds to the scope, but in some cases decides to delete a portion of the work. Typically, in the specifications, the owner reserves the right to make changes in scope, usually on a predetermined basis of negotiation—such as unit prices. If the change amounts to a substantial portion of the value of the contract, the contractor may choose to refuse it, particularly if he had bid on a basis very favorable to the owner. On the other hand, in certain circumstances, the contractor welcomes change orders if the unit prices of the material changing are favorable to him.

2. *Change in materials, method, or equipment.* In some cases, the change order process covers changes in materials such as paint quality, roofing materials, or even major items such as structural material. Usually, these changes are for the convenience of the contractor, and may result in a credit to the owner. Changes of this type would include the value engineering incentive changes wherein the contractor recommends changes in materials or procedures which will save money, and share the results of these changes.

The owner may either on the advice of his designers or for special reasons recommend a change in equipment or method of erection as specified in the specification. The purpose of the change may be to take advantage of special situations, or to avoid a specific problem. For instance, in the case of spray painting with heavy equipment, during the course of an elevator strike, one owner shifted to a type of paint requiring less equipment which permitted the painters to walk up to their work, thus avoiding a delay in the finishing work.

3. *Changing conditions.* The contractor may encounter a change in conditions due to strikes, weather, or subsurface conditions not previously determined. In certain cases, these problem areas are anticipated, and the change order pays for the change in basic concept. This is often the case when foundation conditions are assumed but unknown, and the owner, rather than paying for the entire uncertainty, agrees to pay within the contract a basic price for foundation work, and pays an extra fee for any unusual rock, dewatering, or other anticipated, but unidentified conditions.

Certain other conditions, including strike, Acts of God, and accidents may be the responsibility of the contractor or the owner, depending upon the way in which the specifications have been written. Even in cases in which the contractor is obligated to cover the costs of such delays, a change may be requested to extend the schedule.

4. *Changes due to omissions in the contract.* Particularly in contracts where several contractors are working on the same site, each with a prime contract to the owner—and therefore no cross-contractual commitments—the individual contractors may work through their interpretation of the limitations of their contract, and certain work may go uncovered. In these situations, the owner often attempts to enforce the contract, insisting that the work is indeed covered. However, an airtight contract is not easy to anticipate during the design phase, and often the owner is obliged to issue change orders to cover the gaps in specified work. The role of the inspection team in avoiding or mitigating this type of change order is quite important. The inspection team must be alert to preclude the owner having to pay for work twice. If, in the interpretation of the designers and/or the inspection team, certain work is included in the contract specifications, then it should not be included in any subsequent change orders. Unfortunately, specifications are often somewhat ambiguous, so that each side can interpret them to their own advantage. There are cases in which the inspection team may recommend, even in the face of an ambiguous specification, that the contractor for various reasons including good attitude and effect upon the balance of the work, that a change order be issued. For instance, for a questionable change order amounting to a few hundred dollars, delay in settlement could well hold up other work involving many thousands of dollars, leaving the owner open to a legitimate claim from the other contractors.

CHANGE ORDER NEGOTIATIONS

The request for a change order may be initiated by either party. A contractor, when he encounters a situation which he feels is not included in his scope of work, desires a change in material or equipment, or intends for some other reason to work beyond his contractual obligations, must notify the owner of the situation and request advice. The site inspection team is usually the first to be aware of this type of request, and may insist that the contractor is not entitled to any such change order. Care must be taken in the form and substance of this refusal. For instance, a contractor may make a request for a change order in writing, but be directed to do the job. Even if informed that no change order is authorized, the contractor can use the direction as a change order authorization if, subsequently, interpretation of the specifications confirms that the work was indeed not covered by the specifications. The inspection team must take care not to *direct work* in the field unless they are absolutely certain either that the work should be accomplished or that the contractor is definitely obligated. Even when the inspection team believes that work should be accomplished, they should take care to receive change order authorization through proper channels.

It is important that the inspection team recognize that the owner often has strict rules, even legislative *restraints* for public agencies, which preclude the issuance of change orders beyond a certain point. The purpose of this is clear, so that lower levels of management do not undertake work beyond their authority without prior approval. For this reason, many states and cities require specific legislative or comptroller approval for changes aggregating more than 5% of the contract amount. In most cases, the functional authority in charge of the project has the prerogative to issue changes up to a figure such as 5%.

The owner initiates change orders when the designers recommend, and have accepted, certain additions or deletions to the basic contract scope. Negotiations for this type of change may be very smooth, or may involve substantial negotiation. For instance, the contractor may point out that the change itself is not the total cost or time impact, and that other portions of the project are affected. Accordingly, he will seek relief for the entire impact of the change in scope, and not a narrow cost-only for the addition or deletion.

In most cases, whether the change order is initiated by the owner or the contractor, the contractor upon request is directed to submit an estimate of the cost of the change, in the form of proposal.

The contractor's *proposal* for performance of the change order should be presented in preliminary form for review by the inspection team and/or the designers. It should include the units of work involved according to the contractor's estimate. In most contracts, the contract bid includes authorized units and an approved cost per unit submitted by the contractor prior to the notice to proceed. These are used to price out the change order proposal. The contractor may submit different prices per unit for additions and deductions, but generally the specification precludes these having a spread of more than 15%.

In some cases, the *unit price* categories do not adequately cover the entire change order. Again, the contractor is usually required to present a work breakdown—and to present a lump sum price. If this price is mutually agreeable, it may be accepted by the owner. In reviewing the contractor's price proposal, the inspection team and/or designer should perform their own cost estimate to validate the units utilized, and confirm the scope of work. Where agreement cannot be reached in terms of price on the basis of either units or lump sum, the owner may direct the contractor to perform on a force account basis. Again, the contractor usually is required to accept this direction as a part of the specifications. Specifications usually also specify the amount of allowable mark-up on the standard union rates or skilled labor rates as appropriate. Materials are reimbursed at cost plus an agreed-upon percentage for overhead and profit.

Figure 1 is an approval form for change orders used by Pennsylvania. Figure 2 is the form used to record force account work, material, and equipment; while Fig. 3 is an affidavit that the force account statements are true and correct.

Once the change order is processed and approved, it becomes as part of the contract, and separate identification by change order number is continued only for the purpose of accounting and tracking payments.

Form No. 118-D   Contractor's Name_____

Contract No._____   Sheet_____of_____

APPLICATION FOR PAYMENT OF CHANGE ORDERS, Period Ending_____

(1) Change Order No.	(2) Amount ( ) Denotes Credit	(3) Form 107 Approval Date	(4) % Completed This Period	(5) Amount Payable This Period ( ) Denotes Credit	(6) % Completed To Date	(7) Amount Paid To Date
Transfer to Recapitulation Sheet Line No. 3					X X X X X X X X X X X X	

Fig. 1 Change Order Payment Approval Form. *Source: Pennsylvania.*

When a proposed change order is substantial in size, the owner may consider whether or not to issue a change order, or go into competitive bidding. In certain cases, public agencies are almost required to go into the public bidding cycle. However, in most cases, the inspection team can demonstrate in cooperation with the designer that multiple contracts on a single site are inevitably expensive. Where the owner has awarded prime contracts placing contractors with like capability in close proxim-

CONSTRUCTION	Date .... Contract No. ....
	Today's Work ....
FORCE ACCOUNT	....
18th and Herr Streets	....
Harrisburg, Pennsylvania	Date Authorized ....
DAILY WORK RECORD	By ....

LABOR

Employee	Occupation	Hours Worked	Rate	Total

MATERIAL, EQUIPMENT AND INVOICES

Quantity and Unit	Description	Unit Price	Total

Signed ....
Contractor's Supervisor

Signed ....
Inspector

Fig. 2 Daily Work Record. *Source: Pennsylvania.*

18TH AND HERR STREETS
HARRISBURG, PENNSYLVANIA

Contract No. _____

_____

_____

COMMONWEALTH OF PENNSYLVANIA:

                                    : SS

COUNTY OF   :

_____ , being duly sworn

according to law, deposes and says that he is authorized to and does make

this affidavit on behalf of the Contractor on the above numbered contract,

and that the attached statement is a true and correct statement of the cost

of all the labor and material actually used or employed in the performance

of the "Force-Account Work" under said contract.

                              _____

Sworn to and subscribed:

before me this   day:

of   196 :

_____
      Notary Public
My Commission Expires:

Fig. 3 Work Force Account Affidavit. *Source: Pennsylvania.*

ity, there are inevitably coordination problems, and interferences. Each contractor has potential grounds for authorized change orders as a result of these implied delays.

TIME EXTENSIONS

A time extension is a change. For many years, emphasis on change order negotiations has been upon cost. Generally, the owner has looked to the contractor to absorb all time delays as a result of the change order, while the contractor has generally expected the owner to allow him a day-for-day time extension on the basis of new work. Neither position is correct. If a project has been scheduled on a network basis, there is inherent in the system a means of evaluating the time impact of changes. The owner and the contractor should each be willing to accept the time impact in this context.

The effect of the change order should be evaluated by the inspection team or the designer, whichever is in charge of evaluating the network scheduling system. If the path affected has float, the float can be assumed to be the province of the owner or the contractor (depending upon which point of view is pursued). In most cases, compromise is the rule—and float remaining should be shared. However, if the contractor has utilized float along a path prior to the negotiations for the change order, then realistically the owner must accept this utilization, and can share only in the remaining float. It is really to the owner's best interest to recognize time implications and accept them as part of the change order, rather than pursuing a head-in-the-sand approach which pretends that there is no time impact. In some instances, legislative procedures or administrative procedures for a governmental body may preclude negotiations for time as well as money. This is very unfortunate, since it can result in the negotiation for cost on perhaps several hundred change orders for a large project, and then ultimately a gross negotiation for time. Experience indicates that the owner inevitably will lose more than he should when time is negotiated by lawyers, rather than professionals in the design-construct field. Changes do cause delay, either of themselves, or because they dilute the effort of the contractor. The owner should be ready to pay in terms of time as well as money. Conversely, when a change order is evaluated, and an unacceptable time delay results, then the owner should be ready to pay additional money to buy back the time lost. Another way of looking at this is that the contractor is being required to accelerate his efforts, perhaps in an inefficient manner. In some cases, the owner may come to a compromise solution in which he accepts part of a facility for initial use, extending the acceptance period for the final turnover. The importance of careful study of all sides of schedule delays is that time is money to everybody, and recognition of this earlier in the cycle produces better results for all.

Figure 4 is an example of a daily work log as kept by the California Department of Highways. These records do not automatically result in time extensions. The actual time extension permitted, if any, is a function of the specifications. At one time, any unusual weather was an excuse for

FORM
R-146
1-62

JOB STAMP

CONTRACT No.

WEEKLY STATEMENT OF WORKING DAYS

REPORT NO. _____

To___ J. J. Jones Construction Co. _____ Contractor

The following statement shows the number of working days charged to your contract for the week ending ___ Sept. 18 ___ 19 75

STATE OF CALIFORNIA
DEPARTMENT OF PUBLIC WORKS
DIVISION OF HIGHWAYS

Date	Day	Weather, Weather Conditions or Other Conditions (Note 1)	Working Day	Unworkable Days Caused by Weather	Days Not Worked Conditions other than Weather
Feb. 4	Mon.	Clear	1		
5	Tue.	Rain		1	
6	Wed.	Clear - grade too wet		1	
7	Thur.	Clear - R/W delay	1		1
8	Fri.	Clear - R/W delay	1		1
Days this week			3	2	2
Days previously reported			32	20	9
Total days to date			35	22	11

TIME EXTENSIONS	CCO Numbers (Note 2)	Days Approved CCO	Days Approved Other	Days Unapproved CCO	Days Unapproved Other
Approved to date		4	0		
Requested, approval pending	9			2	2
To be considered					
Total days to date		4	0	2	2

COMPUTATION OF EXTENDED DATE FOR COMPLETION	No. of Days	Numbered Day (Note 4)	Date
1. Date Contract approved by Attorney General		95	6-11-75
2. Working days specified in contract	140		
3. COMPUTED DATE FOR COMPLETION (if all days specified are workable)		235	1-3-76
4. Total time extension days approved to date	4		
5. Total unworkable days to date	22		
6. Sub Total	26		
7. REVISED DATE FOR COMPLETION (Note 3) (Line 3 plus Line 6)		261	2-8-76
8. Revised working days for contract (Line 2 plus Line 4)	144		
9. Total time extensions anticipated but not yet approved	4		
10. Total working days to date	35		
11. PROBABLE WORKING DAYS REMAINING (if all requested days pending are approved) (Line 8 plus Line 9 minus Line 10)	113		
12. EXTENDED DATE FOR COMPLETION (if all requested days pending are approved) (Line 7 plus Line 9)		265	2-14-76

REMARKS: (Continue on reverse)

Sept. 14, 15, 16; Utility Co. lowering 8" gas main across R/W, Sta. 360.
Roadway excavation completed to point where all material remaining
must haul through this area. Present status of utility work prevents
hauling through, forcing suspension of roadway excavation operations.
Approval of two-day extension of time is requested. Utility company
expects to complete Sept. 17.

The Contractor will be allowed fifteen (15) days in which to protest in writing the correctness of the statement; otherwise the statement shall be deemed to have been accepted by the Contractor as correct.
Note: All footnotes are on reverse side.

CONTRACTOR

Sam Brown
Resident Engineer

Fig. 4 Daily Work Log. *Source: California Department of Highways.*

time extensions. Today many designers use specifications which point out
that the contractor shall anticipate *normal* working conditions for weather.
In the mid-Atlantic region, a contractor can get from the Weather Bureau a
forecast for the season, indicating average days of snow, rain, below
freezing, etc. The job weather record, then, would be used to demonstrate

the total number of bad weather days, and in turn, the only consideration for time extension would the the net additional bad weather days. However, to the contractor's favor, net days extension might be amplified by comparing their time of occurrence with the project schedule. For instance, a two-week delay in November could well throw preparation of a subgrade into a hard freeze period which would result in a delay of many additional days.

Similarly, delays in the autumn could also push pipe testing into a period of time wherein the fluids are subject to freezing, and the testing process would materially slow down. In such cases, the causative reasons for the delay could result in substantial costs either to the owner or to the contractor, depending on who was at fault. Also, in negotiating a change order, the owner's team must keep in mind the cumulative effect of changes and delays, rather than viewing each as an abstract topic.

The contractor may formally request an extension of time change order. Figure 5 is the form utilized by Pennsylvania.

PAYMENT OF CHANGE ORDERS

As indicated previously, the payment of change orders is essentially handled as part of the overall progress payment for the contracts. However, the incremental nature of the change orders makes it necessary to keep account of them on a consecutive numerical basis. Figure 6 is the form used by Pennsylvania as an Appendix to the payment requisition for that portion of the work which involves completed change orders.

CLAIMS

Claims differ from extra work in that they are usually made on the basis of a claim of an extra scope of work which has not been authorized or recognized as an extra by the owner. In many cases, the owner may agree with the claim, but because of certain statutory regulations, the contractor is forced to submit a claim, rather than receive an authorization for extra work. For example, in the City of New York, if damage occurs to work already in place, and the contractor claims that the work was in place for an unreasonably long period of time so that he was not any longer required to protect his own work (which is a standard of most New York City specifications), City personnel might recognize the validity of the claim, but could not issue an order for payment. The reason is that the New York City Administrative Code requires any

```
                            CONSTRUCTION          Project No.......Contract No.......
                                                  Request No.......Date..............
The General State Authority     REQUEST FOR       Project Title.....................
    18th & Herr Streets                           Institution.......................
Harrisburg, Pennsylvania  17120  EXTENSION OF TIME  Location..........................
                                 CHANGE ORDER ONLY  Liquidated Damages.........per day
```

CONTRACTOR'S REQUEST & AFFIDAVIT

...., Contractor, herein requests The General
State Authority to grant an extension of..................days for the completion of Contract
No. GSA...........from the original/extended* contract completion date of.....................
to...................the extended contract completion date. This request is made pursuant to
Sections 8.3 and 12.2 of the General Conditions of the Contract and is without prejudice to
the contractor's rights under Section 6.2 of the General Conditions.

...., being sworn according to law
deposes and says that he is the..
of..
the Contractor; that he is authorized to make this affidavit on behalf of the Contractor; that
the following facts are true and correct; that the work under the above captioned contract was
not completed within the time limit specified by the contract; and, that the failure to complete
such work on or before the contract completion date was due to the following reasons outlined
on an attached sheet.

If a controversy should arise for any reason following the presentation of this request for an
"extension of time", it is agreed that any additional substantiating proof or evidence required
to support this request shall be furnished.

Sworn to and subscribed before me this

..........day of............, 19.....
                                                ....
                                                        Signature
....
          Notary Public

Strike out inapplicable words.

          RECOMMENDATIONS OF PROFESSIONAL, DISTRICT ENGINEER AND DIRECTOR OF CONSTRUCTION

....on my knowledge of the construction performed under this contract:

application is/is not recommended: for.......days
                                                ....
                                                Signature - Professional

Application is/is not recommended: for.......days
                                                ....
                                                Signature - District Engineer

application is/is not recommended: for.......days
                                                ....
                                                Signature - Director of Construction

                    ACTION OF EXECUTIVE DIRECTOR
                    Contract No. GSA...........

Approved/Disapproved

....
                                                ....
                                                Executive Director

Fig. 5 Extension of time change order. *Source: Pennsylvania.*

change in the scope of work which is linked to a claim of delay requires
that the entire claim be put through the claim procedures.

There are other cases in which the owner may have a claim against the
contractor, either for work put in which was not in accordance with the
quality level required by the specifications, or other problems—such as
delay. In most cases, the owner is not able to press his claim for delay
unless the specifications lay out a basis for liquidated damages.

	CONSTRUCTION	Project No. _____ _____
		Project Title _____
The General State Authority		Location _____
18th and Herr Streets	Change Order No. _____	
Harrisburg, Penna. 17120	Contract No. _____	Date Prepared: _____
	Additional/Deductible Cost _____	Date Forwarded to Contractor _____

PROPOSAL BY CONTRACTOR

1. Contractor agrees to furnish and/or delete labor and material in order to complete the construction included within the scope of this Change Order to contract plans and specifications for the net additional/deductible cost of $ _____ on a Force Account basis at an amount of not to exceed $_____ . In accordance with the attached cost breakdown. (delete part not applicable).

2. Contractor further agrees that the completion date for construction as contained in the subject contract shall not be changed by approval of this Change Order without the submission of a request for an extension of time to the Authority on forms provided for such purpose.

Signed _____ Date _____
Contractor

RECOMMENDATION OF ARCHITECT/ENGINEER

1. Construction change was requested by _____,
date: _____, to _____.

2. The specific description of this construction change is as follows:

3. The following drawings are involved in this Change Order:
   a. Supplemental drawings are attached as follows:

   b. Sheet and detail of contract drawings to which this Change Order refers are as follows:

*4. Revisions to paragraph and page number of contract specifications affected by this Change Order are as follows:

5. This Change Order did/did not result from an error/omission by the Architect/Engineer in the contract plans or specifications.

* USE EXTRA SHEET IF NECESSARY
7/65   PAGE 1 OF 2

Fig. 6 Change Order Form (2 sheets). *Source: Pennsylvania.*

Contract No. _____ Change Order No. _____ Amount _____

6. Summary of correspondence which created this Change Order is as follows:

7. Prices and quantities of contractor's cost breakdown have been checked and are acceptable.

8. Approval of this Change Order is recommended.

Signed _____ Date _____
Architect/Engineer

RECOMMENDATION OF DISTRICT ENGINEER

1. Architect's description of scope of work has been reviewed
   Additional description is as follows: (if necessary)

2. Prices of contractor's cost breakdown have been checked and are acceptable.

3. Quantities of contractor's cost breakdown have been field measured and are acceptable.

4. This Change Order did did not result from an error or omission by the Architect/Engineer in the contract plans or specifications. This statement is/is not in agreement with that of Architect/Engineer and justification therefor is/is not attached.

5. Additional items of correspondence relative to this Change Order are as follows:

6. Approval is/is not recommended.

Signed _____ Date _____
District Engineer

STATEMENT OF COMPTROLLER

1. Funds for payment of this Change Order are available.

Signed _____ Date _____
Comptroller

RECOMMENDATION OF DIRECTOR OF CONSTRUCTION

1. Approval is/is not recommended

Signed _____ Date _____
Director of Construction

APPROVED:
Signed _____ Date _____
Deputy Executive Director

Original — Master File   Contractor   District Office   Job Site   Accounting   Architect

Rev. 10/69

Fig. 6 (*continued*)

However, if the owner's claim is based on unforeseen or unexpected circumstances—and these circumstances were within the control of the contractor—then the owner's claim may have merit, and may be successfully pressed.

The role of inspector, in either case, is to insist upon performance in accordance with the specifications and the contract between the parties. Where there is deviation from the contract, the work records should show the deviation, and amounts—with amounts described by either time cards for labor trades, or lists of materials used, or both.

If the inspection team does its job of *documentation,* the arbitrator, whether through negotiation or litigation, has a much better base for establishing the actual cost of the claim.

The traditional approach for the owner's inspector is to studiously avoid the facts and admit to nothing, or attempt to admit to nothing. This not only is unacceptable ethically, but almost inevitably boomerangs on the owner. Where the quantities have been recognized and identified, the claim can be limited to actual loss and actual cost. However, where after the fact, a contractor is able to press a claim, and the inspection team ignored the situation in the field, the amount of the claim is often set on the basis of the plaintiff's legal team's ability to negotiate and debate. Settlements achieved in this way are usually severalfold what they might be if the inspection team had been prepared to recognize and accept a legitimate claim.

On the other hand, when the inspection team is maintaining documentation on areas which may be in dispute, the documentation should be kept as a proprietary item, or if it is accomplished in concert with the contractors, then it should be made clear that the maintenance of time and material records is for the purpose only of future resolution, and does not of itself admit that the work is either an extra or suitable claim.

CLAIMS FOR TIME

Time is money, and yet most change orders fail to include a suitable evaluation of the effect upon schedule. The section on change orders has discussed the technique for evaluating time impact of changes, and failure to do so will ultimately cost the owner. Here again, most owners use the ostrich approach—assuming that if the contractor accepts a financial arrangement for a change, he will be willing not to press for a delay in the end of the contract. In a few cases, it works out this way, but only in those few cases where the contractor finishes ahead of schedule, because, after all, the additional work almost inevitably has some delaying effect.

In large projects, it is not unusual to have more than 100 change orders, sometimes several hundred. If these have not been evaluated for time delay on an individual basis, it becomes an almost impossible task to evaluate them en masse at the end of the project. The effect is to make the question of value of delay one of negotiation—and one in which the facts are limited and the art of negotiation holds forth. Here again, the owner tends to lose—and to pay much more than he should. On the other hand, the contractor is also uninhibited by the facts—and usually is unsatisfied with any adjudicated amount—with the ultimate effect of having everyone dissatisfied.

In one complicated case, the owner and contractor had sued and countersued, each claiming that the delays of more than one year on a two-year project had cost substantial sums of money in lost services, as well as increased overhead for the contractor. Each was claiming several million dollars in damages from the other. The owner's project manager charged with the responsibility for representing the owner used CPM to do a post evaluation. This was accomplished by assembling an arrow diagram representing the sequence of work in the project based upon historical fact as disclosed by the work logs, inspector's daily records, progress reports and other documentation. A comprehensive diagram was made in sufficient detail to show the effect of relatively minor disruptions, as well as major changes. In the course of preparing the diagram, a reasonable time for having accomplished each work element was utilized, and then this was extended to match the actual time, by date, at each milestone by adding an increment of delay due to contractor's delay, and an increment due to the owner.

The completed network was totally critical, with no float. Dates compared with actual dates of progress taken from field reports. The network was then calculated by computer, all contractor delays were set to zero, and an earlier than actual completion date was calculated. The difference between this and the actual completion date then represented those delays which were (in the opinion of the owner's engineer) due to the contractor.

A second calculation was made in which the contractor's delays were reinserted and the owner's delays were set at zero. This theoretical earlier date compared with the actual conclusion date showed a block of time which was due to the owner's delays. However, the sum of contractor's delays plus owner's delays did not equal the total delay calculated by setting both the contractor's and the owner's delay times to zero and recalculating. This showed that there was overlapping, and the delay by one did not necessarily cause immediate or sequential delay by the other. While this was intuitively recognized, there had been no previous way of quantifying how much delay each side had caused.

Many of the owner's delays were due to slow recycling of shop drawings. This tardiness was for the very best reasons—because the owner's professional staff was trying to give a careful review—but in many cases, this kindness was a disservice to the contractor. If a shop drawing is to be refused, refuse it. Don't make them wait for re-engineering, which they may not accept anyway.

Roughly speaking, on this project, about a 30% increment of the overall delay was due to the owner, and 90% was due to the contractor, with an overlap indicated therefore of about 20% within which delay by one did not affect the other.

The role of the inspector's documentation was significant, and members of the inspection team were called into the networking effort required to reestablish the historical realities. If a viable CPM plan had been approved and utilized on the project, it would have been possible to achieve about the same results with much less effort. CPM had been used on the project, but quite ineffectually—and the CPM schedules were never ultimately approved—because the contractor's own planning group failed to develop a CPM which reflected what their own field forces were to accomplish.

3. PROGRESS PAYMENTS

One of the most important functions of the inspection team is the approval of progress payments. For the contract which is progressing at appropriate pace, with good attitude on the part of the contractors, the expeditious handling of progress payments requisitions results in a fast return of invested money to the contractor—and therefore lower interest costs.

On the project which is not going well, the approval of progress payments is one of the best ways for the inspection team to maintain control—and the contractor's interest.

There is a legal concern in regard to overpayment. The contractor who advances too far beyond the actual work accomplished in terms of his progress payments, should he choose to or be forced to default, will leave the owner in jeopardy of losing the overinvestment or overpayment.

PROGRESS PAYMENT BASIS

The basis of progress payments is, in all cases, an estimate of the work actually accomplished. This measurement may be by actual field observation of weights and quantities, by acceptance of contractor statements, or a combination. The method and accuracy of work measurement varies with the type of contract. There are many variations of construction contracts, but most break down into the following:

1. Fixed price, lump sum
2. Fixed price per unit quantity
3. Cost plus fixed fee

In the *lump sum* contract, which is usually competitively bid—with sealed confidential bids—the contractor selected is usually the one with the lowest bid price, qualified only by the fact that the contractor has demonstrated to be a responsible individual as proven by both his record and his ability to be bonded for the amount of the bid.

Shortly after the acceptance of the successful bid, the contractor is required to submit a cost breakdown. This is a detailed breakdown of the project costs by specified categories. Usually, the owner details the major categories, and the contractor has flexibility to select subcategories. The purpose of the project breakdown is to facilitate approval of progress payments. Usually, the designer assists the owner in the approval of the job breakdown. In some cases, the inspection team may also have the opportunity to participate. The owner and his representatives are trying to avoid front end loading of the payment schedule, which is a fairly general practice whereby the contractor weights his early activities more heavily, so that he is able to pull his investment out of the job as it progresses. The owner counters this not only by careful review of the job cost breakdown, but also by the fact that traditionally payments are on a monthly basis, and therefore, the contractor has invested that month's work, plus whatever approval on payment cycle time lag the owner imposes, which is usually a month or more. Beyond this, the owner generally reserves a retainer of 10%, in some cases 5%, of the progress payments as a contingency against problems or quality. The theory is that the owner withholds the contractor's profit until the job is completed.

The more detailed the breakdown, the more readily it can be reviewed. It is increasingly common practice to require the cost breakdown in accordance with the CPM breakdown. Figure 1 is a standard cost breakdown for a small road project. Where CPM is used, the breakdown should represent the cost by activity. This not only avoids keeping two sets of books—one for time and the other for cost—but also avoids out-of-sequence claims for progress payments. (Without the CPM sequence information, such claims may be made inadvertently, or even in a few cases, deliberately.)

In many heavy construction jobs, such as major highways, *lump sum unit price* bids are often taken. A typical bid of this type presents a categorical breakdown of units, such as: excavation—earth; excavation —rock; backfill—gravel; and paving per cubic yard in place. Perhaps 25 to 50 categories are specified, and estimates are given of the amounts for each category. The contractor, then, presents a lump sum estimate, but specifies a set fee per unit. The actual work is measured, and payment is made upon the quantities of work accomplished, regardless of the quantities listed by the designer. If a careful quantity survey has been

STATE OF CALIFORNIA
DEPARTMENT OF PUBLIC WORKS
DIVISION OF HIGHWAYS
PUBLIC WORKS BUILDING, SACRAMENTO

FORM WH-11.

IN_____

	ITEMS	QUANTITIES		BID	AMOUNT
	Bid bond, cash or certified check accompanying bid				11959 25
1	Clear and grub			11,200	11200
2	Develop water supply			2000	2000
3	Apply water	725 MGal		2 50	1812 50
4	Roadway excavation	39000 CY		50	19500
5	Structure excavation	290 CY		3	870
6	Structure Backfill	240 CY		4	960
7	Finish roadway			500	500
8	Class 1 aggregate Subbase	11350 Ton		2	22700
9	Class 2 aggregate base	5100 Ton		2 50	12750
10	Aggregate (type B AC)	4050 Ton		5 50	22275
11	Paving asphalt (AC)	205 Ton		35	7175
12	Liquid asphalt SC-2 (on ct.)	8 Ton		45	360
13	Class A concrete (minor struct)	6 CY		150	900
14	Class A concrete (struct.)	109 CY		65	7085
15	Bar reinforcing steel	12300 Lbs		15	1845
16	18" cmp (16 ga.)	334 LF		5	1670
17	36" bit. ctd. paved inv. cmp (12ga)	120 LF		18	2160
18	Property fence (type WM)	5600 LF		55	3080
19	16' property fence gates	3 Ea		75	225
20	Markers	70 Ea.		7 50	525

MADE BY_____ CHECKED BY_____ | | | TOTALS | | 119592 50 |

Fig. 1 Sample bid breakdown by units. *Source: California Department of Highways.*

accomplished by the designer, and too many unforeseen conditions do not occur, the final price should be close to the quantities upon which the bid was based.

In responding to this type of request for proposal, the contractor may apply strategy in the cost breakdown. Naturally, the unit costs bid, when extended by the quantities listed by the designer, result in a fixed price, for terms of comparison only. However, the contractor may utilize his own estimating and cost information to arrive at a total proposal price, and then may revise the units in anticipation of the actual job conditions. Knowing that the bid evaluators will review the unit prices (which become the basis for adding or deducting to the base fee in direct proportion to the actual quantities of work) the contractor may do his own take-off to attempt to determine where the designer may be over or under-estimating quantities. One tactic is to present a very favorable unit price where the contractor thinks there will really be less of a category than is listed. For those categories where the contractor expects substantially greater amounts of work, he may quote a unit price favorable to himself. This will result in a higher profit margin should the quantities run over the base estimate by the owner.

The inspection team has little to do with the establishment of prices in this type of contract. However, they play a vital role in the determination of the actual quantities of work accomplished. Payment is on the basis of units of count, length, area, volume, weight, or, for some items, lump sum. The unit of measurement for any particular category is given in the specifications and the unit breakdown list. In the field, quantities are measured and calculated to a degree of accuracy consistent with the value of the item.

Quantities of some items, such as paving, fencing, and sewer lines are measured in terms of their size and length or area. Measurement of volume by vehicle loads is not a recommended method of determining quantity, since the loading factors vary with the equipment, weather, and operators. Usually, excavation or fill are measured by weight on a truck scale. If by load, a bulking factor of 15–20% should be deducted.

When work has been paid for in a previous estimate, and loses value through damage, loss, theft, or failure to function, the value lost should be deducted from the following monthly requisition.

In certain cases, the owner's regulations permit payment for *materials delivered* to the job and stored for future installation. Generally, only a set percentage of the value of the material is paid as a progress payment. Figure 2 is an example of a request for payment for materials on hand, including an affidavit that the materials are exclusively for the project and have been delivered.

In some instances, a specific list of materials eligible for inclusion on progress payments when delivered to the job site is listed in the special provisions of the specifications.

The inspection team should carefully check out any materials claimed for payment when allowed under the special provisions of the specifications. If this is not done, it is quite possible that nonspecification material, or even incorrect material, may be delivered to the job site and paid for. The inspection team should also be concerned with the methods of unloading, and the storage facilities when the material is sensitive to handling or the elements. Security is also a consideration, and the contractor is usually required to provide fencing, watchmen, and other security measures.

In some cases, credit may be given when the material is shipped to a specified warehouse or holding area. The inspection team should make a point of verifying the delivery of major items.

In *cost-plus fixed-fee* work, the contractor may prepare an estimated bid, but only the fee portion is considered fixed. With few exceptions, the contractor is not responsible for escalation of prices, or even overreaching the estimate. In some cases, the owner may negotiate a contract which is paid on a cost-plus basis up to a certain limit or upset figure. This is a combination of the cost-plus fixed-fee and fixed-price contracts.

REQUISITIONS

Figure 3 is a project payment estimate which includes the breakdown shown in Fig. 1, and an estimate of the quantity of work accomplished and the value of that work for payment purposes.

Figure 4 is the next payment estimate, reflecting progress in the month, followed by Figs. 5 and 6, which show the third monthly period and the semi-final estimate at the end of the fourth month.

Figure 7 shows a separate project record item sheet for a unit price task (apply water), while Fig. 8 shows a similar item sheet for cubic yards of structural excavation.

Note, also, the method of handling change order authorizations shown on Line 14.

PAYMENT REQUISITION

The specification should include the format for requisition for progress payments. An agency which does substantial amounts of construction will generally have its own form, as will government units.

Figure 9 is the form for payment requisition used in Pennsylvania. A prerequisite to the preparation of this form is the project cost breakdown,

Form 135, and because of its importance no requisition is approved until the breakdown has been submitted and accepted. Naturally, the items listed in the requisition must be the same as the project work breakdown terminology.

Other forms are used to list materials on the job site, if the contract permits payment for these. This type of form is shown in Fig. 10. Actual bills of sale must be attached to this form which is appended to the application for payment.

A summary application for payment (Fig. 11) is submitted summarizing information on the stored materials and progress sheets. Also included is a summary of change order status.

The application is signed first by the contractor and then by the inspection team, and, if appropriate, by the design representative.

Signing should not be automatic on the part of either the inspection team or the design representative. In particular, the inspection team is certifying the quantities. Accuracy within a reasonable amount is important. However, the units reported should be considered within the context of their value. Certainly, greater care should be used in certifying the amounts for a high value item, rather than a lesser unit price item, or an item which has relatively small quantities and a medium or low value unit price.

RETENTION

It is usual practice for the specifications to call for retention by the owner of a portion of the progress payments as a contingency against uncompleted or incorrect work. The retention is usually between 5% and 10%. Usually, after the project is 50% complete, the contractor may request a reduction in the retainage. Figure 12 is the form used by Pennsylvania. The significant section is the co-agreement by the surety on the contract bond to the request for reduction in retainage. This agreement should erase the legal liability imposed upon the owner or his inspection team by premature approval of percentage of work. The Pennsylvania DGS considers a reduction to 4% if the percentage of completion is between 50% and 80%, thence to 2% if between 80% and 95%. Above 95%, if the project is on schedule, a reduction in retainage to 1% would be approved. This 1% retainage is the absolute minimum, while punch list items remain open.

Retainage is withheld also on change orders.

REQUEST FOR PAYMENT FOR MATERIALS ON HAND

DATE..

CO., RTE., P.M...................................

CONTRACT NO.................................

To:...
Resident Engineer

From:...
Contractor

In accordance with the provisions of Section 9-1.06 of the Standard Specifications, request is made for payment as "Materials on Hand" for the following materials:

ITEM NO.	QUANTITY	MATERIAL DESCRIPTION	VALUE	TYPE OF SUBSTAN-TIATING EVIDENCE OF PURCHASE AT-TACHED	WHERE STORED *

AFFIDAVIT:

The materials listed above have been purchased exclusively for use on the above referenced project. The material is separated from other like materials and is physically identified as our property for use only on Contract... The State may enter upon the premises for the purposes set forth in Section 6 of the Standard Specifications for inspection, checking or auditing, or for any other purpose as you consider necessary. It is expressly understood and agreed that this information and affidavit is furnished to the State for the purpose of obtaining payment for the above materials before they are delivered to, or incorporated into, the project described above, and that the storage thereof at the location shown is subject to, and under the control of, the State. A revised form showing the current status of the value of materials for which payment is being requested will be submitted each estimate period.

—————————————————————————
Contractor

* When stored at a location other than on the jobsite or at a fabricator's yard, a warehouse receipt for the material issued in the name of the State shall accompany the request for payment. In case the storage location (other than the jobsite or fabricator's yard) is the Contractor's property, the area containing the material to be paid for shall be fenced off and posted to indicate that the material within the fenced area is under the control of the State.

INSTRUCTIONS TO CONTRACTOR:
Submit original and two copies to Resident Engineer not later than one week prior to the end of the estimate period. Attach evidence of purchase (and warehouse receipt when required) to original.

INSTRUCTIONS TO RESIDENT ENGINEER:
Forward duplicate to Materials and Research Department.
Forward triplicate to District or Bridge Department.

Fig. 2 Request for payment of materials on hand. _Source: California Department of Highways._

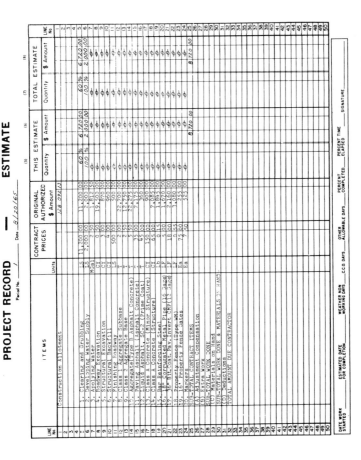

Fig. 3 Project payment estimate. Source: *California Department of Highways.*

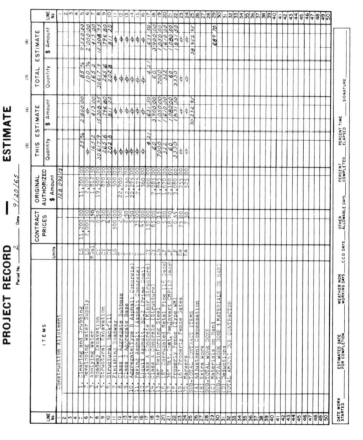

Fig. 4 Progress payment estimate. *Source: California Department of Highways.*

PROJECT RECORD — ESTIMATE

Period No. 3   Date 10/20/65

LINE No	ITEMS	Units	CONTRACT PRICES	ORIGINAL AUTHORIZED $ Amount	THIS ESTIMATE Quantity	THIS ESTIMATE $ Amount	TOTAL ESTIMATE Quantity	TOTAL ESTIMATE $ Amount	LINE No
1	Construction Allotment			128,262/2					1
2									2
3									3
4									4
5	1. Clearing and Grubbing	LS	11,200 00	11,200 00	15 %	1,680 00	100 %	11,200 00	5
6	2. Developing Water Supply	LS	2,000 00	2,000 00	-0-	-0-	100 %	2,000 00	6
7	3. Applying Water	MGal	2 50	1,812 50	484.3	1,210 75	649.5	1,623 75	7
8	4. Roadway Excavation	CY	0	19,100 00	8,927.17	2,464 85		7,961 80	8
9	5. Structural Excavation	CY	3 00	1,870 00	65.5		327.6		9
10	6. Structural Backfill	CY	4 00	960 00	65.9	263 60	268.1	1,074 80	10
11	7. Finishing Roadway	LS	500 00	500 00	-0-		-0-		11
12	8. Class 1 Aggregate Subbase	T	2 00	22,700 00	14,883.9	29,767 80	14,883.9	29,767 80	12
13	9. Class 2 Aggregate Base	T	3 50	22,225 00	5,066		5,066	17,731 00	13
14	10. Aggregate (Type B Asphalt Concrete)	T	5 00	22,375 00	-0-		-0-		14
15	11. Paving Asphalt (Asphalt Concrete)	T	35 00	2,125 00	-0-		-0-		15
16	12. Liquid Asphalt, SC-2 (Prime Coat)	T	45 00	360 00	-0-		-0-		16
17	13. Class A Concrete (Minor Structure)	CY	120 00	7,080 00	4.48	537 00	6.69	1,003 00	17
18	14. Class A Concrete (Structure)	CY	70 00	1,805 00	45.27	3,168 92	107.37	6,122 00	18
19	15. Bar Reinforcing Steel	Lb	0 15	1,670 00	4,050	607 50	1,030	1,657 50	19
20	16. 18" Corrugated Metal Pipe (16 Gage)	LF	5 00	2,180 00	270		372	1,670 00	20
21	17. Bit. Coat. Pav. Invert CMP (12 Gage)	LF	18 00	3,195 00	-240	1,260 00	130	2,340 00	21
22	18. Property Fence (Type WM)	LF	1 05		-0-	-84 00	350	367 50	22
23	19. 16' Property Fence Gates	Ea	75 00	255 00	-0-		-0-		23
24	20. Markers	Ea	2 50	555 00	-0-		-0-		24
25	SUB-TOTAL CONTRACT ITEMS					55,435 05		94,388 00	25
26	(A) Adjustment Compensation								26
27	(B) Extra Work								27
28	SUB-TOTAL WORK DONE								28
29	(C) Materials on Hand							278 27	29
30	SUB-TOTAL WORK DONE & MATERIALS ON HAND								30
31	Deductions								31
32	TOTAL AMOUNT DUE CONTRACTOR								32
33									33
34									34
35									35
36									36
37									37
38									38
39									39
40									40
41									41
42									42
43									43
44									44
45									45
46									46
47									47
48									48
49									49
50									50

DATE WORK STARTED   ESTIMATED DATE FOR COMPLETION   WEATHER NON WORKING DAYS   CCO DAYS   OTHER ALLOWABLE DAYS   PERCENT COMPLETED   PERCENT TIME ELAPSED

Fig. 5 Third monthly payment estimate. *Source: California Department of Highways.*

PROJECT RECORD — ESTIMATE

Period No. _4_   Date _1.8165_

Semi-final or final estimate

LINE No	ITEMS	Units	CONTRACT PRICES	ORIGINAL AUTHORIZED $ Amount	THIS ESTIMATE Quantity	THIS ESTIMATE $ Amount	TOTAL ESTIMATE Quantity	TOTAL ESTIMATE $ Amount	LINE No
1	Construction Allotment			128,072,63					1
2									2
3									3
4									4
5								11,232,00	5
6	1. Clearing and Grubbing	LS	11,200,00	11,200,00	0	0	120	2,000,00	6
7	2. Developing Water Supply	LS	2,000,00	2,000,00	0	0	100%	721,05	7
8	3. Applying Water	MGal	2,50	1,812,50	392	980,00	6981	19,762,80	8
9	4. Roadway Excavation	CY	50	19,890,00	0	0	39,525,6	4,050,00	9
10	5. Structural Excavation	CY	3,00	660,00	0	0	1211	1,054,40	10
11	6. Structural Backfill	CY	4,00	960,00	0	0	2681	500,00	11
12	7. Finishing Roadway	LS	500,00	500,00	100%	500,00	100%	29,167,80	12
13	8. Class 1 Aggregate Subbase	T	2,00	22,700,00	0	0	14,883,9	6,175,00	13
14	9. Class 2 Aggregate Base	T	2,50	23,125,00	0	0	3086	20,269,70	14
15	11. Paving Asphalt (Asphalt Concrete)	T	34,00	7,175,00	216,81	7,388,55	3865,4	7,388,55	15
16	12. Liquid Asphalt, SC-2 (Prime Coat)	T	35,00	360,00	6,55	294,72	216,81	294,72	16
17	13. Class A Concrete (Minor Structure)	CY	150,00	900,00	0	0	6,29	1,003,50	17
18	14. Class A Concrete (Structure)	CY	65,00	1,885,00	0	0	107,57	6992,05	18
19	15. Bar Reinforcing Steel	Lb	15	1,670,00	0	0	11,030	1,650,00	19
20	16. 18" Corrugated Metal Pipe (16 gage)	LF	5,00	2,160,00	0	0	322	1,610,00	20
21	17. 36" St.Coat.Pav. Invert CMP(12 gage	LF	18,00	1,080,00	0	0	130	2,340,00	21
22	19. Property Fence (Type WM)	LF	55		2353	1,294,15	5433	2,977,65	22
23	20. Property Fence Gates	Ea	75,00		5	375,00	5	375,00	23
24	30. Markers	Ea	7,50	525,00	72	540,00	72	540,00	24
25	SUB-TOTAL CONTRACT ITEMS					30,959,95		125,134,90	25
26	(A) Adjustment Compensation								26
27	(B) Extra Work Done								27
28	SUB-TOTAL WORK DONE								28
29	(C) Materials on Hand								29
30	SUB-TOTAL WORK DONE & MATERIALS ON HAND								30
31	(D) Deductions								31
32	TOTAL AMOUNT DUE CONTRACTOR								32
33									33
34									34
35									35
36									36
37									37
38									38
39									39
40									40
41									41
42									42
43									43
44									44
45									45
46									46
47									47
48									48
49									49
50									50

DATE WORK STARTED ___   ESTIMATED DATE FOR COMPLETION ___   WEATHER NON WORKING DAYS ___   CCO DAYS ___   OTHER ALLOWABLE DAYS ___   PERCENT COMPLETED ___   PERCENT TIME ELAPSED ___   SIGNATURE ___

Fig. 6 Semi-final estimate, fourth month. *Source: California Department of Highways.*

ITEM (1) Apply Water 125% ____
UNIT (2) M. Gal. PRICE (3) $ 2.50 75% ____ (4)

PROJECT RECORD — ITEM SHEET

POSTING REFERENCE 3

LINE NO.	DATE	DESCRIPTION OF ENTRY AND OR	REF.	THIS ESTIMATE (5) Quantity	THIS ESTIMATE (6) $ Amount	TOTAL ESTIMATE (7) Quantity	TOTAL ESTIMATE (8) $ Amount	(9) ✓ CHANGE	AUTHORIZED CHANGE (10) Quantity + or −	AUTHORIZED CHANGE (11) $ Amount + or −	TOTAL AUTHORIZED (12) Quantity	TOTAL AUTHORIZED (13) $ Amount		LINE NO.
1	8/13	Preliminary Estimate									725	1812.50		1
2	8/20	Estimate No. 1			⇐		⇐				725	1812.50		2
3														3
4	8/21	Daily Water Report No. 1		15.4										4
5	8/22	" " " " 2		10.5										5
6	8/23	" " " " 3		22.5										6
7	9/7	" " " " 4		34.6										7
8	9/8	" " " " 5		20.6										8
9	9/9	" " " " 6		5.6										9
10	9/12	" " " " 7		19.6										10
11	9/13	" " " " 8		16.8										11
12	9/14	" " " " 9		8.4										12
13	9/15	" " " " 10		11.2										13
14	9/20	Estimate No. 2		165.2	413.00	165.2	413.00				725	1812.50		14
15														15
16	9/21	Daily Water Report No. 11		23.1										16
17	9/22	" " " " 12		35.2										17
18	9/23	" " " " 13		31.7										18
19	9/24	" " " " 14		53.0										19
20	9/27	" " " " 15		18.3										20
21	9/28	" " " " 16		11.3										21
22	9/29	" " " " 17		37.8										22
23	9/30	" " " " 18		28.1										23
24	10/1	" " " " 19		30.0										24
25	10/4	" " " " 20		19.6										25
26	10/5	" " " " 21		17.4										26
27	10/6	" " " " 22		14.0										27
28	10/7	" " " " 23		30.4										28
29	10/8	" " " " 24		5.6										29
30	10/12	Scale Sheet #9-1 Aggr. Base		6.2										30
31	10/13	" " #9-2 " "		8.7										31
32	10/13	Daily Water Report No. 25		11.2										32
33	10/14	Scale Sheet #9-3 Aggr. Base		9.4										33
34	10/14	Daily Water Report No. 26		14.0										34
35	10/15	" " " " 27		14.0										35
36	10/15	Scale Sheet #9-4 Aggr. Base		8.6										36
37	10/18	" " #9-5 " "		12.6										37
38	10/18	Daily Water Report No. 28		11.2										38
39	10/19	" " " " 29		16.8										39
40	10/19	Scale Sheet #9-6 Aggr. Base		10.1										40
41	10/20	Estimate No. 3		484.3	1,210.75	649.5	1623.75				725	1812.50		41
42														42
43	10/20	Daily Water Report No. 30		16.8										43
44	10/21	" " " " 31		22.4										44
45	10/21	Quantity Balance	E4-2						−36.3	−90.75				45
46	11/8	Estimate No. 4		39.2	98.00	688.7	1721.75		−36.3	−90.75	688.7	1,721.75		46
47														47
48														48
49														49
50														50
51														51
52														52
53														53
54														54
55														55

FORM R-55

Fig. 7 Project record by item. *Source: California Department of Highways.*

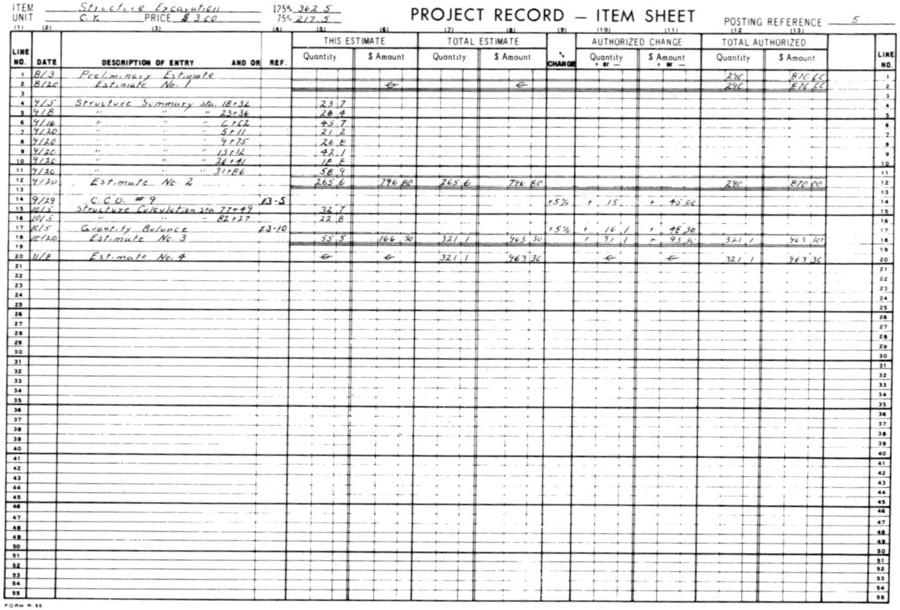

ITEM Structure Excavation 125% 362.5
UNIT C.Y. PRICE $3.00 75% 217.5

PROJECT RECORD — ITEM SHEET

POSTING REFERENCE 5

LINE NO. (1)	DATE (2)	DESCRIPTION OF ENTRY AND OR (3)	REF. (4)	THIS ESTIMATE Quantity (5)	THIS ESTIMATE $ Amount (6)	TOTAL ESTIMATE Quantity (7)	TOTAL ESTIMATE $ Amount (8)	% CHANGE (9)	AUTHORIZED CHANGE Quantity + or — (10)	AUTHORIZED CHANGE $ Amount + or — (11)	TOTAL AUTHORIZED Quantity (12)	TOTAL AUTHORIZED $ Amount (13)		LINE NO.
1	8/3	Preliminary Estimate									290	870.00		1
2	8/20	Estimate No. 1			0		0				290	870.00		2
3														3
4	9/5	Structure Summary sta. 18+32		23.7										4
5	9/8	" " " 23+36		28.4										5
6	9/16	" " " 6+62		45.7										6
7	9/20	" " " 5+11		21.2										7
8	9/20	" " " 9+75		26.8										8
9	9/20	" " " 13+12		42.1										9
10	9/20	" " " 26+41		18.8										10
11	9/20	" " " 31+86		58.9										11
12	9/20	Estimate No. 2		265.6	796.80	265.6	796.80				290	870.00		12
13														13
14	9/29	C.C.O. #9	E3-5					+5%	+ 15	+ 45.00				14
15	10/5	Structure Calculation sta. 77+49		32.7										15
16	10/5	" " " 82+27		22.8										16
17	10/5	Quantity Balance	E3-10					+5%	+ 16.1	+ 48.30				17
18	10/20	Estimate No. 3		55.5	166.50	321.1	963.30		+ 31.1	+ 93.30	321.1	963.30		18
19														19
20	11/8	Estimate No. 4		0	0	321.1	963.30		0	0	321.1	963.30		20
21														21
22														22
23														23
24														24
25														25
26														26
27														27
28														28
29														29
30														30
31														31
32														32
33														33
34														34
35														35
36														36
37														37
38														38
39														39
40														40
41														41
42														42
43														43
44														44
45														45
46														46
47														47
48														48
49														49
50														50
51														51
52														52
53														53
54														54
55														55

FORM R-55

Fig. 8 Project record by item and unit quantities. *Source: California Department of Highways.*

Contractor's Name_____

Contract No._____   Sheet_____of_____

APPLICATION FOR PAYMENT OF MATERIALS/EQUIPMENT INCORPORATED
(EXCLUSIVE OF CHANGE ORDERS OR STORED MATERIALS), Period Ending_____

(1) Form 135 Item No.	(2) Item Description - Form 135	(3) Total Units Form 135	(4) Unit Price Form 135	(5) Units Completed This Period	(6) Amount Earned This Period	(7) Units Completed To Date	(8) Remarks
X X X X X X X X X X X X Transfer to Recapitulation Sheet Line No. 2						X X X X X X X X X X X X X X X X	

GSA FORM NO. 8
CDR NO. 118C

Fig. 9 Payment requisition. *Source: Pennsylvania.*

Contractor's Name _____

Contract No. _____   Sheet _____ of _____

APPLICATION FOR PAYMENT OF STORED MATERIALS, PERIOD ENDING _____

(1) Form 197-B Approval Date	(2) Item No. from Form 135	(3) Extended Price from Form 135	(4) Extended Whole-sale Price from Form 197-B	(5) Amount of Labor and Overhead	(6) % L & O Completed This Period	(7) L & O Payable This Period	(8) % L & O Completed To Date	(9) Amount L & O Paid To Date
Transfer to Recap Sheet - Line No. 4 (Original Submission Only)				X X X X X X X X X X X X X X X X X X X X X X X X X X X X X X X X				
X X X X X X X X X X X X X X X X X X		Transfer to Recap Sheet Line No. 5					X X X X X X X X X X X	

GSA FORM NO. 8
CDR NO. 118B

Fig. 10 Materials at job site. *Source: Pennsylvania.*

18th and Herr Streets
Harrisburg, Pennsylvania

Institution or Facility:_____

District No._____

Location:_____

APPLICATION FOR PAYMENT (INVOICE) OF COMPLETED
WORK ON CONTRACT NO._____
FOR PERIOD ENDING:_____

Contractor's Name and Address

INVOICE NO._____
Sheet_____ of_____

Professional's Name and Address:

CONTRACT AWARD DATE_____
TOTAL CONTRACT AWARD$_____
AMENDED CONTRACT AMOUNT $_____

RECAPITULATION OF CONTRACTOR'S APPLICATION FOR PAYMENT

(1) PREVIOUS REQUESTS FOR PAYMENT TO DATE (Period ending_____) $_____

(2) Labor and Material completed this period (118-C) $_____

(3) Change Orders partially/wholly completed
     this period (118-D) $_____

(4) Stored Materials approved this period
     (Material Only) (118-E) $_____

(5) Labor & Overhead for Stored Materials
     this period (118-E) $_____

(6) TOTAL REQUEST FOR PAYMENT THIS PERIOD (Lines (2) thru (5)) $_____

(7) TOTAL REQUEST FOR PAYMENT TO DATE (Period ending_____) $_____

(8) Less_____ % retained of Line (7) $_____

(9) TOTAL REQUESTS FOR PAYMENT LESS RETAINAGE $_____

(10) Less amount previously paid $_____

(11) TOTAL AMOUNT NOW PAYABLE $_____

I CERTIFY THAT THIS IS A TRUE AND CORRECT APPLICATION FOR PAYMENT (INVOICE) OF WORK DONE AND
MATERIAL INCORPORATED:

Signed_____ Date_____
      (Contractor)

Distribution:
   G.S.A. Comptroller
   Professional (2)   Signed_____ Date_____
   Contractor   (Resident Inspector)
   District Engineer
   Resident Inspector   Signed_____ Date_____
                                   (Professional)

                             Signed_____ Date_____
                                   (District Engineer)

Fig. 11 Application for completed work payment. *Source: Pennsylvania.*

CONSTRUCTION	Project No. .... Contract No. .. ..
	Project Title ....
Request for Reduction or	Location ....
Discontinuance of	....
Retained Percentage	Date ....

18th and Herr Streets
Harrisburg, Pennsylvania

The Contractor, ... hereby, requests that the percentage of the estimates retained by The General State Authority under the

subject contract be reduced to .... %/discontinued.

....
Signature   Date

The Surety on the Contract Bond, ... ,
approves the above request.

Power of attorney   By ....
must be attached   Attorney-in-Fact   Date
to original copy

Approval is/is not recommended. The present percentage of completion as of ....
(date)

is .. .. % and the present percentage of elapsed time as of .... is .. .. %.
(date)

....
District Engineer   Date

Approval is/is not recommended.

....
Director of Construction   Date

Approved/Disapproved.

.. ..
Executive Director   Date

Distribution: Contractor; Surety; Director of Construction; District Engineer; Comptroller; Central File

Fig. 12 Request for reduction of retained percentage. *Source: Pennsylvania.*

4. DOCUMENTATION

An inspection team is, and should be, results oriented. However, there are occasions when investigation of job mishaps, accidents, or other fact-finding requirements require an exhaustive utilization of special portions of the job records. Another situation which requires use of the documentation is the preparation of claims, or defense against claims. The nature of the factfinding process will vary with the job situation, so that prior experience alone is not sufficient to use as a guide in setting up job documentation.

Simply stated, the job records should document every significant activity which occurs throughout the project. Usually, forms or work logs exist and are required for each of these significant steps. The purpose of a documentation plan is the insurance that a comprehensive set of documentation is maintained, and also in some instances that overlapping records are trimmed down to the minimum necessary.

ORGANIZATION

The organization of the files to maintain documentation is a key to minimizing the red tape involved in keeping records, as well as using them. Further, well-planned organization is indispensable to the retrieval of necessary information.

One key to information retrieval is the breakdown of the project into common categories. For instance, Fig. 1 is a table of a typical project file index category breakdown for the California Department of Highways. Where an organization is responsible for many projects, as in this case, file folders and loose-leaf binders can be preprinted with category numbers. Where a category does not apply to a particular project, it can be left blank. Extra numbers are provided for special categories to suit special project conditions.

The *index system* is organized by main subject categories, combined

FILE INDEX CATEGORY NUMBERS AND HEADINGS

Category No.

1. Project Personnel
2. Project Office Equipment and Supplies
3. Equipment and Personnel Reports
4. Service Contracts (Project Office Utilities and Services)
5. General Correspondence
6. Safety
7. Public Relations
8.
9. } Extra category numbers if required
10.
11. Information Furnished at Start of Project
12. Contractor
13. Signs and Striping
14. Photograph Record
15. Accident Reports
16. Utilities Agreements
17. Utility Work Performed
18. Agreements
19.
20.
21.
22. } Extra category numbers if required
23.
24.
25. Federal Requirements
26. Progress Schedule
27. Weekly Statement of Working Days
28. Weekly Newsletter
29. Materials Report and Preliminary Tests
30. Basement Soil Test Results
31. Notice of Materials to be Used (Form R-30)
32. Notice of Materials to be Sampled (Form R-28)
33. Notice of Materials to be Furnished (Form T-608)
34. Copies of Sample Identification
35. Mix Calculations
36. Plant and Spread Records
37. Field Lab Control Tests
38. District and Headquarters Lab Control Tests
39. Progress and Final Tests
40. Field Lab Assistant Reports
41. Bridge Report of Shipment of Material
42. Bridge Materials Tests
43. Bridge Concrete Pour Records
44. Bridge Diaries
45. Resident Engineers' Diaries
46. Contract Item Diaries
47. Structures
48. Contract Item Quantity Documents
49. Contract Change Orders
50. Adjustment of Compensation Calculations
51. Materials on Hand
52. Charges to Total Contract Allotment
53. Credit to Contract
54. Deductions from Payment to Contractor
55.
56.
57. } Extra category numbers if required
58.
59. Bridge Estimate Data
60. Project Record Item Sheets
61. Project Record—Estimate and Contingency—Status Sheet
62. Possible Claims
63. Report of Completion
64. Graphic Portrayal of Test Results

ALPHABETICAL LISTING OF CATEGORIES

Category No.

Accident Reports 15
Agreements 18
Basement Soil Test Results............................... 30
Charges to Total Contract Allotment 52
Contract Change Orders.................................. 49
Contract Item Diaries................................... 46
Contract Item Quantity Documents....................... 48
Contractor 12
Control Tests, District and Headquarters Lab 38
Control Tests, Field Lab................................. 37
Copies of Sample Identification.......................... 34
Credit to Contract 53
Deductions from Payment to Contractor.................. 54
Diaries, Contract Item................................... 46
Diaries, Resident Engineers'.............................. 45
District and Headquarters Lab Control Tests............. 38
Equipment and Personnel Reports........................ 3
Federal Requirements.................................... 25
Field Lab Assistant Reports.............................. 40
Field Lab Control Tests.................................. 37
General Correspondence 5
Graphic Portrayal of Test Result........................ 64
Information Furnished at Start of Project................ 11
Materials on Hand....................................... 51
Materials Report and Preliminary Tests.................. 29
Mix Calculations 35
Notice of Materials to be Furnished (Form T-608) 33
Notice of Materials to be Sampled (Form R-28) 32
Notice of Materials to be Used (Form R-30) 31
Photograph Record 14
Plant and Spread Records................................ 36
Possible Claim 62
Preliminary Tests, Materials Report and................. 29
Progress and Final Tests................................. 39
Progress Schedule 26
Project Office Equipment and Supplies................... 2
Project Personnel 1
Project Record—Estimate and Contingency— Status Sheet 61
Project Record Item Sheet............................... 60
Public Relations 7
Report of Completion 63
Resident Engineers' Diaries.............................. 45
Safety 6
Sample Identification, Copies of......................... 34
Service Contracts 4
Signs and Striping 13
Spread Records, Plant and............................... 36
Structures 47
Utilities Agreements 16
Utility Work Performed.................................. 17
Weekly Newsletter 28
Weekly Statement of Working Days...................... 27

Fig. 1 File index headings. *Source: California Department of Highways.*

RELATIVE COMPACTION SUMMARY

INDEX_____   MATERIAL_____

TEST NO.	DATE	STA.	LOCATION	ELEV. BELOW F.G.	MOISTURE OPT.	MOISTURE FIELD	R.C.	REMARKS

TEST RESULT SUMMARY

MINIMUM FREQUENCY

INDEX NO. _____

MATERIAL _____

DATE SAMPLED	TEST NUMBER	SAMPLE LOCATION	TEST RESULTS			PRODUCTION QUANTITY REPRESENTED	REMARKS

Fig. 2 Compaction summary tests. *Source: California Department of Highways.*

with an alphabetical cross index of subject matter. The file index should be posted prominently in an easily accessible location near the files and records, and should be distributed to the inspection team and administrative assistants.

The categories are generally self-explanatory, and the level of material contained is a function of the specific project and the organization responsible for inspection. In some categories, the material maintained is voluminous, and it is appropriate to maintain summary indices of the material contained in the file. Figure 2 is an example of compaction summary tests, indexing the actual test reports in the file.

Figure 3 is a sample form for maintenance of current information on materials on hand. This is particularly important on projects where the owner pays on the basis of materials delivered to the job site. The first five columns are used for recording and summarizing the release of material. This serves as a control of the quantity of material released to the job, and facilitates checking of requisitions for payment.

As material releases are received, information is entered in Columns 2, 3, and 4. As the material is received on the job or inventory, an appropriate entry is made in Column 5. Since there are no columns for keeping a running total of materials released or received, at requisition time, an estimated number of work completed is entered in Column 1 after the last entry in either Column 4 or 5. Columns 4 and 5 are totaled and the total is used to provide a check to indicate that the materials claimed as in place or delivered to the job are in accordance with documentation.

These forms are typical of those used by the California Highways Department. Figure 4 is a summary of some of the typical standard forms utilized by that organization. It is typical, and by no means all inclusive.

MATERIALS ON HAND & RELEASE RECORD

ITEM: _Bar Rein. Steel_   NUMBER _15_

ENGR. EST. _0.15_ UNIT PRICE _12,300_ QUANTITY _1845.00_ AMOUNT

R-30 RECD. _8-6-65_   R-28 RECD. _8-11-65_   INVOICE UNIT PRICE _0.10_

(1) EST. NO.	(2) DATE OF RELEASE	(3) LOT NUMBER	(4) QUANTITY RELEASED	(5) MATERIAL RECEIVED ON PROJECT		(6) AMOUNT PAID AS ITEM FROM COL. 7 & 8 FORM R-55		(7) MATERIAL ON HAND COL. 5 - 6	(8) CONTRACTOR'S REQUEST FOR PAYMENT FORM R-54		(9) TOTAL PAYMENT COL. 6 + 8		(10) TOTAL AUTHORIZED COL. 12 & 13 FORM R-55	
				DATE	QUANTITY	QUANTITY	AMOUNT	QUANTITY	QUANTITY	AMOUNT	QUANTITY	AMOUNT	QUANTITY	AMOUNT
	8-13	7701	4000	8-16	4000									
	8-16	7702	5000	8-17	5000									
1	8-20		9000		9000	-0-	-0-	9000	9000	900.00	9000	900.00	11 300	1695.00
	9-2	7703	3000	9-6	3000									
2	9-20		12000		12000	7000	1050.00	5000	5000	500.00	12000	1550.00	11330	1699.50
										This exceeds this				So
										this	Would	have	to	be
										reduced	to	4330		

SAMPLE ONLY

FORM HC-148

Fig. 3 Material Status Form. *Source: California Department of Highways.*

TABLE OF FORMS USED BY CONSTRUCTION DEPARTMENT
DISTRIBUTION AND NUMBER OF COPIES

Form Number	Name of Form	Headquarters				District				Time and/or Frequency
		Accounting Department	Office Engineer	Construction Department	Materials & Research	District Engineer	Resident Engineer	Contractor	Sampler	
HA-202	Cash Expenditure Voucher	2				1				End of month.
HA-302	Travel Expense Claim	2				1				End of month.
HA-423	Local Request					1				
HA-1126	Shipping Record					1				
HA-1226	Receiving Record					2	1			
HA-523	Transfer Request					2	1			
EX-5 (Rev.)	Contract Change Order		3		1	1	1	2		
HC-10 (Rev.)	Res. Engr. Daily Report					b	1			Daily
HC-10A	Asst. Res. Engr. Daily Report					1	1			
HC-13 (Rev.)	Load Slip					1	1	1		With each load when required
HC-24 (Rev.)	Daily Extra Work Report					1	1	1		End of day
HC-29	Report of Inspection of Materials				1	1	1			As soon as inspection is made
HC-30	Notice of Material to be used			1	1	1	1		1	As soon as source is known
HC-33	Daily Record of Platform Scale Weights				1	1	1	1		
HC-35	Monthly Report of Day Labor Work		1			1	2			Daily
HC-49	R.E. Report of Assignment		1	1		1	1	1		End of Month
HC-54 (Rev.)	Request for Payment-Materials on Hand			1	1	1				Upon Assignment
HC-55	Project Record Item Sheet						1			Contractor submits to Res. Engr.
HC-55-1	Project Record-Contingency Status						1			Daily
HC-56	Project Record-Estimate		1	1		1	1			Monthly with estimate
HC-113 (Rev.)	Daily Water Report		1	1		1	1			Monthly
HC-146	Weekly Statement of Working Days			1		1	1	1		Make entries daily, submit at end of week
HC-147	Progress Schedule						1			Contractor submits to Res. Engr.
HC-247	Construction Progress						1			Monthly
HMRT-101	Sample Identification Cards				2	1			1	With Sample
HMRT-101R	Record Sample Identification Card								1	With Sample
HMRT-297	Relative Compaction Test				2	1			1	With Sample
HMRT-502	Field Sample of Concrete				2	1	1			As required
HMRT-518	Cement Sample Card				1	1			1	With Sample
HMRT-3096D	Pavement Core Record				1		1	1		Daily entries-submit when coring completed

Fig. 4 Summary of standard forms. *Source: California Department of Highways.*

FIELD NOTES

The inspection team must document its observations in the field. In many cases, these are documented in the form of work logs, which can be maintained on a chronological basis, with entries preferably in ink on a day-by-day basis. For these reports, handwriting is more convenient, and perhaps even preferred. If field log notes are typed, they should be signed by the inspectors on a daily basis.

Forms can assist in the preparation of the daily report. For instance, Fig. 5 is a daily construction report form utilized by Pennsylvania. The format provides a convenient method of recording personnel and equipment on the project by contractor category. It would be particularly appropriate for a unit price or cost-plus job.

Figure 6 is a daily extra work report utilized by the California Highways Department, including extensions of unit costs. This, in effect,

		CONSTRUCTION				Project No. _____
18th and Herr Streets Harrisburg, Pennsylvania		**DAILY CONSTRUCTION REPORT**				Location _____ DISTRICT Date _____

Temperature		Weather	Weather Description			Institution _____ or Department _____
7:30	1:00	Clear	Rain			
9:00	4:30	Cloudy	Snow			

Class of Work	Foreman (F) Engineer (E) Supt. (S)	Mechanics	Laborers	Misc.	No. of pcs. equipment	Remarks
I. GENERAL CONTRACTOR (Name)						
A. Excavation & Backfill						
B. Foundation						
C. Steel Erection (structural)						
D. Masonry Work						
E. Rough Interior						
F. Roofing						
G. Finished Interior						
H.						
I.						
J.						
II. HEATING CONTRACTOR (Name)						
A. Boilers						
B. Roughing-in (Mech. work)						
C. Equipment Installation						
D. Air-conditioning & Vent.						
E. Pipe Work						
F.						
III. PLUMBING CONTRACTOR (Name)						
A. Plumbing & Pipe Work						
B. Roughing-in (Mech. Equip.)						
C. Equipment Installation						
D.						
E.						
IV. ELECTRICAL CONTRACTOR (Name)						
A. Conduit. & Etc.						
B. Equip. Instl.(Switch Gear, Etc.)						
C. Fixture Installation						
D.						
E.						
V. OTHER CONTRACTORS						
TOTAL						

Distribution: Construction Division Harrisburg;   District Office; Field Office.

(Over)

Fig. 5 Daily construction report. *Source: Pennsylvania.*

TO BE FILLED IN BY INSPECTOR IN CHARGE

I. WORK STOPPAGE:

A. Strikes:

Date Work Stopped	Trade	Date Settled	How Settled

Reason ....

(If Jurisdictional dispute give detailed report)

B. Material:

Shortage (what material and reason for shortage) ....

....

....

C. List all delays caused by contractor ....

....

....

D. I .... any delay in construction .... (If YES

yes   no

what was authorized?) ....

E. Temporary Heat

	Type	Location	Hours	Furnished by
....	....	1.	....	1. ....
....	....	2.	....	2. ....
....	....	3.	....	3. ....

F. Visitors:
....

II. DAILY TIME REPORT

A. WORKING PERSONNEL	Started Work	Quit Work	Overtime Hours
SIGNATURE	....	....	....
SIGNATURE	....	....	....
SIGNATURE	....	....	....
SIGNATURE	....	....	....
SIGNATURE	....	....	....
SIGNATURE	....	....	....
SIGNATURE	....	....	....
SIGNATURE	....	....	....
SIGNATURE	....	....	....
SIGNATURE	....	....	....

B. ABSENTEES	Type of Leave
Name	
Name	
Name	
Name	
Name	

Signed ....

Res. Insp. or Insp. In Chge.

Fig. 5 (*continued*)

DAILY EXTRA WORK REPORT   32   5
1,800.00
DATE PERFORMED_____   502.75
DATE OF REPORT_____   515.36
1,018.11

DESCRIPTION OF WORK Removing Slide Right of Sta. 342+50

(Northern Territory Only)

#		HOURLY RATE	HOURS	EXTENDED AMOUNTS
LABOR				
2	Laborers	3 325	16	53 20
2	Laborers, Overtime	4 99	4	19 96
1	Loader Operator	4 57	8	36 56
1	Loader Operator, Overtime	6 855	2	13 71
	(Note: Subsistence based on			
	8-hour days only)			
	SUB-TOTAL			123 43
	ADDED PERCENTAGE—(SEE SPECIAL PROVISIONS) 24%			29 62
	SUBSISTENCE NO 24/8 @ 4.50			13 50
	TRAVEL EXPENSE NO. @ $			
	OTHER			
	TOTAL COST OF LABOR		**A**	166 55
EQUIPMENT				
1	Loader, Crawler, Caterpillar 933	4 75	10	47 50
2	Dump Trucks, 8.3 C.Y.	10 46	20	209 20
2	Dump Truck Drivers, Overtime (1/2 Rate)	2 03	4	8 12
	+ 6.5% on premium wage per Special Provisions			0 53
2	Dump Truck Drivers' Subsistence			
	2 Days @ $4.50			9 00
MATERIALS AND/OR **WORK** DONE BY OTHER THAN CONTRACTOR'S FORCES				
	TOTAL COST OF EQUIPMENT AND MATERIALS		**B**	274 35

The above record is complete and correct

+ 20% ON LABOR COST   (A)   33 31
+ 15% ON EQUIPMENT AND MATERIALS COST   (B)   41 15
TOTAL THIS REPORT   515 36

Signature
Contractor's Representative

Signature
Resident Engineer

FOR DISTRICT USE ONLY
Pd. on Est. No._____
Checked by_____

Orig. to Dist. Engr.

Fig. 6 Daily extra work report. *Source: California Department of Highways.*

becomes a change order—and can be utilized only within the framework of the contract. (In general, the field inspection team is precluded from negotiating *any* extra work. However, in many cases, for certain specified areas or types of work, the field has preauthorization to undertake work within certain limits, and at prenegotiated fees.)

The form also provides a convenient location for notations on strikes, materials, delays (contractor or owner), and other salient project features. On the same form is a brief daily time report for the inspection team. This would only be functional for a smaller project, and preferably should be separated from the specific job reports, although the combined report would provide a vehicle for identification of those personnel actually on the job site in the event that some formal testimony is required in the future.

Job conferences attended by prime contractors, the designers, or the inspection team should be documented. The resident engineer or inspector should assign one or more of the team to take notes and prepare job conference minutes. The distribution of these minutes should be a function of the topic covered, and the implications of delay or other problems. The purpose of job conference documentation is to confirm agreements, as well as to establish areas which were disagreed upon. Wherever possible, particularly where agreements or substantial time extension or cost increase implications exist, the affected contractors should countersign the minutes.

CONTRACTOR DOCUMENTATION

The contractor is also required to document various specific items throughout the project. Usually, his field staff will also maintain progress notes and logs.

One of the first forms to be submitted by the contractor is a request for approval of materials and/or subcontractors. (See Fig. 7.)

Figure 8, representing the contract breakdown sheet submitted by the contractor, is a form which is so important to the progress of the job that it is rarely, if ever, forgotten. However, other forms required for legal protection are more formal in nature, rather than functional, and may easily be neglected. The inspection team should cooperate with the owner and/or designer to agree upon a check-off list to ensure that these more legalistic documents, such as Fig. 9, are submitted as appropriate.

FINAL INSPECTIONS

All of the prior experience and documentation of the inspection staff should be concentrated in the conduct of final inspections. These may be

18th and Herr Streets Harrisburg, Pennsylvania	CONSTRUCTION Request for Approval of Materials and/or Subcontractors	Project No.......Contract No...... Project Title.................... Location......................... .... Date............................

Contractor................................   Professional...............................

Street Address...........................   Street Address.............................

City.....................................   City.......................................

Material or Work	Type of Approval	Name and Address	Remarks (Leave blank-for Prof. use only)

Contractor Professional District Engineer Resident Inspector Dir. of Constr. (2) Asst. Dir. of Eng./Arch. Bur. of Sales & Use Tax	The items listed are tentatively approved, subject to the sub- sequent approval of any required samples or shop drawings. These approvals are given with the understanding that all specification requirements shall be met. ....Date........... Signature of Professional

Fig. 7 Approval of Materials/Subcontractors. *Source: Pennsylvania.*

for the purpose of approving partial completion, with a view toward utilization of part of the facility, or final inspection.

The inspection for acceptance should be requested in writing by the contractor, and scheduled by the inspection team. Further, appropriate parties should be informed, so that they may participate. Typical of the

| Form No. GSA 135 10-62 | CONSTRUCTION | Project No. .... Contract No. .... |
| The General State Authority
18th and Herr Streets
Harrisburg, Pennsylvania | CONTRACT
BREAKDOWN SHEET | Title ....
Institution .... |

ITEM NO.	ITEM DESCRIPTION	UNIT	QUANTITY	UNIT PRICE	EXTENDED PRICE

TOTAL CONTRACT AMOUNT $

I or we hereby certify that the above is a true and correct breakdown including all materials, accessories, labor insurance, etc., per contract requirements.

....
Contractor

....
Signature of Authorized Representative

....
Title   Date

DISTRIBUTION		APPROVED:
Contractor Architect/Engineer Asst. Dir. of Const.	District Engineer (2) Comptroller Insurance Unit Central File	

....
For Asst. Director of Construction   Date

Fig. 8 Contract breakdown sheet. *Source: Pennsylvania.*

CONSTRUCTION	Project No. .... Contract No. ....
	Project Title ....

18th and Herr Streets
Harrisburg, Pennsylvania

PAYROLL AFFIDAVIT **CONTRACTORS AFFIDAVIT** **STATEMENT OF SURETY COMPANY**	Institution ....
	Location ....

PAYROLL AFFIDAVIT

Commonwealth of Pennsylvania

County of ....

I, ... , Title: .... , being duly sworn according to law, depose and say that I am the employee of .... hereinafter called "contractor", who supervises the payments of the employees of said employer; that I am authorized on behalf of said contractor to make this affidavit; that I certify that said contractor upon the above The   State   contract has paid a wage scale in strict compliance with the provisions of the Prevailing Minimum Wage Predetermination, as determined by the Secretary of Labor and Industry, and set forth in the contract above referred to, and I further certify that the contractor is not receiving or requiring, or will not receive or require, directly or indirectly, from any employee, any refund of any such wage or wages.

Sworn to and subscribed before me this

................day of ...., 19 ....

NOTARY PUBLIC ....

My Commission expires ....

CONTRACTORS AFFIDAVIT

Commonwealth of Pennsylvania

County of ....

The undersigned contractor, being duly sworn according to law, deposes and says that all labor, material, outstanding claims and indebtedness of whatsoever nature arising out of the performance of the above contract with The   State   have been paid, except such as are included in a list attached hereto.

And the aforesaid contractor does hereby release The   State   and the Commonwealth of Pennsylvania and/or its employees and agents from any claim for labor and materials furnished and any claims arising through the performance of the above contract, except the amount of ....($ ....) due under the final estimate.

Sworn to and subscribed before me this

................day of ...., 19 ....
                                                                                                    Contractor

NOTARY PUBLIC ....

My Commission expires ....

STATEMENT OF SURETY COMPANY

In accordance with the provisions of the above contract dated.................... between The   State   and .... contractor, of ...., the ...., company of .... , surety on the bond of ...., after a careful examination of the books and records of said contractor or after receipt of an affidavit from contractor, which examination or affidavit satisfies this company that all claims for labor, materials and equipment rental have been satisfactorily settled, except such as are included in a list attached hereto, hereby approves of the final payment of the said.., contractor, and by these presents witnesseth that payment to the contractor of the final estimate shall not relieve the Surety Company of any of its obligations to The   State   or have any claim for labor and materials furnished as set forth in the said Surety Company's Bond.

In witness whereof, the said Surety Company has hereunto set its hand and seal this .... day of .... , 19 ....

(Corporate Seal)   ....
      .   Surety Company

Witness .... By ....
                                                              Attorney-in-Fact

Power of Attorney Must Be Attached

Fig. 9 Payroll Affidavit. *Source: Pennsylvania.*

	FOREMAN	MECHANICS	LABORERS	PRIME OR SUB.	CONTRACTOR	WORK IN PROGRESS
SITE CLEARING						
DEMOLITION						
EXCAVATION						
STORM SEWER						
EXT. PAVING						
FENCES/GATES						
LANDSCAPE						
CONCRETE/REINF.						
C.M.U.						
BRICKWORK						
STR. STEEL						
STR. JOIST						
MTL. DECK						
MISC. MTL.						
ORNAM. MTL.						
CARPENTRY						
ARCH WOODWORK						
WATERPROOFING						
INSULATION						
METAL SIDING						
ROOFING						
SEALERS						
H.M. FRAMES						
H.M. DOORS						
SPECIAL DOORS						
WINDOWS						
HARDWARE						
GLASS/GLAZE						
MTL. LATH, ETC.						
PLASTERING						
DRYWALL						
TILE/CERAMIC						
TILE, CEILING						
FLOORING						
PAINTING						
TOILET PART.						
LOUVERS/PANELS						
FLAG POLE						
DIR./SIGNS						
WIRE PART.						
SCALES						
TEL. BOOTH						
TOILET ACCES.						
SPRAY BOOTH						
COMPACTOR						
DOCK EQUIP.						
ENTR. MATS						
MONO. & HOIST						
M.V. LIFTS						
PLUMBING						
H.V.A.C.						
ELEC.						
MECHANIZATION						

SIG:

CONSTRUCTION MANAGERS
O'BRIEN-KREITZBERG & ASSOC., INC.

LOCATION & PROJECT NO.	A.M.	TEMP.	WEATHER	DATE:
G.M.F. & V.M.F.	P.M.	TEMP.	WEATHER	M T W T F S
SOUTHEASTERN PENNSYLVANIA	DAILY PROGRESS REPORT			SHT. OF
10932-78-A-0006				

Fig. 10 Daily Progress Report (two page), *Source: O'Brien-Kreitzberg & Assoc. Inc.*

ADDITIONAL COMMENTARY
EXTRA WORK IN PROGRESS
DIFFICULTIES (LABOR PROBLEMS, ACCIDENTS, ETC.)
ORDERS ISSUED
UNSAT. WKMNSHP.
DISPUTES & ACTION TAKEN
VISITORS
GENERAL

LOCATION & PROJECT NO.	DAILY PROGRESS REPORT	DATE:
G.M.F. & V.M.F. SOUTHEASTERN PENNSYLVANIA 109320-78-A-0006		M T W T F S
		SHT. OF

Fig. 10 (*continued*)

GENERAL CONTRACTOR: SOMERS CONSTRUCTION CO., INC.

ALERT AREAS ( ) NONE
                ( ) SEE BELOW/ATTACHED

CONSTRUCTION ACTIVITY/ACTIVITIES:

DISTRIBUTION:
U.S.P.S. (WASH.)
O-K-A (FIELD)
O-K-A (OFFICE)   SIG.

	CONSTRUCTION MANAGER O'BRIEN-KREITZBERG & ASSOC., INC.
LOCATION & PROJECT NO. G.M.F. & V.M.F. SOUTHEASTERN PENNSYLVANIA 109320-78-A-0006	WEEKLY PROGRESS REPORT NO. ____
	WEEK ENDING | SHT. OF

Fig. 11 Weekly Progress Summary. *Source: O'Brien-Kreitzberg & Assoc. Inc.*

FOLLOW UP QUALITY CONTROL REPORT FOR MASONRY
UNITED STATES POST OFFICE, WHITE PLAINS, NEW YORK

CONTRACT NUMBER: 109320-78-V-0031   DATE: _____

CONTRACTOR:

WORK FORCE:

SUPT: FOREMAN: LAB: MAS: OTHER:

DESCRIPTION AND LOCATION OF WORK:

THE FOLLOWING ITEMS WERE CHECKED FOR CONTRACT COMPLIANCE:

Masonary mortar [ ]   Brick [ ]   Block [ ]   Reinforcing [ ]   Alignment of masonary [ ]   Weather conditions acceptable [ ]   Masonary joints [ ]   Flashings installed [ ]   Checked frames, in inserts for location and plumbers, etc. [ ]   Fill frames with mortar [ ]   Place Rigid insulation in exterior walls [ ]   Cleaning of Masonry [ ]   Scaffold and general safety [ ]

DEFICIENCES NOTED:

CORRECTIVE ACTION TAKEN:

REMARKS:

The above report is correct to the best of my knowledge.

SIGNED: _____
                              QUALITY CONTROL SUPERVISOR

Fig. 12 Quality Control Report. *Source: O'Brien-Kreitzberg & Assoc. Inc.*

GENERAL	DAMAGED	DEFECT.	MISSING	DIRTY	INCOMP.		DAMAGED	DEFECT.	MISSING	DIRTY	INCOMP.		DAMAGED	DEFECT.	MISSING	DIRTY	INCOMP.
Access Door						Countertop						Partition					
Base						Curtain Track						Peg Board					
Bench						Door						Plaster					
Blind						Door Buck						Plinth					
Borrow. Lt.						File Cabinet						Shelf					
Casework						Filler Piece						Sign					
Ceiling						Fire Ext. Cab.						Tele. Alcove					
Ceiling Molding						Floor						Therapy Grid					
Caulking						Folding Part.						Transfer Grille					
Chaulkboards						Glass						Wainscot					
Closet						Handrail						Wall					
Clothes Bar						Locker #						Window					
Convect. Encl.						Marble						Window Frame					
Corner Guard						Paint						Weatherproof					
						Filler Pieces											
DOOR HARD.																	
Dead Lock						Emerg. Stop						Mutes					
Door Bumpers						Floor Pivot						Panic Device					
Door Checks						Hinge						Pull Plate					
Door Closer						Kick Plate						Push Plate					
Door Glass						Lock Set						Rollor Latch					
Door Stay						Louver						Saddle					
Door Knob																	
PLUMBING																	
Bath Tub						Grab Bar						Shelf Mirrors					
Coat Hooks						Hand Dryer						Soap Holder					
Console						Hose Connect'n						Toilet Pap. Hold.					
Clothing Hook						Ident. Cap.						Towel Rack					
Dispen. Cab.						Oxy.-Vac.						Urinal					
Diver. Valve						Piping						Water Closet					
Emergen. Shower						Racks											
Floor Drains						Shower											
Floor Grading						Sink											
PLUMBING HARD.																	
Aerator						Faucet						Mixing Valve					
Drain						Flushometer						Trap					
Escut. Plate						Foot Valve						Vacuum Breaker					
						Knee Valve						Soap Dish					
HVAC																	
Air Duct						Disch. Plenum						Piping Insul.					
Ceiling Grille						Drip Pans						Refrigerator					
Control Val.						Environ. Rooms						Sprinkler Head					
Convec. Grille						Lint Screen						Thermostat					
Diffuser						Piping											
ELECTRICIAL																	
Clocks						Intercom						Phone Jack					
Circuit Breakers						Light						Receptacle					
Code Card						Light Fixtures						Recept. Plates					
Ele. Panel & Dir.						Night Light						Sign (Operation)					
Fire Alarm Box						Nurses Call						Switch					
Gong						P.A. Speaker						Switch Plate					
						N.C. Pull Cord						Watch Tour Sys.					

Data Punched		C.P.I.D. 95-6		Date Repunched	
Room No.	Gen. Rms.		Sheet No.		HO-145
REMARKS:					

Fig. 13 Hospital Punch List Form. *Source: O'Brien-Kreitzberg & Assoc. Inc.*

Figure 10 is a two-page daily Field Progress Report developed by O'Brien-Kreitzberg and Associates Inc. for a U.S. Postal Service project. Figure 11 is the format for the Weekly Report, an abstract of the daily reports.

QUALITY CONTROL

A number of federal agencies have adopted a procedure calling for quality control inspection by the contractor's own supervision. Figure 12 is a sample form for masonry work.

Figure 13 is a sample form used in final punch listing for a hospital project. A copy of the sheet is provided to the contractor and a second copy posted in the area.

FINAL INSPECTIONS

All of the prior experience and documentation of the inspection staff should be concentrated in the conduct of final inspections. These may be: for the purpose of approving partial completion, with a view toward utilization of part of the facility, or final inspection.

The inspection for acceptance should be requested in writing by the contractor, and scheduled by the inspection team. Further, appropriate parties should be informed, so that they may participate. Typical of the parties to be notified, in writing, would be: the owner; other prime contractors; using agency; the surety company of each prime contractor; federal cognizant agencies (where federal funding programs have been utilized); and the insurance company for the owner.

If all or part of the project (where partial inspection was requested) has been substantially completed, the inspection team should prepare a certificate of substantial completion similar to Fig. 14.

Before the owner may occupy any or all of the building, a *certificate of occupancy* must be issued. Where the federal or state government is building for its own utilization, the C.O. would be similar to Fig. 15. In private work, the C.O. is normally issued by the local authority with cognizant authority over building permits. In certain cases, an additional approval or certificate of occupany is required by the State Bureau of

| 18th and Herr Streets
Harrisburg, Pennsylvania 17120 | Certificate of
Substantial
Completion | Contract No._____
Contractor's Name: |

1. Job Description and Location:

2. Date of Inspection:

3. State responsibilities of the Contractor for maintenance, heat and utilities:

4. The attached list of uncompleted items, if applicable and necessary, shall be completed not later than:

5. The Work on the above Contract is Substantially complete as of:

_____   _____
Professional's Signature   Date

6. The above is acknowledged and accepted:

_____   _____
Contractor's Signature   Date

_____   _____
Resident Inspector's Signature   Date

_____   _____
District Engineer's Signature   Date

Fig. 14 Certificate of Substantial Completion. *Source: Pennsylvania.*

CONSTRUCTION	Project No. .... Con't Nos. ....
	Building ....
CERTIFICATE	or
of	Facility ....
OCCUPANCY	Institution ....

18th and Herr Streets
Harrisburg. Pennsylvania

In accordance with the permission granted by The General State Authority, this Institution has completely partially* occupied the above project contracts* since ....

....

If partial, a detailed description of the occupancy is as follows:

This action was taken with the understanding that construction will be completed in accordance with the plans and specifications.

Signature ....

Title ....

Date ....

* Delete inapplicable words

The Institution has been granted permission to occupy the project or portion thereof as described above.

Signature ....
District Engineer

Date ....

This occupancy is approved by the Using Agency.

Signature ....
Head of Department

Department ....

Date ....

Fig. 15 Certificate of Occupancy. *Source: Pennsylvania.*

| | CONSTRUCTION | Project No. _____ |
| 18th and Herr Streets
Harrisburg, Pennsylvania | Report of Defective
Materials
or
Workmanship | Contract No. _____
Project Title _____
Location _____

Date _____ |

Contract No.	Contractor	Defective Condition	Action to Secure Correction

Inspection Date: Report Date:

_____   _____   _____

District Engineer

Fig. 16 Report of Defective Materials or Workmanship. *Source: Pennsylvania.*

	CONSTRUCTION	Project No._____
18th and Herr Streets Harrisburg, Pennsylvania	Acceptance of Materials and Workmanship	Contract No._____ Project Title_____ Location_____ ——————————————— Date_____

The construction completed under the contracts comprising the subject project has been inspected in connection with the Bonds against Defective or Inferior Materials or Workmanship furnished by the respective contractors.

Accordingly, it has been determined that all materials installed and workmanship performed under the following contracts are acceptable to date:

CONTRACT NO. _____ CONTRACTOR _____

_____

Signature

_____

Title

Fig. 17 Acceptance of Materials and Workmanship. *Source: Pennsylvania.*

Labor for manufacturing facilities, or an appropriate Department of Safety where hazardous materials or equipment is to be utilized.

As part of the *final inspection,* the inspection team should prepare a list of work to be accomplished prior to final acceptance. The value of the list should be offset against any request to pay retainage on the project, and usually final acceptance of the punch list or list of items to be corrected is a prerequisite to payment of the retainage.

In most cases, the contractor is called upon to warrant the workmanship and materials for a period of about one year. Pennsylvania performs a warranty inspection during the eleventh month, noting any defective items, and accepting the balance as indicated in Figs. 16 and 17.

DOCUMENTATION

Forms alone do not make successful documentation. Inspection without documentation is self-limiting, and the results of the inspection may be useless to the owner.

APPENDIX A

This appendix contains a listing of the current ASTM standards which apply to construction. The listing is first by standard number, and then by functional category. The list is taken directly from the table of contents of the 1980 two-volume collection of construction standards published by the American Society for Testing and Materials (ASTM), 18th Edition (1980).

CONTENTS

ASTM Standards Used in Building Codes

Volume I: A 6–B 587
Volume II: C 4–F 476

In the serial designations prefixed to the following titles, the number following the dash indicates the year of original issue as tentative or of adoption as standard or, in the case of revision, the year of last revision. Thus, standards adopted or revised during the year 1980 have as their final number, 80. A letter following this number indicates more than one revision during that year, that is, 80a indicates the second revision in 1980, 80b the third revision, etc. Tentatives are identified by the letter T. Standards that have been reapproved without change are indicated by the year of last reapproval in parentheses as part of the designation number, for example, (1980).

Since the standards in this book are arranged in alphanumeric sequence, no page numbers are given in this contents.

A	6–79b	Spec. for General Requirements for Rolled Steel Plates, Shapes, Sheet Piling, and Bars for Structural Use
A	20–79b	Spec. for General Requirements for Steel Plates for Pressure Vessels
A	27–79	Spec. for Mild- to Medium-Strength Carbon-Steel Castings for General Application
A	29–79	Spec. for Steel Bars, Carbon and Alloy, Hot-Rolled and Cold-Finished, General Requirements for
A	31–76	Spec. for Boiler Rivet Steel and Rivets
A	36–77a	Spec. for Structural Steel
A	47–77	Spec. for Malleable Iron Castings

A 48–76	Spec. for Gray Iron Castings
A 53–79	Spec. for Pipe, Steel, Black and Hot-Dipped, Zinc-Coated Welded and Seamless
A 74–75	Spec. for Cast Iron Soil Pipe and Fittings
A 82–79	Spec. for Cold-Drawn Steel Wire for Concrete Reinforcement
A 90–69(1978)	Tests for Weights of Coating on Zinc-Coated (Galvanized) Iron or Steel Articles
A 105–79	Spec. for Forgings, Carbon Steel, for Piping Components
A 106–79b	Spec. for Seamless Carbon Steel Pipe for High-Temperature Service
A 108–79	Spec. for Steel Bars, Carbon, Cold-Finished, Standard Quality
A 109–72(1979)	Spec. for Steel, Carbon, Cold-Rolled Strip
A 120–79	Spec. for Pipe, Steel, Black and Hot-Dipped Zinc-Coated (Galvanized) Welded and Seamless for Ordinary Uses
A 126–73(1979)	Spec. for Gray Iron Castings for Valves, Flanges, and Pipe Fittings
A 134–74	Spec. for Electric-Fusion (Arc)-Welded Steel Plate Pipe (Sizes 16 in. and Over)
A 135–79	Spec. for Electric-Resistance-Welded Steel Pipe
A 139–74	Spec. for Electric-Fusion (Arc)-Welded Steel Pipe (Sizes 4 in. and over)
A 148–73(1979)	Spec. for High-Strength Steel Castings for Structural Purposes
A 153–78	Spec. for Zinc Coating (Hot-Dip) on Iron and Steel Hardware
A 167–77	Spec. for Stainless and Heat-Resisting Chromium-Nickel Steel Plate, Sheet, and Strip
A 176–79	Spec. for Stainless and Heat-Resisting Chromium Steel Plate, Sheet, and Strip
A 177–69(1979)	Spec. for High-Strength Stainless and Heat-Resisting Chromium-Nickel Steel Sheet and Strip
A 178–79b	Spec. for Electric-Resistance-Welded Carbon Steel Boiler Tubes
A 179–79	Spec. for Seamless Cold-Drawn Low-Carbon Steel Heat-Exchanger and Condenser Tubes
A 181–77	Spec. for Forgings, Carbon Steel, for General Purpose Piping
A 182–79a	Spec. for Forged or Rolled Alloy-Steel Pipe Flanges, Forged Fittings, and Valves and Parts for High-Temperature Service
A 184–79	Spec. for Fabricated Deformed Steel Bar Mats for Concrete Reinforcement
A 185–79	Spec. for Welded Steel Wire Fabric for Concrete Reinforcement
A 192–79	Spec. for Seamless Carbon Steel Boiler Tubes for High-Pressure Service
A 193–79a	Spec. for Alloy-Steel and Stainless Steel Bolting Material for High-Temperature Services
A 194–79	Spec. for Carbon and Alloy Steel Nuts for Bolts for High-Pressure and High-Temperature Service
A 197–79	Spec. for Cupola Malleable Iron
A 199–79a	Spec. for Seamless Cold-Drawn Intermediate Alloy-Steel Heat-Exchanger and Condenser Tubes
A 202–78	Spec. for Pressure Vessel Plates, Alloy Steel, Chromium-Manganese-Silicon
A 203–79a	Spec. for Pressure Vessel Plates, Alloy Steel, Nickel
A 204–79a	Spec. for Pressure Vessel Plates, Alloy Steel, Molybdenum

A 209–79a   Spec. for Seamless Carbon-Molybdenum Alloy-Steel Boiler and Super-heater Tubes

A 211–75   Spec. for Spiral-Welded Steel or Iron Pipe

A 213–79b   Spec. for Seamless Ferritic and Austenitic Alloy-Steel Boiler, Super-heater, and Heat-Exchanger Tubes

A 214–79   Spec. for Electric-Resistance-Welded Carbon Steel Heat-Exchanger and Condenser Tubes

A 216–77   Spec. for Carbon-Steel Castings Suitable for Fusion Welding for High-Temperature Service

A 217–79   Spec. for Martensitic Stainless Steel and Alloy Steel Castings for Pressure-Containing Parts Suitable for High-Temperature Service

A 225–79a   Spec. for Pressure Vessel Plates, Alloy Steel, Manganese-Vanadium

A 226–79   Spec. for Electric-Resistance-Welded Carbon Steel Boiler and Super-heater Tubes for High-Pressure Service

A 234–79a   Spec. for Piping Fittings of Wrought Carbon Steel and Alloy Steel for Moderate and Elevated Temperatures

A 239–73(1978)   Test for Locating the Thinnest Spot in a Zinc (Galvanized) Coating on Iron or Steel Articles by the Preece Test (Copper Sulfate Dip)

A 240–79   Spec. for Heat-Resisting Chromium and Chromium-Nickel Stainless Steel Plate, Sheet, and Strip for Fusion-Welded Unfired Pressure Vessels

A 242–79   Spec. for High-Strength Low-Alloy Structural Steel

A 249–79b   Spec. for Welded Austenitic Steel Boiler, Superheater, Heat-Exchanger, and Condenser Tubes

A 250–79a   Spec. for Electric-Resistance-Welded Carbon-Molybdenum Alloy-Steel Boiler and Superheater Tubes

A 252–77a   Spec. for Welded and Seamless Steel Pipe Piles

A 254–79   Spec. for Copper Brazed Steel Tubing

A 256–46(1976)   Compression Testing of Cast Iron

A 263–79   Spec. for Corrosion-Resisting Chromium Steel Clad Plate, Sheet, and Strip

A 264–79   Spec. for Stainless Chromium-Nickel Steel Clad Plate, Sheet, and Strip

A 265–79   Spec. for Nickel and Nickel-Base Alloy Clad Steel Plate

A 266–78   Spec. for Forgings, Carbon Steel, for Pressure Vessel Components

A 268–79a   Spec. for Seamless and Welded Ferritic Stainless Steel Tubing for General Service

A 269–79a   Spec. for Seamless and Welded Austenitic Stainless Steel Tubing for General Service

A 270–79   Spec. for Seamless and Welded Austenitic Stainless Steel Sanitary Tubing

A 276–79a   Spec. for Stainless and Heat-Resisting Steel Bars and Shapes

A 278–75   Spec. for Gray Iron Castings for Pressure-Containing Parts for Temperatures Up to 650° F (345° C)

A 283–79   Spec. for Low and Intermediate Tensile Strength Carbon Steel Plates, Shapes and Bars

A 285–78   Spec. for Pressure Vessel Plates, Carbon Steel, Low- and Intermediate-Tensile Strength

A 297–79a	Spec. for Heat-Resistant Iron-Chromium and Iron-Chromium-Nickel Alloy Castings for General Application
A 299–79b	Spec. for Pressure Vessel Plates, Carbon Steel, Manganese-Silicon
A 302–79b	Spec. for Pressure Vessel Plates, Alloy Steel, Manganese-Molybdenum and Manganese-Molybdenum-Nickel
A 307–78	Spec. for Carbon Steel Externally Threaded Standard Fasteners
A 308–78	Spec. for Steel, Sheet, Cold-Rolled, Long Terne Coated
A 311–79	Spec. for Stress Relieved Cold-Drawn Carbon Steel Bars Subject to Mechanical Property Requirments
A 312–79a	Spec. for Seamless and Welded Austenitic Stainless Steel Pipe
A 320–79a	Spec. for Alloy-Steel Bolting Materials for Low-Temperature Service
A 321–79	Spec. for Steel Bars, Carbon, Quenched and Tempered
A 322–79a	Spec. for Hot-Worked Alloy Steel Bars
A 325–79	Spec. for High-Strength Bolts for Structural Steel Joints, Including Suitable Nuts and Plain Hardened Washers
A 328–75a	Spec. for Steel Sheet Piling
A 331–74(1979)	Spec. for Steel Bars, Alloy, Cold-Finished
A 333–79	Spec. for Seamless and Welded Steel Pipe for Low-Temperature Service
A 334–79	Spec. for Seamless and Welded Carbon and Alloy-Steel Tubes for Low-Temperature Service
A 335–79a	Spec. for Seamless Ferritic Alloy Steel Pipe for High-Temperature Service
A 336–79	Spec. for Alloy Steel Forgings for Pressure and High Temperature Parts
A 350–79	Spec. for Forgings, Carbon and Low-Alloy Steel, Requiring Notch Toughness Testing for Piping Components
A 351–79	Spec. for Austenitic Steel Castings for High-Temperature Service
A 352–79	Spec. for Ferritic Steel Castings for Pressure-Containing Parts Suitable for Low-Temperature Service
A 353–79a	Spec. for Pressure Vessel Plates, Alloy Steel, Nine Percent Nickel Double-Normalized and Tempered
A 354–79	Spec. for Quenched and Tempered Alloy Steel Bolts, Studs, and Other Externally Threaded Fasteners
A 358–79a	Spec. for Electric-Fusion-Welded Austenitic Chromium-Nickel Alloy Steel Pipe for High-Temperature Service
A 361–76	Spec. for Steel Sheet, Zinc-Coated (Galvanized) by the Hot-Dip Process for Roofing and Siding
A 366–72(1979)	Spec. for Steel, Carbon, Cold-Rolled Sheet, Commercial Quality
A 368–76	Spec. for Stainless and Heat-Resisting Steel Wire Strand
A 369–79a	Spec. for Ferritic Alloy Steel Forged and Bored Pipe for High-Temperature Service
A 370–77	Methods and Definitions for Mechanical Testing of Steel Products
A 372–78	Spec. for Carbon and Alloy Steel Forgings for Thin-Walled Pressure Vessels
A 376–79b	Spec. for Seamless Austenitic Steel Pipe for High-Temperature Central-Station Service
A 377–79	Spec. for Gray Iron and Ductile Iron Pressure Pipe
A 387–79b	Spec. for Pressure Vessel Plates, Alloy Steel, Chromium-Molybdenum

A 395–77	Spec. for Ferritic Ductile Iron Pressure-Retaining Castings for Use at Elevated Temperatures
A 403–79	Spec. for Wrought Austenitic Stainless Steel Piping Fittings
A 405–79a	Spec. for Seamless Ferritic Alloy-Steel Pipe Specially Heat Treated for High-Temperature Service
A 409–79a	Spec. for Welded Large Outside Diameter Light-Wall Austenitic Chromium-Nickel Alloy Steel Pipe for Corrosive or High-Temperature Service
A 412–75	Spec. for Stainless and Heat-Resisting Chromium-Nickel-Manganese Steel Plate, Sheet, and Strip
A 414–79	Spec. for Carbon Steel Sheets for Pressure Vessels
A 416–74	Spec. for Uncoated Seven-Wire Stress-Relieved Strand for Prestressed Concrete
A 420–79	Spec. for Piping Fittings of Wrought Carbon Steel and Alloy Steel for Low-Temperature Service
A 421–77	Spec. for Uncoated Stress-Relieved Wire for Prestressed Concrete
A 423–79a	Spec. for Seamless and Electric Welded Low-Alloy Steel Tubes
A 424–73	Spec. for Steel Sheet for Porcelain Enameling
A 426–79	Spec. for Centrifugally Cast Ferritic Alloy Steel Pipe for High-Temperature Service
A 430–79	Spec. for Austenitic Steel Forged and Bored Pipe for High-Temperature Service
A 434–76	Spec. for Steel Bars, Alloy, Hot-Rolled or Cold-Finished, Quenched and Tempered
A 435–75	Spec. for Straight-Beam Ultrasonic Examination of Steel Plates for Pressure Vessels
A 437–77	Spec. for Alloy-Steel Turbine-Type Boiling Material Specially Heat Treated for High-Temperature Service
A 441–79	Spec. for High-Strengh Low-Alloy Structural Manganese Vanadium Steel
A 442–79b	Spec. for Pressure Vessel Plates, Carbon Steel, Improved Transition Properties
A 444–78	Spec. for Steel Sheet, Zinc-Coated (Galvanized) by the Hot-Dip Process for Culverts and Underdrains
A 446–76	Spec. for Steel Sheet, Zinc Coated (Galvanized) by the Hot-Dip Process, Structural (Physical) Quality
A 449–78a	Spec. for Quenched and Tempered Steel Bolts and Studs
A 450–79a	Spec. for General Requirements for Carbon, Ferritic Alloy, and Austenitic Alloy Steel Tubes
A 451–79	Spec. for Centrifugally Cast Austenitic Steel Pipe for High-Temperature Service
A 452–79	Spec. for Centrifugally Cast Austenitic Steel Cold-Wrought Pipe for High-Temperature Service
A 453–78	Spec. for Bolting Materials, High-Temperature, 50 to 120 ksi Yield Strength, with Expansion Coefficients Comparable to Austenitic Steels
A 455–79b	Spec. for Pressure Vessel Plates, Carbon Steel, High Strength Manganese

A 463–77	Spec. for Steel Sheet, Cold-Rolled, Aluminum-Coated Type 1
A 473–79	Spec. for Stainless and Heat-Resisting Steel Forgings
A 476–70(1976)	Spec. for Ductile Iron Castings for Paper Mill Dryer Rolls
A 478–76	Spec. for Chromium-Nickel Stainless and Heat-Resisting Steel Weaving Wire
A 479–79	Spec. for Stainless and Heat-Resisting Steel Bars and Shapes for Use in Boilers and Other Pressure Vessels
A 480–79	Spec. for General Requirements for Flat-Rolled Stainless and Heat-Resisting Steel Plate, Sheet, and Strip
A 484–79	Spec. for General Requirements for Stainless and Heat-Resisting Wrought Steel Products (Except Wire)
A 487–79a	Spec. for Steel Castings Suitable for Pressure Service
A 489–72(1979)	Spec. for Carbon Steel Eyebolts
A 490–79	Spec. for Quenched and Tempered Alloy Steel Bolts for Structural Steel Joints
A 496–78	Spec. for Deformed Steel Wire for Concrete Reinforcement
A 497–79	Spec. for Welded Deformed Steel Wire Fabric for Concrete Reinforcement
A 500–78	Spec. for Cold-Formed Welded and Seamless Carbon Steel Structural Tubing in Rounds and Shapes
A 501–76	Spec. for Hot-Formed Welded and Seamless Carbon Steel Structural Tubing
A 502–76	Spec. for Steel Structural Rivets
A 505–78	Spec. for General Requirements for Steel Sheet and Strip, Alloy, Hot-Rolled and Cold-Rolled
A 506–73	Spec. for Steel Sheet and Strip, Alloy, Hot-Rolled and Cold-Rolled, Regular Quality
A 507–73	Spec. for Steel Sheet and Strip, Alloy, Hot-Rolled and Cold-Rolled, Drawing Quality
A 508–79	Spec. for Quenched and Tempered Vacuum-Treated Carbon and Alloy Steel Forgings for Pressure Vessels
A 510–77	Spec. for General Requirements for Wire Rods and Coarse Round Wire, Carbon Steel
A 514–77	Spec. for High-Yield Strength, Quenched and Tempered Alloy Steel Plate, Suitable for Welding
A 515–79b	Spec. for Pressure Vessel Plates, Carbon Steel, for Intermediate- and Higher-Temperature Service
A 516–79b	Spec. for Pressure Vessel Plates, Carbon Steel, for Moderate- and Lower-Temperature Service
A 517–79a	Spec. for Pressure Vessel Plates, Alloy Steel, High-Strength, Quenched and Tempered
A 522–76	Spec. for Forged or Rolled 8 and 9% Nickel Alloy Steel Flanges, Fittings, Valves, and Parts for Low-Temperature Service
A 524–79a	Spec. for Seamless Carbon Steel Pipe for Atmospheric- and Lower Temperatures
A 525–79	Spec. for Steel Sheet, Zinc-Coated (Galvanized) by the Hot-Dip Process, General Requirements

A 526–71(1975)   Spec. for Steel Sheet, Zinc-Coated (Galvanized) by the Hot-Dip Process, Commercial Quality

A 527–71(1975)   Spec. for Steel Sheet, Zinc-Coated (Galvanized) by the Hot-Dip Process, Lock-Forming Quality

A 528–71(1975)   Spec. for Steel Sheet, Zinc-Coated (Galvanized) by the Hot-Dip Process, Drawing Quality

A 529–75   Spec. for Structural Stell with 42 000 psi (290 MPa) Minimum Yield Point (½ in. (12.7 mm) Maximum Thickness)

A 530–79a   Spec. for General Requirements for Specialized Carbon and Alloy Steel Pipe

A 533–79a   Spec. for Pressure Vessel Plates, Alloy Steel, Quenched and Tempered, Manganese-Molybdenum and Manganese-Molybdenum-Nickel

A 537–79   Spec. for Pressure Vessel Plates, Heat-Treated, Carbon-Manganese-Silicon Steel

A 538–77   Spec. for Pressure Vessel Plates, Alloy Steel, Precipitation Hardening (Maraging), 18 Percent Nickel

A 540–77a   Spec. for Alloy-Steel Bolting Materials for Special Applications

A 541–79a   Spec. for Steel Forgings, Carbon and Alloy, Quenched and Tempered, for Pressure Vessel Components

A 542–79   Spec. for Pressure Vessel Plates, Alloy Steel, Quenched and Tempered, Chromium-Molybdenum

A 543–79a   Spec. for Pressure Vessel Plates, Alloy Steel, Quenched and Tempered Nickel-Chromium-Molybdenum

A 553–79a   Spec. for Pressure Vessel Plates, Alloy Steel, Quenched and Tempered 8 and 9 Percent Nickel

A 556–79   Spec. for Seamless Cold-Drawn Carbon Steel Feedwater Heater Tubes

A 557–79   Spec. for Electric-Resistance-Welded Carbon Steel Feedwater Heater Tubes

A 562–79a   Spec. for Pressure Vessel Plates, Carbon Steel, Manganese-Titanium for Glass or Diffused Metallic Coatings

A 563–78a   Spec. for Carbon and Alloy Steel Nuts

A 568–74   Spec. for Steel, Carbon and High-Strength Low-Alloy Hot-Rolled Sheet, Hot-Rolled Strip and Cold-Rolled Sheet, General Requirements

A 569–72(1979)   Spec. for Steel, Carbon (0.15 Maximum, Percent) Hot-Rolled Sheet and Strip, Commercial Quality

A 570–79   Spec. for Hot-Rolled Carbon Steel Sheet and Strip, Structural Quality

A 571–71(1976)   Spec. for Austenitic Ductile Iron Castings for Pressure-Containing Parts Suitable for Low-Temperature Service

A 572–79   Spec. for High-Strength Low-Alloy Columbium-Vanadium Steels of Structural Quality

A 573–77   Spec. for Structural Carbon Steel Plates of Improved Toughness

A 576–79   Spec. for Steel Bars, Carbon, Hot-Rolled, Special Quality

A 577–77   Spec. for Ultrasonic Angle-Beam Examination of Steel Plates

A 578–78   Spec. for Straight-Beam Ultrasonic Examination of Plain and Clad Steel Plates for Special Applications

A 580–79   Spec. for Stainless and Heat-Resisting Steel Wire

A 586–68(1976)   Spec. for Zinc-Coated Steel Structural Strand

A 588–79a — Spec. for High-Strength Low-Alloy Structural Steel with 50 000 psi Minimum Yield Point to 4 in. Thick

A 590–72a(1977) — Spec. for Pressure Vessel Plates, Alloy Steel, Precipitation Hardening (Maraging), 12 Percent Nickel

A 591–77 — Spec. for Steel Sheet, Cold-Rolled, Electrolytic Zinc-Coated

A 592–74(1979) — Spec. for High-Strength Quenched and Tempered Low-Alloy Steel Forged Fittings and Parts for Pressure Vessels

A 595–74 — Spec. for Steel Tubes, Low-Carbon, Tapered for Structural Use

A 599–77 — Spec. for Steel, Sheet, Cold-Rolled, Tin-Coated by Electrodeposition

A 603–70(1974) — Spec. for Zinc-Coated Steel Structural Wire Rope

A 605–72(1977) — Spec. for Pressure Vessel Plates, Alloy Steel, Quenched and Tempered, Nickel-Cobalt-Molybdenum-Chromium

A 606–75 — Spec. for Steel Sheet and Strip, Hot-Rolled and Cold-Rolled, High-Strength, Low-Alloy, with Improved Corrosion Resistance

A 607–75 — Spec. for Steel Sheet and Strip, Hot-Rolled and Cold-Rolled, High-Strength, Low-Alloy Columbium and/or Vanadium

A 608–79 — Spec. for Centrifugally Cast Iron-Chromium-Nickel High-Alloy Tubing for Pressure Application at High Temperatures

A 611–72(1979) — Spec. for Steel, Cold-Rolled Sheet, Carbon, Structural

A 612–79b — Spec. for Pressure Vessel Plates, Carbon Steel, High Strength, for Moderate- and Lower-Temperature Service

A 615–79 — Spec. for Deformed and Plain Billet-Steel Bars for Concrete Reinforcement

A 616–79 — Spec. for Rail-Steel Deformed and Plain Bars for Concrete Reinforcement

A 617–79 — Spec. for Axle-Steel Deformed and Plain Bars for Concrete Reinforcement

A 618–74 — Spec. for Hot-Formed Welded and Seamless High-Strength Low-Alloy Structural Tubing

A 632–79 — Spec. for Seamless and Welded Austenitic Stainless Steel Tubing (Small-Diameter) for General Service

A 633–79a — Spec. for Normalized High-Strength Low-Alloy Structural Steel

A 635–74 — Spec. for Steel Sheet and Strip, Carbon, Hot-Rolled Commercial Quality, Heavy-Thickness Coils (Formerly Plate)

A 641–71a(1975) — Spec. for Zinc-Coated (Galvanized) Carbon Steel Wire

A 663–77 — Spec. for Merchant Quality Hot-Rolled Carbon Steel Bars Subject to Mechanical Property Requirements

A 666–72(1979) — Spec. for Austenitic Stainless Steel, Sheet, Strip, Plate, and Flat Bar for Structural Applications

A 668–79a — Spec. for Steel Forgings, Carbon and Alloy, for General Industrial Use

A 671–77 — Spec. for Electric-Fusion-Welded Steel Pipe for Atmospheric and Lower Temperatures

A 672–79 — Spec. for Electric-Fusion-Welded Steel Pipe for High-Pressure Service at Moderate Temperatures

A 674–74(1979) — Rec. Practice for Polyethylene Encasement for Gray and Ductile Cast Iron Pipe for Water or Other Liquids

A 675–79 — Spec. for Steel Bars and Bar Size Shapes, Carbon, Hot-Rolled, Special Quality, Subject to Mechanical Property Requirements

A	678–75	Spec. for Quenched and Tempered Carbon Steel Plates for Structural Applications
A	691–79	Spec. for Carbon and Alloy Steel Pipe, Electric-Fusion-Welded for High-Pressure Service at High Temperatures
A	704–74	Spec. for Welded Steel Plain Bar or Rod Mats for Concrete Reinforcement
A	706–79	Spec. for Low-Alloy Steel Deformed Bars for Concrete Reinforcement
A	722–75	Spec. for Uncoated High-Strength Steel Bar for Prestressing Concrete
A	733–76	Spec. for Welded and Seamless Carbon Steel and Austenitic Stainless Steel Pipe Nipples
A	755–78a	Spec. for General Requirements for Steel Sheet Zinc-Coated (Galvanized) by the Hot-Dip Process and Coil-Coated for Roofing and Siding
A	767–79	Spec. for Zinc-Coated (Galvanized) Bars for Concrete Reinforcement
B	26–80	Spec. for Aluminum-Alloy Sand Castings
B	32–76	Spec. for Solder Metal
B	36–77	Spec. for Brass Plate, Sheet, Strip, and Rolled Bar
B	42–80	Spec. for Seamless Copper Pipe, Standard Sizes
B	43–80	Spec. for Seamless Red Brass Pipe, Standard Sizes
B	68–80	Spec. for Seamless Copper Tube, Bright Annealed
B	75–80	Spec. for Seamless Copper Tube
B	88–80	Spec. for Seamless Copper Water Tube
B	101–78	Spec. for Lead-Coated Copper Sheets
B	108–80	Spec. for Aluminum-Alloy Permanent Mold Castings
B	121–77	Spec. for Leaded Brass Plate, Sheet, Strip, and Rolled Bar
B	134–78	Spec. for Brass Wire
B	135–80	Spec. for Seamless Brass Tube
B	152–79	Spec. for Copper Sheet, Strip, Plate, and Rolled Bar
B	209–80	Spec. for Aluminum-Alloy Sheet and Plate
B	210–79	Spec. for Aluminum-Alloy Drawn Seamless Tubes
B	211–79a	Spec. for Aluminum-Alloy Bars, Rods, and Wire
B	221–79a	Spec. for Aluminum-Alloy Extruded Bars, Rods, Wire, Shapes, and Tubes
B	234–77	Spec. for Aluminum-Alloy Drawn Seamless Tubes for Condensers and Heat Exchangers
B	241–79	Spec. for Aluminum-Alloy Seamless Pipe and Seamless Extruded Tube
B	247–79	Spec. for Aluminum-Alloy Die and Hand Forgings
B	249–79	Spec. for General Requirements for Wrought Copper and Copper-Alloy Rod, Bar, and Shapes
B	250–80	Spec. for General Requirements for Wrought Copper-Alloy Wire
B	251–76	Spec. for General Requirements for Wrought Seamless Copper and Copper-Alloy Tube
B	258–65(1976)	Spec. for Standard Nominal Diameters and Cross-Sectional Areas of Awg Sizes of Solid Round Wires Used as Electrical Conductors
B	280–80	Spec. for Seamless Copper Tube for Air Conditioning and Refrigeration Field Service
B	302–76	Spec. for Threadless Copper Pipe
B	306–80	Spec. for Copper Drainage Tube (DWV)

B	308–80	Spec. for Aluminum-Alloy Standard Structural Shapes, Rolled or Extruded
B	313–79	Spec. for Aluminum-Alloy Round Welded Tubes
B	316–79	Spec. for Aluminum-Alloy Rivet and Cold Heading Wire and Rods
B	370–77	Spec. for Copper Sheet and Strip for Building Construction
B	429–79	Spec. for Aluminum-Alloy Extruded Structural Pipe and Tube
B	447–80	Spec. for Welded Copper Tube
B	483–78	Spec. for Aluminum-Alloy Drawn Tubes for General Purpose Applications
B	547–78a	Spec. for Aluminum-Alloy Formed and Arc-Welded Round Tube
B	585–80	Spec. for Seamless Copper-Alloy Water Tube
B	586–80	Spec. for Welded Copper-Alloy Water Tube
B	587–80	Spec. for Welded Brass Tube
C	4–62(1975)	Spec. for Clay Drain Tile
C	5–59(1974)	Spec. for Quicklime for Structural Purposes
C	12–77	Rec. Practice for Installing Vitrified Clay Pipe Lines
C	14–79	Spec. for Concrete Sewer, Storm Drain, and Culvert Pipe
C	22–77	Spec. for Gypsum
C	25–79	Chemical Analysis of Limestone, Quicklime, and Hydrated Lime
C	27–70(1976)	Classification of Fireclay and high-Alumina Refractory Brick
C	28–76a	Spec. for Gypsum Plasters
C	29–78	Test for Unit Weight and Voids in Aggregate
C	31–69(1975)	Making and Curing Concrete Test Specimens in the Field
C	32–73(1979)	Spec. for Sewer and Manhole Brick (Made from Clay or Shale)
C	33–78	Spec. for Concrete Aggregates
C	34–62(1975)	Spec. for Structural Clay Load-Bearing Wall Tile
C	35–76	Spec. for Inorganic Aggregates for Use in Gypsum Plaster
C	36–78	Spec. for Gypsum Wallboard
C	37–76	Spec. for Gypsum Lath
C	39–72(1979)	Test for Compressive Strength of Cylindrical Concrete Specimens
C	40–73	Test for Organic Impurities in Fine Aggregates for Concrete
C	42–77	Obtaining and Testing Drilled Cores and Sawed Beams of Concrete
C	50–78	Sampling, Inspection, Packing, and Marking of Lime and Limestone Products
C	51–71(1976)	Def. of Terms Relating to Lime and Limestone (As Used by the Industry)
C	52–54(1977)	Spec. for Gypsum Partition Tile or Block
C	55–75	Spec. for Concrete Building Brick
C	56–71(1976)	Spec. for Structural Clay Non-Load-Bearing Tile
C	57–57(1978)	Spec. for Structural Clay Floor Tile
C	59–76	Spec. for Gypsum Casting and Molding Plaster
C	61–76	Spec. for Gypsum Keene's Cement
C	62–75a	Spec. for Building Brick (Solid Masonry Units Made from Clay or Shale)
C	64–72(1977)	Spec. for Refractories for Incinerators and Boilers
C	67–78	Sampling and Testing Brick and Structural Clay Tile
C	73–75	Spec. for Calcium Silicate Face Brick (Sand-Lime Brick)
C	76–79	Spec. for Reinforced Concrete Culvert, Storm Drain, and Sewer Pipe
C	78–75	Test for Flexural Strength of Concrete (Using Simple Beam with Third-Point Loading)

C	79–78	Spec. for Gypsum Sheathing Board
C	85–66(1973)	Test for Cement Content of Hardened Portland Cement Concrete
C	87–69(1975)	Test for Effect of Organic Impurities in Fine Aggregate on Strength of Mortar
C	90–75	Spec. for Hollow Load-Bearing Concrete Masonry Units
C	91–78	Spec. for Masonry Cement
C	94–78a	Spec. for Ready-Mixed Concrete
C	97–47(1977)	Tests for Absorption and Bulk Specific Gravity of Natural Building Stone
C	99–52(1976)	Test for Modulus of Rupture of Natural Building Stone
C	105–47(1976)	Spec. for Ground Fire Clay as a Refractory Mortar for Laying-Up Fireclay Brick
C	109–77	Test for Compressive Strength of Hydraulic Cement Mortars (Using 2-in. or 50-mm Cube Specimens)
C	110–76a	Physical Testing of Quicklime, Hydrated Lime, and Limestone
C	126–71(1976)	Spec. for Ceramic Glazed Structural Clay Facing Tile, Facing Brick, and Solid Masonry Units
C	129–75	Spec. for Non-Load-Bearing Concrete Masonry Units
C	131–76	Test for Resistance to Abrasion of Small Size Coarse Aggregate by Use of the Los Angeles Machine
C	136–76	Test for Sieve or Screen Analysis of Fine and Coarse Aggregates
C	138–77	Test for Unit Weight, Yield, and Air Content (Gravimetric) of Concrete
C	140–75	Sampling and Testing Concrete Masonry Units
C	141–67(1978)	Spec. for Hydraulic Hydrated Lime for Structural Purposes
C	143–78	Test for Slump of Portland Cement Concrete
C	144–76	Spec. for Aggregate for Masonry Mortar
C	145–75	Spec. for Solid Load-Bearing Concrete Masonry Units
C	150–78a	Spec. for Portland Cement
C	155–70(1976)	Classification of Insulating Firebrick
C	167–64(1976)	Test for Thickness and Density of Blanket- or Batt-Type Thermal Insulating Materials
C	170–50(1976)	Test for Compressive Strength of Natural Building Stone
C	171–69(1975)	Spec. for Sheet Materials for Curing Concrete
C	172–71(1977)	Sampling Fresh Concrete
C	183–78	Sampling Hydraulic Cement
C	190–77	Test for Tensile Strength of Hydraulic Cement Mortars
C	192–76	Making and Curing Concrete Test Specimens in the Laboratory
C	206–76	Spec. for Finishing Hydrated Lime
C	207–76	Spec. for Hydrated Lime for Masonry Purposes
C	208–72	Spec. for Insulating Board (Cellulosic Fiber), Structural and Decorative
C	209–72	Testing Insulating Board (Cellulosic Fiber), Structural and Decorative
C	212–60(1975)	Spec. for Structural Clay Facing Tile
C	216–79	Spec. for Facing Brick (Solid Masonry Units Made from Clay or Shale)
C	222–78	Spec. for Asbestos-Cement Roofing Shingles
C	223–78	Spec. for Asbestos-Cement Siding
C	231–78	Test for Air Content of Freshly Mixed Concrete by the Pressure Method
C	260–77	Spec. for Air-Entraining Admixtures for Concrete
C	270–80a	Spec. for Mortar for Unit Masonry

C 279–79	Spec. for Chemical-Resistant Masonry Units
C 287–77	Spec. for Chemical-Resistant Sulfur Mortar
C 296–78	Spec. for Asbestos-Cement Pressure Pipe
C 301–79	Testing Vitrified Clay Pipe
C 309–74	Spec. for Liquid Membrane-Forming Compounds for Curing Concrete
C 315–78c	Spec. for Clay Flue Linings
C 317–76	Spec. for Gypsum Concrete
C 318–78	Spec. for Gypsum Formboard
C 330–77	Spec. for Lightweight Aggregates for Structural Concrete
C 331–77	Spec. for Lightweight Aggregates for Concrete Masonry Units
C 332–77a	Spec. for Lightweight Aggregates for Insulating Concrete
C 377–66(1977)	Spec. for Precast Reinforced Gypsum Slabs
C 387–77a	Spec. for Packaged, Dry, Combined Materials for Mortar and Concrete
C 404–76	Spec. for Aggregates for Masonry Grout
C 406–58(1976)	Spec. for Roofing Slate
C 411–61(1975)	Test for Hot-Surface Performance of High-Temperature Thermal Insulation
C 412–79	Spec. for Concrete Drain Tile
C 425–77	Spec. for Compression Joints for Vitrified Clay Pipe and Fittings
C 428–78	Spec. for Asbestos-Cement Nonpressure Sewer Pipe
C 442–78	Spec. for Gypsum Backing Board
C 443–79	Spec. for Joints for Circular Concrete Sewer and Culvert Pipe, Using Rubber Gaskets
C 444–79	Spec. for Perforated Concrete Pipe
C 471–76	Chemical Analysis of Gypsum and Gypsum Products
C 472–73(1978)	Physical Testing of Gypsum Plasters and Gypsum Concrete
C 473–76a	Physical Testing of Gypsum Board Products, Gypsum Lath, Gypsum Partition Tile or Block, and Precast Reinforced Gypsum Slabs
C 474–67(1978)	Testing Joint Treatment Materials for Gypsum Wallboard Construction
C 475–64(1975)	Spec. for Joint Treatment Materials for Gypsum Wallboard Construction
C 476–71(1976)	Spec. for Grout for Reinforced and Nonreinforced Masonry
C 494–79	Spec. for Chemical Admixtures for Concrete
C 495–77a	Test for Compressive Strength of Lightweight Insulating Concrete
C 496–71	Test for Splitting Tensile Strength of Cylindrical Concrete Specimens
C 503–79	Spec. for Exterior Marble
C 508–78a	Spec. for Asbestos-Cement Underdrain Pipe
C 514–77	Spec. for Nails for the Application of Gypsum Wallboard
C 516–75	Spec. for Vermiculite Loose Fill Insulation
C 530–70(1975)	Spec. for Structural Clay Non-Load-Bearing Screen Tile
C 532–66(1974)	Spec. for Structural Insulating Formboard (Cellulosic Fiber)
C 549–73(1979)	Spec. for Perlite Loose Fill Insulation
C 557–73(1978)	Spec. for Adhesives for Fastening Gypsum Wallboard to Wood Framing
C 564–70(1976)	Spec. for Rubber Gaskets for Cast Iron Soil Pipe and Fittings
C 567–71(1977)	Test for Unit Weight of Structural Lightweight Concrete
C 577–68(1978)	Test for Permeability of Refractories
C 587–68(1978)	Spec. for Gypsum Veneer Plaster
C 588–78	Spec. for Gypsum Base for Veneer Plasters

C 595–79	Spec. for Blended Hydraulic Cements
C 612–77	Spec. for Mineral Fiber Block and Board Thermal Insulation
C 615–68(1977)	Spec. for Granite Building Stone
C 616–68(1977)	Spec. for Sandstone Building Stone
C 617–76	Capping Cylindrical Concrete Specimens
C 618–78	Spec. for Fly Ash and Raw or Calcined Natural Pozzolans for Use As a Mineral Admixture in Portland Cement Concrete
C 629–68(1977)	Spec. for Slate Building Stone
C 630–78	Spec. for Water-Resistant Gypsum Backing Board
C 631–70(1975)	Spec. for Bonding Compounds for Interior Plastering
C 635–78	Spec. for Metal Suspension Systems for Acoustical Tile and Lay-In Panel Ceilings
C 636–76	Rec. Practice for Installation of Metal Ceiling Suspension Systems for Acoustical Tile and Lay-In Panels
C 644–78	Spec. for Asbestos-Cement Nonpressure Small-Diameter Sewer Pipe
C 645–76	Spec. for Non-Load (Axial) Bearing Steel Studs, Runners (Track), and Rigid Furring Channels for Screw Application of Gypsum Board
C 646–78	Spec. for Steel Drill Screws for the Application of Gypsum Board to Light-Gage Steel Studs
C 652–77	Spec. for Hollow Brick (Hollow Masonry Units Made from Clay or Shale)
C 666–77	Test for Resistance of Concrete to Rapid Freezing and Thawing
C 683–76	Test for Compressive and Flexural Strength of Concrete Under Field Conditions
C 700–78a	Spec. for Vitrified Clay Pipe, Extra Strength, Standard Strength, and Perforated
C 707–76	Spec. for Gypsum Board Substrate for Floor or Roof Assemblies
C 739–77	Spec. for Cellulosic Fiber (Wood-Base) Loose-Fill Thermal Insulation
C 744–73(1979)	Spec. for Prefaced Concrete and Calcium Silicate Masonry Units
C 754–79	Spec. for Installation of Steel Framing Members to Receive Screw-Attached Gypsum Wallboard, Backing Board, or Water-Resistant Backing Board
C 780–80	Preconstruction and Construction Evaluation of Mortars for Plain and Reinforced Unit Masonry
C 841–76	Spec. for Installation of Interior Lathing and Furring
C 842–76	Spec. for Application of Interior Gypsum Plaster
C 843–76	Spec. for Application of Gypsum Veneer Plaster
C 844–79	Spec. for Application of Gypsum Base to Receive Gypsum Veneer Plaster
C 847–77	Spec. for Metal Lath
D 5–73(1978)	Test for Penetration of Bituminous Materials
D 25–73	Spec. for Round Timber Piles
D 52–62(1976)	Spec. for Wood Paving Blocks for Exposed Platforms, Pavements, Driveways, and Interior Floors Exposed to Wet and Dry Conditions
D 56–79	Test for Flash Point by Tag Closed Tester
D 86–78	Distillation of Petroleum Products
D 92–78	Test for Flash and Fire Points by Cleveland Open Cup
D 93–79	Test for Flash Point by Pensky-Martens Closed Tester
D 143–52(1978)	Testing Small Clear Specimens of Timber

D 146–78a	Sampling and Testing Bitumen-Saturated Felts and Fabrics Used in Roofing and Waterproofing
D 173–80	Spec. for Bitumen-Saturated Cotton Fabrics used in Roofing and Waterproofing
D 198–76	Static Tests of Timbers in Structural Sizes
D 224–75	Spec. for Smooth-Surfaced Asphalt Roll Roofing (Organic Felt)
D 225–65(1978)	Spec. for Asphalt Shingles Surfaced with Mineral Granules
D 226–77	Spec. for Asphalt-Saturated Organic Felt Used in Roofing and Waterproofing
D 227–78	Spec. for Coal-Tar Saturated Organic Felt Used in Roofing and Waterproofing
D 228–69(1978)	Testing Asphalt Rool Roofing, Cap Sheets, and Shingles
D 245–74	Establishing Structural Grades and Related Allowable Properties for Visually Graded Lumber
D 249–73	Spec. for Asphalt Roll Roofing (Organic Felt) Surfaced with Mineral Granules
D 250–77	Spec. for Asphalt-Saturated Asbestos Felts Used in Roofing and Waterproofing
D 312–78	Spec. for Asphalt Used in Roofing
D 323–79	Test for Vapor Pressure of Petroleum Products (Reid Method)
D 371–58(1970)	Spec. for Wide-Selvage Asphalt Roll Roofing (Organic Felt) Surfaced with Mineral Granules
D 374–79	Tests for Thickness of Solid Electrical Insulation
D 396–79	Spec. for Fuels Oils
D 449–79	Spec. for Asphalt Used in Dampproofing and Waterproofing
D 450–78	Spec. for Coal-Tar Bitumen Used in Roofing, Dampproofing, and Waterproofing
D 461–77	Testing Felt
D 491–79	Spec. for Asphalt Mastic Used in Waterproofing
D 529–76	Rec. Practice for Accelerated Weathering Test of Bituminous Materials
D 568–77	Test for Rate of Burning and/or Extent and Time of Burning of Flexible Plastics in a Vertical Position
D 570–77	Test for Water Absorption of Plastics
D 621–64(1976)	Tests for Deformation of Plastics Under Load
D 635–77	Test for Rate of Burning and/or Extent and Time of Burning of Self-Supporting Plastics in a Horizontal Position
D 648–72(1978)	Test for Deflection Temperature of Plastics Under Flexural Load
D 698–78	Tests for Moisture Density Relations of Soils and Soil-Aggregate Mixtures Using 5.5-lb (2.49-kg) Rammer and 12-in. (305-mm) Drop
D 994–71(1977)	Spec. for Preformed Expansion Joint Filler for Concrete (Bituminous Type)
D1037–78	Evaluating the Properties of Wood-Base Fiber and Particle Panel Materials
D1101–59(1976)	Test for Integrity of Glue Joints in Structural Laminated Wood Products for Exterior Use
D1143–74	Testing Piles Under Axial Compressive Load
D1194–72(1977)	Test for Bearing Capacity of Soil for Static Load on Spread Footings
D1226–70(1976)	Spec. for Asphalt Insulating Siding Surfaced with Mineral Granules

D1227–79	Spec. for Emulsified Asphalt Used as a Protective Coating for Built-up Roofs
D1327–78	Spec. for Bitumen-Saturated Woven Burlap Fabrics Used in Roofing and Waterproofing
D1527–77	Spec. for Acrylonitrile-Butadiene-Styrene (ABS) Plastic Pipe Schedules 40 and 80
D1556–64(1974)	Test for Density of Soil in Place by the Sand-Cone Method
D1557–78	Test for Moisture Density Relations of Soils and Soil-Aggregate Mixtures Using 10-lb (4.54-kg) Rammer and 18-in. (457-mm) Drop
D1586–67(1974)	Penetration Test and Split Barrel Sampling of Soils
D1751–73(1978)	Spec. for Preformed Expansion Joint Fillers for Concrete Paving and Structural Construction (Nonextruding and Resilient Bituminous Types)
D1752–67(1978)	Spec. for Preformed Sponge Rubber and Cork Expansion Joint Fillers for Concrete Paving and Structural Construction
D1761–77	Testing Mechanical Fasteners in Wood
D1785–76	Spec. for Poly(Vinyl Chloride)(PVC) Plastic Pipe, Schedules 40, 80, and 120
D1861–77	Spec. for Homogeneous Bituminized Fiber Drain and Sewer Pipe
D1862–77	Spec. for Laminated-Wall Bituminized Fiber Drain and Sewer Pipe
D1863–77	Spec. for Mineral Aggregate Used on Built-Up Roofs
D1869–78	Spec. for Rubber Rings for Asbestos-Cement Pipe
D1929–77	Test for Ignition Properties of Plastics
D2104–74	Spec. for Polyethylene (PE) Plastic Pipe, Schedule 40
D2164–65(1977)	Tests for Structural Insulating Roof Decks
D2235–79	Spec. for Solvent Cement for Acrylonitrile-Butadiene-Styrene (ABS) Plastic Pipe and Fittings
D2239–74	Spec. for Polyethylene (PE) Plastic Pipe (SDR-PR)
D2241–78	Spec. for Poly(Vinyl Chloride) (PVC) Plastic Pipe (SDR-PR)
D2277–75	Spec. for Fiberboard Nail-Base Sheathing
D2282–77	Spec. for Acrylonitrile-Butadiene-Styrene (ABS) Plastic Pipe (SDR-PR)
D2312–77	Spec. for Perforated Homogeneous Bituminized Fiber Pipe for Septic Tank Disposal Fields
D2313–77	Spec. for Perforated Laminated-Wall Bituminized Fiber Pipe for Septic Tank Disposal Fields
D2321–74	Rec. Practice for Underground Installation of Flexible Thermoplastic Sewer Pipe
D2464–76	Spec. for Threaded Poly(Vinyl Chloride) (PVC) Plastic Pipe Fittings, Schedule 80
D2465–73(1979)	Spec. for Threaded Acrylonitrile-Butadiene-Styrene (ABS) Plastic Pipe Fittings, Schedule 80
D2466–78	Spec. for Poly(Vinyl Chloride) (PVC) Plastic Pipe Fittings, Schedule 40
D2467–76a	Spec. for Socket-Type Poly(Vinyl Chloride)(PVC) Plastic Pipe Fittings, Schedule 80
D2487–69(1975)	Classification of Soils for Engineering Purposes
D2513–78	Spec. for Thermoplastic Gas Pressure Pipe, Tubing, and Fittings
D2517–73(1979)	Spec. for Reinforced Epoxy Resin Gas Pressure Pipe and Fittings
D2555–78	Establishing Clear-Wood Strength Values

D2559–76	Spec. for Adhesives for Structural Laminated Wood Products for Use Under Exterior (Wet Use) Exposure Conditions
D2564–79	Spec. for Solvent Cements for Poly(Vinyl Chloride) (PVC) Plastic Pipe and Fittings
D2609–74	Spec. for Plastic Insert Fittings for Polyethylene (PE) Plastic Pipe
D2661–78	Spec. for Acrylonitrile-Butadiene-Styrene (ABS) Plastic Drain, Waste, and Vent Pipe and Fittings
D2662–78	Spec. for Polybutylene (PB) Plastic Pipe (SDR-PR)
D2665–78	Spec. for Poly(Vinyl Chloride) (PVC) Plastic Drain, Waste and Vent Pipe and Fittings
D2666–75	Spec. for Polybutylene (PB) Plastic Tubing
D2672–78	Spec. for Bell-End Poly(Vinyl Chloride) (PVC) Pipe
D2729–78	Spec. for Poly(Vinyl Chloride) (PVC) Sewer Pipe and Fittings
D2737–74	Spec. for Polyethylene (PE) Plastic Tubing
D2740–76	Spec. for Poly(Vinyl Chloride) (PVC) Plastic Tubing
D2750–77a	Spec. for Acrylonitrile-Butadiene-Styrene (ABS) Plastic Utilities Conduit and Fittings
D2751–77a	Spec. for Acrylonitrile-Butadiene-Styrene (ABS) Sewer Pipe and Fittings
D2822–75	Spec. for Asphalt Roof Cement
D2823–75	Spec. for Asphalt Roof Coatings
D2824–76	Spec. for Aluminum-Pigmented Asphalt Roof Coatings
D2829–76	Rec. Practice for Sampling and Analysis of Built-Up Roofs
D2843–77	Measuring the Density of Smoke from the Burning or Decomposition of Plastics
D2846–79	Spec. for Chlorinated Poly(Vinyl Chloride) (CPVC) Plastic Hot- and Cold-Water Distribution Systems
D2852–77	Spec. for Styrene-Rubber (SR) Plastic Drain Sewer Pipe and Fittings
D2855–78	Rec. Practice for Making Solvent-Cemented Joints with Poly(Vinyl Chloride) (PVC) Pipe and Fittings
D2859–76	Test for Flammability of Finished Textile Floor Covering Materials
D2898–77	Test for Accelerated Weathering on Fire-Retardant-Treated Wood, for Fire Testing
D2899–74	Establishing Design Stresses for Round Timber Piles
D2915–74	Evaluating Allowable Properties for Grades of Structural Lumber
D2925–70(1976)	Measuring Beam Deflection of Reinforced Thermosetting Plastic Pipe Under Full Bore Flow
D2996–71(1977)	Spec. for Filament-Wound Reinforced Thermosetting Resin Pipe
D2997–71(1977)	Spec. for Centrifugally Cast Reinforced Thermosetting Resin Pipe
D3000–73	Spec. for Polybutylene (PB) Plastic Pipe (SDR-PR) Based on Outside Diameter
D3024–78	Spec. for Protein-Base Adhesives for Structural Laminated Wood Products for Use Under Interior (Dry Use) Exposure Conditions
D3033–79	Spec. for Type PSP Poly(Vinyl Chloride) (PVC) Sewer Pipe and Fittings
D3034–78	Spec. for Type PSM Poly(Vinyl Chloride) (PVC) Sewer Pipe and Fittings
D3035–74	Spec. for Polyethylene (PE) Plastic Pipe (SDR-PR) Based on Controlled Outside Diameter
D3036–73	Spec. for Poly(Vinyl Chloride) (PVC) Plastic Line Couplings, Socket-Type
D3140–72(1977)	Rec. Practice for Flaring Polyolefin Pipe and Tubing

D3200–74   Spec. and Methods for Establishing Recommended Design Stresses for Round Timber Construction Poles

D3261–78   Spec. for Butt Heat Fusion Polyethylene (PE) Plastic Fittings for Polyethylene (PE) Plastic Pipe and Tubing

D3378–74T   Spec. for Asphalt-Saturated and Coated Asbestos Felt Base Sheet Used in Roofing

D3462–76   Spec. for Asphalt Shingles Made From Glass Mat and Surfaced with Mineral Granules

D3689–78   Testing Individual Piles Under Static Axial Tensile Load

D3747–79   Spec. for Emulsified Asphalt Adhesives for Adhering Roof Insulation

E   8–79a   Tension Testing of Metallic Materials

E   72–77   Conducting Strength Tests of Panels for Building Construction

E   73–74   Testing Truss Assemblies

E   84–79a   Test for Surface Burning Characteristics of Building Materials

E   90–75   Rec. Practice for Laboratory Measurement of Airborne-Sound Transmission Loss of Building Partitions

E 108–78   Fire Tests of Roof Coverings

E 119–79   Fire Tests of Building Construction and Materials

E 136–79   Test for Behavior of Materials in a Vertical Tube Furnace at 750°C

E 152–78   Fire Tests of Door Assemblies

E 154–68(1975)   Testing Materials for Use as Vapor Barriers Under Concrete Slabs and as Ground Cover in Crawl Spaces

E 160–50(1975)   Test for Combustible Properties of Treated Wood by the Crib Test

E 163–79   Fire Tests of Window Assemblies

E 196–74   Load Tests of Floors and Flat Roofs

E 283–73   Test for Rate of Air Leakage Through Exterior Windows, Curtain Walls, and Doors

E 329–77   Rec. Practice for Inspection and Testing Agencies for Concrete Steel and Bituminous Materials as Used in Construction

E 336–77   Rec. Practice for Measurement of Airborne Sound Insulation in Buildings

E 413–73   Classification for Determination of Sound Transmission Class

E 447–74   Tests for Compressive Strength of Masonry Prisms

E 477–73   Testing Duct Liner Materials and Prefabricated Silencers for Acoustical and Airflow Performance

E 489–73   Test for Tensile Strength Properties of Steel Truss Plates

E 492–77   Laboratory Measurement of Impact Sound Transmission Through Floor-Ceiling Assemblies Using the Tapping Machine

E 514–74   Test for Water Permeance of Masonry

E 518–76   Tests for Flexural Bond Strength of Masonry

E 519–74   Test for Diagonal Tension (Shear) in Masonry Assemblages

E 547–75   Test for Water Penetration of Exterior Windows, Curtain Walls, and Doors by Cyclic Static Air Pressure Differential

E 557–77   Rec. Practice for Architectural Application and Installation of Operable Partitions

E 564–76   Static Load Test for Shear Resistance of Framed Walls for Buildings

E 648–78   Test for Critical Radiant Flux of Floor Covering Systems Using a Radiant Heat Energy Source

| E 662–79 | Test for Specific Optical Density of Smoke Generated by Solid Materials |
| E 476–76 | Test for Security of Swinging Door Assemblies |

LIST BY SUBJECTS

ASTM Standards Used or Referenced in Building Codes

Volume I: A 6-B 587
Volume II: C 4-F 476

Since the standards in this book are arranged in alphanumerical sequence, no page numbers are given in this list by subjects. Italicized references are for information only and are not located in this publication. These standards are available in the part of the *Annual Book of ASTM Standards* indicated. Each is available as a separate reprint from ASTM Headquarters, 1916 Race St., Philadelphia, Pa. 19103.

ACOUSTICAL MATERIALS

Specification for:

| C 635–78 | Metal Suspension Systems for Acoustical Tile and Lay-in Panel Ceilings |
| C 532–66(1974) | Structural Insulating Formboard (Cellulosic Fiber) |

Test methods for:

E 477–73	Duct Liner Materials and Prefabricated Silencers for Acoustical and Airflow Performance
E 492–77	*Impact Sound Transmission Through Floor-Ceiling Assemblies Using the Tapping Machine, Laboratory Measurement of (see Part 18)*
C 384–77	*Impedance and Absorption of Acoustical Materials by the Impedance Tube Method (see Part 18)*
C 423–77	*Sound Absorption and Sound Absorption Coefficients by the Reverberation Room Method (see Part 18)*
C 367–78	*Strength Properties of Prefabricated Architectural Acoustical Materials (see Part 18)*

Practices for:

E 336–77	Airborne Sound Insulation in Buildings, Measurement of
C 636–76	Installation of Metal Ceiling Suspension Systems for Acoustical Tile and Lay-in Panels
E 90–75	Laboratory Measurement of Airborne-Sound Transmission Loss of Building Partitions
E 557–77	*Architectural Application and Installation of Operable Partitions (see Part 18)*

Classification for:

| E 413–73 | Determination of Sound Transmission Class |

ADMIXTURES

Specifications for:

C	260–77	Air-Entraining Admixtures for Concrete
C	494–79	Chemical Admixtures for Concrete
C	618–78	Fly Ash and Raw or Calcined Natural Pozzolans for Use As a Mineral Admixture in Portland Cement Concrete
C	*593–76a*	*Fly Ash and Other Pozzolans for Use with Lime (see Part 13)*

Methods of Sampling and Testing:

C	*311–77*	*Fly Ash or Natural Pozzolans for Use as a Mineral Admixture in Portland Cement Concrete (see Part 14)*

AGGREGATES

Specifications for:

C	144–76	Aggregate for Masonry Mortar
C	404–76	Aggregates for Masonry Grout
C	33–78	Concrete Aggregates
C	35–76	Inorganic Aggregates for Use in Gypsum Plaster
C	331–77	Lightweight Aggregates for Concrete Masonry Units
C	332–77a	Lightweight Aggregates for Insulating Concrete
C	330–77	Lightweight Aggregates for Structural Concrete
C	*897–78*	*Aggregate for Job-Mixed Portland Cement-Based Plasters (see Part 13)*

Test Methods for:

C	87–69(1975)	Effect of Organic Impurities in Fine Aggregate on Strength of Mortar
C	131–76	Resistance to Abrasion of Small Size Coarse Aggregate by Use of the Los Angeles Machine
C	136–76	Sieve or Screen Analysis of Fine and Coarse Aggregates
C	29–78	Unit Weight and Voids in Aggregate
C	*142–78*	*Clay Lumps and Friable Particles in Aggregates (see Part 14)*
C	*123–69(1975)*	*Lightweight Pieces in Aggregate (see Parts 14, 15)*
C	*117–80*	*Materials Finer than (75-μm) No. 200 Sieve in Mineral Aggregates by Washing (see parts 14, 15)*
C	*40–79*	*Organic Impurities in Fine Aggregates for Concrete (see Part 14)*
C	*289–71(1976)*	*Potential Reactivity of Aggregates (Chemical Method) (see Part 14)*
C	*535–69(1975)*	*Resistance to Abrasion of Large Size Coarse Aggregate by Use of the Los Angeles Machine (see Parts 14, 15)*
D	*75–71(1978)*	*Sampling Aggregates (see Parts 14, 15, 19)*
C	*88–76*	*Soundness of Aggregates by Use of Sodium Sulfate or Magnesium Sulfate (see Parts 14, 15)*
C	*127–77*	*Specific Gravity and Absorption of Coarse Aggregate (see Parts 14, 15)*
C	*128–79*	*Specific Gravity and Absorption of Fine Aggregate (see Parts 14, 15)*
C	*70–79*	*Surface Moisture in Fine Aggregate (see Part 14)*

ALUMINUM

Specification for:

B	211–79a	Aluminum-Alloy Bars, Rods, and Wires
B	247–79	Aluminum-Alloy Die and Hand Forgings
B	210–79	Aluminum-Alloy Drawn Seamless Tubes
B	234–77	Aluminum-Alloy Drawn Seamless Tubes for Condensers and Heat Exchangers
B	483–78	Aluminum-Alloy Drawn Tubes for General Purpose Applications
B	221–79a	Aluminum-Alloy Extruded Bars, Rods, Wire, Shapes, and Tubes
B	429–79	Aluminum-Alloy Extruded Structural Pipe and Tube
B	547–78a	Aluminum-Alloy Formed and Arc-Welded Round Tube
B	108–80	Aluminum-Alloy Permanent Mold Castings
B	316–79	Aluminum-Alloy Rivet and Cold Heading Wire and Rods
B	313–79	Aluminum-Alloy Round Welded Tubes
B	26–80	Aluminum-Alloy Sand Castings
B	241–79	Aluminum-Alloy Seamless Pipe and Seamless Extruded Tube
B	209–80	Aluminum-Alloy Sheet and Plate
B	308–80	Aluminum-Alloy Standard Structural Shapes, Rolled or Extruded
B	32–76	Solder Metal
B211M–79		*Aluminum-Alloy Bar, Rod, and Wire [Metric] (see Part 7)*
B247M–80		*Alluminum Alloy Die and Hand Forgings [Metric] (see Part 7)*
B	*85–76*	*Aluminum-Alloy Die Castings (see Part 7)*
B210M–80		*Aluminum-Alloy Drawn Seamless Tubes [Metric] (see Part 7)*
B221M–79		*Aluminum-Alloy Extruded Bars, Rods, Wire, Shapes, and Tubes [Metric] (see Part 7)*
B209M–80		*Aluminum-Alloy Sheet and Plate [Metric] (see Part 7)*

Practice for:

B	*449–67(1972)*	*Chromate Treatments on Aluminum (see Parts 7, 9)*

ASBESTOS-CEMENT PRODUCTS

Specifications for:

C	428–78	Asbestos-Cement Nonpressure Sewer Pipe
C	644–78	Asbestos-Cement Nonpressure Small-Diameter Sewer Pipe
C	296–78	Asbestos-Cement Pressure Pipe
C	222–78	Asbestos-Cement Roofing Shingles
C	223–78	Asbestos-Cement Siding
C	508–78a	Asbestos-Cement Underdrain Pipe
D	1869–78	Rubber Rings for Asbestos-Cement Pipe
C	*551–78*	*Asbestos-Cement Fiberboard Insulating Panels (see Part 16)*
C	*659–78*	*Asbestos-Cement Plastic-Foam Core Insulating Panels (see Part 16)*
C	*746–78*	*Corrugated Asbestos-Cement Sheets for Bulkheading (see Part 16)*
C	*220–77*	*Flat Asbestos-Cement Sheets (see Part 16)*

Specifications and Tests for:

C	*221–77*	*Corrugated Asbestos-Cement Sheets (see Part 16)*

BITUMINIZED FIBER PIPE

Specifications for:

D 1861–77	Homogeneous Bituminized Fiber Drain and Sewer Pipe
D 1862–77	Laminated-Wall Bituminized Fiber Drain and Sewer Pipe
D 2312–77	Perforated Homogeneous Bituminized Fiber Pipe for Septic Tank Disposal Fields
D 2313–77	Perforated Laminated-Wall Bituminized Fiber Pipe for Septic Tank Disposal Fields
D 2417–77	*Perforated Laminated-Wall Bituminized Fiber Pipe for General Drainage (see Part 15)*

BRICK

Specifications for:

C	62–75a	Building Brick (Solid Masonry Units Made from Clay or Shale)
C	73–75	Calcium Silicate Face Brick (Sand-Lime Brick)
C	126–71(1976)	Ceramic Glazed Structural Clay Facing Tile, Facing Brick, and Solid Masonry Units
C	55–75	Concrete Building Brick
C	216–79	Facing Brick (Solid Masonry Units Made from Clay or Shale)
C	105–47(1976)	Ground Fire Clay as a Refractory Mortar for Laying-Up Fireclay Brick
C	652–77	Hollow Brick (Hollow Masonry Units made from Clay or Shale)
C	32–73(1979)	Sewer and Manhole Brick (Made from Clay or Shale)
C	*410–60(1978)*	*Industrial Floor Brick (see Part 16)*
C	*64–72(1977)*	*Refractories for Incinerators and Boilers (see Part 17)*

Methods for:

C	780–80	Preconstruction and Construction Evaluation of Mortars for Plain and Reinforced Unit Masonry
C	67–78	Sampling and Testing Brick and Structural Clay Tile

Classification of:

C	27–70(1976)	Fireclay and High-Alumina Refractory Brick
C	155–70(1976)	Insulating Firebrick

BUILDING CONSTRUCTIONS

Test Methods for:

E	447–74	Compressive Strength of Masonry Prisms
E	648–78	Critical Radiant Flux of Floor Covering Systems using a Radiant Heat Energy Source
E	492–77	Laboratory Measurement of Impact Sound Transmission Through Floor-Ceiling Assemblies Using the Tapping Machine
E	196–74	Load Tests of Floors and Flat Roofs
E	72–77	Panels for Building Construction, Conducting Strength Tests of

E	283–73	Rate of Air Leakage Through Exterior Windows, Curtain Walls, and Doors
E	564–76	Static Load Test for Shear Resistance of Framed Walls for Buildings
E	489–73	Tensile Strength Properties of Steel Truss Plates
E	73–74	Truss Assemblies, Testing
E	547–75	Water Penetration of Exterior Windows, Curtain Walls, and Doors by Cyclic Static Air Pressure Differential
E	283–73	*Air Leakage Through Exterior Windows, Curtain Walls, and Doors, Rate of (see Part 18)*
E	541–77	*Criteria for Agencies Engaged in System Analysis and Compliance Assurance for Manufactured Building (see Part 18)*
E	699–79	*Criteria for Evaluation of Agencies Involved in Testing, Quality Assurance, and Evaluating Building Components in Accordance with Test Methods Promulgated by ASTM Committee E-6 (see Part 18)*
E	632–78	*Developing Short-Term Accelerated Tests for Prediction of the Service Life of Building Components and Materials (see Part 18)*
E	576–79	*Dew/Frost Point of Sealed Insulating Glass Units in Vertical Position (see Part 18)*
E	577–76	*Dimensional Coordination of Rectilinear Building Parts and Systems (see Part 18)*
E	529–75	*Flexural Tests on Beams and Girders for Building Construction (see Part 18)*
E	546–79	*Frost Point of Sealed Insulating Glass Units (see Part 18)*
E	719–80	*Indentation of Building Materials Surfaces Under Concentrated Loads As a Measure of Serviceability (see Part 18)*
E	695–79	*Measuring Relative Resistance of Wall, Floor, and Roof Constructions to Impact Loading (see Part 18)*
E	455–76	*Static Load Testing of Framed Floor or Roof Diaphragm Constructions for Buildings (see Part 18)*
E	330–79	*Structural Performance of Exterior Windows, Curtain Walls, and Doors by Uniform Static Air Pressure Difference (see Part 18)*
E	605–77	*Thickness and Density of Sprayed Fire-Resistive Material Applied to Structural Members (see Part 18)*
E	331–70(1975)	*Water Penetration of Exterior Windows, Curtain Walls, and Doors by Uniform Static Air Pressure Differential (see Part 18)*
E	405–70(1975)	*Wear Testing Rotary Operators for Windows (see Part 18)*

Practice for:

E	336–77	Airborne Sound Insulation in Buildings, Measurement of
E	557–77	Architectural Application and Installation of Operable Partitions
D	3740–78	*Evaluation of Agencies Engaged in Testing and/or Inspection of Soil and Rock as Used in Construction (see Part 19)*
E	548–79	*Generic Criteria for Use in the Evaluation of Testing and Inspection Agencies (see Parts 10, 14)*
E	241–77	*Increasing Durability of Building Constructions Against Water-Induced Damage (see part 18)*
E	575–76	*Reporting Data from Structural Tests of Building Constructions, Elements, Connections, and Assemblies (see Part 18)*

CEMENT

Specifications for:

C	595–79	Blended Hydraulic Cements
C	91–78	Masonry Cement
C	150–78a	Portland Cement
C	897–78	*Aggregate for Job-Mixed Portland Cement-Based Plasters (see Part 13)*
C	226–75	*Air-Entraining Additions for Use in the Manufacture of Air-Entraining Portland Cement (see Part 13)*
C	196–77	*Expanded or Exfoliated Vermiculite Thermal Insulating Cement (see Part 18)*
C	932–80	*Surface-Applied Bonding Agents for Exterior Plastering (see Part 13)*
C	933–80	*Welded Wire Lath (see Part 13)*

Methods for:

C	114–80	*Chemical Analysis of Hydraulic Cement (see Part 13)*
C	183–78	*Sampling Hydraulic Cement (see Parts 13, 15)*

Test Methods for:

C	109–77	Compressive Strength of Hydraulic Cement Mortars (Using 2-in. 50-mm Cube Specimens)
C	184–76	*Fineness of Hydraulic Cement by the No. 100 (150-μm) and 200 (75-μm) Sieves (see Part 13)*
C	204–79a	*Fineness of Portland Cement by Air Permeability Apparatus (see Part 13)*
C	115–79b	*Fineness of Portland Cement by the Turbimeter (see Part 13)*
C	348–80	*Flexural Strength of Hydraulic Cement Mortars (see Part 13)*
C	452–75	*Potential Expansion of Portland Cement Mortars Exposed to Sulfate (see Part 13)*
C	183–78	*Sampling Hydraulic Cement (see Parts 13, 15)*
C	190–77	*Tensile Strength of Hydraulic Cement Mortars (see Part 13)*
C	191–79	*Time of Setting of Hydraulic Cement by Vicat Needle (see Part 13)*
C	266–77	*Time of Setting of Hydraulic Cement by Gillmore Needles (see Part 13)*
C	807–75	*Time of Setting of Hydraulic Cement Mortar by Vicat Needle (see Part 13)*

CLAY PIPE AND TILE

Specifications for:

C	126–71(1976)	Ceramic Glazed Structural Clay Facing Tile, Facing Brick, and Solid Masonry Units
C	4–62(1975)	Clay Drain Tile
C	315–78c	Clay Flue Linings
C	425–77	Compression Joints for Vitrified Clay Pipe and Fittings
C	700–78a	Vitrified Clay Pipe, Extra Strength, Standard Strength, and Perforated
C	212–60(1975)	Structural Clay Facing Tile

C	57–57(1978)	Structural Clay Floor Tile
C	34–62(1975)	Structural Clay Load-Bearing Wall Tile
C	56–71(1976)	Structural Clay Non-Load-Bearing Tile
C	530–70(1975)	Structural Clay Non-Load-Bearing Screen Tile

Methods of:

C	67–78	Sampling and Testing Brick and Structural Clay Tile
C	301–79	Vitrified Clay Pipe

Practice for:

C	12–77	Installing Vitrified Clay Pipe Lines

CONCRETE

Specifications for:

C	260–77	Air-Entraining Admixtures for Concrete
C	494–79	Chemical Admixtures for Concrete
C	309–74	Liquid Membrane-Forming Compounds for Curing Concrete
C	387–77a	Packaged, Dry, Combined Materials for Mortar and Concrete
C	444–79	Perforated Concrete Pipe
D	994–71(1977)	Preformed Expansion Joint Filler for Concrete (Bituminous Type)
D	1751–73(1978)	Preformed Expansion Joint Filler for Concrete Paving and Structural Construction (Nonextruding and Resilient Bituminous Types)
C	94–78a	Ready-Mixed Concrete
C	171–69(1975)	Sheet Materials for Curing Concrete
C412M–80		*Concrete Drain tile [Metric] (see Part 16)*
C	685–80	*Concrete Made by Volumetric Batching and Continuous Mixing (see Part 14)*
C	14M–80	*Concrete Sewer, Storm Drain, and Culbert Pipe [Metric] (see Part 16)*
C	881–78	*Epoxy-Resin-Base Bonding Systems for Concrete (see Part 14)*
C443M–80		*Joints for Circular Concrete Sewer and Culvert Pipe, Using Rubber Gaskets [Metric] (see Part 16)*
C	470–79a	*Molds for Forming Concrete Test Cylinders Vertically (see Part 14)*
C444M–80		*Perforated Concrete Pipe [Metric] (see Part 16)*

Test Methods for:

C	231–78	Air Content of Freshly Mixed Concrete by the Pressure Method
C	617–76	Capping Cylindrical Concrete Specimens
C	85–66(1973)	Cement Content of Hardened Portland Cement Concrete
C	683–76	Compressive and Flexural Strength of Concrete Under Field Conditions
C	39–72(1979)	Compressive Strength of Cylindrical Concrete Specimens
C	495–77a	Compressive Strength of Lightweight Insulating Concrete
C	78–75	Flexural Strength of Concrete (Using Simple Beam with Third-Point Loading)
C	31–69(1975)	Making and Curing Concrete Test Specimens in the Field
C	192–76	Making and Curing Concrete Test Specimens in the Laboratory

E	154–68(1975)	Materials for Use as Vapor Barriers Under Concrete Slabs and as Ground Cover in Crawl Spaces
C	42–77	Obtaining and Testing Drilled Cores and Sawed Beams of Concrete
D	1752–67(1978)	Preformed Sponge Rubber and Cork Expansion Joint Fillers for Concrete Paving and Structural Construction
C	666–77	Resistance of Concrete to Rapid Freezing and Thawing
C	172–71(1977)	Sampling Fresh Concrete
C	143–78	Slump of Portland Cement Concrete
C	496–71	Splitting Tensile Strength of Cylindrical Concrete Specimens
C	567–71(1977)	Unit Weight of Structural Lightweight Concrete
C	138–77	Unit Weight, Yield, and Air Content (Gravimetric) of Concrete
A	767–79	Zinc-Coated (Galvanized) Bars for Concrete Reinforcement
C	779–76	*Abrasion Resistance of Horizontal Concrete Surfaces (see Part 14)*
C	173–78	*Air Content of Freshly Mixed Concrete by the Volumetric Method (see Part 14)*
C	233–78	*Air-Entraining Admixtures for Concrete (see Part 14)*
C	882–78	*Bond Strength of Epoxy-Resin Systems Used with Concrete (see Part 14)*
D	98–77a	*Calcium Chloride (see Parts 14, 15, and 19)*
C	234–71(1977)	*Comparing Concretes on the Basis of the Bond Developed With Reinforcing Steel (see Part 14)*
C	116–68(1980)	*Compressive Strength of Concrete Using Portions of Beams Broken in Flexure (see Part 14)*
C	827–78	*Early Volume Change of Cementitious Mixtures (see Part 14)*
C	883–80	*Effective Shrinkage of Epoxy-Resin Systems Used with Concrete (see Part 14)*
C	293–79	*Flexural Strength of Concrete (Using Simple Beam with Center-Point Loading) (see Part 14)*
C	341–79	*Length Change of Drilled or Sawed Specimens of Cement Mortar and Concrete (see Part 14)*
C	174–49(1975)	*Measuring Length of Drilled Concrete Cores (see Part 14)*
C	884–78	*Thermal Compatibility Between Concrete and an Epoxy-Resin Overlay (see Part 14)*

Practices for:

C	823–75	*Examination and Sampling of Hardened Concrete in Constructions (see Part 14)*
C	801–75	*Mechanical Properties of Hardened Concrete Under Triaxial Loads, Determining the (see Part 14)*
C	457–71	*Microscopical Determination of Air-Void Content and Parameters of the Air-Void System in Hardened Concrete (see Part 14)*

CONCRETE PIPE

Specifications for:

C	412–79	Concrete Drain Tile
C	14–79	Concrete Sewer, Storm Drain, and Culvert Pipe

C	443–79	Joints for Circular Concrete Sewer and Culvert Pipe, Using Rubber Gaskets
C	76–79	Reinforced Concrete Culvert, Storm Drain, and Sewer Pipe
C	654–79	*Porous Concrete Pipe (see Part 16)*
C	478–79	*Precast Reinforced Concrete Manhole Sections (see Part 16)*
C 76M–80		*Reinforced Concrete Culvet Storm Drain, and Sewer Pipe [Metric] (see Part 16)*
C	655–77	*Reinforced Concrete D-Load Culvert, Storm Drain and Sewer Pipe (see Part 16)*
C	507–79	*Reinforced Concrete Elliptical Culvert Storm Drain, and Sewer Pipe (see Part 16)*
C	361–78	*Reinforced Concrete Low-Head Pressure Pipe (see Part 16)*

Definition of Terms Relating to:

C	822–80	*Concrete Pipe and Related Products (see Part 14)*

CONCRETE MASONRY UNITS

Specifications for:

C	55–75	Concrete Building Brick
C	90–75	Hollow Load-Bearing Concrete Masonry Units
C	129–75	Non-Load-Bearing Concrete Masonry Units
C	744–73(1979)	Prefaced Concrete and Calcium Silicate Masonry Units
C	145–75	Solid Load-Bearing Concrete Masonry Units
C	139–73(1979)	*Concrete Masonry Units for Construction of Catch Basins and Manholes (see Part 16)*

Methods of:

C	140–75	Sampling and Testing Concrete Masonry Units
E	488–76	*Strength of Anchors in Concrete and Masonry Elements (see Part 18)*

COPPER AND COPPER BASE ALLOYS

Specifications for:

B	36–77	Brass Plate, Sheet, Strip, and Rolled Bar
B	134–78	Brass Wire
B	306–80	Copper Drainage Tube (DWV)
B	370–77	Copper Sheet and Strip for Building Construction
B	152–79	Copper, Sheet, Strip, Plate, and Rolled Bar
B	250–80	General Requirements for Wrought Copper-Alloy Wire
B	249–79	General Reqirements for Wrought Copper and Copper-Alloy Rod, Bar, and Shapes
B	251–76	General Requirements for Wrought Seamless Copper and Copper-Alloy Tube
B	101–78	Lead-Coated Copper Sheets
B	121–77	Leaded Brass Plate, Sheet, Strip, and Rolled Bar
B	135–80	Seamless Brass Tube

B	585–80	Seamless Copper-Alloy Water Tube
B	42–80	Seamless Copper Pipe, Standard Sizes
B	75–80	Seamless Copper Tube
B	68–80	Seamless Copper Tube, Bright Annealed
B	280–80	Seamless Copper Tube for Air Conditioning and Refrigeration Field Service
B	88–80	Seamless Copper Water Tube
B	43–80	Seamless Red Brass Pipe, Standard Sizes
B	587–80	Welded Brass Tube
B	586–80	Welded Copper-Alloy Water Tube
B	447–80	Welded Copper Tube
B	506–78	*Copper-Clad Stainless Steel Sheet and Strip for Building Construction (see Part 6)*
B250M–79		*General Requirements for Wrought Copper-Alloy Wire [Metric] (see Part 6)*
B249M–79		*General Requirements for Wrought Copper and Copper Alloy Rod, Bar, and Shapes [Metric] (see Part 6)*
B251M–79		*General Requirements for Wrought Seamless Copper and Copper-Alloy Tube [Metric] (see Part 6)*
B	227–70(1976)	*Hard-Drawn Copper-Clad Steel Wire (see Parts 3, 6)*
B	315–80	*Seamless Copper Alloy Pipe and Tube (see Part 6)*
B	188–80	*Seamless Copper Bus Pipe and Tube (see Part 6)*
B	302–76	*Threadless Copper Pipe (see Part 6)*
B	543–80	*Welded Copper and Copper Alloy Heat Exchanger Tube (see Part 6)*
B	248–76a	*Wrought Copper and Copper-Alloy Plate, Sheet, Strip, and Rolled Bar, General Requirements (see Part 6)*

ELECTRODEPOSITED METALLIC COATINGS

Specifications for:

A	164–71	*Electrodeposited Coatings of Zinc on Steel (see Parts 3 and 9)*
A	165–71	*Electrodeposited Coatings of Cadmium on Steel (see Parts 3 and 9)*
B	454–76	*Mechanically Deposited Coatings of Cadmium and Zinc on Ferrous Metals (see Parts 4 and 9)*

Methods for:

B	530–75(1980)	*Measurement of Coating Thicknesses by the Magnetic Method: Elecrodeposited Nickel Coatings on Magnetic and Nonmagnetic Substrates (see Part 9)*
B	499–75(1980)	*Measurement of Coating Thicknesses by the Magnetic Method: Nonmagnetic Coatings on Magnetic Basis Metals (se Part 9)*
B	487–79	*Measurement of Metal and Oxide Coating Thicknesses by Microscopical Examination of a Cross Section (see Part 9)*
B	504–70(1976)	*Measurement of the Thickness of Metallic Coatings by the Coulometric Method (see Part 9)*
B	244–79	*Measurement of Thickness of Anodic Coatings on Aluminum and of Other Nonconductive Coatings on Nonmagnetic Basis Metals with Eddy-Current Instruments (see Parts 7 and 9)*

Practice for:
 B   201–68   *Chromate Coatings on Zinc and Cadmium Surfaces, Testing* (*see Part 9*)

ELECTROMAGNETIC

Practices for:
 E   309–77   *Eddy-Current Examination of Steel Tubular Products Using Magnetic Saturation* (*see Part 11*)
 E   426–76   *Electromagnetic* (*Eddy-Current*) *Testing of Seamless and Welded Tubular Products, Austenitic Stainless Steel and Similar Alloys* (*see Part 11*)
 E   243–80   *Electromagnetic* (*Eddy-Current*) *Testing of Seamless Copper and Copper-Alloy Tubes* (*see Parts 6, 11*)
 E   215–67(1979)   *Electromagnetic Testing of Seamless Aluminum-Alloy Tube, Standardizing Equipment for* (*see Part 11*)

ELECTRICAL CONDUCTORS

Specification for:
 B   258–65(1976)   Standard Nominal Diameters and Cross-Sectional Areas of Awg Sizes of Solid Round Wires Used as Electrical Conductors

FASTENERS

Specifications for:
 F   541–77   *Alloy Steel Eyebolts* (*see Part 4*)
 A   574–80   *Alloy Steel Socket-head Cap Screws* (*see Part 4*)
 A   614–79   *Bolting Material for Nuclear and Other Special Applications, Special Requirements for* (*see Parts 1 and 45*)
 F   436–78   *Hardened Steel Washers* (*see Part 4*)
 A   687–79   *High-Strength Nonheaded Steel Bolts and Studs* (*see Part 4*)
 F   468–80   *Nonferrous Bolts, Hex Cap Screws, and Studs for General Use* (*see Parts 4, 6, 7, and 8*)
 F   467–80   *Nonferrous Nuts for General Use* (*see Parts 4, 6, 7, and 8*)
 F   432–78   Roof and Rock Bolts and Accessories (*see Part 4*)

Definitions of Terms Relating to:
 F   547–77   *Nails for Use with Wood and Wood-Base Materials* (*see Parts 4 and 22*)

FIRE TESTS

Test Methods for:
 E   136–79   Behavior of Materials in a Vertical Tube Furnace at 750° C
 E   160–50(1975)   Combustible Properties of Treated Wood by the Crib Test

E 119–79   Fire Tests of Building Construction and Materials
E 152–78   Fire Tests of Door Assemblies
E 108–78   Fire Tests of Roof Coverings
E 163–79   Fire Tests of Window Assemblies
E 662–79   Specific Optical Density of Smoke Generated by Solid Materials
E   84–79   Surface Burning Characteristics of Building Materials
E   69–50(1975)   Combustible Properties of Treated Wood by the Fire-Tube Apparatus (see Part 18)

Definitions of Terms Relating to:
E   176–80   Fire Tests of Building Construction and Materials (see Part 18)

FLAMMABILITY TESTS

Test Methods for:
D 2859–76   Flammability of Finished Textile Floor Covering Materials
D   92–78   Flash and Fire Points by Cleveland Open Cup
D   93–79   Flash Point by Pensky-Martens Closed Tester
D   56–79   Flash Point by Tag Closed Tester
D 568–77   Rate of Burning and/or Extent and Time of Burning of Flexible Plastics in a Vertical Position
D 777–74   Flammability of Treated Paper and Paperboard (see Part 20)
E 162–79   Surface Flammability of Materials Using a Radiant Heat Energy Source (see Part 18)

FUEL, OIL, SOLVENTS

Test Methods for:
D   86–78   Distillation of Petroleum Products
D 396–79   Fuel Oils
D 323–79   Vapor Pressure of Petroleum Products (Reid Method)
D 2156–65(1975)   Smoke Density in Flue Gases From Burning Distillate Fuels (see Part 24)

GLASS AND GLASS PRODUCTS

Method for:
C 240–72(1979)   Testing Cellular Glass Insulating Block (see Part 18)

GYPSUM AND PLASTERING

Specifications for:
C 557–73(1978)   Adhesives for Fastening Gypsum Wallboard to Wood Framing
C 844–79   Application of Gypsum Base to Receive Gypsum Veneer Plaster
C 843–76   Application of Gypsum Veneer Plaster

C	842–76	Application of Interior Gypsum Plaster
C	631–70(1975)	Bonding Compounds for Interior Plastering
C	22–77	Gypsum
C	442–78	Gypsum Backing Board
C	588–78	Gypsum Base for Veneer Plasters
C	707–76	Gypsum Board Substrate for Floor or Roof Assemblies
C	59–76	Gypsum Casting and Molding Plaster
C	317–76	Gypsum Concrete
C	318–78	Gypsum Formboard
C	61–76	Gypsum Keene's Cement
C	37–76	Gypsum Lath
C	52–54(1977)	Gypsum Partition Tile or Block
C	28–76a	Gypsum Plasters
C	79–78	Gypsum Sheathing Board
C	587–68(1978)	Gypsum Veneer Plaster
C	36–78	Gypsum Wallboard
C	841–76	Installation of Interior Lathing and Furring
C	754–79	Installation of Steel Framing Members to Receive Screw-Attached Gypsum Wallboard, Backing Board, or Water-Resistant Backing Board
C	475–64(1975)	Joint Treatment Materials for Gypsum Wallboard Construction
C	847–77	Metal Lath
C	514–77	Nails for the Application of Gypsum Wallboard
C	645–76	Non-Load (Axial) Bearing Steel Studs, Runners (Track), and Rigid Furring Channels for Screw Application of Gypsum Board
C	377–66(1977)	Precast Reinforced Gypsum Slabs
C	646–78	Steel Drill Screws for the Application of Gypsum Board to Light-Gage Steel Studs
C	630–78	Water-Resistant Gypsum Backing Board
C	893–78	*Type G Steel Screws for the Application of Gypsum Board to Gypsum Board (see Part 13)*
C	894–78	*Type W Screws for the Application of Gypsum Board to Wood Framing (see Part 13)*

Methods for:

C	471–76	Chemical Analysis of Gypsum and Gypsum Products
C	473–76a	Physical Testing of Gypsum Board Products, Gypsum Lath, Gypsum Partition Tile or Block, and Precast Reinforced Gypsum Slabs
C	472–73(1978)	Physical Testing of Gypsum Plasters and Gypsum Concrete
C	474–67(1978)	Testing Joint Treatment Materials for Gypsum Wallboard Construction
C	471–76	*Gypsum and Gypsum Products (see Part 13)*

Definitions of Terms Relating to:

C	11–80a	*Ceilings and Walls (see Part 13)*

IRON

Specifications for:

A	74–75	Cast Iron Soil Pipe and Fittings

A 197–79	Cupola Malleable Iron
A 476–70(1976)	Ductile Iron Castings for Paper Mill Dryer Rolls
A 395–77	Ferritic Ductile Iron Pressure-Retaining Castings for Use at Elevated Temperatures
A 377–79	Gray Iron and Ductile Iron Pressure Pipe
A 48–76	Gray Iron Castings
A 278–75	Gray Iron Castings for Pressure-Containing Parts for Temperatures up to 650° F (345° C)
A 126–73(1979)	Gray Iron Castings for Valves, Flanges, and Pipe Fittings
A 47–77	Malleable Iron Castings
C 564–70(1976)	Rubber Gaskets for Cast Iron Soil Pipe and Fittings
A 211–75	Spiral-Welded Steel or Iron Pipe
A 308–78	Steel, Sheet, Cold-Rolled, Long Terne Coated
A 338–61(1977)	*Malleable Iron Flanges, Pipe Fittings, and Valve Parts for Railroad, Marine, and Other Heavy Duty Service at Temperatures Up to 650° F (345° C) (see Part 2)*

Practice for:

| A 256–46(1976) | Compression Testing of Cast Iron |
| A 674–74(1979) | Polyethylene Encasement for Gray and Ductile Cast Iron Pipe for Water or Other Liquids |

LIME

Specifications for:

C 207–76	Hydrated Lime for Masonry Purposes
C 141–67(1978)	Hydraulic Hydrated Lime for Structural Purposes
C 5–59(1974)	Quicklime for Structural Purposes
C 206–76	Finishing Hydrated Lime

Methods of:

C 25–79	Chemical Analysis of Limestone, Quicklime, and Hydrated Lime
C 110–76a	Physical Testing of Quicklime, Hydrated Lime, and Limestone
C 50–78	Sampling, Inspection, Packing, and Marking of Lime and Limestone Products

Definitions of Terms Relating to:

| C 51–71(1976) | Lime and Limestone (As Used by the Industry) |

LIQUID PENETRANT

Practice for:

| E 165–80 | *Liquid Penetrant Inspection Method (see Part 11)* |

Definitions of Terms Relating to:

| E 270–78 | *Liquid Penetrant Inspection (see Part 11)* |

Reference Photographs for:

| E 433–71(1980) | *Liquid Penetrant Inspection (see Part 11)* |

MAGNESIUM AND MAGNESIUM ALLOY BARS, RODS, SHAPES, TUBES, AND WIRES

Specification for:
  B   107–76   *Magnesium-Alloy Extruded Bars, Rods, Shapes, Tubes, and Wires (see Part 7)*

MASONRY MORTAR

Specifications for:
  C   144–76   Aggregate for Masonry Mortar
  C   279–79   Chemical-Resistant Masonry Units
  C   287–77   Chemical-Resistant Sulfur Mortar
  C   476–71(1976)   Grout for Reinforced and Nonreinforced Masonry
  C   270–80a   Mortar for Unit Masonry

Definitions of Terms Relating to:
  C   51–71(1976)   Lime and Limestone (As Used by the Industry)

Test Methods for:
  C   267–77   *Chemical Resistance of Mortars (see Part 16)*

MASONRY PERFORMANCE

Test Methods for:
  E   519–74   Diagonal Tension (Shear) in Masonry Assemblages
  E   518–76   Flexural Bond Strength of Masonry
  E   514–74   Water Permeance of Masonry

MECHANICAL TESTING

Test Method for:
  E   18–79   *Rockwell Hardness and Rockwell Superficial Hardness of Metallic Materials (see Part 10)*

Definition of Terms Relating to:
  E   6–76   *Methods of Mechanical Testing (see Parts 10, 35)*

NATURAL BUILDING STONE

Specification for:
  C   503–79   Exterior Marble
  C   615–68(1977)   Granite Building Stone
  C   406–58(1976)   Roofing Slate
  C   616–68(1977)   Sandstone Building Stone
  C   629–68(1977)   Slate Building Stone

Test Methods:

C	97–47(1977)	Absorption and Bulk Specific Gravity of Natural Building Stone
C	170–50(1976)	Compressive Strength of Natural Building Stone
C	99–52(1976)	Modulus of Rupture of Natural Building Stone
C	*241–51(1976)*	*Abrasion Resistance of Stone Subjected to Foot Traffic (see Part 19)*
C	*120–52(1976)*	*Flexure Testing of Slate (Modulus of Rupture, Modulus of Elasticity) (see Part 19)*
C	*121–48(1976)*	*Water Absorption of Slate (see Part 19)*
C	*217–58(1976)*	*Weather Resistance of Natural Slate (see Part 19)*

NICKEL AND NICKEL ALLOY PIPE AND TUBE

Specification for:

B	*165–75*	*Nickel-Copper Alloy (UNS NO4400) Seamless Pipe and Tube (see Part 8)*

PAINT

Specification for:

D	*962–66(1973)*	*Aluminum Pigments, Powder and Paste, for Paints (see Part 28)*

PLASTICS

Test Methods:

D	648–72(1978)	Deflection Temperature of Plastics Under Flexural Load
D	621–64(1976)	Deformation of Plastics Under Load

Test Methods for:

D 2843–77	Density of Smoke from the Burning or Decomposition of Plastics
D 1929–77	Ignition Properties of Plastics
D 635–77	Rate of Burning and/or Extent and Time of Burning of Self-Supporting Plastics in a Horizontal Position
D 570–77	Water Absorption of Plastics
D 757–77	*Incandescence Resistance of Rigid Plastics in a Horizontal Position (see Part 35)*
D 1433–77	*Rate of Burning and/or Extent and Time of Burning of Flexible Thin Plastic Sheeting Supported On a 45-Deg Incline (see Part 35)*
D 543–67(1972)	*Resistance of Plastics to Chemical Reagents (see Part 35)*
D 1242–56(1975)	*Resistance to Abrasion of Plastic Materials (see Part 35)*
D 1502–60(1974)	*Transverse Load of Corrugated Reinforced Plastic Panels (see Part 36)*

Practice for:

D 2299–68(1977)	*Relative Stain Resistance of Plastics, Determining (see Parts 3, 5)*
D 756–78	*Weight and Shape Changes of Plastic Under Accelerated Service Conditions, Determination of (see Part 35)*

Definition of Terms Relating to:
  D 883–80a   *Plastics (see Parts 34, 35, 36)*

PLASTIC PIPE AND FITTINGS

Specifications for:

D 2661–78	Acrylonitrile-Butadiene-Styrene (ABS) Plastic Drain, Waste, and Vent Pipe and Fittings
D 2282–77	Acrylonitrile-Butadiene-Styrene (ABS) Plastic Pipe (SDR-PR)
D 1527–77	Acrylonitrile-Butadiene-Styrene (ABS) Plastic Pipe, Schedules 40 and 80
D 2750–77a	Acrylonitrile-Butadiene-Styrene (ABS) Plastic Utilities Conduit and Fittings
D 2751–77a	Acrylonitrile-Butadiene-Styrene (ABS) Sewer Pipe and Fittings
D 2672–78	Bell-End Poly(Vinyl Chloride) (PVC) Pipe
D 3261–78	Butt Heat Fusion Polyethylene (PE) Plastic Fittings for Polyethylene (PE) Plastic Pipe and Tubing
D 2997–71(1977)	Centrifugally Cast Reinforced Thermosetting Resin Pipe
D 2846–79	Chlorinated Poly(Vinyl Chloride) (CPVC) Plastic Hot- and Cold-Water Distribution Systems
D 2996–71(1977)	Filament-Wound Reinforced Thermosetting Resin Pipe
D 2609–74	Plastic Insert Fittings for Polyethylene (PE) Plastic Pipe
D 2662–78	Polybutylene (PB) Plastic Pipe (SDR-PR)
D 3000–73	Polybutylene (PB) Plastic Pipe (SDR-PR) Based on Outside Diameter
D 2666–75	Polybutylene (PB) Plastic Tubing
D 2104–74	Polyethylene (PE) Plastic Pipe, Schedule 40
D 2239–74	Polyethylene (PE) Plastic Pipe (SDR-PR)
D 3035–74	Polyethylene (PE) Plastic Pipe (SDR-PR) Based on Controlled Outside Diameter
D 2737–74	Polyethylene (PE) Plastic Tubing
D 2241–78	Poly(Vinyl Chloride) (PVC) Plastic Pipe (SDR-PR)
D 2665–78	Poly(Vinyl Chloride) (PVC) Plastic Drain, Waste and Vent Pipe and Fittings
D 3036–73	Poly(Vinyl Chloride) (PVC) Plastic Line Couplings, Socket-Type
D 2466–78	Poly(Vinyl Chloride) (PVC) Plastic Pipe Fittings, Schedule 40
D 1785–76	Poly(Vinyl Chloride) (PVC) Plastic Pipe, Schedules 40, 80, and 120
D 2740–76	Poly(Vinyl Chloride) (PVC) Plastic Tubing
D 2729–78	Poly(Vinyl Chloride) (PVC) Sewer Pipe and Fittings
D 2517–73(1979)	Reinforced Epoxy Resin Gas Pressure Pipe and Fittings
D 2467–76a	Socket-Type Poly(Vinyl Chloride) (PVC) Plastic Pipe Fittings, Schedule 80
D 2235–79	Solvent Cement for Acrylonitrile-Butadiene-Styrene (ABS) Plastic Pipe and Fittings
D 2564–79	Solvent Cements for Poly(Vinyl Chloride) (PVC) Plastic Pipe and Fittings
D 2852–77	Styrene-Rubber (SR) Plastic Drain Sewer Pipe and Fittings
D 2513–78	Thermoplastic Gas Pressure Pipe, Tubing, and Fittings

D 2465–73(1979)	Threaded Acrylonitrile-Butadiene-Styrene (ABS) Plastic Pipe Fittings, Schedule 80
D 2464–76	Threaded Poly(Vinyl Chloride) (PVC) Plastic Pipe Fittings, Schedule 80
D 3033–79	Type PSP Poly(Vinyl Chloride) (PVC) Sewer Pipe and Fittings
D 3034–78	Type PSM Poly(Vinyl Chloride) (PVC) Sewer Pipe and Fittings
D 2133–78	*Acetal Resin Injection Molding and Extrusion Materials (see Part 36)*
D 2680–79a	*Acrylonitrile-Butadiene-Styrene (ABS) Composite Sewer Piping (see Part 34)*

Specifications for:

F 443–77	*Bell-End Chlorinated Poly(Vinyl Chloride) (CPVC) Plastic Pipe Schedule 40 (see Part 34)*
D 1503–73(1978)	*Cellulose Acetate Butyrate (CAB) Plastic Pipe, Schedule 40 (see Part 34)*
F 441–77	*Chlorinated Poly(Vinyl Chloride) (CPVC) Plastic Pipe Schedules 40 and 80 (see Part 34)*
F 442–77	*Chlorinated Poly(Vinyl Chloride) (CPVC) Plastic Pipe (SDR-PR) (see Part 34)*
F 405–77a	*Corrugated Polyethylene (PE) Tubing and Fittings (see Part 34)*
F 477–76	*Elastomeric Seals (Gaskets) for Joining Plastic Pipe (see Part 34)*
D 788–78	*Methacrylate Molding and Extrusion Compounds (see Part 36)*
D 789–78a	*Nylon Injection Molding and Extrusion Materials (see Part 36)*
F 546–77	*Perfluoro(Ethylene-Propylene)Copolymer (FEP) Plastic-Lined Ferrous Metal Pipe and Fittings (see Part 34)*
D 1248–78	*Polyethylene Plastics Molding and Extrusion Materials (see Parts 34, 36)*
D 2447–74	*Polyethylene (PE) Plastic Pipe, Schedules 40 and 80 Based on Outside Diameter (see Part 34)*
F 423–75	*Polytetrafluoroethylene (PTFE) Plastic Lined Ferrous Metal Pipe and Fittings (see Part 34)*
F 491–77	*Poly(Vinylidene Fluoride) (PVDF) Plastic-Lined Ferrous Metal Pipe and Fittings (see Part 34)*
F 545–77	*PVC and ABS Injected Solvent Cemented Plastic Pipe Joints (see Part 34)*
F 512–79	*Smooth-Wall Poly(Vinyl Chloride) (PVC) Conduit and Fittings for Underground Installation (see Part 34)*
D 2468–76	*Socket-Type Acrylonitrile-Butadiene-Styrene (ABS) Plastic Pipe Fittings, Schedule 40 (see Part 34)*
F 438–77	*Socket-Type Chlorinated Poly(Vinyl Chloride) (CPVC) Plastic Pipe Fittings Schedule 40 (see Part 34)*
F 439–77	*Socket-Type Chlorinated Poly(Vinyl Chloride) (CPVC) Plastic Pipe Fittings Schedule 80 (see Part 34)*
F 493–79	*Solvent Cements for Chlorinated Poly(Vinyl Chloride) (CPVC) Plastic Pipe and Fittings (see Part 34)*
D 3138–79	*Solvent Cements for Transition Joints Between Acrylonitrile-Butadiene-Styrene (ABS) and Poly(Vinyl Chloride) (PVC) Nonpressure Piping Components (see Part 34)*

D 3122–78	*Solvent Cements for Styrene-Rubber Plastic Pipe and Fittings (see Part 34)*
D 1457–78	*TFE-Fluorocarbon Resin Molding and Extrusion Material (see Part 36)*
F 409–77	*Thermoplastic Accessible and Replaceable Plastic Tube and Tubular Fittings (see Part 34)*
F 480–80	*Thermoplastic Water Well Casing Pipe and Couplings Made in Standard Dimension Ratios (SDR) (see Part 34)*
F 437–77	*Threaded Chlorinated Poly(Vinyl Chloride) (CPVC) Plastic Pipe Fittings Schedule 80 (see Part 34)*
D 1694–79	*Threads for Reinforced Thermosetting Resin Pipe (see Part 34)*

Test Methods for:

D 2925–70(1976)	Measuring Beam Deflection of Reinforced Thermosetting Plastic Pipe Under Full Bore Flow
D 2122–76	*Determining Dimensions of Thermoplastic Pipe and Fittings (see Part 34)*
D 2924–79	*External Pressure Resistance of Reinforced Thermosetting Plastic Pipe (see Part 34)*
D 757–77	*Incandescence Resistance of Rigid Plastics in a Horizontal Position (see Part 35)*
D 635–77	*Rate of Burning and/or Extent and Time of Burning of Self-Supporting Plastics in a Horizontal Position (see Part 35)*
D 1929–77	*Ignition Properties of Plastics (see Part 35)*
D 2843–77	*Measuring the Density of Smoke from the Burning or Decomposition of Plastics (see Part 35)*
F 492–77	*Propylene and Polypropylene (PP) Plastic-Lined Ferrous Metal Pipe and Fittings (see Part 34)*
D 543–67(1978)	*Resistance of Plastics to Chemical Reagents (see Part 35)*
D 1242–56(1975)	*Resistance of Plastic Materials to Abrasion (see Part 35)*
D 1502–60(1974)	*Transverse Load of Corrugated Reinforced Plastic Panels (see Part 36)*
D 2774–72(1978)	*Underground Installation of Thermoplastic Pressure Piping (see Part 34)*
D 570–77	*Water Absorption of Plastics (see Part 35)*
D 756–78	*Weight and Shape Changes of Plastics Under Accelerated Service Conditions, Determination of (see Part 35)*

Classification for:

D 2310–80	*Machine-Made Reinforced Thermosetting Resin Pipe (see Part 34)*

Practices for:

D 3140–72(1977)	Flaring Polyolefin Pipe and Tubing
D 2855–78	Making Solvent-Cemented Joints with Poly(Vinyl Chloride) (PVC) Pipe and Fittings
D 2321–74	Underground Installation of Flexible Thermoplastic Sewer Pipe
D 2299–68(1977)	*Determining Relative Stain Resistance of Plastics (see Part 35)*
D 2657–79	*Heat Joining of Polyolefin and Fittings (see Part 34)*

F	481–76	*Installation of Thermoplastic Pipe and Corrugated Tubing In Septic Tank Leach Fields (see Part 34)*
F	402–80	*Safe Handling of Solvent Cements and Primers Used for Joining Thermoplastic Pipe and Fittings (see Part 34)*
F	449–76	*Subsurface Installation of Corrugated Thermoplastic Tubing for Agricultural Drainage or Water Table Control (see Part 34)*

Definition of Terms Relating to:
| F | 412–80 | *Plastic Piping Systems (see Part 34)* |

RADIOGRAPHS

Reference Radiographs for:
E	155–79	*Aluminum and Magnesium Castings, Inspection of (see Parts 7, 11)*
E	505–75(1979)	*Aluminum and Magnesium Die Castings (see Parts 7, 11)*
E	280–75(1979)	*Heavy-Walled (4½ to 12-in. (114 to 305-mm)) Steel Castings (see Parts 2, 11)*
E	186–75(1979)	*Heavy-Walled (2 to 4½-in. (51 to 114-mm)) Steel Castings (see Parts 2, 11)*
E	272–75(1979)	*High-Strength Copper-Base and Nickel-Copper Alloy Castings (see Parts 6, 11)*
E	192–75(1979)	*Investment Steel Castings for Aerospace Applications (see Parts 2, 11)*
E	446–78(1979)	*Steel Castings up to 2 in. (51 mm) in Thickness (see Parts 2, 11)*
E	390–75(1979)	*Steel Fusion Welds (see Parts 2, 11)*

REFRACTORIES

Specification for:
| C | 64–72(1977) | Refractories for Incinerators and Boilers |

Test Method for:
| C | 577–68(1978) | Permeability of Refractories |

Classification of:
| C | 401–77 | *Castable Refractories (see Part 17)* |

Definition of Terms Relating to:
| C | 71–73(1978) | *Refractories (see Part 17)* |

ROOFING

Specifications for:
D 2824–76	Aluminum-Pigmented Asphalt Roof Coatings
D 449–79	Asphalt Used in Dampproofing and Waterproofing
D 312–78	Asphalt Used in Roofing

D 1226–70(1976)   Asphalt Insulating Siding Surfaced with Mineral Granules
D 491–79   Asphalt Mastic Used in Waterproofing
D 249–73   Asphalt Roll Roofing (Organic Felt) Surfaced with Mineral Granules
D 2822–75   Asphalt Roof Cement
D 2823–75   Asphalt Roof Coatings
D 3378–74T   Asphalt-Saturated and Coated Asbestos Felt Base Sheet Used in Roofing
D 250–77   Asphalt-Saturated Asbestos Felt Used in Roofing and Waterproofing
D 226–77   Asphalt-Saturated Organic Felt Use in Roofing and Waterproofing
C 644–78   Asbestos-Cement Nonpressure Small-Diameter Sewer Pipe
D 3462–76   Asphalt Shingles Made from Glass Mat and Surfaced with Mineral Granules

Specifications for:

D 225–65(1978)   Asphalt Shingles Surfaced with Mineral Granules
D 1327–78   Bitumen-Saturated Woven Burlap Fabrics Used in Roofing and Waterproofing
D 450–78   Coal-Tar Bitumen Used in Roofing, Dampproofing, and Waterproofing
D 227–78   Coal-Tar Saturated Organic Felt Used in Roofing and Waterproofing
D 3747–79   Emulsified Asphalt Adhesive for Adhering Roof Insulation
D 1227–79   Emulsified Asphalt Used as a Protective Coating for Built-Up Roofs
D 1863–77   Mineral Aggregate Used on Built-Up Roofs
D 2312–77   Perforated Homogeneous Bituminized Fiber Pipe for Septic Tank Disposal Fields
D 224–75   Smooth-Surfaced Asphalt Roll Roofing (Organic Felt)
D 371–58(1970)   Wide-Selvage Asphalt Roll Roofing (Organic Felt) Surfaced with Mineral Granules
D 173–80   Bitumen-Saturated Cotton Fabrics Used in Roofing and Waterproofing
D 3468–80   *Liquid-Applied Neoprene and Chlorosulfonated Polyethylene Used in Roofing and Waterproofing (see Part 15)*

Test Methods for:

D 228–69(1978)   Asphalt Roll Roofing, Cap Sheets, and Shingles
D   5–73(1978)   Penetration of Bituminous Materials
D 146–78a   Sampling and Testing Bitumen-Saturated Felts and Fabrics Used in Roofing and Waterproofing
D 461–77   Testing Felt
D 3409–75   *Adhesion of Asphalt Roof Cement to Damp, Wet, or Underwater Surfaces (see Part 15)*
D   86–78   *Distillation of Petroleum Products (see Parts 15, 23, 29)*
D 3461–77   *Softening Point of Pitches (Mettler Cup and Ball Method) (see Part 15)*

Practices for:

D 529–76   Accelerated Weathering Test of Bituminous Materials

D 2829-76   Sampling and Analysis of Built-Up Roofs
D 3423-78   *Emulsified Coal-Tar Pitch (Mineral Colloid Type), Application of*
              (see Part 15)

SEALANTS

Specifications for:

C 509-79   *Cellular Elastomeric Performed Gasket and Sealing Material (see*
              Parts 18, 38)
C 542-76   *Lock-Strip Gaskets (see Parts 18, 38)*
C 570-72(1978) *Oil and Resin Base Caulking Compound for Building Construction*
              (see Part 18)
C 564-70(1976) *Rubber Gaskets for Cast Iron Soil Pipe and Fittings (see Parts 2, 18,*
              38)
C 510-77   *Staining and Color Change of Single- or Multicomponent Joint*
              Sealants (see Part 18)

Test Methods for:

C 719-79   *Adhesion and Cohesion of Elastomeric Joint Sealants Under Cyclic*
              Movement (see Part 18)
C 732-77   *Aging Effects of Artificial Weathering on Latex Sealing Compounds*
              (see Part 18)
C 712-72(1978) *Bubbling of One-Part, Elastomeric Solvent-Release Type Sealants*
              (see Part 18)
C 733-72(1977) *Determining the Volume Shrinkage of Latex Sealing Compounds*
              (see Part 18)
C 731-72(1977) *Extrudability, after Package Aging, of Latex Sealing Compounds*
              (see Part 18)
C 711-72(1978) *Low-Temperature Flexibility and Tenacity of One-Part, Elastomeric,*
              Solvent-Release Type Sealants (see Part 18)
C 734-77   *Low-Temperature Flexibility of Latex Sealing Compounds After*
              Artificial Weathering (see Part 18)
C 713-72(1978) *Slump of an Oil Base Knife Grade Channel Glazing Compound (see*
              Part 18)
C 679-71(1977) *Tack-Free Time of Elastomeric Type Joint Sealants (see Part 18)*
C 718-72(1978) *UV-Cold Box Exposure of One-Part, Elastomeric, Solvent-Release*
              Type Sealants (see Part 18)
C 681-73(1979) *Volatility of Oil- and Resin-Based, Knife-Grade, Channel Glazing*
              Compounds (see Part 18)

SOILS

Test Methods for:

D 1194-72(1977)   Bearing Capacity of Soil for Static Load on Spread Footings
D 2487-69(1975)   Classification of Soils for Engineering Purposes
D 1556-64(1974)   Density of Soil in Place by the Sand-Cone Method
D 3689-78   Individual Piles Under Statis Axial Tensile Load, Testing

Test Methods for:

D 698–78	Moisture Density Relations of Soils and Soil Aggregate Mixtures Using 5.5-lb (2.49-kg) Rammer and 12-in. (305-mm) Drop
D 1557–78	Moisture Density Relations of Soils and Soil-Aggregate Mixtures Using 10-lb (4.54-kg) Rammer and 18-in. (457-mm) Drop
D 1586–67(1974)	Penetration Test and Split Barrel Sampling of Soils
D 1143–74	Piles Under Axial Compressive Load
D 3385–75	*Infiltration Rate of Soils in Field Using Double-Ring Infiltrometers (see Part 19)*
D 3017–78	*Moisture Content of Soil and Soil-Aggregate in Place by Nuclear Methods (Shallow Depth) (see Part 19)*
D 422–63 (1972)	*Particle-Size Analysis of Soils (see Part 19)*
D 424–59 (1971)	*Plastic Limit and Plasticity Index of Soils*

STEEL

Steel Bolting and Rivet Material

Specifications for:

A 193–79a	Alloy-Steel and Stainless Steel Bolting Materials for High-Temperature Services
A 320–79a	Alloy-Steel Bolting Materials for Low-Temperature Service
A 540–77a	Alloy-Steel Bolting Materials for Special Applications
A 437–77	Alloy-Steel Turbine-Type Bolting Material Specially Heat-Treated for High-Temperature Service
A 31–76	Boiler Rivet Steel and Rivets
A 453–78	Bolting Materials, High-Temperature, 50 to 120 ksi Yield Strength, with Expansion Coefficients Comparable to Austenitic Steels
A 563–78a	Carbon and Alloy Steel Nuts
A 194–79	Carbon and Alloy Steel Nuts for Bolts for High-Pressure and High-Temperature Service
A 307–78	Carbon Steel Externally Threaded Standard Fasteners
A 489–72(1979)	Carbon Steel Eyebolts
A 325–79	High Strength Bolts for Structural Steel Joints Including Suitable Nuts and Plain Hardened Washers
A 490–79	Quenched and Tempered Alloy Steel Bolts for Structural Steel Joints
A 354–79	Quenched and Tempered Alloy Steel Bolts, Studs, and Other Externally Threaded Fasteners
A 449–78a	Quenched and Tempered Steel Bolts and Studs
A 502–76	Steel Structural Rivets
A325 M–79	*High-Strength Bolts for Structural Steel Joints, Including Suitable Nuts and Plain Hardened Washers [Metric] (see Part 4)*

Steel Castings

Specifications for:

A 372–78	Carbon and Alloy Steel Forgings for Thin-Walled Pressure Vessels

A 216–77   Carbon-Steel Castings Suitable for Fusion Welding for High-Temperature Service
A 297–79a   Heat-Resistant Iron-Chromium and Iron-Chromium-Nickel Alloy Castings for General Application
A 148–73(1979)   High-Strength Steel Castings for Structural Purposes
A 217–79   Martensitic Stainless Steel and Alloy Steel Castings for Pressure-Containing Parts Suitable for High-Temperature Service
A   27–79   Mild- to Medium-Strength Carbon-Steel Castings for General Application
A 387–79b   Pressure Vessel Plates, Alloy Steel, Chromium-Molybdenum
A 508–79   Quenched and Tempered Vacuum-Treated Carbon and Alloy Steel Forgings for Pressure Vessels
A 487–79a   Steel Castings Suitable for Pressure Service

Test Methods and Definitions for:
A 370–77   Mechanical Testing of Steel Products

Concrete Reinforcement Steel

Specifications for:
A 617–79   Axle-Steel Deformed and Plain Bars for Concrete Reinforcement
A   82–79   Cold-Drawn Steel Wire for Concrete Reinforcement
A 615–79   Deformed and Plain Billet-Steel Bars for Concrete Reinforcement

Specifications for:
A 496–78   Deformed Steel Wire for Concrete Reinforcement
A 184–79   Fabricated Deformed Steel Bar Mats for Concrete Reinforcement
A 706–79   Low-Alloy Steel Deformed Bars for Concrete Reinforcement
A 616–79   Rail-Steel Deformed and Plain Bars for Concrete Reinforcement
A 722–75   Uncoated High Strength Steel Bar for Prestressing Concrete
A 416–74   Uncoated Seven-Wire Stress-Relieved Strand for Prestressed Concrete
A 421–77   Uncoated Stress-Relieved Wire for Prestressed Concrete
A 497–79   Welded Deformed Steel Wire Fabrics for Concrete Reinforcement
A 704–74   Welded Steel Plain Bar or Rod Mats for Concrete Reinforcement
A 185–79   Welded Steel Wire Fabric for Concrete Reinforcement
A615M–79   *Deformed and Plain Billet-Steel Bars for Concrete Reinforcement [Metric] (see Part 4)*

Steel Pipe and Tube

Specifications for:
A 430–79   Austenitic Steel Forged and Bored Pipe for High-Temperature Service
A 452–79   Centrifugally Cast Austenitic Steel Cold-Wrought Pipe for High-Temperature Service
A 451–79   Centrifugally Cast Austenitic Steel Pipe for High-Temperature Service
A 426–79   Centrifugally Cast Ferritic Alloy Steel Pipe for High-Temperature Service

A 608–79	Centrifugally Cast Iron-Chromium-Nickel High-Alloy Tubing for Pressure Application at High Temperatures
A 500–78	Cold-Formed Welded and Seamless Carbon Steel Structural Tubing in Rounds and Shapes
A 254–79	Copper Brazed Steel Tubing
A 139–74	Electric-Fusion (Arc)-Welded Steel Pipe (Sizes 4 in. and over)
A 134–74	Electric-Fusion (Arc)-Welded Steel Plate Pipe (Sizes 16 in. and Over)
A 358–79a	Electric-Fusion-Welded Austenitic Chromium-Nickel Alloy Steel Pipe for High-Temperature Service
A 250–79a	Electric-Resistance-Welded Carbon-Molybdenum Alloy-Steel Boiler and Superheater Tubes
A 226–79	Electric-Resistance-Welded Carbon Steel Boiler and Superheater Tubes for High-Pressure Service
A 178–79b	Electric-Resistance-Welded Carbon Steel Boiler Tubes
A 557–79	Electric-Resistance-Welded Carbon Steel Feedwater Heater Tubes
A 214–79	Electric-Resistance-Welded Carbon Steel Heal-Exchanger and Condenser Tubes
A 135–79	Electric-Resistance-Welded Steel Pipe
A 369–79a	Ferritic Alloy Steel Forged and Bored Pipe for High-Temperature Service
A 182–79a	Forged or Rolled Alloy-Steel Pipe Flanges, Forged Fittings, and Valves and Parts for High-Temperature Service
A 522–76	Forged or Rolled 8 and 9% Nickel Alloy Steel Flanges, Fittings, Valves, and Parts for Low-Temperature Service
A 181–77	Forgings, Carbon Steel, for General Purpose Piping
A 350–79	Forgings, Carbon and Low Alloy Steel, Requiring Notch Toughness Testing for Piping Components
A 105–79	Forgings, Carbon Steel, for Piping Components
A 450–79a	General Requirements for Carbon, Ferritic Alloy, and Austenitic Alloy Steel Tubes
A 530–79a	General Requirements for Specialized Carbon and Alloy Steel Pipe
A 501–76	Hot-Formed Welded and Seamless Carbon Steel Structural Tubing
A 618–74	Hot-Formed Welded and Seamless High-Strength Low-Alloy Structural Tubing
A 53–79	Pipe, Steel, Black and Hot-Dipped, Zinc-Coated Welded and Seamless
A 120–79	Pipe, Steel, Black and Hot-Dipped Zinc-Coated (Galvanized) Welded and Seamless for Ordinary Uses
A 420–79	Piping Fittings of Wrought Carbon Steel and Alloy Steel for Low-Temperature Service
A 234–79a	Piping Fittings of Wrought Carbon Steel and Alloy Steel for Moderate and Elevated Temperatures
A 423–79a	Seamless and Electric Welded Low-Alloy Steel Tubes
A 312–79a	Seamless and Welded Austenitic Stainless Steel Pipe
A 270–79	Seamless and Welded Austenitic Stainless Steel Sanitary Tubing
A 269–79a	Seamless and Welded Austenitic Stainless Steel Tubing for General Service
A 632–79	Seamless and Welded Austenitic Stainless Steel Tubing (Small-Diameter) for General Service

A 334–79   Seamless and Welded Carbon and Alloy-Steel Tubes for Low-Temperature Service

Specifications for:

A 268–79a	Seamless and Welded Ferritic Stainless Steel Tubing for General Service
A 333–79	Seamless and Welded Steel Pipe for Low-Temperature Service
A 376–79b	Seamless Austenitic Steel Pipe for High-Temperature Central-Station Service
A 209–79a	Seamless Carbon-Molybdenum Alloy-Steel Boiler and Superheater Tubes
A 192–79	Seamless Carbon Steel Boiler Tubes for High-Pressure Service
A 524–79a	Seamless Carbon Steel Pipe for Atmospheric and Lower Temperatures
A 106–79b	Seamless Carbon Steel Pipe for High-Temperature Service
A 556–79	Seamless Cold-Drawn Carbon Steel Feedwater Heater Tubes
A 199–79a	Seamless Cold-Drawn Intermediate Alloy-Steel Heat-Exchanger and Condenser Tubes
A 179–79	Seamless Cold-Drawn Low-Carbon Steel Heat-Exchanger and Condenser Tubes
A 335–79a	Seamless Ferritic Alloy Steel Pipe for High-Temperature Service
A 405–79a	Seamless Ferritic Alloy-Steel Pipe Specially Heat Treated for High-Temperature Service
A 213–79b	Seamless Ferritic and Austenitic Alloy-Steel Boiler, Superheater, and Heat Exchanger Tubes
A 210–79a	Seamless Medium-Carbon Steel Boiler and Superheater Tubes
A 211–75	Spiral-Welded Steel or Iron Pipe
A 595–74	Steel Tubes, Low-Carbon, Tapered for Structural Use
A 733–76	Welded and Seamless Carbon Steel and Austenitic Stainless Steel Pipe Nipples
A 252–77a	Welded and Seamless Steel Pipe Piles
A 249–79b	Welded Austenitic Steel Boiler, Superheater, Heat-Exchanger, and Condenser Tubes
A 409–79a	Welded Large Outside Diameter Light-Wall Austenitic Chromium-Nickel Alloy Steel Pipe for Corrosive or High-Temperature Service
A 403–79	Wrought Austenitic Stainless Steel Piping Fittings
A 192–79	Seamless Carbon Steel Boiler Tubes for High Pressure Service
A 702–74	*Steel Fence Posts and Assemblies, Hot-Rolled, for Field and Line-Type Fencing (see Part 36)*

Sheet and Strip Steel

Specifications for:

A 336–79	Alloy Steel Forgings for Pressure and High Temperature Parts
A 571–71(1976)	Austenitic Ductile Iron Castings for Pressure Containing Parts Suitable for Low-Temperature Service
A 414–79	Carbon Steel Sheets for Pressure Vessels
A 263–79	Corrosion-Resisting Chromium Steel Clad Plate, Sheet, and Strip
A 507–73	Drawing Quality Hot-Rolled and Cold-Rolled Alloy Steel Sheet and Strip

A	266–78	Forgings, Carbon Steel, for Pressure Vessel Components
A	480–79	General Requirements for Flat-Rolled Stainless and Heat-Resisting Steel Plate, Sheet, and Strip
A	505–78	General Requirements for Hot-Rolled and Cold-Rolled Alloy Steel Sheet and Strip
A	20–79b	General Requirements for Steel Plates for Pressure Vessels
A	240–79	Heat-Resisting Chromium and Chromium-Nickel Stainless Steel Plate, Sheet, and Strip for Fusion-Welded Unfired Pressure Vessels
A	592–74(1979)	High-Strength Quenched and Tempered Low-Alloy Steel Forged Fittings and Parts for Pressure Vessels
A	177–69(1979)	High-Strength Stainless and Heat-Resisting Chromium-Nickel Steel Sheet and Strip
A	612–79b	High-Strength Steel Plates for Pressure Vessels for Moderate- and Lower-Temperature Service
A	514–77	High-Yield Strength, Quenched and Tempered Alloy Steel Plate, Suitable for Welding
A	570–79	Hot-Rolled Carbon Steel Sheet and Strip, Structural Quality
A	202–78	Pressure Vessel Plates, Alloy Steel, Chromium-Manganese-Silicon
A	517–79a	Pressure Vessel Plates, Alloy Steel, High-Strength, Quenched and Tempered
A	302–79b	Pressure Vessel Plates, Alloy Steel, Manganese-Molybdenum and Manganese-Molybdenum-Nickel
A	225–79a	Pressure Vessel Plates, Alloy Steel, Manganese-Vanadium
A	204–79a	Pressure Vessel Plates, Alloy Steel, Molybdenum
A	203–79a	Pressure Vessel Plates, Alloy Steel, Nickel
A	590–72a(1977)	Pressure Vessel Plates, Alloy Steel, Precipitation Hardening (Maraging), 12 Percent Nickel
A	542–79	Pressure Vessel Plates, Alloy Steel, Quenched and Tempered, Chromium-Molybdenum
A	553–79a	Pressure Vessel Plates, Alloy Steel, Quenched and Tempered 8 and 9 Percent Nickel

Specifications for:

A	543–79a	Pressure Vessel Plates, Alloy Steel, Quenched and Tempered Nickel-Chromium-Molybdenum
A	605–72(1977)	Pressure Vessel Plates, Alloy Steel, Quenched and Tempered, Nickel-Cobalt-Molybdenum-Chromium
A	515–79b	Pressure Vessel Plates, Carbon Steel, for Intermediate- and Higher-Temperature Service
A	516–79b	Pressure Vessel Plates, Carbon Steel, for Moderate- and Lower-Temperature Service
A	612–79b	Pressure Vessel Plates, Carbon Steel, High Strength, for Moderate- and Lower-Temperature Service
A	455–79b	Pressure Vessel Plates, Carbon Steel, High Strength Manganese
A	442–79b	Pressure Vessel Plates, Carbon Steel, Improved Transition Properties
A	285–78	Pressure Vessel Plates, Carbon Steel, Low- and Intermediate-Tensile Strength

A 299-79b	Pressure Vessel Plates, Carbon Steel, Manganese-Silicon
A 562-79a	Pressure Vessel Plates, Carbon Steel, Manganese-Titanium for Glass or Diffused Metallic Coatings
A 412-75	Stainless and Heat-Resisting Chromium-Nickel-Manganese Steel Plate, Sheet, and Strip
A 167-77	Stainless and Heat-Resisting Chromium-Nickel Steel Plate, Sheet, and Strip
A 176-79	Stainless and Heat-Resisting Chromium Steel Plate, Sheet, and Strip
A 276-79a	Stainless and Heat-Resisting Steel Bars and Shapes
A 473-79	Stainless and Heat-Resisting Steel Forgings
A 368-76	Stainless and Heat-Resisting Steel Wire Strand
A 264-79	Stainless Chromium-Nickel Steel Clad Plate, Sheet and Strip
A 434-76	Steel Bars, Alloy, Hot-Rolled or Cold-Finished, Quenched and Tempered
A 108-79	Steel Bars, Carbon, Cold-Finished, Standard Quality
A 568-74	Steel, Carbon and High-Strength Low-Alloy Hot-Rolled Sheet, Hot-Rolled Strip and Cold-Rolled Sheet, General Requirements
A 366-72(1979)	Steel, Carbon, Cold-Rolled Sheet, Commercial Quality
A 109-72(1979)	Steel, Carbon, Cold-Rolled Strip
A 569-72(1979)	Steel, Carbon (0.15 Maximum, Percent) Hot-Rolled Sheet and Strip, Commercial Quality
A 668-79a	Steel Forgings, Carbon and Alloy, for General Industrial Use
A 541-79a	Steel Forgings, Carbon and Alloy, Quenched and Tempered, for Pressure Vessel Components
A 507-73	Steel Sheet and Strip, Alloy, Hot-Rolled and Cold-Rolled, Drawing Quality
A 505-78	Steel Sheet and Strip, Alloy, Hot-Rolled and Cold-Rolled, General Requirements for
A 506-73	Steel Sheet and Strip, Alloy, Hot-Rolled and Cold-Rolled, Regular Quality
A 607-75	Steel Sheet and Strip, Hot-Rolled and Cold-Rolled, High Strength, Low-Alloy Columbium and/or Vanadium
A 606-75	Steel Sheet and Strip, Hot-Rolled and Cold-Rolled, High-Strength, Low-Alloy with Improved Corrosion Resistance
A 463-77	Steel Sheet, Cold-Rolled, Aluminum-Coated Type I
A 591-77	Steel Sheet, Cold-Rolled, Electrolytic Zinc-Coated
A 308-78	Steel Sheet, Cold-Rolled, Long Terne Coated
A 599-77	Steel Sheet, Cold-Rolled, Tin-Coated by Electrodeposition
A 424-73	Steel Sheet for Porcelain Enameling
A 444-78	Steel Sheet, Zinc-Coated (Galvanized) by the Hot Dip Process for Culverts and Underdrains
A 578-78	Straight-Beam Ultrasonic Examination of Plain and Clad Steel Plates for Special Applications
A 577-77	Ultrasonic Angle-Beam Examination of Steel Plates
B 409-77	*Nickel-Iron-Chromium Alloy Plate, Sheet, and Strip (see Part 8)*
A568M-77	*Steel, Carbon, and High-Strength Low-Alloy Hot-Rolled Sheet and Cold-Rolled Sheet, General Requirements [Metric] (see Part 3)*
A109M-77	*Steel, Carbon, Cold-Rolled Strip, [Metric] (see Part 3)*

Structural Steel

Specifications for:

A	666–72(1979)	Austenitic Stainless Steel, Sheet, Strip, Plate and Flat Bar for Structural Applications
A	351–79	Austenitic Steel Castings for High-Temperature Service
A	478–76	Chromium-Nickel Stainless and Heat-Resisting Steel Weaving Wire
A	184–79	Fabricated Deformed Steel Bar Mats for Concrete Reinforcement

Specifications for:

A	352–79	Ferritic Steel Castings for Pressure-Containing Parts Suitable for Low-Temperature Service
A	6–79b	General Requirements for Rolled Steel Plates, Shapes, Sheet Piling, and Bars for Structural Use
A	29–79	General Requirements for Steel Bars, Carbon and Alloy, Hot-Rolled and Cold-Finished
A	484–79	General Requirements for Stainless and Heat-Resisting Wrought Steel Products (Except Wire)
A	510–77	General Requirements for Wire Rods and Coarse Round Wire, Carbon Steel
A	572–79	High-Strength Low-Alloy Columbium-Vanadium Steels of Structural Quality
A	441–79	High-Strength Low-Alloy Structural Manganese Vanadium Steel
A	242–79	High-Strength Low-Alloy Structural Steel
A	588–79a	High-Strength Low-Alloy Structural Steel with 50 000 psi Minimum Yield Point to 4 in. Thick
A	570–79	Hot-Rolled Carbon Steel Sheets and Strip, Structural Quality
A	322–79a	Hot-Worked Alloy Steel Bars
A	283–79	Low and Intermediate Tensile Strength Carbon Steel Plates, Shapes and Bars
A	663–77	Merchant Quality Hot-Rolled Carbon Steel Bars Subject to Mechanical Property Requirements
A	265–79	Nickel and Nickel-Base Alloy Clad Steel Plate
A	633–79a	Normalized High-Strength Low-Alloy Structural Steel
A	353–79a	Pressure Vessel Plates, Alloy Steel, Nine Percent Nickel Double-Normalized and Tempered
A	538–77	Pressure Vessel Plates, Alloy Steel, Precipitation Hardening (Maraging), 18 Percent Nickel
A	533–79a	Pressure Vessel Plates, Alloy Steel, Quenched and Tempered, Manganese-Molybdenum and Manganese-Molybdenum-Nickel
A	537–79	Pressure Vessel Plates, Heat-Treated, Carbon-Manganese-Silicon Steel
A	434–76	Steel Bars, Alloy, Hot-Rolled or Cold-Finished, Quenched and Tempered
A	321–79	Quenched and Tempered Carbon Steel Bars
A	678–75	Quenched and Tempered Carbon Steel Plates for Structural Applications
A	576–79	Special Quality Hot-Rolled Carbon Steel Bars

A 479–79	Stainless and Heat-Resisting Steel Bars and Shapes for Use in Boilers and Other Pressure Vessels
A 580–79	Stainless and Heat-Resisting Steel Wire
A 331–74(1979)	Steel Bars, Alloy, Cold-Finished
A 675–79	Steel Bars and Bar Size Shapes, Carbon, Hot-Rolled, Special Quality Subject to Mechanical Property Requirements
A 611–72(1979)	Steel, Cold-Rolled Sheet, Carbon, Structural
A 606–75	Steel, Hot-Rolled and Cold-Rolled Sheet and Strip, High Strength, Low Alloy with Improved Corrosion Resistance
A 635–74	Steel Sheet and Strip, Carbon, Hot-Rolled Commercial Quality, Heavy-Thickness Coils (Formerly Plate)
A 607–75	Steel Sheet and Strip, Hot-Rolled and Cold-Rolled, High Strength, Low-Alloy Columbium and/or Vanadium
A 328–75a	Steel Sheet Piling
A 435–75	Straight-Beam Ultrasonic Examination of Steel Plates for Pressure Vessels
A 311–79	Stress Relieved Cold-Drawn Carbon Steel Bars Subject to Mechanical Property Requirements
A 573–77	Structural Carbon Steel Plates of Improved Toughness
A 36–77a	Structural Steel
A 529–75	Structural Steel with 42 000 psi (290 MPa) Minimum Yield Point (½ in. (12.7 mm) Maximum Thickness)
A 276–79a	*Stainless and Heat-Resisting Steel Bars and Shapes (see Part 5)*
A 108–79	*Steel Bars, Carbon, Cold-Finished, Standard Quality (see Part 5)*
A635M–77	*Steel Sheet and Strip, Carbon, Hot-Rolled Commercial Quality, Heavy-Thickness Coils (Formerly Plate) [Metric] (see Part 3)*
A510M–77	*Wire Rods and Coarse Round Wire, Carbon Steel, General Requirements [Metric] (see Part 3)*

Practice for:

A 700–78	*Packaging, Marking, and Loading Methods for Steel Products for Domestic Shipments (see Parts 1, 3, 4, 5,)*

Specifications for:

A 591–77	Steel Sheet, Cold-Rolled, Electrolytic Zinc-Coated
A 755–78a	Steel Sheet Zinc-Coated (Galvanized) by the Hot-Dip Process and Coil-Coated for Roofing and Siding, General Requirements
A 526–71(1975)	Steel Sheet, Zinc-Coated (Galvanized) by the Hot-Dip Process, Commercial Quality
A 528–71(1975)	Steel Sheet, Zinc-Coated (Galvanized) by the Hot-Dip Process, Drawing Quality
A 525–79	Steel Sheet, Zinc-Coated (Galvanized) by the Hot-Dip Process, General Requirements
A 361–76	Steel Sheet, Zinc-Coated (Galvanized) by the Hot-Dip Process for Roofing and Siding
A 527–71(1975)	Steel Sheet, Zinc-Coated (Galvanized) by the Hot-Dip Process, Lock-Forming Quality
A 446–76	Steel Sheet, Zinc-Coated (Galvanized) by the Hot-Dip Process, Structural (Physical) Quality

A 641–71a(1975) Zinc-Coated (Galvanized) Carbon Steel Wire
A 586–68(1976) Zinc-Coated Steel Structural Strand
A 603–70(1974) Zinc-Coated Steel Structural Wire Rope
A 153–78 Zinc Coating (Hot-Dip) on Iron and Steel Hardware
A 164–71 *Electrodeposited Coatings of Zinc on Steel (see Parts 3 and 9)*
A 123–78 *Zinc (Hot-Galvanized) Coatings on Products Fabricated from Rolled, Pressed, and Forged Steel Shapes, Plates, Bars and Strip (see Part 3)*

Test Methods for:
A 239–73(1978) Locating the Thinnest Spot in a Zinc (Galvanized) Coating on Iron or Steel Articles by the Preece Test (Copper Sulfate Dip)
A 90–69(1978) Weight of Coating on Zinc-Coating (Galvanized) Iron or Steel Articles

TESTING AGENCIES

Practices for:
E 329–77 Inspection and Testing Agencies for Concrete Steel and Bituminous Materials as Used in Construction
E 699–79 *Criteria for Evaluation of Agencies Involved in Testing, Quality Assurance, and Evaluation Building Components in Accordance with Test Methods Promulgated by ASTM Committee E-6 (see Part 18)*
E 543–76 *Determining the Qualification of Nondestructive Testing Agencies (see Part 11)*
D 3740–78 *Evaluation of Agencies Engaged in Testing and/or Inspection of Soil and Rock Used in Construction (see Part 19)*
E 548–79 *Generic Criteria for Use In the Evaluation of Testing and Inspection Agencies (see Parts 10, 41)*

TESTING METHODS, GENERAL

Methods of:
F 476–76 Security of Swinging Door Assemblies
E 8–79a Tension Testing of Metallic Materials
D 374–79 Thickness of Solid Electrical Insulation, Tests for

Specification for:
E 11–70(1977) *Wire-Cloth Sieves for Testing Purposes (see Part 4)*

THERMAL INSULATING MATERIALS

Specifications for:
C 739–77 Cellulosic Fiber (Wood-Base) Loose-Fill Thermal Insulation
C 612–77 Mineral Fiber Block and Board Thermal Insulations

C 549–73(1979)   Perlite Loose Fill Insulation
C 208–72   Insulating Board (Cellulosic Fiber), Structural and Decorative
C 532–66(1974)   Structural Insulating Formboard (Cellulosic Fiber)
C 516–75   Vermiculite Loose Fill Insulation
C 551–78   *Asbestos-Cement Fiberboard Insulating Panels (see Part 16)*
C 533–72   *Calcium Silicate Block and Pipe Thermal Insulation (see Part 18)*
C 552–79   *Cellular Glass Block and Pipe Thermal Insulation (see Part 18)*
C 640–69(1976)   *Corkboard and Cork Pipe Insulation for Low-Temperature Thermal Insulation (see Part 18)*
C 517–71(1979)   *Diatomaceous Earth Block and Pipe Thermal Insulation (see Part 18)*

Specifications for:
C 553–70(1977)   *Mineral Fiber Blanket and Felt Insulation (Industrial Type) (see Part 18)*
C 592–70(1979)   *Mineral Fiber Blanket Insulation and Blanket-Type Pipe Insulation (Metal-Mesh Covered) (see Part 18)*
C 547–77   *Mineral Fiber Preformed Pipe Insulation (see Part 18)*
C 195–77   *Mineral Fiber Thermal Insulating Cement (see Part 18)*
C 728–72   *Perlite Thermal Insulation Board (see Part 18)*
C 578–69   *Preformed, Block-Type Cellular Polystyrene Thermal Insulation (see Part 18)*
C 534–77   *Preformed Flexible Elastomeric Cellular Thermal Insulation in Sheet and Tubular Form (see Part 18)*
C 591–69   *Rigid Preformed Cellular Urethane Thermal Insulation (see Part 18)*
C 720–72(1979)   *Spray-Applied Fibrous Thermal Insulation for Elevated Temperature (see Part 18)*
C 656–79a   *Structural Insulating Board, Calcium Silicate (see Part 18)*

Practices for:
C 680–71   *Heat Gain or Loss, and Surface Temperatures of Insulated Pipe and Equipment Systems by the Use of a Computer Program, Determination of (see Part 18)*
C 585–76   *Inner and Outer Diameters of Rigid Thermal Insulation for Nominal Sizes of Pipe and Tubing (NPS System) (see Part 18)*
C 687–71   *Thermal Resistance of Low-Density Fibrous Loose Fill-Type Building Insulation, Determination of (see Part 18)*
C 677–71(1977)   *Use of a Standard Reference Sheet for the Measurement of the Time-Averaged Vapor Pressure in a Controlled Humidity Space (see Part 18)*
C 727–72(1978)   *Use of Reflective Insulation in Building Constructions (see Part 18)*

Test Methods for:
C 411–61(1975)   Hot-Surface Performance of High-Temperature Thermal Insulation
C 209–72   Insulating Board (Cellulosic Fiber), Structural and Decorative
C 167–64(1976)   Thickness and Density of Blanket- or Batt-Type Thermal Insulating Materials
C 692–77   *Evaluating the Influence of Wicking-Type Thermal Insulations on*

		the Stress Corrosion Cracking Tendency of Austenitic Stainless Steel (see Part 18)
C	447–76	Maximum Use Temperature of Preformed Homogeneous Thermal Insulations (see Part 18)
C	390–79	Sampling Preformed Thermal Insulation (see Part 18)
C	335–79	Steady-State Heat Transfer Properties of Horizontal Pipe Insulations (see Part 18)
C	177–76	Steady-State Thermal Transmission Properties by Means of the Guarded Hot Plate (see Part 18)
C	691–78	Steady-State Thermal Transmission Properties of Nonhomogeneous Pipe Insulation Installed Horizontally (see Part 18)
C	236–66 (1971)	Thermal Conductance and Transmittance of Built-Up Sections by Means of the Guarded Hot Box (see Part 18)
C	421–77	Tumbling Friability of Preformed Block-Type Thermal Insulation (see Part 18)

ULTRASONIC TECHNIQUES

Methods of:

E	273–68 (1974)	Ultrasonic Inspection of Longitudinal and Spiral Welds of Welded Pipe and Tubing (see Part 11)

Practices for:

E	127–75	Fabricating and Checking Aluminum Alloy Ultrasonic Standard Reference Blocks (see Part 11)
E	428–71 (1980)	Fabrication and Control of Steel Reference Blocks Used in Ultrasonic Inspection (see Part 11)
E	214–68 (1979)	Immersed Ultrasonic Testing by the Reflection Method Using Pulsed Longitudinal Waves (see Part 11)
E	317–79	Performance Characteristics of Ultrasonic Pulse-Echo Testing Systems Without the use of Electronic Measurement Instruments, Evaluating (see Part 11)
E	164–74	Ultrasonic Contact Examination of Weldments (see Part 11)
E	213–79	Ultrasonic Inspection of Metal Pipe and Tubing (see Part 11)
E	114–75	Ultrasonic Pulse-Echo Straight Beam Testing by the Contact Method (see Part 11)
E	113–67 (1974)	Ultrasonic Testing by the Resonance Method (see Part 11)
E	494–75 (1980)	Ultrasonic Velocity in Materials, Measuring (see Part 11)

Definitions of Terms Relating to:

E	500–74	Ultrasonic Testing (see Part 11)

WATER VAPOR

Test Methods for:

E	96–66 (1972)	Water Vapor Transmission of Materials in Sheet Form (see Parts 18, 20, 21, 35)

WOOD AND WOOD PRESERVATIVES

Wood

Specifications for:

D 2559-76	Adhesives for Structural Laminated Wood Products for Use Under Exterior (Wet Use) Exposure Conditions
D 2277-75	Fiberboard Nail-Base Sheathing
D 3200-74	Establishing Recommended Design Stresses for Round Timber Construction Poles
D 25-79	Round Timber Piles
D 1031-59(1976)	*Creosoted End-Grain Wood Block Flooring for Interior Use (see Part 22)*
D 52-62(1976)	*Wood Paving Blocks for Exposed Platforms, Pavements, Driveways, and Interior Floors Exposed to Wet and Dry Conditions (see Part 22)*
D 1324-60(1977)	*Modified Wood*

Methods of:

D 2555-78	Establishing Clear-Wood Strength Values
D 2899-74	Establishing Design Stresses for Round Timber Piles
D 245-74	Establishing Structural Grades and Related Allowable Properties for Visually Graded Lumber
D 2915-74	Evaluating Allowable Properties for Grades of Structural Lumber
D 1037-78	Evaluating the Properties of Wood-Base Fiber and Particle Panel Materials
D 198-76	Static Tests of Timbers in Structural Sizes
D 1101-59(1976)	Test for Integrity of Glue Joints in Structural Laminated Wood Products for Exterior Use
D 1761-77	Testing Mechanical Fasteners in Wood
D 143-52(1978)	Testing Small Clear Specimens of Timber
D 2164-65(1977)	Tests for Structural Insulating Roof Decks
D 3502-76	*Moisture Absorption of Compressed Wood Products (see Part 22)*
E 661-78	*Performance of Wood and Wood-Based Floor and Roof Sheathing Under Concentrated Static and Impact Loads (see Part 18)*
D 3501-76	*Plywood in Compression, Testing (see Part 22)*
D 3500-76	*Plywood in Tension, Testing (see Part 22)*
D 3503-76	*Swelling and Recovery of Compressed Wood (see Part 22)*
E 160-80	*Test for Combustible Properties of Treated Wood by the Crib Test (see Part 18)*
D 3499-76	*Toughness of Plywood (see Part 22)*

Practice for:

D 3737-78	*Establishing Stresses for Structural Glued-Laminated Timber (Glulam) Manufactured from Visually Graded Lumber (see Part 22)*

Wood Preservatives

Specifications for:

D 3024–78	Protein-Base Adhesives for Structural Laminated Wood Products for Use Under Interior (Dry Use) Exposure Conditions
D 1624–71 (1976)	*Acid Copper Chromate (see Part 22)*
D 1625–71 (1976)	*Chromated Copper Arsenate (see Part 22)*
D 1032–76	*Chromated Zinc Chloride (see Part 22)*
D 390–80	*Coal-Tar Creosote for the Preservative Treatment of Piles, Poles, and Timbers for Marine Land, and Fresh Water Use (see Part 22)*
D 1034–76	*Fluor-Chrome-Arsenate-Phenol (see Part 22)*
D 1272–56 (1976)	*Pentachlorophenol (see Part 22)*

Test Methods for:

D 2898–77	Accelerated Weathering on Fire-Retardant Treated Wood, for Fire Testing

Test Methods for:

D 3225–73 (1979)	*Low-Boiling Hydrocarbon Solvent for Oil-Borne Preservatives (see Part 22)*
D 2016–74	*Moisture Content of Wood (see Part 22)*
D 3224–73 (1979)	*Water Solubility of Auxiliary Solvent for Wood Preserving Solutions (see Part 22)*
D 2718–76	*Plywood in Rolling Shear (Shear in Plane of Piles) Testing (see Part 22)*
D 2719–76	*Polywood in Shear Through-the-Thickness (see Part 22)*

APPENDIX B

This appendix contains an extract from the 1981 Catalog of American National Standards. This 206-page catalog has two listings, by topic and by standard. The complete catalog is available from the American National Standards Institute (ANSI), 1430 Broadway, New York, NY 10018.

This extract is from the topic list. The ANSI list includes appropriate ASTM standards. Since these already appear in Appendix A, they are omitted from Appendix B.

ABBREVIATIONS

The list below provides the full names of sponsor organizations whose acronyms and designations are used in the ANSI catalog to identify many American National Standards. The year date shown in the sponsor's designation indicates the latest edition of the standard approved by ANSI. In some instances, that edition may have been succeeded by a subsequent document that is pending ANSI approval. Information on such subsequent editions may be obtained by writing directly to the sponsor.

AAMA	Architectural Aluminum Manufacturers Association
AAMI	Association for the Advancement of Medical Instrumentation
AAR	Association of American Railroads
AATCC	American Association of Textile Chemists and Colorists
ACI	American Concrete Institute
ADA	American Dental Association
AFMBA	Anti-Friction Bearing Manufacturers Association, Inc
AGMA	American Gear Manufacturers Association
AHAM	Association of Home Appliance Manufacturers
ANS	American Nuclear Society
API	American Petroleum Institute
ARI	Air-Conditioning and Refrigeration Institute
ASAE	American Society of Agricultural Engineers
ASHRAE	American Society of Heating, Refrigerating and Air-Conditioning Engineers
ASLE	American Society of Lubricating Engineers

ASME	American Society of Mechanical Engineers
ASQC	American Society for Quality Control
ASSE	American Society of Sanitary Engineers
ASTM	American Society for Testing and Materials
AWS	American Welding Society
AWWA	American Water Works Association
BIFMA	Business and Institutional Furniture Manufacturers Assocation
CAPPA	Crusher and Portable Plant Association
CEMA	Conveyor Equipment Manufacturers Association
CGA	Compressed Gas Association
DCDMA	Diamond Core Drill Manufacturers Association
EIA	Electronic Industries Association
FCI	Fluid Controls Institute
FIPS	Federal Information Processing Standards Publication
HPMA	Hardwood Plywood Manufacturers Association
IEEE	Institute of Electrical and Electronics Engineers
IES	Illuminating Engineering Society
IIAR	International Institute of Ammonia Refrigeration
IME	Institute of Makers of Explosives
IPC	Institute of Printed Circuits
IPCEA	Insulated Power Cable Engineers Association
ISA	Instrument Society of America
ISDSI	Insulated Steel Door Systems Institute
ISO	International Organization for Standardization
ITE	Institute of Traffic Engineers
MPTA	Mechanical Power Transmission Association
MSS	Manufacturers Standardization Society
NAAMM	National Association of Architectural Metal Manufacturers
NBS	National Bureau of Standards
NCCLS	National Committee for Clinical Laboratory Standards
NEMA	National Electrical Manufacturers Association
NFPA	National Fire Protection Association
NFSA	National Fertilizer Solutions Association
NILECJ	National Institute of Law Enforcement and Criminal Justice
NLGI	National Lubricating Grease Institute
NMA	National Microfilm Association
NWMA	National Woodwork Manufacturers Association
RETMA	Radio-Electronics-Television Manufacturers Association
RMA	Rubber Manufacturers Association
RVIA	Recreational Vehicle Industry Assocation
RWMA	Resistance Welder Manufacturers Association
SAE	Society of Automotive Engineers
SAMA	Scientific Apparatus Makers Association
SMA	Screen Manufacturers Association (also designated as SMR and SMS)
SPR	Simplified Practice Recommendation
TAPPI	Technical Association of the Pulp and Paper Industry
UL	Underwriters Laboratories
Vol Prod Std	Voluntary Product Standard

LEGEND

A diamond (♦) indicates new and revised standards approved since last (1980) issue of the Catalog.

A triangle (▲) indicates that the standard is not yet available and price will be announced at a later date.

A dagger (†) indicates Standards Institute publications to which quantity and member discounts apply.

R indicates reaffirmed as up-to-date—no change.

BLASTING

† Ventilation and Safe Practices of Abrasive Blasting Operations, ANSI Z9.4-1979.

BOILER-FURNACES

Fuel Oil-Fired Multiple Burner Boiler-Furnaces, Explosion Prevention, ANSI/NFPA 85D-1978.   [Z287.3]

♦ Furnace Explosions in Fuel Oil- and Natural Gas-Fired Watertube Boiler-Furnaces with One Burner, Prevention of, ANSI/NFPA 85-1976.   [Z287.1]

Furnace Implosions in Multiple Burner Boiler-Furnaces, Prevention of, ANSI/NFPA 85G-1980.

Gas-Fired Multiple Burner Boiler-Furnaces, Explosion Prevention, ANSI/NFPA 85B-1978.   [Z287.2]

♦ Pulverized Coal-Fired Multiple Burner Boiler-Furnaces, Explosion Prevention, ANSI/NFPA 85E-1980.   [Z287.4]

Oil-Fired Boiler Assemblies, Safety Standard for, ANSI/UL 726-1975.   [Z96.3]

♦ Qualifications and Duties of Personnel Engaged in ASME Boiler and Pressure Vessel Code, Section III, Division 1 and 2, Certifying Activities, ANSI/ASME N626.3-1978.

BUILDING AREAS

† Areas in Hospitals and Related Facilities, Methods of Determining, ANSI Z65.4-1959(R1973).

† Areas in Industrial Buildings, Methods of Determining, ANSI Z65.5-1962(R1973).

† Areas in Public Buildings, Methods of Determining, ANSI Z65.3-1958(R1973).

† Areas in School Buildings, Methods of Determining, ANSI Z65.2-1958(R1973).

♦ Floor Areas in Office Buildings, Method of Measuring, ANSI Z65.1-1980.

BOILERS AND PRESSURE VESSELS

see also Appliances, Gas-Burning; ASME Boiler and Pressure Vessel Code; Gas Equipment, Industrial

♦ Electric Boilers, Safety Standard for, ANSI/UL 834-1980, ▲

Human Occupancy, Safety Standard for Pressure Vessels for (includes supplement ANSI/ASME PVHO-1a-1979), ANSI/ASME PVHO-1-1977.

Human Occupancy, Safety Standard for Pressure Vessels for (supplement to ANSI/ASME PVHO-1-1977), ANSI/ASME PVHO-1a-1979.

BUILDING CONSTRUCTION

see also Construction and Demolition

Building Construction and Demolition Operations, ANSI/NFPA 241-1975.   [A10.21]

BUILDING MATERIALS

see also Building Construction; Specific Materials

♦ Wood and Fire Retardant Coatings for Building Materials, Fire Retardant Impregnated, ANSI/NFPA 703-1979.

CONCRETE

see also Mortars; Refractory Materials

Consolidated of Concrete, Practice for, ANSI/ACI 309-72(1978).   [A185.1]

Evaluation of Strength Test Results of Concrete, Recommended Practice for, ANSI/ACI 214-77.   [A146.1]

Measuring, Mixing, Transporting, and Placing of Concrete, Practice for, ANSI/ ACI 304-73(1978).   [A186.1]

Normal and Heavy Weight Concrete, Practice for Selecting Proportions for, ANSI/ACI 211.1-77.   [A167.1]

Preparation of Notation for Concrete, ANSI/ACI 104-71.   [A188.1]

Reinforced Concrete, Building Code Requirements for, ANSI/ACI 318-77.
    [A89.1]

Reinforced Gypsum Concrete, Specifications for ANSI A59.1-1968(R1972).

Selecting Proportions for No-Slump Concrete, Practice for, ANSI/ACI 211.3-75.
    [A167.2]

♦ Selecting Proportions for Structural Lightweight Concrete, Practice for, ANSI/ ACI 211.2-1977.   [A164.1]

Shotcrete, Specification for Materials, Proportioning, and Application of, ANSI/ACI 506.2-77.

Shotcreting, Practice for, ANSI/ACI 506-1966(1978).   [A141.1]

Shrinkage-Compensating Concrete, Recommended Practice for Use of, ANSI/ ACI 223-77.

♦ Structural Concrete for Buildings, Specifications for, ANSI/ACI 301-79.
    [A138.1]

Structural Plain Concrete, Building Code Requirements for, ANSI/ACI 322-71.
    [A169.1]

CONCRETE CONSTRUCTION

see also Construction and Demolition; Masonary Units

Bins, Silos, and Bunkers for Storing Granular Materials, Recommended Practice for, Design and Construction of Concrete, ANSI/ACI 313-77.

♦ Bonding to Hardened Concrete with a Multi-Component Epoxy Adhesive, Producing Skid-Resistant Surface with a Multi-Component Epoxy System, and Repairing with Epoxy Mortars (includes ANSI/ACI 503.1, 503.2, 503.3 and 503-4), ANSI/ACI 503-79.

Concrete Floor and Slab Construction, Practice for, ANSI/ACI 302-69.
    [A157.1]

♦ Concrete Formwork, Recommended Practice for, ANSI/ACI 347-1978. [A145.1]

Concrete Highway Bridge Deck Construction, Practice for, ANSI/ACI 345-74.
    [A89.2]

Concrete Inspection, Practice for, ANSI/ ACI 311-75.   [A188.2]

Construction of Concrete Pavements and Concrete Bases, Practice for, ANSI/ ACI 316-74.   [A144.2]

† Forms for Two-Way Concrete Joist Floor and Roof Construction, ANSI A48.2-1978.

† Joint Construction, Forms for One-Way Concrete, ANSI A48.1-1978.

♦ Masonry Construction, Specification for Concrete, ANSI/ACI 531-79.

Nuclear Safety Related Concrete Structures, Code Requirements for, ANSI/ ACI 349-76.

♦ Piers, Construction of End Bearing Drilled, ANSI/ACI 336.1-1979.

♦ Structures, Building Code Requirements for Concrete Masonry, ANSI/ACI 531-1979.

CONCRETE CURING

see also Concrete

Atmospheric Pressure Steam Curing of Concrete, Practice for, ANSI/ACI 517-70.   [A172.1]

Curing Concrete, Practice for, ANSI/ACI 308-71(1978).   [A168.1]

CONCRETE REINFORCEMENT

Steel Spirals for Reinforced Concrete Columns, ANSI/SPR R53-63. [A38.1]

CONDUCTORS
Non-Specular Surface Finish on Bare Overhead Aluminum Conductors, ANSI C7.69-1976.

CONDUIT AND DUCTS
♦ Duct and Fittings for Underground Installation, PVC and ABS Plastic Communications, ANSI/ NEMA TC10-1978.
♦ Duct for Underground Installation, Extra-Strength PVC Plastic Utilities, ANSI/ NEMA TC8-1978.
♦ Duct for Underground Installation, Fittings for ABS and PVC Plastic Utilities, ANSI/ NEMA TC9-1978.
♦ Duct, Corrugated Polyolefin Coilable Plastic Utilities, ANSI/ NEMA TC5-1978.
♦ Duct, Smooth-Wall Coilable Polyethylene Electrical Plastic, ANSI/ NEMA TC7-1978.
Electrical Metallic Tubing, Safety Standard for, ANSI/ UL 797-1977. [C33.98]
† Electrical Metallic Tubing, Zinc Coated, Specification for, ANSI C80.3-1977.
Fiber Conduit, Safety Standard for, ANSI/ UL 543-1972. [C33.37]
♦ Fittings and Supports for Conduit and Cable Assemblies, ANSI/ NEMA FB1-1977, ▲
Flexible Metal Conduit, Safety Standard for, ANSI/ UL 1-1979. [C33.92]
♦ Flexible Nonmetallic Tubing, Safety Standard for, ANSI/ UL 3-1797. [C33.15]
♦ PVC and ABS Plastic Utilities Duct for Underground Installaton, ANSI/ NEMA TC-6-1978. [C130.2]
† Rigid Aluminum Conduit, Specification for, ANSI C80.5-1977.
♦ Rigid Nonmetallic Conduit, Safety Standard for, ANSI/ UL 651-1980, ▲ [C33.91]
† Rigid Steel Conduit, Enamelled, Specification for, ANSI C80.2-1977.
† Rigid Steel Conduit, Zinc Coated, Specification for, ANSI C80.1-1977.

♦ Steel Conduits, Liquid-Tight Flexible, ANSI/ UL 360-1980.

CONNECTIONS, ELECTRIC
† Highly Reliable Soldered Connections in Electronic and Electrical Applications, Criteria for Inspection for (includes 2 * 2 Color Transparencies), ANSI C83.86-1966(R1978).
Solderless Wrapped Electrical Connections, ANSI/ EIA RS-280-B-1977. [C83.72]

CONNECTORS, ELECTRIC
see also Printed Circuits
Cable Connectors for Audio Facilities Requirements for, ANSI/ EIA RS-297-B-1956. [C83.12]
Connectors for Use between Aluminum or Aluminum-Copper Overhead Conductors, ANSI/ NEMA CC3-1973.[C119.4]
Connectors, Electrical Flat Cable Type, Industry Standard for, ANSI/ EIA RS-429-1976, ANSI/ IPC FC-218B. [C83.110]
Connectors, Electrical, Printed-Wiring Board, ANSI/ EIA RS-406-1973 (R1979). [C83.88]
Electric Power Connectors for Substations, ANSI/ NEMA CC1-1972. [C119.3]
Electrical Connectors, Small Contact Standard for, ANSI/ EIA RS-380-A-1978. [C83.64]
Precision Coaxial Connectors, ANSI/ IEEE 287-1968. [C16.43]
Precision Coaxial Connectors for CATV Application (75 Ohms), ANSI/ EIA RS-403-1973. [C83.100]
† Sealed Insulated Underground Connector Systems Rated 600 Volts, ANSI C119.1-1974.
Separable Insulated Connectors for Power Distribution Systems above 600 V, ANSI/ IEEE 386-1976. [C119.2]
Wire Connectors and Soldering Lugs, Safety Standard for, ANSI/ UL 486-1969. [C33.5]

CONNECTORS, HOSE
see also Fire Fighting Equipment
Pigtails and Flexible Hose Connectors for LP-Gas, Safety Standard for, ANSI/ UL 569-1975.   [B126.1]

CONSTRUCTION AND DEMOLITION
see also Building Construction
† Asphalt Pavement Construction, Safe Operating Practice for, ANSI A10.17-1975.
† Ceramic Tile, Terrazzo, and Marble Work, Safety Requirements for, ANSI A10.20-1977.
† Concrete Construction and Masonry Work, Safety Requirements for, ANSI A10.9-1970.
† Demolition, Safety Requirements for, ANSI A10.6-1969.
† Dredging, Safety Requirements for, ANSI A10.15-1974.
† Floor and Wall Openings, Flat Roofs, Stairs, Railings, and Toeboards, Construction Safety Requirements for Temporary, ANSI A10.18-1977.
† Hoists, Safety Requirements for Rope-Guided and Non-Guided Workmen's, ANSI A10.22-1977.
† Material Hoists, Safety Requirements for, ANSI A10.5-1975.
† Personnel Hoists, Safety Requirements for, ANSI A10.4-1975.
† Powder Actuated Fastening Systems, Safety Requirements for, ANSI A10.3-1977.
† Safety Belts, Harnesses, Lanyards, Lifelines, and Drop Lines for Construction and Industrial Use, Requirements for, ANSI A10.14-1975.
♦ Safety nets Used During Construction, Repair, and Demolition Operations, ANSI A10.11-1979.
† Scaffolding, Safety Requirements for, ANSI A10.8-1977.
† Steel Erection, Safety Requirements for, ANSI A10.13-1978.
† Temporary and Portable Space Heating Devices and Equipment Used in the Construction Industry, Safety Requirements for, ANSI A10.10-1970.
† Transportation, Storage, Handling, and Use of Commercial Explosives and Blasting Agents in the Construction Industry, Safety Requirements for, ANSI A10.7-1970.

DOORS AND FRAMES
see also Windows
Aluminum Sliding Glass Doors, Specifications for, ANSI/ AAMA 402.9-1977.   [A134.2]
Aluminum Sliding Screen Doors, Specifications for, ANSI/SMA 2005-1975.   [A201.2]
Aluminum Storm Doors, Specifications for, ANSI/ AAMA 1102.7-1977.   [A134.4]
Aluminum Swinging Screen Doors, Specifications for, ANSI/SMA 3001-1975.   [A201.3]
Dimensional Standard for Insulated Steel Door Systems, ANSI/ISDSI 100-1979.
Door, Drapery, Gate, Louver, and Window Operators and Systems, Safety Standard for, ANSI/ UL 325-1979.
Fire Door Frames, Safety Standard for, ANSI/UL 63-1970.   [A155.1]
Fire Doors and Windows, ANSI/NFPA 80-1977.   [A2.7]
Fire Resistance for Vault and File Room Doors, Tests for, ANSI/ UL 155-1979.   [A153.1]
♦ Fire Tests of Door Assemblies, Methods of, ANSI/ ASTM E152-80.
Overhead Type Doors, Specifications for Sectional, ANSI A216.1-1977.
Pedestrian Doors, Power Operated, ANSI A156.10-1979.
Ponderosa Pine Doors, ANSI/NWMA I.S. 5-73.   [A200.2]
Security of Swinging Door Assemblies, Test for, ANSI/ ASTM F476-76.
Sliding Hardware for Standard, Horizontally Mounted Tin-Clad Fire Doors, Safety Standard for, ANSI/UL 14B-1978.   [A143.1]

Standard Steel Doors, Frames, Anchors, Hinge Reinforcings and Exit Device Reinforcings, Performance Test for, ANSI A151.1-1969.

Steel Door Systems, Installation Standard for Insulated, ANSI/ISDSI 102-1979.

Steel Doors and Steel Door Frames, Nomenclature for, ANSI A123.1-1974.

♦ Steel Surfaces for Steel Doors and Frames, Test Procedure and Acceptance Criteria for Prime Painted, ANSI A224.1-1980.

▲

Swinging Hardware for Standard Tin-Clad Fire Doors Mounted Singly and in Pairs, Safety Standard for, ANSI/UL 14C-1978.   [A133.1]

♦Tin-Clad Fire Doors, ANSI/UL 10A-1979.   [A142.1]

Wood Sliding Patio Doors, ANSI/ NWMA I.S. 3-70.   [A200.4]

ELECTRIC EQUIPMENT

Electrical Equipment Maintenance, Recommended Practice for, ANSI/NFPA 70B-1977.   [C132.1]

♦Hazardous Location, Intrinsically Safe Apparatus and Associated Apparatus for Use in Class I, II, and III, Division 1, ANSI/UL/NFPA 4913-1979.

Purged and Pressurized Enclosures for Electrical Equipment in Hazardous Locations, ANSI/NFPA 496-1974.
                                        [C106.1]

ELECTRIC LINE CONSTRUCTION

† Anchor Rods and Nuts for Overhead Line Construction, Threaded Galvanized Ferrous Strand-Eye, ANSI C135.2-1979.

† Bolts and Nuts for Overhead Line Construction, Galvanized Ferrous Eye, ANSI C135.4-1979.

† Bolts and Nuts for Overhead Line Construction, Galvanized Steel, ANSI C135.1-1979.

♦† Cable Racks and Cable Rack Hooks, Galvanized Ferrous Underground, ANSI C135.35-1980.

♦†Crossarm Braces, Galvanzied Ferrous, ANSI C135.6-1980. ▲

♦†Crossarm Gains, Galvanized Ferrous, ANSI C135.33-1980.

† Eyenuts and Eyelets, Galvanized Ferrous for Overhead Line Construction, ANSI C135.5-1979.

†Ground Rods for Overhead or Underground Line Construction, Galvanized Ferrous, ANSI C135.30-1979.

†Insulator Pins with Lead Threads for Overhead Line Construction, Galvanized Ferrous Pole-Top, ANSI C135.22-1979.

†Insulator Pins with Lead Threads for Overhead Line Construction, Galvanized Ferrous Bolt-Type, ANSI C135.17-1979.

♦† Spool Insulator Bolts, Galvanized Ferrous Single and Double Upset, ANSI C135. 31-1980.

†Staples with Rolled or Slash Points for Overhead Line Construction, ANSI C135.14-1979.

ELEVATORS

Elevators, Dumbwaiters, Escalators, and Moving Walks, ANSI A17.1-1978.

Elevators, Escalators, and Moving Walks, Practice for the Inspection of (Inspector's Manual), ANSI A17.2-1979.

ENGINES

see also Turbines, Gas

♦ Combustion Engines and Gas Turbines, Stationary, ANSI/NFPA 37-1979.

Performance Test Code—Reciprocating Internal-Combustion Engines, ANSI/ ASME PTC17-1973(R1980).

FITTINGS, FLANGES, AND VALVES

Bronze Pipe Flanges and Flanged Fittings, Class 150 and 300, ANSI B16.24-1979.

Buttwelding Ends, ANSI B16.25-1979.

Cast Bronze Threaded Fittings, Class 125 and 250, ANSI B16.15-1978.

Cast Copper Alloy Fittings for Flared Copper Tubes, ANSI B16.26-1975.

Cast Copper Alloy Solder Joint Drainage Fittings—DWV, ANSI B16.23-1976.

Cast Copper Alloy Solder Joint Fittings for Sovent Drainage Systems, ANSI B16.32-1979.

Cast Copper Alloy Solder Joint Pressure Fittings, ANSI B16.8-1978.

Cast Iron Pipe Flanges and Flanged Fittings, Class 25, 125, 250, and 800, ANSI B16.1-1975.

Cast-Iron Threaded Drainage Fittings, ANSI B16.12-1977.

Cast-Iron Threaded Fittings, Class 125 and 250, ANSI B16.4-1977.

♦ Commercial Seat Tightness to Safety Relief Valves with Metal-to-Metal Seats, ANSI/API 527-1978.   [B147.1]

Constant-Level Oil Valves, Safety Standard for, ANSI/UL 352-1977. [B127.1]

Control Valve Seat Leakage, Quality Control Standard for, ANSI/FCI 70-2-1976.   [B16.104]

Control Valve Sizing Equations, ANSI/ISA S75.01-1977.

♦ Ductile Iron Fittings, 3-Inch through 24-Inch, for Gas, ANSI A21.14-1979.

Electrically Operated Valves for Use in Hazardous Locations, Class I, Groups A, B, C, and D, and Class II, Groups E, F, and G, Safety Standard for, ANSI/UL 1002-1977.   [C33.83]

Face-to-Face and End-to-End Dimensions of Ferrous Valves, ANSI B16.10-1973.

Factory-Made Wrought Steel Buttwelding Fittings, ANSI B16.9-1978.

Ferrous Pipe Plugs, Bushings, and Locknuts with Pipe Threads, ANSI B16.14-1977.

♦ Fittings, Class 150 and 300, Ductile Iron Pipe Flanges and Flanged, ANSI B16.42-1979.

Fittings, 3 in through 48 in. for Water and Other Liquids, Gray-Iron and Ductile-Iron, ANSI A21.10-77, ANSI/AWWA C110-77.

Flanged Steel Safety Relief Valves, ANSI/API 526-1969 (2nd ed).   [B146.1]

♦ Flanges (Nominal Pipe Size 26 to 60, Inclusive: Classes 75, 150, and 300) Large-Diameter Carbon Steel, ANSI/API 605 (2 Edition)—1978.

Flanges for Water Works Service, 4 in. through 144 in. Steel, ANSI/AWWA C207-78.

♦ Forged Steel Fittings, Socket-Welding, and Threaded, ANSI B16.11-1980.

Gas Shut-Offs and Valves in Gas Distribution Systems, Manually Operated Thermoplastic, ANSI B16.40-1977.

Gas Values in Gas Distribution Systems Whose Maximum Allowable Operating Pressure Does Not Exceed 125 psig (8.6 bar, gage), Large Manually Operated Metallic, ANSI B16.38-1978.

Joints, Grooved and Shouldered Type, ANSI/AWWA C606-78.

Malleable-Iron Threaded Figgings, Class 150 and 300, ANSI B16.3-1977.

♦ Marine Through-Hull Fittings and Sea-Valves, Safety Standard for, ANSI/UL 1121-1979.   [Z254.1]

Measuring Electrical Characteristics of Solenoid Valves, Test Conditons and Procedures for, ANSI/FCI 75-1-1979.

Non-Ferrous Pipe FLanges, 150, 300, 400, 600, 900, 1500, and 2500 lb. ANSI B16.31-1971.

Nonmetallic Flat Gaskets for Pipe Flanges, ANSI B16.21-1978.

Performance Test Code—Safety and Relief Valves, ANSI/ASME PTC25.3-1976.   [PTC25.3]

Pipe Connectors for Flammable Liquids and Combustible Liquids and LP-Gas, Safety Standard for, ANSI/UL 567-1978.   [B148.1]

Pipe Unions (Class 150, 250 and 300) Mallable Iron Threaded, ANSI B16.39-1977.

Ring-Joint Gaskets and Grooves for Steel Pipe Flanges, ANSI B16.20-1973.

Small Manually Operated Metallic Gas Valves, Sizes 1/2 to 2 Inch, in Gas Distribution Systems Whose Maximum Allowable Operating Pressure Does Not Exceed 60 PSIG or 125 PSIG, ANSI B16.33-1973.

Steel Orifice Flanges, Class 300, 600, 900, 1500, and 2500 (includes supplement ANSI B16.36a-1979), ANSI B16.36-1975.

Steel Orifice Flanges, Class 300, 600, 900, 1500, and 2500 (supplement to ANSI B16.36-1975), ANSI B16.36a-1979.

Steel Pipe Flanges and Flanged Fittings, Including Ratings for Class 150, 300, 400, 600, 900, 1500 and 2500, ANSI B16.5-1977.

Steel Valves, Flanged and Butt-Welding End, ANSI B16.34-1977.

♦ Valve Bodies, Uniform Face-to-Face Dimensions for Flanged Globe Style Control, ANSI/ISA S75.03-1979.

♦ Valves, Face-to-Face Dimensions of Flangeless Control, ANSI/ISA S75.04-1979.

♦ Valves, Hydrostatic Testing of Control, ANSI B16.37-1980.

♦ Valves, Rubber-Seated Butterly, ANSI/AWWA C504-80.

♦ Valves, Spring Loaded Lift Disc Check, ANSI/FCI 74-1-1979.

♦ Valves, 3 through 12 NPS, for Water and Sewage Systems, Resilient-Seated Gate, ANSI/AWAW C509-80.

♦ Valves, 3 through 48 Inch NPS, for Water and Sewage Systems, Gate, ANSI/AWWA C500-80.

FURNACES

Industrial Furnaces, Using Special Atmospheres, ANSI/NFPA 86C-1977.   [Z294.3]

Industrial Furnaces—Design, Location, and Equipment, ANSI/NFPA 86B-1974.   [Z294.1]

♦ Oil-Fired Central Furnaces, Safety Standard for, ANSI/UL 727-1980. [Z96.1]

Oil-Fired Floor Furnaces, Safety Standard for, ANSI/UL 729-1975. [Z96.4]

Oild-Fired Wall Furnaces, Safety Standard for, ANSI/UL 730-1978. [Z96.5]

† Performance Requirements for Oil Powered Central Furnaces, ANSI Z91.1-1972.

♦ Vacuum Atmosphere Industrial Furnaces, ANSI/NFPA 86D-1979.

FUSES AND HOLDERS
see also Switchgear

Class H Fuses, Safety Standard for, ANSI/UL 198B-1975.   [C33.114]

Fuseholders, Safety Standard for, ANSI/UL 512-1975.   [C33.10]

Fuses, Safety Standard for, ANSI/UL 198-1971. ▲   [C33.42]

High-Interrupting Capacity Fuses, Current-Limiting Types, Safety Standard for, ANSI/UL 198.2-1974.   [C33.99]

† Low-Voltage Cartridge Fuses 600 Volts or Less, ANSI C97.1-1972(R1978).

Plug Fuses, Safety Standard for, ANSI/UL 198F-1975.   [C33.113]

GAGES

Gauges and Indicators, Pressure and Vacuum, Indicating Digital Type, ANSI B40.2-1977.

♦ Gauges—Indicating Dial Type—Elastic Element (Metric), ANSI B40.1M-1979.

Gauges—Pressure and Vacuum, Indicating Dial Type—Elastic Element, ANSI B40.1-1974.

GAS EQUIPMENT, INDUSTRIAL

Boilers, Gas Utilization Equipment in Large (includes supplements ANSI Z83-3a-1972 and Z83.3b-1976), ANSI Z83.3-1971.

Boilers, Gas Utilization Equipment in Large (supplement to ANSI Z83.3-1971), ANSI Z83.3b-1976.

Boilers, Gas Utilization Equipment in Large (supplement to ANSI Z83.3-1971), ANSI Z83.3a-1972.

Gas-Fired Construction Heaters, ANSI Z83.7-1974.

♦ Heaters, Direct Gas-Fired Make-Up Air, ANSI Z83.4-1980, ▲

Infrared Heaters, Gas-Fired (includes supplement ANSI Z8.6a-1975), ANSI Z83.6-1974.

Infrared Heaters, Gas-Fired (supplement to ANSI Z83.6-1974), ANSI Z83.6a-1975.

Infrared Heaters, Gas-Fired (supplement to ANSI Z83.6-1974), ANSI Z83.6b-1979.

GLAZING MATERIAL

♦ Burglary-Resisting Glazing Material, Safety Standard for, ANSI/UL 972-1978.
[SE4.5]

Glazing Compounds for Back Bedding and Face Glazing of Metal Sash, Specification for, ANSI/ASTM C669-75.

† Glazing Material Used in Buildings, Performance Specifications and Methods of Test for Safety (supplement to ANSI Z97.1-1975), ANSI Z97.1a-1977.

† Glazing Material Used in Buildings, Performance Specifications and Methods of Test for Safety (includes supplement ANSI Z97.1a-1977), ANSI Z97.1-1975.

† Glazing Materials for Glazing Motor Vehicles Operating on Land Highways, Safety Code for Safety (includes supplement ANSI Z26.1a-1980), ANSI Z26.1-1977.

♦ Glazing Materials for Glazing Motor Vehicles Operating on Land Highways, Safety Code for Safety (supplement to ANSI Z26.1-1977), ANSI Z26.1a-1980.

Volatility of Oil- and Resin-Based, Knife-Grade, Channel Glazing Compounds, Test for, ANSI/ASTM C681-73(1979).

HARDWARE

♦ Architectural Door Trim, ANSI A156.6-1979.

Auxiliary Locks and Associated Products, ANSI A156.5-1978.

Butts and Hinges, ANSI A156.1-1976.

Cabinet Hardward, ANSI A156.9-1975.

Cabinet Locks, ANSI A156.11-1978.

♦ Door Controls (Closers), ANSI A156.4-1980, ▲

Door Controls (Overhead Holders), ANSI A156.8-1974.

♦ Door Hardware, Sliding and Folding, ANSI A156.14-1980.

Exit Devices, ANSI A156.3-1978.

♦ Locks and Latches, Interconnected, ANSI A156.12-1979.

Locks and Lock Trim, ANSI A156.2-1976.

Template Hinges, ANSI A156.7-1972.

HEATERS

♦ Air Heaters, Electric, ANSI/UL 1025-1980, ▲

Electric Baseboard Heating Equipment, Safety Standard for, ANSI/UL 1042-1978.
[C33.95]

Electric Dry Bath Heaters, Safety Standard for, ANSI/UL 875-1975. [C33.75]

Electric Heaters for Use in Hazardous Locations, Class I, Groups A, B, C, and D, and Class II, Groups E, F, and G, Safety Standard for, ANSI/UL 823-1977.
[C33.48]

♦ Electric Oil Heaters, Safety Standard for, ANSI/UL 574-1980. [C33.44]

Electric Space-Heating Equipment, Safety Standard for, ANSI/UL 573-1972.
[C33.12]

Electric Water Bed Heaters, Safety Standard for, ANSI/UL 1445-1978.

Household Electric Storage-Tank Water Heaters, Safety Standard for, ANSI/ UL 174-1977.
[C33.87]

Liquid Fuel-Burning Heating Appliances for Mobile Homes and Recreational Vehicles, Safety Standard for, ANSI/UL 307A-1978.
[A147.1]

Oil-Fired Storage Tank Water Heaters, Safety Standard for, ANSI/UL 732-1975.
[Z95.3]

Oil-Fired Unit Heaters, Safety Standard for, ANSI/UL 731-1974. [Z95.2]

Performance Test Code—Closed Feedwater Heaters, ANSI/ASME PTC 12.1-1978.
[PTC12.1]

Sheathed Electrical Resistance Heaters for Nuclear or Other Specialized Service, Specification for, ANSI/ASTM E420-71.

Sheathed Heating Elements, Safety Standard for, ANSI/UL 1030-1975. [C33.108]

HEATING AND AIR CONDITIONING

Air Conditioners, Central Cooling, Safety Standard for, ANSI/UL 465-1978. [B144.1]

Air Conditioning and Ventilating Systems, ANSI/NFPA 90A-1976. [B144.2]

Electric Central Air-Heating Equipment, Safety Standard for, ANSI/UL 1096-1973. [C33.104]

Packaged Terminal Air-Conditioners, ANSI/ARI 310-1976.

Thermal Environmental Conditions for Human Occupancy, ANSI/ASHRAE 55-74. [B193.1]

Unitary Air Conditioning and Heat Pump Equipment, Methods of Testing for Rating, ANSI/ASHRAE 37-78.

♦Warm Air Heating and Air Conditioning Systems, Installation of, ANSI/NFPA 90B-1980. [B144.3]

LIFTING DEVICES

Base Mounted Drum Hoists, Safety Code for, ANSI B30.7-1977.

Controlled Mechanical Storage Cranes, ANSI B30.13-1977.

♦Cranes, (Top Running Bridge, Single Girder, Underhung Hoist) Overhead and Gantry, ANSI B30.17-1980, ▲

Crawler, Locomotive, and Truck Cranes, Safety Code for, ANSI B30.5-1968.

Derricks, Safety Code for, ANSI B30.6-1977.

Floating Cranes and Floating Derricks, Safety Code for, ANSI B30.8-1971.

Hammerhead Tower Cranes, ANSI B30.3-1975.

Handling Loads Suspended from Rotor-craft, Safety Standard for, ANSI B30.12-1975.

Hooks, Safety Standard for, ANSI B30.10-1975.

Jacks, Safety Standard for, ANSI B30.1-1975.

Mobile Hydraulic Cranes, Safety Sandard for, ANSI B30.15-1973.

♦Monorails and Underhung Cranes, ANSI B30.11-1980.

Overhead and Gantry Cranes (Top Running Bridge, Multiple Girder) Safety Standard for, ANSI B30.2.0-1976.

Overhead Hoists, Safety Standard for, ANSI B30.16-1973.

Portal, Tower, and Pillar Cranes, Safety Standard for, ANSI B30.4-1973.

Side Boom Tractors, ANSI B30.14-1979.

Slings, Safety Standard for, ANSI B30.9-1971.

LIGHTING

♦Emergency Lighting Equipment, Safety Standard for, ANSI/UL 924-1978.

♦Illuminating Engineering, Nomenclature and Definitions for, ANSI/IES RP-16-1980, ▲

Industrial Lighting, Practice for, ANSI/IES RP7-1979. [A11.1]

♦Lighting Strings, Temporary, ANSI/UL 1088-1979.

Office Lighting, Practice for, ANSI/IES RP 1-1973. [A132.1]

Protective Lighting, Practice for, ANSI/IES RP10-1956. [A85.1]

School Lighting, Guide for, ANSI/IES RP3-1977. [A23.1]

LIGHTING FIXTURES

Electric Lighting Fixtures for Use in Hazardous Locations, Safety Standard for, ANSI/UL 844-1978. [C33.28]

Marine Type Electric Lighting Fixtures, ANSI/UL 595-1980.

Portable Electric Lighting Units for Use in Hazardous Locations, Class I, Groups C and D, and Class II, Group G, Safety Standard for, ANSI/UL 781-1977.

[C33.79]

♦ Swimming Pools, Safety Standard for Underwater Lighting Fixtures for, ANSI/UL 676-1980, ▲

PIPE, CONCRETE

Cast-in-Place Nonreinforced Concrete Pipe, Specifications for, ANSI/ACI 346-70.   [A171.1]

♦ Prestressed Concrete Pressure Pipe, Steel Cylinder Type, for Water and Other Liquids, ANSI/AWWA C301-79.

Reinforced Concrete Pressure Pipe—Steel Cylinder Type, Pretensioned for Water and Other Liquids, ANSI/AWWA C303-78, ▲

PIPE, IRON

Ductile-Iron Pipe, Centrifugally Cast, in Metal Molds or Sand-Lined Molds for Water or Other Liquids, ANSI A21.51-1976, ANSI/AWWA C151-76.

Ductile-Iron Pipe, Centrifugally Cast, in Metal Molds or Sand-Lined Molds for Gas, ANSI A21.52-1976.

Flanged Cast-Iron and Ductile-Iron Pipe with Threaded Flanges, ANSI A21.15-1975, ANSI/AWWA C115-75.

♦ Gray-Iron Pipe Centrifugally Cast in Metal Molds, for Water of Other Liquids, ANSI A21.6-80, ANSI/AWWA C106-80.

Polyethylene Encasement for Gray and Ductile Cast-Iron Piping for Water and Other Liquids, ANSI A21.5-1972, ANSI/AWWA C105-72(1977).

Thickness Design of Cast Iron Pipe, ANSI A21.1-1967, ANSI/AWWA C101-67 (1977).

Thickness Design of Ductile-Iron Pipe, ANSI A21.50-1976, ANSI/AWWA C150-76.

Threaded Cast-Iron Pipe for Drainage, Vent, and Waste Services, ANSI A40.5-1943.

♦ Water Mains and Appurtances, Installation of Gray and Ductile Cast Iron (supplement to ANSI/AWWA C600-77), ANSI/AWWA C600a-80.

Water Mains and Appurtenances, Installation of Gray and Ductile Cast Iron (includes supplement ANSI/AWWA C600a 80), ANSI/AWWA C600-77.

PIPELINE COATINGS AND LININGS

♦ Cement-Mortar Lining for Ductile-Iron and Gray-Iron Pipe and Fittings for Water. ANSI A21.4-80, ANSI/AWWA C104-80.

♦ Cement-Mortar Protective Lining and Coating for Steel Water Pipe—4 in and Larger—Shop Applied, ANSI/AWWA C205-80.

Chemical Resistance of Pipeline Coatings, Test for, ANSI/ASTM G20-77.

Coal-Tar Epoxy Coating System for the Interior and Exterior of Steel Water Pipe, ANSI/AWWA C210-78.

Coal-Tar Protective Coatings and Linings for Steel Water Pipelines—Enamel and Tape—Hot-Applied, ANSI/AWWA C203-78.

Fusion bonded Epoxy Coating for the Interior and Exterior of Steel Water Pipelines, ANSI/AWWA C213-1979.

PIPING AND PIPING SYSTEMS

♦ Chemical Plant and Petroleum Refinery Piping, ANSI B31.3-1980.

Fuel Gas Piping, ANSI B31.2-1968.

Gas Transmission and Distribution Piping Systems, ANSI B31.8-1975.

Identification of Piping Systems, Scheme for the, ANSI A13.1-1975.

♦ Liquid Petroleum Transportation Piping Systems, ANSI/ASME B31.4-1979.

♦ Power Piping, ANSI B31.1-1980.

Power Piping, Corrosion Control for ANSI B31.1 Power Piping Systems; this document contains guidelines that are applicable to existing operating piping systems contained in the scope of ANSI

B31.1 as well as to new construction, ANSI B31 Guide.

Refrigeration Piping (includes supplement ANSI B31.5a-1978), ANSI B31.5-1974.

Refrigeration Piping (supplement to ANSI B31.5-1974), ANSI B31.5a-1978.

PLUGS AND RECEPTACLES

† Attachement Plugs and Receptacles, Dimensions of (includes supplement ANSI C73a-1980), ANSI C73-1973.

♦† Attachment Plugs and Receptacles, Dimensions of (supplement to ANSI C73-1973 Edition), ANSI C73a-1980.

♦ Attachment Plugs and Receptables, Safety Standard for, ANSI/ UL 498-1980, ▲ [C33.77]

Electrode Receptacles for Gas-Tube Signs, Safety Standard for, ANSI/UL 879-1971. [C33.24]

Receptacle-Plug Combinations for Use in Hazardous Locations, Safety Standard for, ANSI/UL 1010-1977. [C33.97]

PLUMBING

Air Gaps in Plumbing systems, ANSI A112.1.2-1942(R1979).

Backwater Valves, ANSI A112.14.1-1975.

Bowls, Tanks and Urinals, Trim for Water-Closet, ANSI A112.19.5-1979.

Diverters for Plumbing Faucets with Hose Spray, Anti-Siphon Type, Residential Applications, Performance Requirements for, ANSI/ASSE 1025-1978.

Enameled Cast Iron Plumbing Fixtures, ANSI A112.19.1M-1979.

♦ Finished and Rough Brass Plumbing Fixture Fittings, ANSI A112.18.1M-1979.

Floor Drains, ANSI A112.21.1-1968 (R1974).

Hand Held Showers, Plumbing Requirements for, ANSI/ASSE 1014-1979.

♦ Home Laundry Equipment, Plumbing Requirements for, ANSI/ASSE 1007, ANSI/AHAM HLW-2PR-1980.
[A197.2]

Hose Connection Vacuum Breakers, Performance Requirements for, ANSI/ ASSE 1011-1970. [A112.1.3]

Hot Water Dispenser, Household Storage Type Electrical, Plumbing Requirements for, ANSI/ASSE 1023-1979.

♦ Household Dishwashers, Plumbing Requirements for, ANSI/ASSE 1006, ANSI/AHAM DW-2PR-1980, ▲ [A197.1]

♦ Household Food Waste Disposer Units, Plumbing Requirements for, ANSI/ ASSE 1008, ANSI/AHAM FWD-2PR-1980, ▲ [A197.3]

Hydrants for Utility and Maintenance Use, ANSI A112.21.3-1976.

Individual Shower Control Valves—Anti-Scald Type, Performance Requirements for, ANSI/ASSE 1016-1979.

Metallic Cleanouts, ANSI A112.36.2-1975.

Pipe-Applied Atmospheric-Type Anti-Siphon Vacuum Breakers, Performance Requirements, Methods of Test for, ANSI/ASSE 1001-1970. [A112.1.1]

♦† Plastic Bathtub Units, ANSI Z124.1-1980.

♦ Plastic Lavatories, ANSI Z124.3-1980.

♦† Plastic Shower Receptors and Shower Stalls, ANSI Z124.2-1980.

Roof Drains, ANSI A112.21.2-1971.

Stainless Steel Plumbing Fixtures (Designed for Residential Use), ANSI A112.19.3-1976.

Steel Plumbing Fixtures, Porcelain Enameled Formed, ANSI A112.19.4-1977.

♦ Supports for Off-the-Floor Plumbing Fixtures for Public Use, ANSI A112.6.1-1979.

Thermostatic Mixing Valves, Self Actuated for Primary Domestic Use, Performance Requirements for, ANSI/ ASSE 1017-1979.

Trap Seal Primer Valves, Performance Requirements for, ANSI/ASSE 1018-1978.

Vacuum Breakers, Anti-Siphon, Pressure Type, Performance Standard for, ANSI/ ASSE 1020-1974. [A112.1.7]

Vitreous China Plumbing Fixtures, ANSI A112.19.2-1973.

Wall Hydrants, Frost Proof Automatic Draining, Backflow Types, Performance Requirements for, ANSI/ASSE 1019-1978.

Water Closet Flush Tank Ball Cocks, Performance Requirements for, ANSI/ASSE 1002-1979.

Water Hammer Arresters, ANSI A112.26.1-1969(R1975).

Water Pressure Reducing Valves, Performance Requirements for, ANSI/ASSE 1003 AND 1003-1-1970.

[A112.26.2]

REFRIGERATION

Capacity Rating of Thermostatic Refrigerant Expansion Valves, Method of Testing for, ANSI/ASHRAE 17. [B60.1]

♦ Coolers, Refrigeration Unit, ANSI/UL 412-1980.

Equipment, Design, and Installation of Ammonia Mechanical Refrigeration Systems, ANSI/IIAR 74-2-1978.

Mechanical Refrigeration Installations on Shipboard, Practice for, ANSI/ASHRAE 26-78. [B59.1]

Mechanical Refrigeration, Safety Code for, ANSI/ASHRAE 15-78. [B9.1]

Number Designation of Refrigerants, ANSI/ASHRAE 34-1978.

Reciprocating Water-Chilling Packages, ANSI/ARI 590-1976.

Refrigerant-Containing Component and Accessories, Nonelectrical, Safety Standard for, ANSI/UL 207-1975. [Z263.1]

♦ Refrigeration and Air-Conditioning Condensing and Compressor Units, Safety Standard for, ANSI/UL 303-1979.
[Z262.1]

Refrigeration Terms and Definitions, ANSI/ASHRAE 12-75. [B53.1]

SCAFFOLDS AND PLATFORMS
see also Construction and Demolition
♦ † Elevating Work Platforms, Manually Propelled, ANSI A92.3-1980.

† Mobile Ladder Stands and Scaffolds (Towers), Manually Propelled, ANSI A92.1-1977.

† Powered Platforms for Exterior Building Maintenance, Safety Code for, ANSI A120.1-1970.

♦ † Self-Propelled Boom-Supported Elevated Work Platforms, ANSI A92.5-1980.

` † Vechicle Mounted Elevating and Rotating Aerial Devices, ANSI A92.2-1979.

† Work Platforms, Self-Propelled Elevating, ANSI A92.6-1979.

STEEL FABRICATION
† Ships Fabricating Structural Steel and Steel Plate, Safety Requirments for, ANSI Z229.1-1973.

SWITCHES
Automatic Transfer Switches, Safety Standard for, ANSI/UL 1008-1976.
[C33.122]

Basic Sensitive Switches, ANSI/EIA RS-437-1-1978.

Switches for Use in Hazardous Locations, Class I, Groups A, B, C, and D, and Class II, Groups E, F, and G. Safety Standard for, ANSI/UL 894-1977.
[C33.29]

SWITCHGEAR
see also Circuit Breakers; Circuit Breakers, High Voltage; Circuit Protectors; Insulators; Relays

Application, Installation, Operation, and Maintenance of High-Voltage Air Disconnecting and Load Interrupter Switches, ANSI C37.35-1976.

Application, Operation, and Maintenance of Automatic Circuit Reclosers, Guide for the, ANSI/IEEE C37.61-1973 (R1979). [C37.61]

† Application, Operation, and Maintenance of Distribution Cutouts and Fuse Links, Secondary Fuses, Distribution Enclosed

Single-Pole Air Switches, Power Fuses,
Fuse Disconnecting Switches, and Ac-
cessories, Guide for, ANSI C37.48-1969
(R1974).

Automatic Circuit Reclosers for Alternating Current Systems, Requirements for,
ANSI/IEEE C37:60-1974.   [C37.60]

TANKS

Design and Construction of Large,
Welded, Low-Pressure Storage Tanks,
Rules for, ANSI/API 620-1978.
[B184.1]
Painting Steel Water-Storage Tanks,
ANSI/AWWA D102-78.
♦Steel Inside Tanks for Oil-Burner Fuel,
Safety Standard for, ANSI/UL 80-1980.
[B110.1]
Steel Underground Tanks for Flammable
and Combustible Liquids, ANSI/UL 58-
1976   [B137.1]
† Ventilation and Operation of Open-Sur-
face Tanks, Practices for, ANSI Z9.1-
1977.
♦ Water Storage, Factory-Coated Bolted
Steel Tanks for, ANSI/AWWA D103-
80.
♦ Water Storage, Welded Steel Tanks for,
ANSI/AWWA D100-1979.
♦† Welded Aluminum-Alloy Field-Erected
Storage Tanks. Specification for, ANSI
B96.1-1980, ▲
Welded Steel Tanks for Oil Storage,
ANSI/API 650-1978.   [B184.2]

TRANSFORMERS

Application of Transformer Connections
in Three-Phase Distribution Systems.
Guide for, ANSI/IEEE C57.105-1978.
Distribution, Power and Regulating Trans-
formers, General Requirements for Liq-
uid-Immersed (see Appendix ANSI
C57.91), ANSI/IEEE C57.12.00-1979,
▲
♦ Distribution, Power and Regulating Trans-
formers, Test Code for Liquid-Immersed,
ANSI/IEEE C57.12.90-1979, ▲

Dry-Type Distribution and Power Trans-
formers, General Requirements for,
ANSI/IEEE C57.12.01-1978.
Dry-Type Distribution and Power Trans-
formers, Test Code for, ANSI/IEEE
C57.12.91-1978.
Dry-Type Transformers for General Ap-
plications, ANSI/NEMA ST 20-1972.
[C89.2]

TURBINES, GAS
Control and Protection Systems, Gas Tur-
bine, ANSI B133.4-1978.
Electrical Equipment, Gas Turbine, ANSI
B133.5-1978.
Engine, Gas Turbine, ANSI B133.2-1977.
Environmental Requirements and Respon-
sibilities, Procurement Standard for Gas
Turbine, ANSI B133.9-1979.
Fuels, Gas Turbine, ANSI B133.7-1977.
Installation Sound Emission, Gas Turbine,
ANSI B133.8-1977.
♦ Performance Test Code—Gas Turbine
Power Plants, ANSI/ASME PTC22-
1974(R1980).
Ratings and Performance, Gas Turbine,
ANSI B133.6-1978.
Terminology, Gas Turbine, ANSI B133.1-
1978.

TURBINES, STEAM
Performance Test Code—Pressure-Con-
trol Systems Used on Steam Turbine-
Generator Units, ANSI/ASME PTC20.
3-1970(1980).
Performance Test Code—Steam Turbines,
ANSI/ASME PTC6-1976.
Performance Test Code—Steam Turbines
(appendix to ANSI/ASME PTC6),
ANSI/ASME PTC6 Appendix A-1974.
[PTC6A]
♦ Performance Test Code—Steam Turbines,
Guidance for Evaluation of Measure-
ment Uncertainty, ANSI/ASME
PTC6R-1974(R1980).   [PTC6R]
♦ Performance Test Code—Steam Turbines,

Simplified Procedures for Routine Tests, ANSI/ASME PTC6S-1974(R1980).
[PTC6S]
Speed and Load Governing Systems for Steam Turbine Generator Units, ANSI/ASME PTC 20.1-1977.

VENTILATION

see also Blasting; Ducts, Air; Exhaust Systems; Nuclear Reactors; Spray Finishing; Tanks; Vaults

Gas Vents, Safety Standard for, ANSI/UL 441-1979   [A131.2]

Low-Temperature Venting Systems, Type L, Safety Standard for, ANSI/UL 641.
[Z125.1]

Natural and Mechanical Ventilation, ANSI/ASHRAE 62-73.   [B194.1]

Power Roof Ventilators, Safety Standard for, ANSI/UL 705-1977.   [C33.89]

†Ventilation Control of Grinding, Polishing, and Buffing Operations, ANSI Z43.1-1966.

†Working in Tanks and Other Confined Spaces, Safety Requirements for, ANSI Z117.1-1977.

WELDING AND CUTTING

♦Arc Welding, Gas Metal, ANSI/AWS C5.6-79.

Bare Carbon Steel Electrodes and Fluxes for Submerged-Arc Welding, Specification for, ANSI/AWS A5.17-77. [W3.17]

Brazing Filler Metal, Specification for, ANSI/AWS A5.8-77.   [W3.8]

Cutting and Welding Processes, ANSI/NFPA 51B-1977.   [Z49.2]

Electric Arc-Welding Apparatus, ANSI/NEMA EW 1-1976.   [C87.1]

♦Filler Metals for Gas Shielded Arc Welding, Low Alloy Steel, ANSI/AWS A5.28-79.

♦Inspection of Welds, Guide for Nondestructive, ANSI/AWS B1.0-77, ▲

Mechanical Testing of Welds, Method for, ANSI/AWS B4.0-77.

Oxygen-Fuel Gas Systems for Welding and Cutting, ANSI/NFPA 51-1977.
[W7.1]

♦Pipe, Welding Plain Carbon Steel, ANSI/AWS D10.12-79.

♦Piping and Tubing, Welding Austenitic Chromium-Nickel Stainless Steel, ANSI/AWS D10.4-79.

♦Piping and Tubing, Welding of Chromium Nickel Steel, ANSI/AWS D10.8-78, ▲

♦Root Pass Welding and Gas Purging, ANSI/AWS D10.11-80.

Sampling Airborne Particulates Generated by Welding and Allied Processes, Method for, ANSI/AWS F1.1-78.

♦Structural Welding Code—Reinforcing Steel, ANSI/AWS D1.4-79

♦Structural Welding Code—Steel, ANSI/AWS D1.1-79.

♦Symbols for Welding and Nondestructive Testing. ANSI/AWS A2.4-79.

♦Transformer-Type Arc-Welding Machines, Safety Standard for, ANSI/UL 551-1980.   [C33.2]

Welding and Cutting, Safety in, ANSI Z49.1-1973.

Welding Earthmoving and Construction Equipment, Specification for, ANSI/AWS D14.3-78.

♦Welding Sheet Steel in Structures, ANSI/AWS D1.3-78.

♦Welding Terms and Definitions, Including Terms for Brazing, Soldering, Thermal Spraying, and Thermal Cutting, ANSI/AWS A3.0-80.

WELDING RODS AND ELECTRODES

♦Aluminum and Aluminum Alloy Arc-Welding Electrodes, Specification for, ANSI/AWS A5.3-80, ▲   [W3.3]

♦Aluminum and Aluminum Alloy Welding Rods and Bare Electrodes, Specification for, ANSI/AWS A5.10-80, ▲ [W3.10]

Bare Low-Alloy Steel Electrodes and Fluxes for Sumberged Arc Welding, Specification for, (including Addenda 1-76). ANSI/AWS A5.23-76.

♦ Carbon Steel Covered Arc-Welding Electrodes, Specification for, ANSI/AWS A5.1-78, ▲   [W3.1]

♦ Carbon Steel Electrodes for Flux-Cored Arc Welding, Specification for, ANSI/AWS A5.20-79, ▲   [W3.20]

♦ Carbon Steel Filler Metal for Gas Shielded Arc Welding, Specification for, ANSI/AWS A5.18-79, ▲   [W3.18]

♦ Composite Surfacing Welding Rods and Electrodes, ANSI/AWS A5.21-80, ▲

Composite Surfacing Welding Rods and Electrodes, Specification for, ANSI/AWS A5.21-70.   [W3.21]

♦ Consumables Used for Electrogas Welding of Carbon and High Strength Low Alloy Steels, ANSI/AWS A5.26-78, ▲

♦ Consumables Used for Electroslag Welding of Carbon and High Strength Low Alloy Steels, ANSI/AWS A5.25-78, ▲

Copper and Copper-Alloy Bare Welding Rods, and Electrodes, Specification for, ANSI/AWS A5.7-77   [W3.7]

Copper and Copper-Alloy, Covered Electrodes, Specification for, ANSI/AWS A5.6-76.   [W3.6]

Corrosion-Resisting Chromium and Chromium-Nickel Steel Bare and Composite Metal Cored and Stranded Arc Welding Electrodes and Welding Rods, ANSI/AWS A5.9-77.   [W3.9]

♦ Corrosion-Resisting Chromium and Chromium-Nickel Steel Covered Welding Electrodes, Specification for, ANSI/AWS A5.4-78   [W3.4]

♦ Electrodes, Automotive Resistance Spot Welding, ANSI/AWS D8.6-77.

Flux Cored Corrosion-Resisting Chromium and Chromium-Nickel Steel Electrodes, Specification for, ANSI/AWS A5.22-74   [W3.22]

♦ Iron and Steel Gas Welding Rods, Specification for, ANSI/AWS A5.2-80, ▲   [W.32]

Low-Alloy Steel Covered Arc-Welding Electrodes, Specification for, ANSI/AWS A5.5-69.   [W3.5]

♦ Magnesium-Alloy Welding Rods and Bare Electrodes, Specification for, ANSI/AWS A5.19-80, ▲   [W3.19]

Nickel and Nickel-Alloy Bare Welding Rods and Electrodes, Specification for, ANSI/AWS A5.14-76.   [W3.14]

Nickel and Nickel-Alloy Covered Welding Electrodes, Specification for, ANSI/AWS A5.11-76.   [W3.11]

♦ Rods, Copper and Copper Alloy Gas Welding, ANSI/AWS A5.27-78.

♦ Solid Surfacing Welding Rods and Electrodes, Specification for, ANSI/AWS A5.13-80, ▲   [W3.13]

Tungsten Arc-Welding Electrodes, Specification for, ANSI/AWS A5.12-69.   [W3.12]

Welding Rods and Covered Electrodes for Welding Cast Iron, Specification for, ANSI/AWS A5.15-69.   [W3.15]

WINDOWS
see also Doors and Frames

♦ Aluminum Combination Vertically-Sliding or Horizontally-Operating Storm Windows for External Application, Specifications for, ANSI/AAMA 1002.10-1980, ▲   [A134.3]

Aluminum Windows, Specifications for, ANSI/AAMA 302.9-1977   [A134.1]

♦ Fire Tests of Window Assemblies, Methods of, ANSI/ASTM E163-80.

♦ Installation of Storm Windows, Replacement Windows, Multi-Glazing Storm Doors and Replacement Doors, Practice for the, ANSI/ASTM E737-80.

Rate of Air Leakage through Exterior Windows, Curtain Walls, and Doors, Test for, ANSI/ASTM E283-73.

♦ Structural Performance of Exterior Windows, Curtain Walls, and Doors under the Influence of Wind Loads, Test for, ANSI/ASTM E330-79.

Water Penetration of Exterior Windows,

Curtain Walls, and Doors by Uniform Static Air Pressure Differential, Test for, ANSI/ASTM E331-70(1975).

Wear Testing Rotary Operators for Windows, ANSI/ASTM E405-70(1975).

Window Assemblies, Test Methods for

Measurement of Forced Entry Resistance, of, ANSI/ASTM F588-79.

†Window Cleaning, Safety Requirements for, ANSI A39.1-1969.

♦Wood Windows, ANSI/NWMA 1.S.2-79,
▲   [A200.1]

APPENDIX C

This appendix lists many of the organizations and associations which can provide reference material and information in regard to the construction process:

1. Organizations of Construction Contractors
2. Organizations of the Design Professions
3. Construction-Material and Equipment Suppliers and Product Research
4. Construction Labor Organizations
5. Coordination and Arbitration
6. Inspection, Specifications, and Cost

1. ORGANIZATIONS OF CONSTRUCTION CONTRACTORS

American Institute of Constructors. Glenbrook Centre, Ste. 511, 1140 N.W. 63rd St., Oklahoma City, OK 73116.

American Road Builders Association. ARBA Bldg., 525 School St., N.W., Washington, DC 20024. (202) 737-5440.

American Subcontractors Association. 402 Shoreham Bldg., Washington, DC 20005. (202) 783-1883.

Associated Builders and Contractors, Inc. Box 8643, Friendship International Airport, MD 21240. (301) 796-0971.

Associated General Contractors of America. 1957 E. St., N.W., Washington, DC 20006. (202) EX 3-2040.

Building Waterproofers Association. 60 E. 42 St., New York. NY 10017

Ceilings and Interior Systems Contractors Association. 1201 Waukegan Rd., Glenview, IL 60025.

Contracting Plasterers & Lathers International Association. 304 Landmark Bldg., 1343 H St., N.W., Washington, DC

Council of Mechanical Specialty Contracting Industries, Inc. 1611 N. Kent St., Arlington, VA 22209.

651

Gypsum Drywall Contractors International. 127 Wacker Dr., Chicago, IL 60606. (312) 263-2770.

Insulation Distributor-Contractors National Association. 1406 Chestnut St., Philadelphia, PA 19102.

International Association of Wall and Ceiling Contractors. 20 E. St., N.W., Washington, DC 20001. (202) 638-1072.

Mason Contractors Association of America, Inc. Suite 750, 5530 Wisconsin Ave., N.W., Washington, DC 20016 (202) 638-1072.

Mechanical Contractors Association of America, Inc. Suite 750, 5330 Wisconsin Ave., N.W., Washington, DC 20016 (202) 654-7960.

National Acoustical Contractors Association. 1201 Waukegan Rd., Glenview, Ill., 60025. (302) 724-7700.

National Association of Elevator Contractors. 4647 Hampton Ave., St. Louis, MO 63109.

National Association of Home Builders of the United States. 1625 L St., N.W., Washington, DC 20036. (202) RE 7-7435.

National Association of Lighting Maintenance Contractors. 2121 Keith Bldg., Cleveland, OH 44115.

National Association of Plumbing-Heating-Cooling Contractors. 1016 20th St., N.W., Washington, DC 20036. (202) 337-1675.

National Association of River & Harbor Contractors. 3900 N. Charles St., Baltimore, MD 21218.

National Constructors Association. 1133 15th St., N.W., Washington, DC 20005. (202) 466-8880.

National Electrical Contractors Association. 1730 Rhode Island Ave., N.W., Washington, DC 20036. (202) 293-2150.

National Environmental Systems Contractors Association (1969) [merger of National Warm Air Heating and Air Conditioning Association (1914) and Air Conditioning & Refrigeration Contractors of America (1946)]. 221 La Salle St., Chicago, IL 60601. (313) 726-6026.

National Erectors Association (1969). Suite 912, 1625 Eye St., N.W., Washington, DC 20006. (202) 223-6534.

National Housing Producers Association. 900 Peachtree St., N.E., Atlanta, GA 30309.

National Remodelers Association (1969) (Formerly NERSICA—National Established Roofing Siding Insulation Contractors Association (1933). 50 E. 42nd St., New York, NY 10017. (212) 687-5224.

National Roofing Contractors Association (1886). 1515 North Harlem Ave., Oak Part, IL 60302. (312) 383-9513.

Painting & Decorating Contractors of America (1884). 2625 W. Peterson Ave., Chicago, IL 60645. (312) 561-2328.

Pipe Line Contractors Association (1948). 2800 Republic National Bank Building, Dallas, TX 75201. (214) 741-6251.

Sheet Metal and Air Conditioning Contractors National Association, Inc. (1943). 1611 N. Kent St., Suite 200, Arlington, VA 22209. (703) 337-3380.

Tile Contractors Association of America, Inc. (1903). 112 N. Alfred, Alexandria, VA 22314. (703) 836-5995.

2. ORGANIZATIONS OF THE DESIGN PROFESSIONS

American Association for Hospital Planning. 122 S. Michigan Ave., Rm. 939, Chicago, IL 60603.

American Association of Hospital Consultants, 1700 K St., N.W., Washington, DC 20036.

American Institute of Architects (1857). 1735 New York Ave., N.W., Washington, DC 20006. (202) EX 307050

American Institute of Consulting Engineers (1910). 345 E. 47th St., New York, NY 10017. (212) 752-6800.

American Institute of Interior Designers. 730 Fifth Ave., New York, NY 10019.

American Institute of Landscape Architects. 501 E. San Juan Ave., Phoenix, AZ 85012.

American Institute of Planners. 917 15th St., N.W., Rm. 800, Washington, DC 20005.

American Society of Civil Engineers (1852). 345 E. 47th St., New York, NY 10017.

American Society of Golf Course Architects. 221 N. LaSalle St., Chicago, IL 60601.

American Society of Heating, Refrigerating and Air-Conditioning Engineers, Inc. (1894). 345 E. 47th St., New York, NY 10017. (212) 752-6800.

American Society of Landscape Architects (1899). 2013 Eye St., N.W., Washington, DC 20006. (202) 659-9550.

American Society of Mechanical Engineers (1880). 345 E. 47th St., New York, NY 10017. (212) PL 2-6800.

American Society of Professional Estimators, Inc. 14918 Burbank Blvd., Van Nuys, CA 91401.

Consulting Engineers Council of the U.S.A. (1956). 1155 15th St., N.W., Washington, DC 20005. (202) 296-1780.

Illuminating Engineering Society (1906). 345 E. 47th St., New York, NY 10017. (212) PL 2-6800.

Industrial Designers Society of America, Inc. 60 W. 57th St., New York, NY 10019.

Institute of Electrical & Electronic Engineers (1884). 345 E. 47th St., New York, NY 10017. (212) PL 2-6800.

International Society of Food Consultants. 2710 N. Salisbury St., West Lafayette, IN 47906.

National Society of Professional Engineers (1934). 2029 K St., N.W., Washington, DC 20006. (202) 337-0420.

Project Management Institute (1969). Drexel Hill, PA 19026. (215) 622-1796.

Society for Advancement of Management. 1472 Broadway, New York, NY 10036. (212) 594-2392.

3. CONSTRUCTION-MATERIAL AND EQUIPMENT SUPPLIERS AND PRODUCT RESEARCH

Acoustical Materials Association (1933). 335 E. 45th St., New York, NY 10017. (202) MU 5-1940.

Adhesive & Sealant Council (1958). 159 N. Dearborn St., Chicago, IL 60601. (312) 332-6333.

Air Conditioning and Refrigeration Institute (1953). 1815 N. Ft. Myer Drive, Arlington, VA 22209. (703) 524-8800.

Air Distribution Institute (1946). 22 W. Monroe St., Chicago, IL 60603. (312) DE 2-7127.

Air Filter Institute (1933). Box 85, Station E, Louisville, KY 40208. (502) ME 7-0336.

Air Moving and Conditioning Association (1963). 205 W. Touhy Ave., Park Ridge, IL 60068. (312) 823-5147.

Air Pollution Control Association (1960). 4400 Fifth Ave., Pittsburgh, PA 15213.

Allied Building Metals Industries (1921). 211 E. 43rd St., New York, NY 10017. (212) OX 7-5551.

Aluminum Association (1933). 750 Third Ave., New York, NY 10017. (212) 972-1800.

Aluminum Siding Association (1956). Tribune Tower, Chicago, IL 60611.

American Association of Nurserymen (1920). 835 Southern Bldg., Washington, DC 20005. (202) RE 7-4060.

American Carpeting Institute (1928). 350 Fifth Ave., New York, NY 10001. (212) 736-2043.

American Concrete Institute (1906). P.O. Box 4754, Redfern Station, Detroit, MI·48219. (313) KE 1-1802.

American Concrete Paving Association. 1211 West 22nd St., Ste. 727, Oakbrook, IL 60523.

American Concrete Pipe Association (1907). 228 N. La Salle St., Chicago, IL 60601. (312) AN 3-6488.

American Forest Institute. 1835 K St., N.W., Washington, DC 20006.

American Forest Products Industries, Inc. (1932). 1816 N St., N.W., Washington, DC 20036. (202) 338-3330.

American Gas Association (1918). 605 Third Ave., New York, NY 10016. (212) 972-5500.

American Hardboard Association (1953). 20 N. Wacker Dr. Chicago, IL 60603. (312) CE 6-8008.

American Home Lighting Institute (1945). 360 N. Michigan Ave., Chicago, IL 60601. (312) CE 6-7796.

American Institute of Kitchen Dealers. 199 Main St., Hackettstown, NJ 07840.

American Institute of Steel Construction (1921). 101 Park Ave., New York, NY 10017. (212) MU 5-7374.

American Institute of Supply Association (1938). 1505 22nd St., N.W., Washington, DC 20037. (202) 387-5055.

American Institute of Timber Construction (1952). 1700 K St., N.W., Washington, DC 20006. (202) 296-3390.

American Iron & Steel Institute (1908). 150 E. 42nd St., New York, NY 10017. (212) OX 705900.

American National Standards Institute (1969) (formerly United States of America Standards Institute). 1430 Broadway, New York, NY 10018. (212) 354-3300.

American Nuclear Society (1955). 244 E. Ogden Ave., Hinsdale, IL 60521.

American Pipe Fittings Association (1938). 60 E. 42nd St., New York, NY 10017.

American Plywood Association (1936). 1109 A St., Tacoma, WA 98401. (206) BR 2-2283.

American Public Works Association (1937). 1313 E. 60th St., Chicago, IL 60637. (312) FA 4-3400.

American Society for Metals (1920). Metals Park, OH 44073. (216) 338-5151.

American Society of Architectural Hardware Consultants (1940). Box 599, Mill Valley, CA 94941. (408) LI 5-1756.

American Society of Concrete Construction. 2510 Dempster St., Des Plaines, IL 60016. ·

American Waterworks Association, Inc. (1881). 2 Park Ave., New York, NY 10016. (212) MU 4-6686.

American Wood Preservers Institute (1904). 1707 L Street, N.W., Washington, DC 20036. (202) 296-1280.

American Zinc Institute (1918). 292 Madison Avenue, New York, NY 10017. (212) OR 9-6020.

Architectural Precast Association. 2201 East 46th St., Indianapolis, IN 46205.

Asbestos-Cement Products Association. 520 Fifth Ave., Rm. 2000, New York, NY 10017.

Asphalt and Vinyl Asbestos Tile Institute (1939). 101 Park Ave., New York, NY 10017. (212) MU 6-3937.

Asphalt Institute (1919). Asphalt Institute Bldg., College Park, MD (301) 927-0422.

Asphalt Roofing Bureau (1919). 752 Third Ave., New York, NY 10017.

Asphalt Roofing Industry Bureau (1919). 50 East 41st St., New York, NY 10017.

Bituminous Pipe Institute. 111 E. Wacker Dr., Chicago, IL 60601.

Builders Hardware Manufacturers Association (1961). 60 E. 42nd St., New York, NY 10017.

Building Stone Institute. 420 Lexington Avenue, New York, NY 10017.

Canvas Products Institute (1912). 224 Endicott Building, St. Paul, MN 55101. (612) 222-2500.

Cast Iron Pipe Research Association (1949). 1826 Jefferson Place, N.W., Washington, DC 20036. (202) 338-9042.

Cellular Concrete Association. 715 Boylston St., Boston, MA 02116.

Central Supply Association (1894). 221 N. La Salle St., Chicago, IL 60601.

Certified Ballast Manufacturers Association (1941). 2120 Keith Bldg., Cleveland, OH 44115. (216) 241-0711.

Clay Products Association (1917). P.O. Box 172, Barrington, IL 60010 (312) 381-3260.

Concrete Pipe Institute. 1501 Wilson Blvd., Arlington, VA 22209.

Concrete Reinforcing Steel Institute (1924). 228 N. La Salle St., Chicago, IL 60601.

Construction Industry Manufacturers Association (1949). Marine Plaza 1700, 111 E. Wisconsin Ave., Milwaukee, WI 53202.

Copper Development Association (1963). 405 Lexington Ave., New York, NY 10017.

Copper Institute (1927). 50 Broadway, New York, NY 10004.

Copper Research Association International (1960). 1271 Avenue of the Americas, New York, NY 10020.

Edison Electric Institute (1933). 750 Third Ave., New York, NY 10017. (212) YU 6-4100.

Electric Heating Association, Inc. 437 Madison Ave., New York, NY 10022.

Expanded Shale, Clay and Slate Institute (1952). 1041 National Press Building, Washington, DC 20004. (202) ST 3-1669.

Fine Hardwoods Association. 666 N. Lakeshore Dr., Chicago, IL 60611.

Flexicore Manufacturers Association. 666 N. Lakeshore Dr., Chicago, IL 60611.

Forest Products Research Society. 2801 Marshall Ct., Madison, WI 53705.

Gas Appliance Manufacturers Association, Inc. (1935). 60 E. 42nd St., New York, NY 10017. (212) MU 2-8743.

Gypsum Association (1930). 201 N. Wells St., Chicago, IL 60606.

Hardwood Plywood Manufacturers Association (1930). 2310 S. Walter Reed Dr., Arlington, VA 22206. (703) 671-6262.

Home Manufacturers Association (1943). 1625 L St., N.W., Washington, DC 20036. (202) 296-3813.

Indiana Limestone Institute of America (1929). 400 E. 7th St., Bloomington, IN 47401. (317) 332-9338.

Industrial Fasteners Institute (1950). 1505 Ohio Building, 1717 E. 9th St., Cleveland, OH 44414.

Institute of Appliance Manufacturers (1933). Ste. 455, 2000 K St., N.W., Washington, DC 20006.

Institute of Heating and Air Conditioning Industries. 5107 W. First St., Los Angeles, CA 90004.

Institute of Masonry Research. 9013 Old Harford Rd., Baltimore, MD 21234.

International Masonry Institute. 825 15 St., N.W., Washington, DC 20005.

Stone Manufacturers Association (1872). Ste. 455, 2000 K St., N.W., Washington, DC 20006. (202) 337-4122.

Instrument Society of America (1945). 530 Wm. Penn Pl., Pittsburgh, PA 15219.

Insulation Board Institute (1932). 111 W. Washington St., Chicago, IL 60602. (312) 782-9542.

Lead Industries Association, Inc. (1928). 292 Madison Ave., New York, NY 10017. (212) OR 9-6020.

Lightning Protection Institute, 2 N. Riverside Dr., Chicago, IL 60606.

Manufacturers Standardization Society of Valve and Fitting Industry (1915). 420 Lexington Ave., New York, NY 10017. (212) LE 2-6146.

Maple Flooring Manufacturers Association (1905). 424 Washington Ave., Oshkosh, WI 54901.

Marble Institute of America (1944). 425 13th St., N.W., Washington, DC 20004. (202) 628-0214.

Metal Buildings Manufacturers Association (1956). 2130 Keith Bldg., Cleveland, OH 44115. (216) 241-7333.

Metal Lath Association (1910). 12703 Triskett Road, Cleveland, OH 44114. (216) 781-3374.

Mobile Housing Association of America (1945). 39 S. La Salle St., Chicago, IL 60603.

National American Wholesale Lumber Association (1893). 180 Madison Ave., New York, NY 10016. (212) LE 2-9161.

National Apartment Owners Association, Inc. (1939). 1401 K St., N.W., Washington, DC 20005. (202) 347-3766.

National Asphalt Pavement Association (1955). 6715 Kenilworth Ave., Riverdale, MD 20840. (301) 779-4880.

National Association of Architectural Metal Manufacturers (1937). 228 N. La Salle St., Chicago, IL 60601. (312) ST 2-4454.

National Association of Building Manufacturers. 1619 Massachusetts Ave., N.W., Washington, DC 20036.

National Association of Decorative Architectural Finishes. 112 N. Alfred St., Alexandria, VA 22314.

National Association of Electrical Distributors (1908). 600 Madison Ave., New York, NY 10022.

National Association of Marble Dealers (1901). 219 E. Island Ave., Minneapolis, MN 55401.

National Association of Marble Producers. P.O. Box 718, Carthage, MO 64836. Marble quarries.

National Association of Mirror Manufacturers (1957). 1225 19th St., N.W., Washington, DC 20036. (202) 296-5339.

National Association of Sheet Metal Distributors (1894). 1900 Arch St., Philadelphia, PA 19103.

National Automatic Sprinkler & Fire Control Association (1914). 277 Park Ave., New York, NY 10017.

National Builders Hardware Association (1934). 1290 Avenue of the Americas, New York, NY 10019.

National Building Granite Quarries Association (1917). North State St., Concord, NH 03301.

National Building Materials Distributors Association (1952). 221 N. La Salle St., Chicago, IL 60601.

National Building Products Association. 120-44 Queens Blvd., Kew Gardens, NY 11415.

National Bureau for Lathing and Plastering (1952). 938 K St., N.W., Washington, DC 20001. (202) 347-3683.

National Ceramic Manufacturers Association. 53 E. Main Street, Moorestown, NJ 20277.

National Cinder Concrete Products Association (1923). Box 196, Newtown Squre, PA 19073.

National Clay Pipe Institute (1942). 1028 Connecticut Ave., N.W., Washington, DC 20036. (202) 296-5270.

National Concrete Masonry Association (1943). 2009 N. 14th St., Arlington, VA 22201. (703) 524-0815.

National Corrugated Steel Pipe Association (1956). 140 S. Dearborn St., Chicago, IL 60603.

National Crushed Stone Association (1918). 1415 Elliot Place, N.W., Washington, DC 20007. (202) 333-1536.

National Electric Sign Association (1944). 10922 S. Western Ave., Chicago, IL 60643.

National Elecrical Manufacturers Association (1926). 155 E. 44th St., New York, NY 10017. (212) MU 2-1500.

National Elevator Manufacturing Industry (1934). 101 Park Ave., New York, NY 10017.

National Forest Products Association (1902). 1619 Massachusetts Ave., N.W., Washington, DC 20036. (202) 332-1050.

National Glass Dealers Association (1948). 1000 Connecticut Ave., N.W., Washington, DC 20036. (202) 296-5680.

National Hardwood Lumber Association (1898). 59 E. Van Buren St., Chicago, IL 60605.

National Home Improvement Council (1955). 11 E. 44th St., New York, NY 10027. (212) PL 1-7178.

National Insulation Manufacturers Association (1941). 441 Lexington Ave., New York, NY 10017. (212) MU 2-3574.

National Kitchen Cabinet Association (1955). 910 Watterson City Office Building, Louisville, KY 40218.

National Lime Association (1902). 4000 Brandywine St., N.W., Washington, DC 20016. (202) 966-3418.

National Limestone Institute (1945). 1315 16th St., Washington, DC 20006. (202) 232-2650.

National Lumber & Building Materials Dealers Association (1916). 1200 18th St., N.W., Washington, DC 20036. (202) FE 8-3770.

National Mineral Wool Insulation Association (1933). 211 E. 51st St., New York, NY 10022. (212) 265-8171.

National Oak Flooring Manufacturers Association (1909). 814 Sterick Building, Memphis, TN 38103. (901) JA 6-5016.

National Oil Fuel Institute (1961). 60 E. 42nd St., New York, NY 10017. (212) TN 7-0260.

National Ornamental Metal Manufacturers Association (1958). Box 183, Timonium, MD 21093.

National Paint, Varnish & Lacquer Association (1887). 1500 Rhode Island Ave., N.W., Washington, DC 20005. (202) HO 2-6272.

National Particleboard Association (1960). 711 14th St., N.W., Washington, DC 20005. (202) 783-8258.

National Plywood Distributors Association (1942). 6809 Canyon Crest Road, S.W., Portland, OR 97225.

National Ready Mixed Concrete Association (1930). 900 Spring St., Silver Spring, MD 20910. (301) 587-1400.

National Sand & Gravel Association (1916). 900 Spring St., Silver Spring, MD 20910 (301) 587-1400.

National Sash & Door Jobbers Association. 20 N. Wacker Drive, Chicago, IL 60606. (312) AN 3-2670.

National Slag Association (1918). 613 Perpetual Building, Washington, DC 20004. (202) ME 8-2040.

National Slate Association (1922). 455 W. 23rd St., New York, NY 10011. (212) 3-5846.

National Swimming Pool Institute (1956). 2000 K St., N.W., Washington, DC 20006. (202) 337-2244.

National Terrazzo and Mosaic Association (1931). 716 Church, Alexandria, VA 22314. (703) 836-6765.

National Woodwork Manufacturers Association (1926). 400 W. Madison Ave., Chicago, IL 60606. (312) 782-6232.

North American Heating & Air Conditioning Wholesalers Association (1947). 1200 W. Fifth Ave., Columbus, OH 43212.

Northern Hardwood and Pine Manufacturers Association (1910). 207 Northern Building, Green Bay, WI 54301. (414) HE 2-9161.

Paint and Wallpaper Association of America, Inc. (1947). 2101 S. Brentwood Blvd., St. Louis, MO 63144. (314) VO 3-4505.

Perlite Institute, Inc. (1949). 45 W. 45th St., New York, NY 10036. (212) CO 5-2145.

Plastics in Construction Council (1964). 250 Park Ave., New York, NY 10017.

Plumbing Brass Institute (1956). 205 W. Wacker Drive, Chicago, IL 60606.

Plumbing-Heating-Cooling Information Bureau (1919). 35 E. Wacker Drive, Chicago, IL 60601. (312) FR 2-7331.

Ponderosa Pine Woodwork Association (1941). 39 S. La Salle St., Chicago, IL 60603.

Portland Cement Association (1916). Old Orchard Road, Skokie, IL 60076.

Prestressed Concrete Institute (1954). 205 W. Wacker Drive, Chicago, IL 60606. (312) 346-4071.

Producers Council, Inc. (1921). 1717 Massachusetts Ave., N.W., Washington, DC 20036. (202) 667-8727. Membership: 250. 52 chapters.

Red Cedar Shingle & Handsplit Shake Bureau (1915). 5510 White Building, Seattle, WA 98101. (206) MA 3-4881.

Reinforced Concrete Research Council (1948). 3129 Civil Engineering Building, Urbana, IL 61801.

Resilient Tile Institute. 101 Park Ave., New York, NY 10017.

Roof Drainage Manufacturers Institute (1950). 221 N. La Salle St., Chicago, IL 60601.

Rubber Manufacturers Association (1915). 444 Madison Ave., New York, NY 10022. (212) PL 5-9200.

Screen Manufacturers Association (1955). 110 N. Wacker Drive, Chicago, IL 60606.

Society of the Plastics Industry, Inc. (1937). 250 Park Ave., New York, NY 10017. (212) MU 7-2675.

Southern Pine Association (1914). Box 5268, New Orleans, LA 70150. (504) 525-7381.

Steel Deck Institute (1939). 9836 W. Roosevelt Road, Westchester, IL 60153. (312) 345-8550.

Steel Door Institute (1955). Keith Building, Cleveland, OH 44115. (216) CH 1-7333.

Steel Joist Institute (1928). 1346 Connecticut Ave., N.W., Washington, DC 20036. (202) AD 4-2244.

Steel Plate Fabricators Institute (1933). 19 S. La Salle St., Chicago, IL 60603. (312) DE 2-1682.

Steel Scaffolding & Shoring Institute (1960). 2130 Keith Bldg. Cleveland, OH 44115. (216) 241-3468.

Steel Service Center Institute (1909). 540 Terminal Tower, Cleveland, OH 44114.

Steel Structures Painting Council (1951). 4400 5th Ave., Pittsburgh, PA 15213.

Steel Window Institute (1920). Keith Building, Cleveland, OH 44115.

Structural Clay Products Institute (1934). 1750 Old Meadow Road, McLean, VA 22101. (703) 893-4010.

Stucco Manufacturers Association (1957). 15926 Kittridge St., Van Nuys, CA 91406.

Tile Council of America (1945). 800 Second Ave., New York, NY 10017. (212) OX 7-9269.

Timber Products Manufacturers (1926). W. 721 Second Ave., Spokane, WA 99204.

Vermiculite Institute (1941). 208 S. La Salle St., Chicago, IL 60604. (312) DE 2-0121.

Water and Wastewater Equipment Manufacturers Association, Inc. (1908). 744 Broad St., Newark, NJ 10102.

Water Pollution Control Federation (1928). 3900 Wisconsin Ave., N.W., Washington, DC 20016. (202) EM 2-4100.

Water Systems Council (1932). 205 W. Wacker Drive, Chicago, IL 60606.

Western Wood Products Asociation (1964). Yeon Building, Portland, OR 97204.

Window Shade Manufacturers Association (1949). 110 N. Wacker Drive, Chicago, IL 60606.

Wire Reinforcement Institute (1930). 5034 Wisconsin Ave., N.W., Washington, DC 20016. (202) 966-4411.

Wood and Synthetic Flooring Institute. 1441 Shermer Rd., Northbrook, IL 60062.

Wood Flooring Institute of America. 201 N. Wells St., Chicago, IL 60606. (312) CE 6-1569.

Wood Kitchen Cabinet Institute (1946). 600 Old Country Rd., Garden City, NY 11530.

Zinc Institute. 292 Madison Avenue, New York, NY 10017.

4. CONSTRUCTION LABOR ORGANIZATIONS

Bricklayers, Masons & Plasterers International Union (1865). 815 15th St., N.W., Wshington, DC 20005. (202) 783-3788.

Brotherhood of Painters, Decorators & Paperhangers of America (1887). Suite 508, 1925 K St., N.W., Washington, DC 20006. (202) 338-4890.

Construction and Building Trades Department AFL-CIO (1908). 815 16th St., N.W., Washington, DC 20006. (202) DI 7-1461.

Granite Cutters International Association (1888). 18 Federal Ave., Quincy, MA 02169. (617) GR 2-0209.

International Association of Bridge, Structural & Ornamental Iron Workers of America (1901). 3615 Oliver St., St. Louis, MO 63108. (314) FR 1-3900.

International Association of Heat & Frost Insulators & Asbestos Workers (1910). 1300 Connecticut Ave., N.W., Washington, DC 20036. (202) HU 3-6288.

International Association of Marble, Stone and Slate Polishers, Rubbers & Sawyers, Tile & Terrazzo Helpers (1902). 821 15th St., N.W., Washington, DC 20005. (202) 347-7414.

International Brotherhood of Boilermakers, Iron Ship Builders, Blacksmiths, Forgers & Helpers (1887). New Brotherhood Building, Kansas City, KS 66101. (913) DR 1-2642.

International Brotherhood of Electrical Workers (1891). 1200 15th St., N.W., Washington, DC 20005. (202) CO 5-8040.

International Union of Elevator Operators (1903). 12 S. 12th St., Philadelphia, PA 19107. (215) 922-2226.

International Union of Operating Engineers (1897). 1125 17th St., N.W., Washington, DC 20036. (202) DI 7-8560.

Laborers International Union of North America (1903). 905 16th St., N.W., Washington, DC 20006. (202) 737-8320.

Operative Plasterers & Cement Masons International Association (1864). 1125 17th St., N.W., Washington, DC 20036. (202) EX 3-6569.

Sheetmetal Workers International Association, 1000 Connecticut Ave., N.W., Washington, DC 20036. (202) 296-5880.

United Association of Journeymen & Apprentices of the Plumbing & Pipe Fitting Industry (1889). 901 Massachusetts Ave., N.W., Washington, DC 20001. (202) NA 8-5823.

United Brotherhood of Carpenters and Joiners of America (1881). 101 Constitution Ave., N.W., Washington, DC 20005. (202) 546-6206.

United Slate, Tile & Composition Roofers, Damp & Waterproof Workers Association (1905). 1125 17th St., N.W., Washington, DC 20036. (202) ME 8-3228.

Wood, Wire and Metal Lathers International Union (1899). 6530 New Hampshire Ave., Takoma Park, MD 20012. (301) 270-1200.

5. COORDINATION AND ARBITRATION

American Arbitration Association (1926). 140 W. 51st St., New York, NY 10020.

Building Officials Conference of America, Inc. (1915). 1313 E. 60th St., Chicago, IL 60637. (312) FA 4-3400.

Building Research Institute, Inc. (1952). 1725 De Sales St., N.W., Washington, DC 20036. (202) 783-4866.

Chamber of Commerce of the United States (1912). (Construction and Community Development Department.) 1615 H St., N.W., Washington, DC 20006. (202) NA 8-2380. Membership: 37,000.

Construction Industry Collective Bargaining Commission (1969). 5220 Labor Dept. Building, Washington, DC (202) 393-2420, ext. 3736.

Construction Industry Foundation (1969). 211 E. 51st St., New York, NY 10022. (212) 593-3152.

Construction Industry Joint Conference (1958). 1012 14th St., N.W., Washington, DC 20005. (202) ST 3-0038.

Construction Writers Association. 202 Homer Building, Washington, DC 20005.

Council of Mechanical Specialty Contracting Industries (1955). 1730 Rhode Island Ave., N.W., Washington, DC 20036. (202) 293-2150.

Council on Industrial Relations (1920). 1200 15th St., N.W., Washington, DC 20005. (202) CO 5-8040.

International Conference of Building Officials (1922). 50 S. Los Robles, Pasadena, CA 91101. (415) 684-1310.

National Association of Women in Construction (1954). 723 Markham St., Little Rock, AR 72201.

6. INSPECTION, SPECIFICATIONS, AND COST

American Association of Cost Engineers (1956). 308 Monongahela Bdg., Morgantown, W. Va. 26505, (304) 296-8444. Cost estimates and cost-control research.

American Council of Independent Laboratories (1937). 1026 17th St., N.W., Washington, DC 20036. (202) EX 3-5674. Applied research and testing methods.

American National Standards Institute (1969). (Formerly United States of America Standards Institute, the predecessor of which was the American Standards Association.) 1430 Broadway, New York, NY 10018. (212) 354-3300. Checks and records standards.

American Society for Testing and Materials (1898). 1916 Race St., Philadelphia, PA 19103. Materials testing.

American Specifications Institute (1921). 134 N. La Salle St., Chicago, IL 60602. Architectural and engineering specifications.

Construction Specifications Institute (1948). 1346 Connecticut Ave., N.W., Washington, DC 20036. (202) 483-6645. Promotes better specification writing.

Construction Surveyors Institute (1926). 420 Lexington Ave., New York, NY 10017. Quantity surveys and cost appraisals.

International Association of Electrical Inspectors (1928). 201 Erie St., Chicago, IL 60611. Membership: 13,000. Electrical inspectors.

National Fire Protection Association (1896). 60 Batterymarch St., Boston, MA 02110. (617) 482-8755. Fire-protection and -prevention information.

Society of Construction Superintendents (1933). 38 Park Row, New York, NY 10038. Contractor supervisors.

Society of Fire Protection Engineers (1950). 60 Batterymarch St., Boston, MA 02110. Fire-protection engineers.

Underwriters' Laboratories, Inc. (1894). 333 Pfingsten Road, Northbrook, IL 60062. (312) 272-8800. Testing devices and materials for safety.

INDEX

Acceleration, 97
Acoustical Treatment, 381
  grids, 381
  plaster, 381
AGC, 24, 131
Aggregates, 114–116, 216
  bituminous, 192–193
Air-conditioners, package, 463
  controls, 463–464
  cooling towers, 463
Air distribution systems, 457–459
  air handling equipment, 457–458
  warm air furnaces, 457
Air filtration, 459–462
  filter banks, 459
  sprays, 460
Air handling equipment, 457–458
  access doors, 458
  alignment, 458
  fans, 458
  filtration, 459
  multizone, 459
  vibration, 458
American Concrete Institute, 230, 252
American National Standards Institute, 39
  Appendix B
American Society Architectural Hardware,
  356
American Society of Testing and Materials,
  39, Appendix A
American Standard for Nursery Stock, 188
Anchors, 96–97
ASME code, 112, 204
Astroturf, 186
Audio visual, 408

Beams, 89–94

Bituminous, 116–117
  aggregate, 193
  bins, 193
  raking, 194
  rolling, 194
  spreader, 193
  temperatures, 194
Boilers, 449–451
  ASME code, 449
  draft fans, 450
  feed, 450
  refractory materials, 450
  safety valves, 450
Bonds, 51–52
Bracing, 175
Brick (see Masonry)
Building licenses, 45
Building permits, 45, 56

Caisson, 171–174
  bell bottom, 172–173, 174
  Chicago, 171
  Corporation, 174
  inspection, 174
Carpentry, 312–328
  creosote, 313
  custom cabinets, 327
  finish, 326–327
  framing, 313
  heavy timber, 322–324
  joists, 316
  millwork, 326–327
  nails, 319–321
  OSHA, 328
  panel, 327
  rough, 312

Carpentry (cont.)
  sheathing, 321–322
  woodwork, 327
Cast-in-place concrete, 241–253
  aggregate, 241, 253–254
  air entrainment, 252
  ASTM, 252
  batching, 243, 245
  cement, 241–244
  cold weather, 246, 251
  conveyor, 250
  curing, 245–246, 251
  freeze/thaw cycles, 241
  handling, 246–250
    screed, 249
    vibration, 248–249
  hydration, 241
  laitance, 253
  PCA, 244, 252
  Portland Cement classification, 243, 244
  troweling, 250–251
  warm weather, 251–252
  water/cement ratio, 241, 242
Caulking, 343
Centrifugal compressors, 451
Change orders, 20, 526–541
Circulating system, 451
Clerk-of-the-Works, 4
Cofferdams, 174–175
Cold formed framing, 306
Columns, 95–96
Compressed air, 440
Concrete, 211–272
  aggregates, 241
  camber, 218
  cleanout, 220
  coverage, 237
  cylinders, 118–119
  deflection, 255–256
  finishes, 256–257
  form deflection, 218
  formwork, 214–229
  honeycomb, 220
  hydraulic loads, 216
  joists, 231–235
  lateral hydraulic pressure, 218
  live load, 216
  magnetite iron ore, 215

  nailable, 256
  National Forest Products Association, 214
  OSHA, 213–214
  paving, 190–199
  pneumatic, 257
  screeds, 249–250
  slip form, 223–225
  snap ties, 220
  temperature reinforcement, 240–241
  test, 117–122
  underwater placement, 247
  vermiculite, 215
  vibration, 217–218
  whalers, 220
  wood, working stress, 218–220
Concrete paving, 195–199
  batch plant, 196
  forms, 196
  reinforcing, 196
Concrete unit masonry, 279–281
  bond beams, 281
  control joints, 280
Construction management, 7
Contractor, attitudes, 9
Cooling towers, 463
Conveying systems, 421–426
  dumbwaiters, 422–423
  elevators, 423–425
  hoists and cranes, 425–426
Cost breakdown, 546, 547, 568
Cost-plus, fixed fee, 546
  payment, materials, 548
  payment, requisitions, 546
  retention, 547
Craftsmen, 11
Cranes, derricks, hoists, 109–110
CSI, 24, 131, 132
Curbs and gutters, 199–200
Curtainwall system, 359–360
  clips, 359
  preglazed, 360
  tolerances, 360

Dampproofing, 333–334
  bituminous, 333
  cementitious, 334
  silicone, 333–334

temperatures, 333
vapor barrier, 334
Derricks, 109–110
Design, 16
  contract documents, 17
  preliminary, 16
  schematic, 16
  working drawings, 17
Designer, 10–11
Detour, 200
Documentation, 558–580
Doors, 346–354
  bronze, 351
  coiling doors and grilles, 354
  frames, 347–348
  hand, 347
  hollow metal, 348
  metal, 346–351
  metal covered, 353
  NAAMM, 346
  National Woodwork Manufacturers
    Association, 351
  sizes (wood), 351, 352
  sliding metal fire doors, 352–353
  special doors, 351–352
  stainless steel, 351
  steel door, 349
  storm and screen, 351
  swing, 348, 349, 350
  Underwriter's Laboratories, 351, 352
  wood, 351
Doors, windows, glass, 344–360
Drawings, 22
  architectural, 22
  mechanical-electrical, 23
Dual temperature system, 457
Duct systems, 460–462
  acoustic liners, 461
  balancing, 461
  bracing, 461
  dampers, 461
  hangers, 461
  high velocity, 460, 462
  lock seam, 461
  manometer, 462
  turning vanes, 461
  velocity, 460
Dumbwaiters, 422–423

Duties, inspector, 5, 6, 7–8

Electric, 465–501
  baseboard heaters, 498
  busways, 473–475
  cable trays, 473
  cathodic protection, 492–493
  circuit breaker, 479, 481
  codes and fees, 468–469
  communication systems, 494–498
    alarm and detection, 494
    clock and program, 498
    fire alarm and detection, 494
    public address, 497
    smoke detector, 494
    telephone, 495
    television, 498
  conductor, 473, 474
  contract drawings, 472
  controls and instrumentation, 500
  duct heaters, 498
  emergency power, 491
  grounding, 486
  lighting, 488–491
  lightning, 491
  motor starters, 478–481
  NEMA, 478, 481
  NFPA, 468, 500
  panelboards, 481
  power transmission, 482–484
    capacitors, 483
    substations, 483
    switchgear, 482
    transformers, 482
  raceways, 472–473
  service and distribution, 484–486
  snow-melting mats, 498
  switches and receptacles, 477
  tests, 469–471
  wires and cables, 475–476
Elevators, 423–425
  cable, 423
  counterweights, 423
  guide rails, 424
  safety shoes, 423
  tracks, 423
Embecco, 153

ENR pile formula, 168
Equal Opportunity, 54–55
Equipment, 402–411
Erection, 81

Fencing, 185–186
Field tests, 112–127
    aggregates, 114–116
    Anteus consolidometer, 114
    bituminous, 116–117
    concrete test, 117–122
    slump test, 117–118
    soil test, 113–114
    weld test, 122–123
Final inspections, 566–567
Finishes, 361–392
    acoustical treatment, 381
    conductive ceramic tile, 379
    concrete, 367
    cornerbead, 367
    grounds, 367
    gypsum, 367–370, 371
    Gypsum Association, 364, 368, 372
    gypsum dry wall, 373–377
    lath, 364–367
    lath, gypsum, 364–366
    lath, metal, 364
    painting, 386–391
    plaster, 367–373
    plaster, perlite, 368
    plaster, vermiculite, 368
    quarry tile, 379
    resilient flooring, 383–384
    special coatings, 385–386
    special flooring, 384–385
    terrazzo, 380–381
    tile work, 377–380
    wood flooring, 381–383
Fire hydrants, 202
Fire protection systems, 446–449
    automatic sprinkler, 447–448
        deluge, 447
        dry pipe, 447–448
        sprinkler heads, 447
        wet pipe, 447–448
    CO_2 systems, 448–449
    standpipe and hose, 449
Fire rating, 269

Fittings, pipe, 435–436
Floor gratings, 309–310
Flooring, 381–386
    cementitious coatage, 385
    coatings, 385–386
    coatings, elastometric, 385
    coatings, fire resistant, 385–386
    elastometric floors, 384
    heavy duty topping, 385
    parquet, 382–383
    plywood block, 383
    resilient, 383–384
    sleepers, 382
    terrazzo, 380–381
    wood strip, 381–382
Food service equipment, 400–401
    bars, 400
    cooking, 400
    dishwashing, 401
    machines, 401
    refrigerators, 401
    serving, 401
Forces, 82–84
Forms, 521–575
    Certificate of Occupancy, 577
    Certificate of Substantial Completion,
        576
    change order, 537, 538
    compaction summary, 560
    completed work payment, 556
    contract breakdown, 568 (units, 544)
    C.O. payment chart, 530
    daily extra work, 565
    daily force account, 531
    daily report, 534, 563–564
    defective materials/workmanship, 578
    extension of time, C.O., 536
    materials at job site, 555
    material status, 561
    materials subcontractor approval, 567
    payment estimate, 549–554
    payment requisition, 555
    payroll affidavit, 569
    progress chart, 521
    reduction of retainage, 557
    request for payments-materials, 548
    work force account affidavit, 532
    work stoppage, 564
Fountains, 184–185

Fuel lines, 203
Fuel piping, 449
Furnishings, 412–415
  art work, 414
  blinds and shades, 414
  cabinets and fixtures, 414
  carpets and mats, 415
  drapery and curtains, 415
  seating, 415
Furnishing and moving, 20–21

Gas piping, 442
  shutoff valves, 442
  vents, 442
General Conditions, 25–30, 33
Glass and glazing, 357–359
  acrylic, 357
  gasket, 357
Glass unit masonry, 281
Grading plan, 177
Ground covers, 188–189
Grout, non-shrink, 153
Gunite, 257
Gypsum, 268–272
  Association, 268, 270
  concrete, 269
  decks, 269–271
  dry wall, 373–377
  unit masonry sheathing, 281
  water-cement ratio, 269

Hoists and cranes, 109–110, 425, 426
Hot water heating systems, 456, 457
  boiler, 456
  circulating, 456
  snow melting, 457
  terminal units, 456
Humidity control, 460

Indiana Limestone Institute of America,
  285
Inspector
  authority, 5
  conduct, 10
  Plans and Specifications, 9–10
  specific duties, 7–8

Insulation, 334–335, 439–440
  asbestos, 335
  conductivity, 439
  cork, 335
  fiberglass, 335
  styrofoam, 335
  urethane, 335
  vapor barrier, 439
Insurance, 52–54
Irrigation system, 184

James Electronics Co., 122
Jeep test, 202
John Hancock Tower, 169
Joists, open web, 305

Laboratory equipment, 409–410
Labor relations, 550
Landscaping, 186–187
Lawns, 187–188
Laws of Motion, 82–83, 97
Library equipment, 408
Light gauge framing, 306

Manholes, 203, 204
Manitowoc, 169, 174
Marine work, 209–210
  dredging, 210
  piling, 209
  protective structures, 209–210
Masonry, 273–287
  ASTM-C144, 275
  brick, 278–279
    back-up, 278
    corbelling, 279
    coursing, 277, 278
    effervescence, 278
    grades, 278
    mortar, 277
    pattern bonds, 278
    placement temperature, 276
    reinforcing, 276
    restoration, 286–287
    soldier course, 279
    story pole, 277
    test panels, 275

Mechanical, 427–464
  air distribution systems, 457–463
  controls, 434–435, 463
  fire protection, 446–449
  Freon, 436
  insulation, 439–440
  liquid heat transfer, 456–457
  pipe and pipe fittings, 435–437
  piping specialties, 437
  plumbing fixtures and trim, 446
  plumbing systems, 443–444
  power, heat generation, 449–451
  refrigeration, 452, 456
  soil and waste piping systems, 444–446
  special piping systems, 440–443
  supports, 437–439
  test, 435–436
  tubing, 435
Mechanics, 81–98
Medical equipment, 410–411
  case work, 410
  dental, 410
  examination room, 410
  incubators, 410
  mortuary, 411
  radiology, 410
  sterilizers, 410
  therapy, 411
Metal deck, 305–306
Metal, miscellaneous, 307
Metals, 288–310
  base plates, 299–300
  bolts, 302–304
  calibration, 303
  connection, 302–303
  cross-bracing, 298
  erection, 298–301
  lateral stability, 297
  OSHA, 293–294, 301–302
  painting, 296–297
  purlins, 297–298
  rivet, 302
  trusses, 297
  welding, 290–295, 304
Metal stairs
  classes and types, 307–309
Metal windows, 354–355
  aluminum, 355
  bronze, 355

stainless steel, 355
Steel Window Institute, 354
Mobilization, 56–57
Moisture protection, 329–343
Moments, 85–87
Mortuary equipment, 411
Motion, 97–98
Motor vehicles, 108–109

NAAMM, 307, 308, 356, 359
National Association Standards for
    Ventilation Systems, 458
National Builder's Hardware Association,
    356–357
National Fire Protection Association, 447, 458
National Forest Products Association, 80
Needle beams, 153

Oakum, 204
Oil storage, 449
Ornamental metal, 310
OSHA, 8, 99, 105, 149–151
  carpentry, 328
  excavations, 165
  fire prevention, 105–106
  fire protection, 105
  first aid, 101
  hand and power tools, 107–108
  inspections, 100
  laser, 71–72
  LPG, 107
  noise, 101–102
  piling, 174
  protective equipment, 103–105
  safety belts, 104
  safety nets, 104
  sanitation, 102
  scaffolding, 75, 78
  temporary heating, 107
  training, 101
  ventilation, 103
  water, 105
OSHA requirements, piling, 174
Owner, 11, 25

Painting, 386–392
  failures, 386–389

inspectors, 386–387
  quality, 388–389
  rejection, 391–392
  sampling, 390
  spray, 390–391
  temperature, 390–391
  thinning, 390
  ventilation, 390
Parking equipment, 408
Paving and surfacing, 190–199
  bituminous, 191, 192–195
  concrete, 191, 195–199
  subgrade, 191–192
Payment, 31–32
  materials, 545, 548
  requisitions, 546, 548–556
  retention, 547, 557
Piling, 165–174
  Armco, 167
  cast-in-place, 167
  load test, 168
  monotube, 167
  pre-cast, pre-stressed, 167
  Raymond, 167
  safety, OSHA, 174
  Steel-H, 166
  steel pipe, 166
  wood, 166
Pipe
  backfill, 179
  bedding, 179–181
  corrugated metal, 179
  crossovers, 181
  thread, 204
Piping
  compressed air, 440
  nitrous oxide, 441
  oxygen, 441
  process pipe, 441
  vaccum piping, 440
  valves, 436, 437
Plans and Specifications, 9
Plants, 188–189
Plaster, 364–373
  aggregates, 368–369
  base, 367
  dryout, 372
  Keene's, 371, 372
  lime, 372

machines, 371
  perlite, 368
  plaster of Paris, 367
  scratch coat, 370
  scratch test, 368–369
  sweatout, 373
  thickness, 371
  three-coat, 370
  water ratio, 370
Playing fields, 186
Plumbing fixtures, 446
  faucets, 446
  fixtures, 446
  traps, 446
  vacuum breakers, 446
Plumbing systems, 443–444
  cold water, 443
  hot water, 443
  refrigerated, 444
Post-tensioned concrete, 255
PPBS, 14
Precast concrete, 259–268
  AACE, 259
  Karl Middendorf, 259
  Kenneth Braselton, 259
  Magnel, 261
Precasting plant, 261
  C. H. Raths, 262
  Lin Tee, 265
  precast mold, 263
  thermal shock, 264
Precast panels:
  connections, 264–265
  handling, 264
Precast structural members, 265
Predesign, 14
Prescon, 259, 261
Pressure, 98
Prestressed anchorage, 267
Prestressed concrete, 259–262, 266
Prestressed tendons, 266
Prison equipment, 409
Programming, 15
Progress payments, 542–557
Project plans, 9
Project structure, 13, 19–20
Project, topical breakdown, 558–560
Punch list, 20
Purchasing, 20

Railroad work, 205–209
American Railway Engineering
Association, 205
ballast, 208
guard rail, 207
ICC regulations, 205
joint bars, 206
roadbed, 205
ties, 207–208
track, 206
Reciprocating compressors, 452–454
auxiliaries, 453
air washers, 454
humidifiers, 454
pumps, 454
Refrigeration, 452–456
Freon, 452
machines, 454–456
Reinforced masonry, 281
Reinforcement, concrete, 229–241
accessories, 236
bending, 240
chairs, 236, 237
CRSI, 229
deformations, 229
delivery, 230
fabrication, 230
grades, 229–230
placement drawings, 231
shop drawings, 230
sizes, 230
splice, 239
stirrup, 234
temperature reinforcement, 240
tie wires, 239
Residential equipment, 409
central vacuum cleaner, 409
kitchen and laboratory cabinets, 409
kitchen equipment, 409
laundry equipment, 409
unit kitchens, 409
Resilient flooring, 383–384
adhesive, 383
conductive, 384
epoxy-marble-chip, 384
filler, 383
sheet, 384
storage, 383

tile, 383
underlayment, 383
Roads and parking, 200
Roofing, 338–340
aggregate, 339–340
built-up, 338
elastic liquid, 340
elastic sheet, 340
hypalon, 340
insulation, 339
roll, 340
sheetmetal, 340
vapor barrier

Safety, 99–111
Sanitary sewer lines, 203
Scaffolding, 75–78
Scaffolding and Shoring Institute, 76
Schedule, 32, 57
construction, 18
design, 17
network, 19
working plan, 19
Scheduling, 505–525
bar graph, 506, 507
California highway, paving chart,
523, 524
change orders, 520–541
computer output, 513
CPM, 513, 516
critical path, 512
delays, shop drawings, 533
direction, 528
documentation, 539
DuPont, 505, 513
i–j, 513, 517
network calculations, 512–513
PERT, 513
plan, 512
prebid CPM, 523
precedence diagramming, 513
progress chart GSA, 520
schedule, 512
time extension, 532
Value Engineering, 527
Screeds, 70, 249, 250
Sealants, 343

Seating, 415
Seed, 187–188
Sheeting, 175–177
    anchors, 177
    blowouts, 177
    lagging, 177
    sheet piling, 176
Sheet metal, 340–342
    downspouts, 342
    gravel stops, 341
    gutters, 342
    metal flashing, 341
    through-wall flashing, 341
    wall flashing, 341
Sheet Piling, 176
Shingles, 335–337
    alignment, 336
    asbestos-cement, 336
    asphalt, 335–336
    clay tiles, 337
    slate, 337
    starter course, 336
    wood, 336–337
Shop drawings, 20
Shoring, 78–80
Shoring, concrete, 175
Shotcrete, 257
Siding, preformed, 337–338
Site, 14–15
Site work, 131–210
    backfill, 159–163
    blasting, 151–152
    blasting, NYC, 149
    blasting, OSHA, 149–152
    blasting, safety, 158–159
    clearing, 146–148
    demolition, 148–149
    dewater, 181–182
    disposal, 147
    drainage, 177–183
    dust, 162
    earthwork, 153–165
    excavations, OSHA, 165
    Moretrench, 181
    moving structure, 152
    rock, 157
    stabilization, 164
    soil poisoning, 164–165

    subdrainage, 163
    topsoil, 154
    utilities, 146, 156, 159, 201–205
Skylights, 342
Slump, 117–118, 252
SMACNA, 458, 460
Snow melting, 457
Sodding, 188
Soil and waste pipe, 444–446
    manhole, 445
    piping, 444–445
    stacks, 445
    traps, 445
    vent, 445
Soil preparation, 187
Special construction, 416–420
Specialties, 393–401
    bar units, 400
    chalkboard, 395
    cooking equipment, 400
    cubicles, hospital, 395–396
    cubicles, office, 396
    directory and bulletin signs, 344
        plaques, 344
        signs, painted, 344
        three-dimensional signs, 344
    dishwashing equipment, 401
    firefighting devices, 397–398
    fireplace equipment, 396
    fireplaces, prefab, 396
    flagpoles, 396–397
    food preparation machines, 401
    food services equipment, 400–401
    food serving units, 401
    lockers, 397
    partitions, demountable, 398
    partitions, folding, 399
    partitions, mesh, 398
    postal, 398
    scales, 399
    shelving, storage, 399
    sun control devices, 399
    tackboard, 395
    telephone booths, 399
    toilet and bath accessories, 400
    toilet and shower compartments, 396
    wardrobe specialties, 400
Specifications, 23

Stage equipment, 408
Standpipe and hose, 449
Steam heating, 451
Steam lines, 203
  anchorages, 203
  contraction, 203
  expansion, 203
Steel Joist Institute, 305
Steel Window Institute, 354
Stone, 284–286
Storm sewer lines, 203
Subsurface piping, 178
Supports, 437–439
  anchors, 438
  hangers, 438
  stress, 438–439
Surface run-off, 178
Surveying, 59–74
  accuracy, 59
  benchmarks, 59, 64, 67
  distances, 65–67
  laser, 70–72, 74
  levels, 64–65
  plumbob, 61
  procedures, 67–69
  tapes, 65
  theodolites, 62–63
  transits, 60, 62–63

Telephone cable, 201
Testing, paving, 191
  laboratory, 191
Testing, pipe, 435–446
  Freon, 436
  hydrostatic, 436
Testing, site utilities, 202
  coating, 202
  "jeep" test, 202
Tile, 377–380
  joints, 378
  mortar, 378
  temperature, 378
  Tile Council of America, 377
  ceramic mosaics, 378–379
  conductive, 379
Topsoil, 187, 188

Trees and shrubs, 189
Tubing connections, 204

Underpinning, 152–153
Utilities, public, 201

Valves, 436–437
Vectors, 82
Vibration, 97
Vibration isolation, 439

Waco Scaffold and Shoring, Co., 76, 79, 82
Walkways, 200
Waste treatment, 444–445
  manhole, 445
  stacks, 445
  traps, 445
  vents, 445
Water lines, sterilized, 203
Waterproofing, 331–333
  bituminous, 332
  liquid, 332
  membrane, 332
  metallic oxide, 332–333
Weather striping, 357
  astragal, 357
Water supply, 443–444
  cold water, 443
  hot water, 443–444
  iced, 444
Welding, 290–295
  American Welding Society, 291
  arc, 290
  gas, 290
  Heliarc, 291
  OSHA regulations, 293–294
Windows, 354–360
  curtainwall, 359–360
  glass, 357–359
  metal, 354–355
  weatherstripping, 357
  wood, 355–356